AF880898

# The Parasite-Stress Theory of Values and Sociality

Randy Thornhill • Corey L. Fincher

# The Parasite-Stress Theory of Values and Sociality

## Infectious Disease, History and Human Values Worldwide

Springer

Randy Thornhill
Department of Biology
The University of New Mexico
Albuquerque, NM, USA

Corey L. Fincher
Department of Psychology
The University of Warwick
Coventry, UK

ISBN 978-3-319-08039-0 ISBN 978-3-319-08040-6 (eBook)
DOI 10.1007/978-3-319-08040-6
Springer Cham Heidelberg New York Dordrecht London

Library of Congress Control Number: 2014942639

© Springer International Publishing Switzerland 2014
This work is subject to copyright. All rights are reserved by the Publisher, whether the whole or part of the material is concerned, specifically the rights of translation, reprinting, reuse of illustrations, recitation, broadcasting, reproduction on microfilms or in any other physical way, and transmission or information storage and retrieval, electronic adaptation, computer software, or by similar or dissimilar methodology now known or hereafter developed. Exempted from this legal reservation are brief excerpts in connection with reviews or scholarly analysis or material supplied specifically for the purpose of being entered and executed on a computer system, for exclusive use by the purchaser of the work. Duplication of this publication or parts thereof is permitted only under the provisions of the Copyright Law of the Publisher's location, in its current version, and permission for use must always be obtained from Springer. Permissions for use may be obtained through RightsLink at the Copyright Clearance Center. Violations are liable to prosecution under the respective Copyright Law.
The use of general descriptive names, registered names, trademarks, service marks, etc. in this publication does not imply, even in the absence of a specific statement, that such names are exempt from the relevant protective laws and regulations and therefore free for general use.
While the advice and information in this book are believed to be true and accurate at the date of publication, neither the authors nor the editors nor the publisher can accept any legal responsibility for any errors or omissions that may be made. The publisher makes no warranty, express or implied, with respect to the material contained herein.

Printed on acid-free paper

Springer is part of Springer Science+Business Media (www.springer.com)

*Thornhill dedicates this book to the loving memories of the women of his family who encouraged his boyhood dreams and visions for his life course: Bessie Mae Norton, Lillie Flora Pickens, Rebecca Pickens, Mattie Polk, Clara Jestenia Thornhill, Mary Teresa Thornhill, and Willie Raiford Thornhill.*

*Fincher dedicates this book to his children, Ashlyn, Peyton and Lane, and his wife, Lisa.*

# Preface

This book grew out of our intensive research collaboration over the last 9 years. The order of the book's authorship could have been determined by a coin flip. Shared interests in the causes of human cultural behavior as well as its variation across regions and times fostered our collaboration. Our shared interests also include the long-standing, unsolved scientific problem in the social and behavioral sciences of the causes of social prejudice and its flipside, equalitarianism. Our book, in part, provides a novel solution to this problem. The central ideas and empirical evidence in the book comprise what we have called "the parasite-stress theory of values" or "the parasite-stress theory of sociality." This theory is a general theory of human culture and of the range of human values, including prejudicial and egalitarian values. Drawing on our research and that of many others, our book presents a new interpretation of human values and their various manifestations in cultural behavior and related group-level phenomena. The theory proposes that, both on the evolutionary time scale and the ecological time scale, humans interfacing with infectious diseases cause many core human values. On both time scales, infectious diseases account for a huge amount of human morbidity and mortality, and hence cause strong natural selection for traits that reduce contact with the diseases and manage their negative effects upon contact. The parasite-stress theory of values provides new and encompassing ways to understand the wide range of regionally variable cultural patterns in the values dimension of collectivism–individualism and the similar values dimension of conservatism–liberalism, as well as patterns across the world in religiosity, personality, sexual behavior, marital systems, cooperative breeding and family organization in general, interpersonal violence, intergroup violence (warfare), and cognitive ability. The theory also reveals how infectious diseases and the values they cause generate geographic variation in governmental systems (e.g., autocracy versus democracy, governmental corruption versus transparency), economic outcomes (e.g., wealth per capita, and wealth inequity), and the creation and diffusion of innovations and technologies. Hence, our book proposes new theories of economics, political science, and a wide range of other human affairs. It also proposes new interpretations, based on the parasite-stress theory of sociality, of the evolution of human reciprocal altruism and human-unique intelligence.

In addition to our interest in cultural variation on the broadest scale across the planet, we share a scientific fascination with the conservative culture of the Old South, the southern USA prior to the region's racial desegregation beginning in the 1960s. Our study of the Old South, and in Thornhill's case, experiences in the Old South, contributed to our interests in the causes of human values. As shown in this book, the highly conservative social life that Thornhill observed as he grew up is similar to that of children in other highly conservative cultures, both throughout history and currently. He was born in 1944 in Alabama, the so-called Heart of Dixie. His natal culture had changed little in basic values over the previous 100 years or more. Some people unfamiliar with the South's history may find this claim about the region's stasis incredible. Historians, however, have shown the region's cultural isolation and temporal constancy in the values held by people into the 1960s and 1970s. The slogan "the South will rise again" refers to the desire of traditionalist southerners to re-establish the culture of the Old South.

Another interest we share is the causes of biodiversity. The evidence we present in this book reveals that parasite adversity and associated preferences or values provide a novel theory of how new cultures and new species arise. We have called this new theory of diversification "the parasite-driven-wedge model." The causes specified in this model may lead commonly to new cultures and species arising side-by-side (i.e., parapatric diversification) from a common ancestor, and lead to new species arising sympatrically. The parasite-driven-wedge also may account for the sympatric origin of human caste social systems.

We share, too, an interest in understanding sociality across all species, not just humans. Although *Homo sapiens* is our primary topic, the book treats how recent knowledge of the interrelationship between infectious diseases, values/preferences and sociality may illuminate topics concerning non-human sociality, especially group cohesiveness, intergroup segmentation, family organization, and dispersal.

Our book is a scientific research monograph and not a survey textbook of the many and diverse topics we treat. Its purpose is to create a theoretical and empirical synthesis based on the parasite-stress theory of sociality of many areas of scholarship that traditionally have been largely or entirely separated. We draw on, analyze and reinterpret many literatures. In drawing on such a range of literatures, we have tried our best to represent them fairly. We appreciate fully that even the ideas we criticize have contributed in an important way to a dialogue among scholars.

We have tried to make our book understandable to all. We explain specialist terms and theoretical, analytical and methodological issues in some detail. Our desire to achieve comprehension of the book by all comes from our view that the parasite-stress theory of values and its empirical support are relevant to the lives and interests of everyone. First, every person has values and may wish to understand them scientifically. Second, the parasite-stress theory is a scientific theory of the ultimate or evolutionary, as well as the ecological or immediate, causes of values. Such a theory can provide the knowledge necessary to change values if this is an ideological goal. Scientific discovery of the causes of prejudice and egalitarianism opens up two opposing paths for those who may use these discoveries to engineer the future cultural course of the human species. One path is to use the causal knowledge to erect liberal culture and its associated democratic values throughout the world. The second path is to use the causal knowledge to erect conservative culture and associated prejudice and inequality. The parasite-stress theory of values and its empirical support do not claim one path is morally superior to the other. This theory and its discovered findings, as with all science, inform about nature as it is, but provide no moral judgment or direction.

We sincerely thank the many colleagues who provided comments on our ideas at various stages of their development and empirical testing. In the last chapter of the book, we discuss the criticisms and comments of many scholars who provided commentaries on our recent *Behavioral and Brain Sciences* article on parasite stress, religiosity and family values. Others also have given us useful input: Paul Andrews, Bram Buunk, Martin Daly, Chris Eppig, Steve Gangestad, Ed Hagen, Ashley Hoben, Jinguang (Andrew) Zhang, Kenneth Letendre, Kamil Luczaj, Damian Murray, Steven Pinker, Scott Reid, Pete Richerson, Mark Schaller, Joy Thornhill-Montoya, Robert Trivers, Josh Tybur, Paul Watson, and the anonymous reviewers of the book manuscript. We thank Anne Rice for formatting the manuscript and other assistance. We thank, too, Meghan Bentz, Djente Jo Fawcett, Parisa Mortaji, Vishal Patel, Abbie Reade, William A. Strickler, Samana Tasnim and Savannah Woodward for various critical assistances. Finally, we are grateful to Dan Colman who allowed us to present his unpublished material on intercollegiate sports-teams discussed in Chap. 12.

Albuquerque, NM, USA — Randy Thornhill
Coventry, UK — Corey L. Fincher

# Contents


**1 Background and Overview of the Book** ........ 1
- 1.1 Introduction ........ 1
- 1.2 Particularist Studies of the South ........ 1
- 1.3 Other Particularistic–Historical Studies ........ 3
- 1.4 Culture Is Acquired Strategically ........ 4
- 1.5 Causes of the Particularistic–Historical Ideology ........ 5
- 1.6 The Scientific Promise of Better Methods ........ 7
- 1.7 The Naturalistic Fallacy ........ 8
- 1.8 How Values Relate to Science ........ 9
- 1.9 Accommodationism ........ 10
- 1.10 Book Overview ........ 11
- 1.11 Summary ........ 17
- References ........ 18

**2 Evolutionary Aesthetics, Values, and Methodology** ........ 21
- 2.1 Introduction ........ 21
- 2.2 Philosophical Aesthetics Is Not Science ........ 21
- 2.3 Biology Encompasses All Life ........ 23
- 2.4 Other Reasons Why Philosophical Aesthetics Failed ........ 24
  - 2.4.1 Common Sense Is Biased ........ 24
  - 2.4.2 Intuitions Vary Because Core Values Vary ........ 25
  - 2.4.3 Personal Introspection Is Incomplete Analytically ........ 26
- 2.5 Scientific Aesthetics ........ 27
- 2.6 Categories of Ultimate Causation ........ 27
- 2.7 By-products ........ 28
- 2.8 Units of Selection and Altruism ........ 29
- 2.9 Individually Selected Altruism ........ 31
- 2.10 A Role for Parasites ........ 33
- 2.11 The Ontogeny of Culture ........ 34
- 2.12 Current Adaptiveness ........ 37
- 2.13 Genes, Environment, and Ontogeny ........ 38

2.14 Human Values Are Not an Objectively Delimited Category ......... 39
2.15 Adaptationism and Special-Purpose Adaptation ......... 40
2.16 Beauty as Truth ......... 41
2.17 Comparative Methodology ......... 42
2.18 A Cultural Phylogenetic Example ......... 47
2.19 Interspecies Comparative Research ......... 49
2.20 Opponents of Evolution Applied to Human Behavior ......... 50
2.21 Summary ......... 53
References ......... 54

**3 The Parasite-Stress Theory of Values** ......... 59
3.1 Introduction ......... 59
3.2 Immunity ......... 59
3.3 Host–Parasite Coevolution ......... 61
3.4 Assortative Sociality: An Aspect of the Behavioral Immune System ......... 62
3.5 The Genesis of Cultures and Species ......... 63
3.6 Conditional Behavioral Immunity ......... 64
3.7 Behavioral Immunity Adaptively Manages False Positives ......... 66
3.8 Brief History of Research Inspired by the Parasite-Stress Theory of Sociality ......... 67
3.9 Summary ......... 75
References ......... 76

**4 Human Values Research Prior to the Parasite-Stress Theory** ......... 83
4.1 Introduction ......... 83
4.2 Collectivism–Individualism Is Conservatism–Liberalism ......... 85
4.2.1 Psychometric Studies ......... 86
4.2.2 Correlates of Collectivism–Individualism ......... 88
4.2.3 Cross-National Findings ......... 88
4.2.4 Interstate USA Findings ......... 102
4.3 Overview of Patterns in Table 4.1 ......... 102
4.4 Additional Differences Between Conservatives and Liberals ......... 103
4.5 Summary ......... 106
References ......... 106

**5 Collectivism–Individualism, Family Ties, and Philopatry** ......... 113
5.1 Introduction ......... 113
5.2 Collectivism–Individualism Across Countries ......... 113
5.2.1 The Origin of the Hypothesis ......... 113
5.2.2 Measures of Collectivism–Individualism ......... 115
5.2.3 Cross-National Parasite Severity ......... 116
5.3 Findings ......... 117
5.4 Nonzoonosis and Collectivism–Individualism Cross-Nationally ......... 120
5.4.1 Methods ......... 120
5.4.2 Findings ......... 122
5.4.3 Severity of Disease Types ......... 124

5.5 Scientific Strengths of the Research ........ 124
5.6 A Caution About Interpreting Correlational Results ........ 125
5.7 Ecological Correlations, the Ecological Fallacy, and Multilevel Modeling ........ 129
5.8 Family Ties Cross-Nationally ........ 130
5.8.1 Methods ........ 131
5.8.2 Findings ........ 132
5.9 United States: Collectivism and Family Ties ........ 134
5.9.1 Methods ........ 134
5.9.2 Results ........ 135
5.10 Summary: Cross-National and USA States ........ 137
5.11 Minkov (2011) ........ 137
5.12 Cashdan and Steele (2013): The Standard Cross-Cultural Sample ........ 138
5.13 Convergence of Evidence ........ 138
5.14 Collectivism and Family Ties: Other Implications ........ 139
5.14.1 Life History ........ 139
5.14.2 Implications for Future Regional Analyses ........ 140
5.14.3 The Demographic Transition ........ 140
5.15 Philopatry ........ 141
5.15.1 Ethnographic Societies ........ 142
5.15.2 Residential Movement Between USA States ........ 144
5.15.3 Overview: Philopatry ........ 145
5.16 Collectivism, Family Ties, and Cooperative Breeding ........ 145
5.17 Reciprocity ........ 147
5.18 Human-Specific Cognitive Ability ........ 149
5.19 Patriotism ........ 151
5.20 Xenophobia ........ 152
5.21 Moral Foundations Theory ........ 152
5.22 Summary ........ 153
5.23 Appendix 1 ........ 156
5.24 Appendix 2 ........ 161
5.25 Appendix 3 ........ 163
References ........ 166

**6 Mating Systems, Mate Choice, Marriage, Sexual Behavior, and Inbreeding** ........ 171
6.1 Introduction ........ 171
6.2 Polygyny ........ 171
6.2.1 Low's Research ........ 171
6.2.2 Collectivisim and Polygyny ........ 172
6.3 Mate Choice ........ 173
6.3.1 Gangestad and Buss' Research ........ 173
6.3.2 Women's Mate-Choice Trade-Off ........ 174
6.3.3 Other Attractiveness Research ........ 175
6.3.4 Overview: Mate-Choice Studies ........ 177

6.4 Marital Divorce ... 178
6.5 Female Sociosexual Orientation ... 179
6.6 Inbreeding ... 180
6.6.1 The Hypothesis and Hoben et al. (2010) ... 180
6.6.2 Hoben's Dissertation ... 183
6.7 Male Sexual Competition ... 184
6.7.1 Out-Group Men as Sexual Competitors ... 184
6.7.2 White-on-Black Violence ... 185
6.8 Disgust and Sexual Arousal ... 187
6.8.1 Women's Sexual Arousal ... 187
6.8.2 Men's Sexual Arousal ... 188
6.9 Sex Differences in Sickness Behavior and Suppression of Sexual Activity ... 188
6.10 Summary ... 189
References ... 190

**7 Personality** ... 195
7.1 Introduction ... 195
7.2 Schaller and Murray (2008) ... 196
7.3 Experimental Studies ... 198
7.4 Thornhill et al. (2010) ... 199
7.5 Personality and Parasite Severity ... 201
7.6 Condition-Dependent Personality ... 201
7.7 Implications for Synthesis and Future Research ... 202
7.8 Summary ... 204
References ... 205

**8 Interpersonal Violence** ... 207
8.1 Introduction ... 207
8.2 Nonlethal Physical Aggression Between Romantic Partners, and Honor Ideology ... 207
8.3 Romantic-Partner Homicide ... 211
8.3.1 Men Killing Female Romantic Partner ... 212
8.3.2 Women Killing Male Romantic Partner ... 213
8.4 Potential Confounders ... 213
8.5 Male-Honor Homicides ... 215
8.5.1 Context ... 215
8.5.2 Hypothesis ... 216
8.5.3 The South's Culture of Male Honor ... 216
8.5.4 Analyses ... 218
8.6 Felony-Related Homicide ... 219
8.7 Male Lifespan ... 219
8.8 Sex Ratio ... 220
8.9 Medical Care Affects Homicide Rates ... 221

8.10 Cross-National Homicide ... 222
8.10.1 Data Issues ... 222
8.10.2 Analyses ... 223
8.11 Family-Honor-Based Homicide ... 224
8.12 Child Maltreatment and Parasite Stress ... 225
8.12.1 Hypothesis ... 225
8.12.2 Evidence ... 226
8.13 Child Maltreatment and Collectivism ... 228
8.14 Overall Summary of Findings ... 229
8.15 Other Tests ... 230
8.16 Prospect for Societal Control of Homicide ... 231
8.17 Summary ... 232
References ... 233

**9 Religiosity** ... 237
9.1 Introduction ... 237
9.2 Religiosity: A Costly Signal of In-Group Allegiance ... 238
9.3 The Parasite-Stress Hypothesis of Religiosity ... 239
9.4 Religiosity and Signal Redundancy ... 240
9.5 Other Hypotheses to Explain Cross-Cultural Differences in Religiosity ... 241
9.6 Methods for Establishing an Empirical Link Across Regions Between Religiosity and Parasite Stress ... 242
9.6.1 Religious Affiliation ... 243
9.6.2 Religious Participation and Value ... 244
9.6.3 In-Group Assortative Sociality Variables for Countries and US States ... 245
9.7 Results ... 245
9.7.1 Cross-National Analyses ... 245
9.7.2 United States Analyses ... 250
9.8 Overview of Findings ... 254
9.9 The Findings' Relationships to Some Other Research Topics ... 254
9.9.1 The USA Is Not an Exception ... 254
9.9.2 Economics ... 255
9.9.3 Secularization ... 255
9.9.4 Ontogeny of Religiosity ... 256
9.9.5 Health ... 257
9.9.6 Geographical Expansions of Religions ... 257
9.9.7 In-Group Assortative Sociality ... 258
9.10 Life History and Assortative Sociality ... 259
9.11 Other Tests of the Parasite-Stress Hypothesis of Religiosity ... 259
9.12 Summary ... 260
References ... 261

**10 Democracy and Other Governmental Systems** 265
10.1 Introduction 265
10.2 Autocracy and Democracy in Political Science 266
10.3 Traditional Theories of Democracy and Associated Empirical Patterns 267
10.4 The Parasite-Stress Hypothesis of Democratization 268
10.5 Thornhill et al. (2009) 273
10.5.1 Measures of Democratization 273
10.5.2 Parasite Stress 275
10.5.3 Measures of Collectivism–Individualism 275
10.5.4 Property Rights 275
10.5.5 Gender Equality 275
10.5.6 Sexual Liberalization 276
10.5.7 Results 276
10.6 Thornhill et al. (2010) 280
10.6.1 Democracy 280
10.6.2 Results 281
10.7 Nonzoonotic and Zoonotic Disease Severity Compared 282
10.8 F-Scale and Autocracy 282
10.9 Authoritarianism in the Standard Cross-Cultural Sample 283
10.10 Cross-National Freedom of the Media 283
10.11 Overview of Findings 284
10.11.1 Democracy 284
10.11.2 Collectivism–Individualism 286
10.11.3 Innovation and Diffusion 287
10.11.4 Property Rights 288
10.11.5 Gender Relations and Women's Sexual Liberties 288
10.12 Political Elections and Parasites 289
10.13 The Western Revolution in Values 291
10.14 Reconstruction, Parasite Stress, and Civil Rights 293
10.15 Early Western Transitions Toward Democracy 294
10.16 Politics and Human Nature 296
10.17 Summary 297
References 298

**11 Economics, Values, and Cognitive Ability** 303
11.1 Introduction 303
11.2 The Parasite-Stress Hypothesis of Economics 305
11.3 Eppig et al. (2010): International IQ Research 307
11.3.1 The Parasite-Stress Hypothesis of Cognitive Ability 307
11.3.2 Methods 308
11.3.3 Results 310
11.3.4 Discussion 311
11.3.5 Trade-Off Mechanisms 312
11.3.6 Other Implications 313
11.3.7 Other Tests 314

11.4 Eppig et al. (2011): US States IQ Research ... 315
11.4.1 Methods ... 315
11.4.2 Results ... 316
11.4.3 Discussion ... 316
11.5 Conclusions from the Two Studies Combined ... 318
11.6 Values and IQ ... 319
11.6.1 International ... 319
11.6.2 US States ... 319
11.7 GDP, IQ, and Collectivism ... 319
11.8 Parasite Stress Mediates the Relationship Between IQ and Economic Productivity ... 320
11.9 Corruption and Value Systems ... 320
11.10 Innovation and Diffusion ... 321
11.10.1 Diffusion in the USA ... 322
11.10.2 Cross-National Technology Adoption ... 325
11.11 Knowledge Unity ... 327
11.12 Economic Prosperity Has a Major Cost ... 327
11.13 Summary ... 328
References ... 328

**12 Wars, Revolutions and Coups, and the Absence of Peace Across the World** ... 333
12.1 Introduction ... 333
12.2 Relevant Aspects of the Parasite-Stress Theory ... 333
12.3 Civil War ... 335
12.3.1 Letendre et al. (2010) ... 335
12.3.2 Collectivism ... 337
12.4 Other Intra-Nation Intergroup Conflicts ... 338
12.5 Non-State Wars ... 339
12.5.1 Methods ... 339
12.5.2 Findings ... 340
12.6 Revolutions and Coups ... 340
12.6.1 Methods ... 340
12.6.2 Findings ... 341
12.7 Peace ... 341
12.7.1 Methods ... 341
12.7.2 Findings ... 341
12.8 Discussion and Conclusions ... 342
12.9 Parasite Stress and Civil Conflict: Further Clarifications ... 342
12.10 Limitation of Our Findings on Intra-Nation Conflicts ... 347
12.11 Prospects for Eliminating Civil Conflicts ... 347
12.12 The American Civil War ... 349
12.13 Team-Sport Competition in the USA ... 349
12.14 Summary ... 350
References ... 351

**13 Biodiversity and the Parasite-Driven Wedge** ........ 353
13.1 Introduction ........ 353
13.2 The Parasite-Driven-Wedge Model ........ 354
13.2.1 Step 1: The Ancestral Culture/Species ........ 354
13.2.2 Step 2: Localized Host–Parasite Races ........ 355
13.2.3 Step 3: Spatially Variable Immunity ........ 355
13.2.4 Step 4: Subdivision of the Ancestral Range ........ 356
13.2.5 Step 5: Divergence ........ 358
13.2.6 Step 6: Coupling of Immunity Traits ........ 359
13.2.7 Step 7: The Wedge Causes New Parapatric Kinds ........ 362
13.2.8 Step 8: Parasites Also Speciate ........ 362
13.2.9 General Prediction ........ 363
13.3 The Parasite-Driven Wedge Creates Ecological Diversification ........ 363
13.4 Language ........ 364
13.4.1 Methods ........ 365
13.4.2 Results ........ 366
13.4.3 Parasite Severity ........ 369
13.4.4 Conclusion ........ 369
13.5 Religion ........ 370
13.5.1 Methods ........ 370
13.5.2 Results ........ 371
13.5.3 Conclusions and Discussion ........ 374
13.6 Studies' Findings and Their Relationship to Other Models of Cultural Diversity ........ 375
13.7 Political Factionalism ........ 377
13.8 Caste Social Systems ........ 378
13.9 Cultural Diversity in Nonhuman Animals ........ 380
13.10 On the Origin of Species ........ 380
13.10.1 The Mystery of Mysteries ........ 380
13.10.2 Speciation Processes ........ 381
13.10.3 Other Research Relevant to the Wedge Model of Parapatric Speciation ........ 384
13.10.4 Parasite-Driven Sympatric Speciation ........ 387
13.11 Summary ........ 388
References ........ 389

**14 Reflections, Criticisms, and Future Research** ........ 395
14.1 Introduction ........ 395
14.2 Creativity and the Temporary Limit of Science ........ 395
14.3 There Is No Such Thing as a Free Lunch ........ 400
14.4 Criticisms of the Parasite-Stress Theory of Sociality ........ 402
14.5 Alternative Hypotheses ........ 403

14.6 Other Criticisms ........ 410
14.7 Additional Future Research ........ 428
14.8 Summary ........ 435
References ........ 436

**Index** ........ 441

# Chapter 1
# Background and Overview of the Book

## 1.1 Introduction

In this chapter we introduce the topic of research on human values as well as our approach to the study of this topic. Also, we give a brief overview of the chapters that follow.

## 1.2 Particularist Studies of the South

A range of scholars, especially historians, psychologists, and sociologists, have generated a voluminous published literature about the Confederate or Deep South USA, the southeastern states of the USA, prior to the federally legislated racial desegregation of the region in the 1960s. This era in the Deep South is commonly called "The Old South." Vandello and Cohen's (1999) subdivision of the USA into cultural regions identifies 11 states with high collectivist (conservative) ideology as comprising the Deep South region: Alabama, Arkansas, Florida, Georgia, Louisiana, Mississippi, North Carolina, South Carolina, Tennessee, Texas, and Virginia. These 11 states are the same ones that seceded from the USA and formed the Confederate States of America from 1861 to 1865. In the literature about the Deep South (hereafter the South), southern hospitality, politeness and manners, emotionally rich Southern American English, marital durability, family duty and honor, female modesty and sexual continence, reverence of hierarchy, elder respect, localism or parochialism, lifelong friendships, hygiene, religiosity, conformity, obedience of norms, and traditionalism are seen often as benevolent values of the region. In contrast, the mental rigidity or dogmatism in the ideologies of sexism with assumed male superiority, classism (elitism) and associated authoritarianism, and racism and other prejudices are considered widely the region's malevolent values, because they are undemocratic/anti-egalitarian. Although the South's culture of male honor is sometimes viewed as a positive cultural feature, male honor seems to be a cause of

R. Thornhill and C.L. Fincher, *The Parasite-Stress Theory of Values and Sociality: Infectious Disease, History and Human Values Worldwide*,
DOI 10.1007/978-3-319-08040-6_1, © Springer International Publishing Switzerland 2014

the high male-on-male homicide rate in the region and thus is discussed sometimes as a morally negative southern value (Nisbett and Cohen 1996).

The customary analysis of the South by its scholars is to describe southern culture in terms of the particulars of southern events without much or any consideration of how the South's value system is related to other value systems across the globe that are similar (conservative) or dissimilar (liberal). For instance, it is often said by historians of the South that the USA Civil War of 1861–1865 was caused by a clash of core values between the two regions involved. Scholarly accounts certainly support the North's liberal attitude of opposing slavery and the South's racism, as well as the South's xenophobia toward the northern invaders and meddlers, as parts of the causal picture. These accounts, however, do not give an encompassing explanation, as all wars are determined, at least in part and fundamentally, by conflicts in moral or ideological systems. Why did southern and northern values differ in the ways they did and what were the causes of these conflicting core values that escalated into interregional aggression and eventually that civil war? Why did the decision to secede from the US arise in the South rather than in the North? Given that civil conflicts are common in some regions of the planet, how can knowledge of their causes inform historians about the US Civil War? A deeper and more final understanding of the US Civil War is achievable by looking both inside and outside the events in the South.

Civil wars are extremely common in some areas of the world, but not in others. As an example, more than two-thirds of the countries in sub-Saharan Africa have seen civil war since 1960 (Blattman and Miguel 2010), but many large regions of the world have had no such wars over the same period of time. Knowledge of the common denominators of civil war onsets in regions and times, as well as those of the absence of civil war in regions and times, is essential for the understanding of any single civil war. This broader or encompassing approach can demonstrate the causes that are common to all civil wars—and thereby illuminate causes of any particular civil war. Encompassing comparative methodology is highly valued in the best of sociological research on civil war, because it provides scientific testing of hypotheses about causes of civil war, wherever and whenever they occur. Strictly particularistic studies lack comparative data from outside their restricted focus of analysis, and thus cannot test for and hence identify a general or a fundamental cause of civil war or of any other cultural feature.

The American Civil War is a very popular topic among book authors. Many of the thousands of books on this war have particularistic points of view about the causes of the war. An additional challenge, however, is to identify the general causes of civil wars and thereby illuminate each of these wars, including the American Civil War. Our effort to meet this challenge is just one of many themes of this book.

Another way to describe the particularistic or atomistic method of research on culture is in terms of the high importance it gives to local history as causation: the South's unique history caused the region's unique culture. As an example, Nisbett and Cohen (1996) use the South's history of male-honor ideology to explain the region's current male-honor culture. Actually, the cultural history of the South is part of the phenomenon to be explained and not an explanation in itself. People's values in a region yield that region's culture and cultural history. Hence, a big question for the history of the South is why these people's values were the way they were but not another way. This is answerable only from evidence of the functional organization of peoples' value systems—i.e., the effects that values are designed by evolution by natural selection to accomplish—and therefore how the people of the South fit into the overall empirical picture of variable value systems across the world and history. Understanding what evolved purpose values serve will illuminate why they exist and why they vary across individuals, time, and geographic regions.

Overall, then, the particularistic–historical method cannot identify the larger causal picture. (See also Daly and Wilson 2010 for a critique of this method.) In the case of the South, the method cannot yield an understanding of the region that is part of an empirically consistent, general scientific theory of human values. Particularistic scholars of the South have generated a large, valuable body of facts that require synthesis into a general causal framework by evolutionary theory and comparative data analysis in order to inform causation of the culture of the Old South in light of causation of cultural variation across the globe. This is one of the goals of this book.

There are relatively few notable exceptions to the particularistic–historical approach in prior studies of the Old South. For example, van den Berghe (1981) discussed the similarities between African-American slavery and servitude in the Old South and the caste systems in other countries. Vandello and Cohen (1999) treated the South as an extreme in a continuum of conservative (collectivist) values across the USA states. But even these important efforts toward engaging bigger issues are descriptive and anecdotal, and hence remain outside a scientifically synthetic and robust general theory of values. More encompassing questions are: Why did strongly hierarchical social systems with strict boundaries between social strata and extreme prejudice against people of a particular color, caste, or hereditary background arise independently in the Old South and Asia, as well as in some other places, but not in other regions? Why did the Old South stand out in the USA in its highly conservative value system and show similarity in values to other conservative regions of the world? Why was the Old South's culture more similar to that of contemporary Guatemala or Syria than to that of contemporary Sweden? Why is the southeastern USA today more conservative than other regions of the same country and other regions of the West? A general scientific theory of values would provide answers to all these questions in a small set of causes common to them all.

## 1.3 Other Particularistic–Historical Studies

The particularistic–historical approach is certainly not restricted to scholars of the South. Its assumptions about culture are held widely by academics and nonacademics alike. This approach assumes to answer the question of how humans obtain their culture or socially learned values and behavior, a major question of cultural anthropology and cultural psychology. Obtaining the answer to this question is fundamental to our book because the book emphasizes a general theory about the causes of the acquisition of culture by people as well as the causes of cultural stasis and cultural change.

In some forms, the particularistic–historical approach relies fundamentally on a view of acquisition and transmission of culture that not only ignores the evolved design of human psychology, but is spiritual as well. Certain traditions in anthropology and related fields view the transmission of values between generations as automatic, inevitable, and passive, resulting from culture itself as an incorporeal force with an inertia that drives it within and across generations in an often-unchanging course. In essence, this view sees culture as a ghost explicable only in terms of itself, and hence there is no need to consider people as decision-makers that affect the adoption and transmission of culture. William Irons (1979) wrote a masterful early critique of this supernatural perspective on culture. (See also similar critiques by Tooby and Cosmides 1992 and Buss 2001.) Recently, Martin Daly and Margo Wilson (2010) discussed the currently widespread advocacy of a similar view by some culture theorists. This nonmaterial view, as Daly and Wilson call it, understands culture as being transmitted by its own inertia—as something that flows along through time independently of any strategic cognition and action of humans in choosing, discarding, and modifying cultural items to meet the problems they face.

## 1.4 Culture Is Acquired Strategically

The alternative view—that individual humans are cultural strategists—is a foundation for this book. It is a commonly held theory of culture among scholars who apply the evolutionary biological theory of human psychology and behavior in their research on culture (e.g., Irons 1979; Alexander 1979a, b; Daly 1982; Flinn and Alexander 1982; Boyd and Richerson 1985; Tooby and Cosmides 1992; Billing and Sherman 1998; Buss 2001; Gangestad et al. 2006; Fincher et al. 2008; Daly and Wilson 2010; Henrich and Henrich 2010; Chudek et al. 2012). Accordingly, as will be explained more fully in Chap. 2, humans are evolved cultural strategists with psychological adaptations placed in the human nervous system by past Darwinian selection favoring individuals who learned cultural items, including ideologies, that ancestrally maximized personal reproductive success, as measured by the number of produced descendant and nondescendant kin. Hence, such psychological adaptations are responsible for, i.e., cause, selective assessment, and use of cultural items by individuals. Such adaptations also guide individual decisions about discarding or retaining and modifying/not modifying cultural items, including values. Moreover, such adaptations determine people's decisions that affect the fate of cultural items that arise de novo within a society or diffuse into a society from another society. Certainly, culture is transmitted between and within generations and between societies. This transmission, however, is caused by historically adaptive and highly discriminative psychological learning adaptations of individuals, not by arbitrariness or by the ghost of cultural inertia. These discriminative adaptations positively bias culture-item adoption and use toward those items that maximize the benefit-to-cost ratio, where benefits and costs are measured in terms of reproductive success of individuals in evolutionarily ancestral environments.

Our view of cultural acquisition does not assume that individual people are always free agents to adopt whatever available cultural items will maximize their reproductive success. Manipulation and/or coercion from parents or other family sources, and from peers, allies, and enemies often become part of the context in which individuals decide among values and other cultural items. Nor do we assume that cultural choices are made primarily with conscious calculation, although this oftentimes is one cause of the selection, retention, and use of cultural items. The role of consciousness, specifically self-awareness, in enculturation is discussed further in Chap. 2.

An irrevocable finding from a long and noble history of research by dedicated and admirable scholars is that a material brain causes all of people's feelings, cognitions, behaviors, and learning. This research has scientifically falsified the many ghosts that have been proposed as causes of human mental activity and behavior. (See the treatments of this research history by Carl Degler (1992) and Steve Pinker (2002).) The ghost of cultural inertia has not yet been purged from the thinking of some researchers and many ordinary people. As demonstrated in this book, the evolutionary theory of people as evolved cultural strategists has profound implications for the understanding of enculturation and of cultural diversity.

## 1.5 Causes of the Particularistic–Historical Ideology

We find the ideology that is the basis of historical–particularistic thinking about culture is scientifically fascinating. Some readers might hypothesize that the notion of a nonmaterial cultural inertia arises from the psychological adaptations that yield spiritual feelings and religious behavior, given that such feelings exist in some form in the minds of essentially all people. (See Atran (2002) and Boyer (2002) for discussions of religiosity as a by-product of psychological adaptations for other purposes.) Certainly, that is a reasonable start, but we suggest there is much more to it. The belief that historical cultural tradition has the overriding influence on us, even determining who we are or become, is an important deduction and core value of many people. We propose that there are, however, predictable individual and group differences in this belief.

Conservatives support and value traditional thinking more than liberals do, as documented in Chaps. 4 and 5. A useful example is the US Marine Corps reverence for the warrior tradition. This value system, or ethos as it is called in the Marines, encourages merging into the stream of tradition—of history—with those brave Marine warriors who came before and, in this way, become that tradition or history. In the South, there is great value placed on tradition, which is manifested in folklore and everyday conversation. An example is the tenacity of the ethos of the American Civil War in the South. We hypothesize that the conveyance in language and other behavior of the value that tradition is one's essence is an honest or truthful signal to observers of fully understanding, believing, and being embedded in conservatism and hence in local in-group well-being, harmony, and goals. It is a way to display that you are not just "whistling Dixie," to use a common phrase in the South. To just whistle Dixie is a dishonest portrayal of the sacred conservative values, and it is a derogatory label for those who are not truly Dixie, i.e., not a product of, faithful to, and merged with southern tradition and history. Contemporary southerners who celebrate the birthday of Robert E. Lee or Stonewall Jackson (two major southern generals in the American Civil War) or fly the flag of the Confederate States are not whistling Dixie.

We are saying that honest signaling of traditionalist values assists conservatives in social navigation in conservative culture. (Honest signaling, the general theory of signals in biology, is treated more fully later in the book, first in Chap. 2.) Our hypothesis predicts that future research will find that those who hold strongly the historical–particularistic perspective on how people come by their culture, whether inside or outside academics, will score right of center on psychometric (questionnaire) measures of conservatism. A major theme and empirical finding of this book is that conservatism is ideological defense against infectious diseases. Thus, we predict, too, that future research will show that the degree of belief in tradition as the basic cause of one's being will correlate positively with scores on questionnaires that measure people's concern about contracting infectious diseases. We predict also that the importance of the historical–particularistic ideology across countries of the world or states of the US will correlate positively with regional severity of infectious diseases. It is already well established scientifically that regional severity of infectious disease is robustly related positively to the importance of traditionalism, a component of conservative ideology, in the value systems across the globe (Chap. 5).

On a related topic, for a long while there has been a schism between researchers who understand culture as a product of the evolved psychology of individual actors, reducible to the strivings of individuals to secure goals that promoted high ancestral reproductive success, and those who see culture as a supra-individual phenomenon that cannot be reduced to individual cognitions and motives (see Irons 1979; Tooby and Cosmides 1992; Buss 2001; Chiu et al. 2010; Daly and Wilson 2010). The view that individuals are designed by evolutionary selection to adopt culture strategically endorses the former view, whereas the particularistic–historical view endorses the latter.

We suggest that this schism is fueled, in part, by core ideological differences pertaining to how the person or self is conceptualized by liberals versus conservatives. As documented and discussed in more detail in Chap. 4, individualists (liberals) see the individual person as an independent agent who has personal goals that are paramount relative to group interests; in sharp contrast, collectivists (conservatives) view the world in interdependent terms: an individual is merged or embedded in the group to the extent that his or her being is only understandable in terms of the group's harmony and goals. The psychologist Donald Campbell is cited often as a proponent of the views that culture (a) exists as a happening very distant or entirely independent of individual behavior and goals and (b) functions for the good of the society, specifically to prevent human activity from becoming overly selfish.

Campbell believed that culture anticipates, controls, and regulates human behavior in order to promote group harmony, cooperation, and survival. (See Chiu et al. (2010) for a useful summary of these ideas.) This way of interdependent or collectivist thinking is similar to that of theists who view god as one who controls human activity and knows fully the best course of action for humans. Campbell anticipated resistance to his ideas from individualists. Chiu et al. (2010) document this and try to promote Campbell's view and its collectivist thinking about culture.

A critical commentator might suggest that our reservations about Campbell's ideas stem from our core values: we are individualists and hence see people in terms of independent agency. The scientific validity of the two views, however, cannot be settled by ideological arguments. In science, only evidence counts. The hypothesis that people strategically adopt culture as a result of psychological adaptation created by past evolutionary selection for maximizing individual reproductive success, as measured by the production of descendant and nondescendant relatives, is supported both by strong theory and empiricism, including the copious evidence about the nature and diversity of culture presented in this book.

Certainly, we are not claiming that the striving of individual people or groups of people does not create physical cultural things that exist external to individuals. The local zoo, the corner church building, the Internet, and this book are examples of such things. Instead, we are saying that all aspects of culture are physical/material and are caused by evolved psychological and behavioral activity of individuals that often includes individuals striving in groups for competitive edge.

The human animal is an evolved group-living organism and hence regardless of core ideology people live and operate socially in groups. The groups established and preferred by conservatives are different than those liberals form and prefer. Simply put, conservatives are in-group specialists whereas liberals are out-group specialists. The social tightness of conservatives in their in-group reflects their high in-group interdependence and priority of in-group goals. Groups of conservatives are relatively durable and strictly bounded by like values. Liberals are less interdependent and hence are said to be individualistic and exhibit an autonomous self-conception and agency. Liberal groups are relatively impermanent and open to diversity (Chaps. 4 and 5).

In sum, we are hypothesizing that the persistence of the beliefs claiming that culture is a group-protective, supra-individual, noncorporeal, passively transmitted phenomenon, with a trajectory independent of evolved human psychology and behavior, are explained by two core values of conservative ideology: traditionalism and the view of self as interdependent.

## 1.6 The Scientific Promise of Better Methods

We have stressed above that the particularistic–historical method, including its foundational mechanism of cultural transmission by inertia, is flawed. Fortunately, better ideas and methods for the scientific study of culture and cultural history are available.

A general perspective on human ideologies/values—what we have called the parasite-stress theory of values—recently has emerged in the scientific literature. It assumes people are functionally designed to learn, adopt, and use culture strategically. The parasite-stress theory of values is a subtheory of the general theory of biology, evolution. In science, the word "theory" has a special meaning: fact-based principles that serve the scientist in making discoveries. Scientific theory, then, is a combination of facts and conceptual proposals that organize facts and discover more facts. The fact-based principles that comprise a scientific theory inspire new hypotheses that are then tested by observations, including observations derived from experiments. Experimentation is not synonymous with scientific testing, but is just one avenue for testing a hypothesis. A hypothesis is a statement about a possible cause of some thing in nature. The cause may be one of deep-time-past history (i.e., evolutionary history) or one acting now or recently; in biology, the former type of causation is referred to as ultimate and the latter as proximate. Hypotheses are either supported or not by observational data addressing a given hypothesis's predictions/empirical implications—those things that must exist in nature if the hypothesis is an actual statement of causation. A hypothesis is supported when it successfully predicts new findings or better predicts old findings, and strongly supported when its discovered findings cannot be explained by alternative hypotheses. Charles Darwin gave biology its general theory, and biologists since Darwin's work have demonstrated evolutionary theory's utility in understanding life, by far the most complex and diverse component of the universe. Evolutionary theory, then, is a fact-based method for discovering the causes of living things and their diverse characteristics.

Only recently has evolutionary theory been applied broadly and in detail to human value systems, resulting in the cornucopia of recent findings presented in this book. A highlight of our scientific careers is participating with colleagues in this new research area. Our book emphasizes the story of this application of evolutionary theory and methods to human values and their diversity across epochs and geography. The parasite-stress theory of values appears to explain why many features of human social life and the value systems that cause them take the same fundamental form in many times and places in the world, but a different fundamental form in other times and places in the world. It seems, too, that the parasite-stress theory of values is a general theory of sociality with the potential to explain some key features of sociality in nonhuman animals. Our book tells that story as well. Given the parasite-stress theory's broad application to social life, this theory is also called the parasite-stress theory of sociality.

## 1.7 The Naturalistic Fallacy

Before proceeding we emphasize that this book is a scientific document and hence its findings in themselves can provide no moral guidance. The findings in the book do not endorse morally any value system nor do they claim that one system is morally superior to another. Science identifies what is, and cannot identify what ought to be. The belief that a fact about the nature of the world gives moral direction—that "is" equates with "ought"—is called the naturalistic fallacy, a fallacy of logic. It is logically erroneous to believe that the universe created by natural processes—processes without supernatural intervention—provides evidence, in these processes or their products, of morality and immorality. The assumption that a supernatural, morality-knowing being(s) created and guides nature, the assumption of most of the people of the world, leads to the conviction that moral truth is to be found in the features and workings of nature. Science is solely concerned with discovering the causes of the effects that are the universe. Science does not identify moral and immoral paths; people identify these paths based on their values. Theism includes the belief that morality and immorality are generated from some supernatural sphere, but, in fact, people, to promote their personal interests—according to evolutionary biology, their inclusive-fitness interests (Chap. 2)—generate morals and belief in spirits. As discussed throughout this book, values, including secular ones, are proximate mechanisms of promoting personal reproductive interests.

The naturalistic fallacy is endorsed by some secularist scholars (e.g., Wilson 1998; Baschetti 2007a, b; Harris 2010). Most notably, Sam Harris (2010) argued in his popular book, *The Moral Landscape: How Science Can Determine Moral Values*, that the naturalistic fallacy is erroneous and that science identifies morality in its findings about the causes of human well-being. However, the interpretation of what promotes well-being depends on a person's values. As shown in later chapters, perceived and valued well-being to conservatives is different than well-being to liberals. For instance, conservatives are most satisfied with tradition, cultural stasis, and unequal treatment of people, whereas liberals find well-being in nontraditional ways and means and egalitarianism. People's perception of well-being will change only if their values change, as we discuss first immediately below and document subsequently throughout our book. Our book scientifically identifies the "is" of "ought," which is not in any way the same as finding ought in is. Our book is about the proximate and ultimate causes of morality. Scientific identification of such causes provides only "is." "Ought" depends on the ideological beliefs of the person, which serve the person in dealing with adversity in her/his niche. In this book, the fundamental adversity that we advocate as important is infectious disease.

## 1.8 How Values Relate to Science

Although the discovery and use of facts about the content of nature as the way to identify morality is not part of science, there are two principled ways in which morals (and values broadly) actually are related to scientific research. As noted earlier, one way stems from the fact that a person's morals and a group's morals have causes. Scientific research is *the* means to identify causation. The scientific study of values, including all aspects of religious ideology, is no different than the scientific study of the giraffe's neck or planetary motion—all these studies pursue knowledge of causation. The second way is that scientific understanding of causation allows the achievement of the moral goals desired and decided by people, but, as we have emphasized, it cannot identify moral goals. If the people of a region decide that an increase in democratization is the correct moral path, then knowledge of the causes of democratic and undemocratic values is the necessary information for implementing changes that will democratize future generations. The evidence we will present in this book indicates that reducing social prejudice and authoritarianism and increasing equalitarianism can be accomplished by emancipating people from infectious diseases. In contrast, if the moral goal of a government is to make authoritarianism, ethnocentrism, and xenophobia more widespread, or even to achieve their extremes in full-blown fascism or genocide, then this goal will be achievable by promoting widespread mortality and morbidity from infectious diseases.

Recent evidence indicates that the majority of researchers investigating the relationship between evolution and human behavior are quite liberal (Tybur et al. 2007). Our personal ideologies are left of center, and therefore consistent with this general pattern. Our approach in understanding the causes of values is scientific, however, and thus value neutral. Given our values, we hope governments and people widely will elect the moral path of liberalism rather than the moral path of conservatism.

## 1.9 Accommodationism

Above we mentioned that, in terms of the scientific goal of elucidating causation, the scientific study of religious ideology is no different than the scientific study of any other feature of nature. We want to be clear about what we mean. There is a widespread notion that science is limited in its application to the universe. According to this ideology, often called accommodation, certain aspects of the universe are off-limits to science because God and other similar deities act not only in mysterious ways, but also in ways that are unknowable because they are supernatural. Accommodationists believe that this allows the compatibility of science and religion. Accommodationists vary in which realms of the universe are designated as scientifically unknowable, but such topics as the deep-time history of the universe, including life's history on earth—evolution, basically—and religion are commonly ones that are deemed off-limits to science. Notably, some accommodationists feel that morality is a purely spiritual realm of human affairs and, as such, can only be addressed by religion—science has nothing to say about it. This is the opinion of Francis Collins, head of the National Institutes of Health. It is also the opinion of some members of the National Academy of Sciences. Many scientists have criticized these opinions on the appropriate grounds that science applies to all features of the universe. Singham (2010) has provided an informative discussion and documentation of contemporary accommodationism and its prevalence in the West.

It should be clear to readers that we are not accommodationists. Science is the avenue without limit for illuminating all aspects of nature. The supernatural does not exist as a material feature of the universe and hence is not part of nature.

Nonexistent phenomena provide no challenge for science and hence are outside science. All feelings, thoughts, and beliefs about the so-called supernatural and the behaviors associated with these feelings and cognitions, however, are as real or natural as the giraffe's neck or the human stomach, and hence are fully subject to scientific analysis. Human feelings, cognitions, and behaviors about spiritual matters are salient components of the central topic of our book, value systems.

Again, our book is a science book and therefore, given it is science, it does not, nor can it, address moral correctness or incorrectness. It follows, then, that we endorse only one version of the notion that science and religion are compatible. Science provides only the "is," including the causes of moral systems; it doesn't and cannot provide any "ought." Therefore, the findings of science do not and cannot threaten one's held ideology, whether it includes spirituality or not. A person's values may change as she/he acquires scientific knowledge, but such change is not based on the knowledge itself identifying a more moral path. The scientific knowledge so gained is just more "is" and "is" never identifies "ought." Gained scientific knowledge, like any information acquisition, simply may contribute to perception and pursuit of an assumed better path for achieving one's interests.

## 1.10 Book Overview

A brief overview of the remainder of the book is as follows. In Chap. 2, the oldest scholarly treatment of human values is discussed and criticized. It involved the study of values within the discipline of aesthetics, a branch of philosophy. It had the limitation of concerning itself fundamentally with attempts by philosophers to discover the causes of morals and moral truths entirely from the scholars' personal aesthetic experiences generated by thinking about alternative values. Chapter 2 also treats the evolutionary methods and statistical methods and assumptions we use throughout the book to study values.

Chapter 3 presents in detail the parasite-stress theory of values. Immediately below, we briefly mention some of the theory's main features. This theory, based on fact-based evolutionary principles as well as supporting evidence of those principles applied to values, claims to inform a long-standing mystery of political science and sociology: the psychological underpinnings and consequent behavior of in-group favoritism and concomitant out-group avoidance and dislike, and their flipside, out-group tolerance, respect, and affiliation. (On the perpetual centrality of these matters in political research and sociological studies, see Frenkel-Brunswik et al. 1947; Frenkel-Brunswik 1948; Adorno et al. 1950; Lipset 1960; Rokeach 1960; Lipset and Raab 1978.) The parasite-stress theory of values proposes that functionally organized psychological traits, or said differently, psychological adaptations, for dealing with in-group and out-group interactions under variable infectious-disease stress account for a huge range of people's values and their associated behaviors currently and historically.

A brief sketch of the parasite-stress theory of values is as follows. This new theory of values explains and synthesizes how egalitarian values and their opposite, prejudicial values, relate to in-group and out-group behavioral preferences, and how these preferences of individuals and groups relate to avoidance and management of infectious diseases. In this book, as in biology typically, parasite, pathogen, and infectious disease are synonyms.

A fundamental feature of the parasite-stress theory of values is that infectious diseases are an important context for evolutionary change by natural selection. In humans, infectious diseases appear to be the most important cause of recent evolution (changes in allelic frequencies) and account for a major portion of contemporary morbidity and mortality. This, too, appears to have been the case in the distant evolutionary ancestral environments of human history. The evolutionary power of infectious diseases in shaping animal, including human, sociality has not been central in the thinking of scholars until the advent of the parasite-stress theory.

Another fundamental feature of the theory is that parasite–host coevolutionary races are localized geographically. This renders in-group members more immunologically similar to one another than to out-group members. As parasite stress increases in a region, out-group members become increasingly dangerous as sources of contagion and in-group members become increasingly valuable for social alliances and support to offset the negative effects of parasites. As parasite stress declines, out-group contacts and transactions become increasingly advantageous to individuals because of the associated benefits of providing new ideas and technologies and broader and more diverse social networks. The psychological underpinnings of interest are psychological adaptations that function to contingently guide feelings, cognition, decision making, and behavior toward one set of values versus another based on individuals' assessment of the local infectious disease stress.

According to the parasite-stress theory, under high parasite stress, conservative values are adopted. They are optimal ideological solutions under high parasite stress because they reduce contact with novel infectious disease through philopatry (the absence of dispersal from the birth locale), in-group favoritism (ethnocentrism), neophobia (etymologically, fear of the new; fear of nontraditional and nonconformist ideas, ways, and means), and out-group avoidance (xenophobia). Moreover, philopatric, neophobic, ethnocentric, and xenophobic values embed enduringly the ideologue in the local traditions and local social network of supportive kin and friends, and thereby function in managing the adverse morbidity and mortality effects caused by infectious agents when the diseases are prevalent in a region.

Conservatism, too, is an engine of ethnogenesis, dividing a region's original culture into new cultures. The conservative values of ethnocentrism, neophobia, xenophobia, and philopatry fractionate the original culture and cause the rise of cultural group boundaries and thereby produce cultural isolation of groups from neighboring groups. The xenophobia of conservative groups leads to conflict with nearby out-groups, including civil war and other forms of within-region intergroup violence (e.g., clan and tribal wars).

In contrast, under low infectious disease stress, liberal values are evoked. These are beneficial because they promote openness toward and pursuit of new ideas, ways, and means (neophilia, etymologically, new loving), even acceptance of those of out-groups. Liberal values also promote dispersal as well as interactions and alliances with out-groups, even those with different values, skin color, language, and background. These benefits of liberal values, however, exceed the costs of liberal values—i.e., contact with novel parasites harbored in out-groups—only under low parasite stress. Liberalism, by way of its values of the equality of people and tolerance of diversity, leads to mixing among groups and even between distinct cultures. The openness and high intergroup contact, emigration, and immigration of liberal groups prevent or retard ethnogenesis. The liberal values of tolerance and respect of out-groups and conflict resolution by diplomacy reduce within-region, intergroup violence.

Chapter 3 also provides an overview of background published research inspired by the parasite-stress theory of values. This research is quite recent, as is the parasite-stress theory itself.

Chapter 4 treats the findings of social scientists over the last 60 years or so that described the values that yield the two ideological poles, as well as all the versions of ideology between the two poles. One pole is right-wing ideology or conservatism, which is very similar to the value system labeled collectivism by cross-cultural psychologists and sociologists. The antipole is the set of values that give what is referred to as left-wing ideology or liberalism, which is similar to the belief system labeled individualism by cross-cultural scholars. Conservatism–liberalism, or said differently collectivism–individualism, is a continuous one-dimensional variable across countries or other regional divisions or among individuals within a geographic region (Vandello and Cohen 1999; Gelfand et al. 2004; Carney et al. 2008; Jost et al. 2009). Thus, high conservatism (collectivism) is low liberalism (individualism), and vice versa.

Beginning with Chap. 5 and extending through Chap. 13, we provide empirical tests of the parasite-stress theory of values. Each of these chapters focuses on a major testing ground.

Chapter 5 shows that parasite stress predicts collectivism–individualism, family ties/embeddedness, and philopatry across regions of the world. High parasite-stress regions are characterized by collectivism and related high philopatry and extended family nepotism, whereas low parasite-stress regions are characterized by individualism, greater dispersal, and nuclear family nepotism. Chapter 5 also treats collectivistic social systems, so-called cooperative breeding, in nonhuman animals. The parasite-stress theory offers a general theory of family life across human cultures and nonhuman animal species.

Finally, Chap. 5 discusses the application of the parasite-stress theory to the evolution of human reciprocal altruism and to human-unique cognitive abilities. In periods of time of relatively high parasite stress, parochial or localized sociality is optimal for maximum reproductive success of individuals. Under low parasite stress, however, a wider extension of social interactions with nonrelatives—what biologists call reciprocity or reciprocal altruism—including interactions with out-groups is optimal. Natural selection in human evolutionary history crafted a contingent psychological adaptation of reciprocity because local conditions change and affect the adaptiveness of in-group and out-group interaction. Variable local parasite stress, including variation within a human generation, creates a context of continual dynamic change in optimal values and associated behaviors. This context, we argue, created important aspects of the social complexity that led to the evolution of the lofty mental capabilities of people.

Chapter 6 shows that parasite stress predicts human mating systems, mate choice behavior, marriage patterns, and sexual behavior across countries. High parasite-stress regions are characterized by salience of a marital partner's physical attractiveness and sometimes by marital inbreeding. Inbreeding appears to keep together genetic factors (alleles) that defend against local parasites, and physical attractiveness certifies sound condition or what biologists call phenotypic and genetic quality, a major component of which is resistance to infectious disease. Across traditional societies, as parasite stress increases, polygynous marriages increase, because fewer men have attractive phenotypic and genetic quality and the social and other resources that the quality provides. Chapter 6 also documents the relationship between parasite stress and the variable of restrictiveness–unrestrictiveness in sexual relationships. As parasite stress increases, peoples', especially women's, sexual continence or restrictiveness also increases. Chapter 6 also discusses how different value systems of men affect their concerns about out-group men as mating threats.

Chapter 7 treats cross-national patterns in personality dimensions, specifically extroversion–introversion and openness–closed-mindedness to new experiences. As parasite stress increases across regions, people defend by becoming more introverted and closed to new experiences and social contacts. The groundwork is established for a new and scientifically superior theory of personality that takes into account recent findings from the application of the parasite-stress theory to personality. In light of the parasite-stress theory, a synthetic perspective on personality that encompasses both nonhuman animal personality traits and human personality traits is indicated.

Chapter 8 treats interpersonal violence in humans in relation to the parasite-stress theory of values. This theory offers a novel and fruitful approach for understanding the major types of this violence. The conventional value of assumed female inferiority to males and its associated ideology that endorses male physical and sexual control of females are components of collectivist ideology, and hence are most developed in high parasite-stress regions. Moreover, the elevated sexual control of women by men under collectivism is expected from the relatively high priority women place on a sire's genetic quality (physical attractiveness) in high parasite-stress settings, which promotes extra-pair mating by women pair-bonded to men of low genetic quality. Thus, rates of male-perpetuated spousal abuse and homicide, which reflect sexual control of romantic partners by men, are predicted to show positive relationships with parasite stress and collectivism across regions. This is true empirically. In fact, as predicted by the parasite-stress theory of values, all major types of adult-on-adult homicide are related positively to parasite stress and collectivism. Furthermore, it is shown that, across regions, lethal and nonlethal violence by parents on children is associated positively with parasite stress, theoretically because high parasite stress generates more offspring of low reproductive value in which parents divest.

Chapter 9 focuses on religiosity, an important component of human value systems, ranging from atheism to high religious adherence and participation. The parasite-stress theory allows a fuller understanding of religiosity than earlier theories. We argue that religiosity is a collectivist value of in-group assortative sociality and boundary formation, and hence a behavioral defense against infectious disease. It functions to promote the infectious disease defenses of in-group embeddedness, the restriction of social interactions to others with similar values, and creation of a boundary between a religious in-group and out-groups. Parasite stress positively predicts religious commitment across countries of the world and the states of the USA.

Chapter 10 shows that the parasite-stress theory of values predicts political systems of democracy and their opposite, autocracy, across the countries of the world. The parasite-stress theory offers a new theory of political systems. The evidence presented indicates that democracy is caused by individualistic values, and autocracy by collectivist values, and by low and high parasite stress, respectively. Also treated in this chapter is the dimension of democratization pertaining to gender relations: gender equality to inequality. Across countries, gender equality and all other aspects of democracy increase as parasite stress declines. This chapter also looks at how relative emancipation from infectious disease accounts for the increase in liberalism and associated democratization that was the Western social revolution of the 1960s, and accounts for the earliest democratic efforts in the West.

Chapter 11 details how the parasite-stress theory provides a new theory of economics. It emphasizes how parasite stress and associated values affect a region's wealth, diffusion of innovations, and governmental transparency and efficacy. Infectious disease reduces the ability of people to do work and evokes ethnocentric values that restrict people's concern to the well-being of their local community, both of which affect negatively a country's economic productivity. Furthermore, parasite stress generates traditionalism, xenophobia, and neophobia, which act against the creation and diffusion of innovations. Moreover, Chap. 11 shows that international variation in governmental corruption appears to be explicable by the parasite-stress theory. As predicted by the theory, across countries, there is a positive relationship between governmental corruption and ineffectiveness and both parasite stress and collectivism. These patterns arise, in part, from the reverence collectivists place on people in authority (authoritarianism), which gives those in authority greater freedom to violate the interests of the populace and impunity when such violations occur. The patterns also arise from the cronyism—in-group favoritism—of leaders and other officials in collectivist regions. Chapter 11 also treats the highly variable cognitive ability of people across regions and its relationship to economic development. Parasite stress negatively affects cognitive ability (IQ) and hence economic development. We argue that parasite stress variation across regions is a fundamental cause of economic stagnation versus progress. Indeed, parasite stress appears to be the strongest known predictor (negative) of economic productivity.

Chapter 12 shows that parasite stress predicts, across countries, frequencies of within-nation warfare of all kinds (large and small civil wars, clan wars, tribal wars) as well as revolutions and coups. As parasite stress increases across regions, so do collectivism and its associated ideological factionalism and fractionalization of groups. We argue that civil conflicts are caused by these factors. The parasite-stress theory of values provides a new theory of civil conflicts.

Chapter 13 addresses evidence for parasite stress as a generator of within-region biodiversity. The chapter demonstrates that parasite prevalence predicts religion richness (religion number) across the globe, as well as global variation in language richness. Chapter 13 also discusses the role of parasite stress in generating diversity in kinds of nonhuman organisms. High infectious-disease stress fractionates nonhuman species just as it does human cultural groups. This chapter includes discussion of a new hypothesis of ethnogenesis (the origin of new cultures) and of species' formation based on the parasite-stress theory.

Chapter 14 reflects on the book's findings. We discuss our hypothesis for why scientific ideas that are empirically encompassing are rare. We treat the various criticisms of the parasite-stress theory of values and of certain empirical findings that we claim support the theory. Future research areas for additional testing of the theory's empirical implications also are discussed.

In this book, we show that the essence of the culture of the Old South is duplicated across the many places of the world wherever infectious disease stresses are high. Conservative cultures arise independently and are maintained across space and time under high infectious disease stress. Evidence implies that the Old South will rise again only if infectious disease stresses increase dramatically in the region; otherwise, it is gone forever. But devout southern traditionalists need not despair, because the cultural duplicates of the Old South remain in high parasite-stress regions across our planet.

There is some overlap and repetition across chapters. This is necessary, in part, for the synthesis we attempt. It allows bringing material discussed earlier into a given chapter's treatment of a topic. Also, we want the individual chapters to be relatively independent so they can be read without knowing the details of other chapters. The structure of the book is such that after reading the first four chapters, then Chaps. 5–13 can be read in any order. Chapter 14 includes an overview and reflections as well as responses to critics that are most easily understood after all the previous chapters are read.

Sections of some of the chapters contain condensed versions of more detailed treatments of specific topics in our recent papers published in the scientific literature. Where appropriate, we direct readers to the papers for details about methods and results. The book also includes many new analyses and ideas of ours that we have not published previously. We make explicit throughout the book the previously published results and the results first to appear in this book in order to not confuse issues of replication of results. Included, too, are the published findings and ideas of the many scholars who have contributed to the literature across the range of topics in our book. This book's combination of overview of previously published material and new material allows the theoretical and empirical synthesis of diverse areas of human affairs as well as aspects of the social life of nonhuman animals in light of the parasite-stress theory of values.

## 1.11 Summary

We criticize the particularistic method of cultural analysis, which relies on a region's specific cultural history to explain why the region's culture is the way it is. A region's cultural history is part of the problem to be explained scientifically and not a causal explanation of the region's culture. Deductions that arise from particularistic analyses are always questionable as general theories because they may not be supported by cultural patterns outside the restricted region of analysis. Hence, particularistic ideas must be examined broadly across regions and times in order to support or refute them scientifically. The culture of the Old South and USA Civil War of 1861–1865 are discussed to illustrate some limitations of particularistic studies. The particularistic method assumes incorrectly the widely accepted view that culture is passively accepted by future generations. Culture, in some forms of this thinking, is an incorporeal or nonmaterial thing that controls human behavior.

A very different perspective on culture sees people as evolved cultural strategists who accept, reject, and modify values and other cultural items using psychological adaptation that is functionally designed for this purpose. The contents of our book support this perspective on enculturation.

We hypothesize that the particularistic interpretation of enculturation derives from a value system that prioritizes tradition and interdependent self-concept—i.e., the value system of collectivism/conservatism. Research studies are mentioned that could test this hypothesis.

We emphasize the meaning of "theory" in science. We discuss briefly the parasite-stress theory of values, which is the general theory used throughout the book to analyze enculturation and cultural diversity. According to this theory, the level of infectious-disease stress in a region evokes people's values. Low parasite stress evokes liberalism/individualism and high parasite stress evokes conservatism/collectivism. This theory is treated in more detail in Chap. 3.

Scientific findings, including those about values in our book, do not endorse or advocate any moral goal. Although the discoveries about values cannot identify moral or immoral activity, they may allow achievement of whatever moral goals are identified by people. The book's empirical findings, discovered by applying the parasite-stress theory of values, indicate that democratic or equalitarian values can be promoted by reducing infectious disease in a region whereas undemocratic values can be promoted by increasing infectious disease.

Accommodationism is the popular opinion in the West that science and morality are compatible because science's realm of explanation does not encompass morality. All features of nature, however, including religion and other aspects of morality, have causes, which mean that they are explicable by science. Science and religion are compatible in one sense only: the causes of values are amoral and hence do not alone threaten or validate one's values.

Brief overviews of some major features of each of the subsequent 13 chapters of the book are provided.

## References

Adorno, T. W., Frenkel-Brunswik, E., Levinson, D. J. et al. (1950). *The Authoritarian Personality.* Harper and Brothers, New York, NY.

Alexander, R. D. (1979a). Evolution and culture. In *Evolutionary Biology and Human Social Behavior: An Anthropological Perspective* (eds. N. A. Chagnon & W. Irons), pp. 59–78. Duxbury Press, North Scituate, MA.

Alexander, R. D. (1979b). *Darwinism and Human Affairs*. University of Washington Press, Seattle, WA.

Atran, S. (2002). *In Gods We Trust: The Evolutionary Landscape of Religion*. Oxford University Press, USA, New York, NY.

Baschetti, R. (2007a). The dawn of science-based moral reasoning. *Medical Hypotheses* 68: 4–8.

Baschetti, R. (2007b). Evolutionary, neurobiological, gene-based solution of the ideological "puzzle" of human altruism and cooperation. *Medical Hypotheses* 69: 241–249.

Blattman, C., & Miguel, E. (2010). Civil war. *Journal of Economic Literature* 48: 3–57.

Billing, J., & Sherman, P. W. (1998). Antimicrobial functions of spices: Why some like it hot. *Quarterly Review of Biology* 73: 3–49.

Boyd, R., & Richerson, P. J. (1985). *Culture and the Evolutionary Process*. Chicago University Press, Chicago.

Boyer, P. (2002). *Religion Explained: The Evolutionary Origins of Religions Thought*. Basic Books, New York, NY.

Buss, D. M. (2001). Human nature and culture: An evolutionary and psychological perspective. *Journal of Personality* 69: 955–978.

Carney, D. R., Jost, J. T., Gosling, S. D. et al. (2008). The secret lives of liberals and conservatives: Personality profiles, interaction styles, and the things they leave behind. *Political Psychology* 29: 807–840.

Chiu, C., Kim, Y. H., & Chaturvedi, A. (2010). Collective Evolution: Revisiting Donald Campbell's Legacy. In *Evolution, Culture, and the Human Mind* (eds. M. Schaller, A. Norenzayan, S. J. Heine et al.), pp. 39–48. Psychology Press, New York, NY.

Chudek, M., Heller, S., Birch, S., et al. (2012). Prestige-biased cultural learning: Bystander's differential attention to potential models influences children's learning. *Evolution and Human Behavior* 33: 46–56.

Daly, M. (1982). Some caveats about cultural transmission models. *Human Ecology* 10: 401.

Daly, M., & Wilson, M. (2010). Cultural inertia, economic incentives and the persistence of "Southern violence." In *Evolution, Culture, and the Human Mind* (eds. M. Schaller, A. Norenzayan, S. Heine et al.), pp. 229–241. Psychology Press, New York, NY.

Degler, C. N. (1992). *In Search of Human Nature: The Decline and Revival of Darwinism in American Social Thought*. Oxford University Press USA, New York, NY.

Fincher, C. L., Thornhill, R., Murray, D. R. et al. (2008). Pathogen prevalence predicts human cross-cultural variability in individualism/collectivism. *Proceedings of the Royal Society of London Biological Sciences* 275: 1279–1285.

Flinn, M. V., & Alexander, R. D. (1982). Cultural theory: The developing synthesis from biology. *Human Ecology* 10: 383–400.

Frenkel-Brunswik, E. (1948). A study of prejudice in children. *Human Relations* 1: 295–306.

Frenkel-Brunswik, E., Levinson, D. J., & Sanford, R. N. (1947). The antidemocracy personality. In *Readings in Social Psychology* (eds. E. E. Maccoby, T. M. Newcomb, & E. L. Hartley), pp. 636–646. Henry Holt and Co., New York, NY.

Gangestad, S. W., Haselton, M. G., & Buss, D. M. (2006). Evolutionary foundations of cultural variation: Evoked culture and mate preferences. *Psychological Inquiry* 17: 75–95.

Gelfand, M. J., Bhawuk, D. P. S., Nishii, L. H. et al. (2004). Individualism and collectivism. In *Culture, Leadership, and Organizations: The GLOBE Study of 62 Societies* (eds. R. J. House, P. J. Hanges, M. Javidan et al.), pp. 437–512. Sage Publications, Thousand Oaks, CA.

Harris, S. (2010). *The Moral Landscape: How Science Can Determine Human Values*. Free Press, New York, NY.

Henrich, J., & Henrich, N. (2010). The evolution of cultural adaptations: Fijian food taboos protect against dangerous marine toxins. *Proceedings of the Royal Society B* 277: 3715–3724.

Irons, W. (1979). Natural selection, adaptation and human social behavior. In *Evolutionary Biology and Human Social Behavior: An Anthropological Perspective* (eds. N. A. Chagnon & W. Irons), pp. 4–38. Duxbury Press, North Scituate, MA.

Jost, J. T., Federico, C. M., & Napier, J. L. (2009). Political ideology: Its structure, functions, and elective affinities. *Annual Review of Psychology* 60: 307–337.

Lipset, S. (1960). *Political Man*. Doubleday, Garden City, NY.

Lipset, S., & Raab, E. (1978). *The Politics of Unreason: Right-wing Extremism in America, 1790–1977.* University of Chicago Press, Chicago.

Nisbett, R. E., & Cohen, D. (1996). *Culture of Honor: The Psychology of Violence in the South*. Westview, Boulder, CO.

Pinker, S. (2002). *The Blank Slate: The Modern Denial of Human Nature*. Viking Penguin, New York.

Rokeach, M. (1960). *The Open and Closed Mind*. Basic Books, Oxford, U.K.

Singham, M. (2010). The new war between science and religion. *The Chronicle of Higher Education Review*, http://chronicle.com/article/The-New-War-Between-Science/65400.

Tooby, J., & Cosmides, L. (1992). Psychological foundations of culture. In *The Adapted Mind: Evolutionary Psychology and the Generation of Culture* (eds. J. Barkow, L. Cosmides, & J. Tooby), pp. 19–136. Oxford University Press, New York, NY.

Tybur, J. M., Miller, G. F., & Gangestad, S. W. (2007). Testing the controversy: An empirical examination of adaptationists' attitudes toward politics and science. *Human Nature* 18: 313–328.

Vandello, J. A., & Cohen, D. (1999). Patterns of individualism and collectivism across the United States. *Journal of Personality and Social Psychology* 77: 279–292.

van den Berghe, P. L. (1981). *The Ethnic Phenomenon*. Elsevier, New York, NY.

Wilson, E. O. (1998). The biological basis of morality. *The Atlantic Monthly* 281: 53–70.

# Chapter 2
# Evolutionary Aesthetics, Values, and Methodology

## 2.1 Introduction

In this chapter, we treat the basic scientific methods and assumptions used throughout the remainder of the book. We criticize the philosophical method as a way of knowing the causes of values. We also treat the fundamentals of scientific investigation. We discuss evolutionary social theory, distinguish proximate and ultimate causation, and elaborate on the topic of enculturation that was introduced in Chap. 1. We discuss and resolve criticisms of comparative methodology, which is the method we emphasize throughout. Lastly, we comment on ideological criticisms of evolutionary theory applied to human affairs.

## 2.2 Philosophical Aesthetics Is Not Science

Philosophical aesthetics, the study of the beautiful and the ugly using methods from philosophy, is an old, broad, and diverse field of scholarship. It deals with such topics as the beauty of values, as well as the beauty of facial and other bodily features, natural landscapes, architecture, scents and tastes, and art forms (for reviews, see Thornhill 1998, 2003). Aesthetics became a distinct discipline within philosophy with G. Baumgarten's *Aesthetica*, published in 1750, but, as documented by the historian Kovach (1974), speculations by scholars about the nature of beauty and ugliness have been made in the West at least since the sixth century BC in Greece. Philosophical aesthetics applies only pure reasoning to a topic to discover the topic's essence or reason for being. Aesthetic philosophers tried to understand many domains of natural beauty and ugliness this way, including the aesthetic valuation of different belief systems.

The philosophical method provides only a first step toward producing knowledge of how the universe is. Although thinking is essential for discovery, a fatal problem with philosophy per se is the lack of the necessary follow-up in testing whether the

R. Thornhill and C.L. Fincher, *The Parasite-Stress Theory of Values and Sociality: Infectious Disease, History and Human Values Worldwide*,
DOI 10.1007/978-3-319-08040-6_2, © Springer International Publishing Switzerland 2014

thinking is supported by evidence or is not. In science, the thinking or hypothesis must be transformed into predictions about what will exist in nature if the thinking is accurate. If, upon observation, data of that existence is absent, the thinking is then demonstrated to be inaccurate, and the investigation must go "back to the drawing board" to attempt another thought that may have worth as evidenced by tested, empirical support. If the thinking, when transformed into predictions about nature, leads to discovery, the idea or hypothesis generated by the thinking is supported and hence not falsified. This procedure continues and leads to facts about nature.

This method of knowing is the scientific method. In our opinions, this method is humankind's most valuable idea. Our ideology is that there is far too much human misery in the world and human misery can be reduced only when its causes are discovered and eliminated. This is where science comes in, because it is the only way to demonstrate causation. Science does not tell us that a moral world is one without genocide, fascism, racism, classism, sexism, and rape, but, through its ability to determine the causes of those happenings, it can facilitate progress toward achieving such a world if its findings are used in relevant social policy. The scientific method is humanity's nascent path to a world more benevolent and egalitarian than ever imagined by liberal idealism.

In the mid to late 1800s, and in biology's most foundational research, Charles Darwin extended scientific methodology to the study of evolutionary history. Darwin's method of historical science is the method used for understanding deep-time historical causes that are unobservable when they cause their effects. Contrary to arguments by some creationists and accommodationists, we can fully know deep-time historical causation, even though we cannot observe it directly as it happened. As Ghiselin (1969) pronounced, the general method of historical science that Darwin invented has "triumphed." It is straightforward and powerful. Consider this: scientific hypotheses conjecture possible causation, to be tested by empirical evaluation of predictions or consequences. This means that scientific hypotheses about evolutionary histories conjecture possible causation, such as common ancestry of different species or processes of Darwinian selection, which acted in the deep-time past. Actual deep-time historical causes have consequences, which are the predictions offered by evolutionary historical hypotheses to be evaluated through empirical research. Darwin's method can penetrate vast stretches of deep-time history to identify causation; it is applied respectfully and productively not only in biology, but also in all other sciences charged with understanding the distant past, including geology and astronomy. In this book, we apply Darwin's method of historical science to illuminate the evolutionary history of people's values.

The scientific method first arose and was widely discussed and applied during the Scientific Revolution of 1500–1700, which, in turn, fueled the Age of Reason between 1600 and the early 1800s. The Enlightenment was the period of 1700–1800 within the Age of Reason. The Age of Reason was a widespread intellectual movement emphasizing rationalist, antiauthoritarian, liberal, humanitarian, and scientific values, and de-emphasizing tradition, religion, and authority as sources of knowledge. This reduction in the salience of religion in people's values was and is called secularism.

Secularist thinking that excludes science is pure old philosophy—speculation without empirical test—and alone cannot give knowledge of the world. As one noted preacher, Martin Luther, put it, "reason is the devil's greatest whore" (see http://en.wikiquote.org/wiki/Martin_Luther). He meant that reason is a powerful lure that can pull one down the path of false and trivial thought (in Luther's religious moral view, the path of sin) (Pelikan and Lehmann 1955–1986). Secularist rationalism has no more claims on truth than religion or other ideology. Regardless of how principled and objective a philosopher's thoughts seem to be, they are trivial until they show claim to empirical ground through testing. The scientific method is the guillotine for thought incapable of discovering the empirical reality of nature.

Aesthetic philosophy, then, is an anachronistic and failed way of knowing because it cannot empirically falsify or verify hypotheses. As a result, it has been replaced by science. Science is the study of cause and effect using the scientific method. The effects of interest to scientists are the world around us as it is. Each of the three major branches of science—biology, chemistry, and physics—is charged with understanding the causes of the effects or features in its research domain.

## 2.3 Biology Encompasses All Life

The adjective "biological" is defined typically as, of or pertaining to life. Biology is the scientific study of all of life's features. Examples of subdisciplines of biology are anthropology, biochemistry, botany, economics, entomology, ethics, genetics, history, linguistics, ornithology, paleontology, parasitology, physiology, political science, psychology, sociology, and all other fields engaged in study of life. The dichotomy of social versus biological made by some is highly erroneous, because many of life's effects are social interactions, which are mediated by environmentally cued computations in brains. Using biological as a synonym of genetic is misinformed profoundly too, because genetics is only one of many subdisciplines of biology. Moreover, genetic causes have no primacy among all the causes of biological features (see below).

Scholars are evolved social animals and hence are socially competitive. Ideas and values, inside and outside scholarly realms, are major tools of social competition (Alexander 1979a, b; Flinn 1997). As a result, many areas of scholarly study of people have arisen because they provide useful means of social competition for their originators and followers. Regardless, all the areas of science dealing with life, including all those dealing with human activity, are necessarily tied together conceptually by, and anchored in, the fact that all life's history on Earth was an evolutionary history involving two distinct categories of evolutionary causation creating each feature or trait of individual organisms: (1) phylogenetic origin of the trait, and (2) the trait's maintenance after its origin. In fact, there is simply no such thing as a nonevolutionary study of humans because humans are evolved animals. Likewise, there is no such thing as a nonbiological study of humans because humans are living things. Hence, all studies of humans are evolutionary and biological studies. These studies only differ in their degree of sophistication in using evolutionary theory to make discoveries about humans. This dimension of difference, however, is huge. Many scholars in research areas investigating human affairs lack an understanding of evolutionary biology and its heuristic power for illuminating human behavior. Ignoring evolutionary theory in research on any topic of life amounts to ignoring the most basic and fundamental knowledge for understanding the topic.

Yet, researchers in the many traditions of human scholarship not informed by evolutionary theory are increasingly realizing that biology's general theory and Darwin's method of historical science are essential for creating new and empirically fruitful ideas and testing those ideas. As an example, the economist Robert Frank recently discussed why this is so for economic scholarship. (See his article in *The New York Times*, July 12, 2009.) Our book is a testament to the utility of the parasite-stress theory of values for informing many fields of human scholarship, and, most importantly, for their empirical and conceptual synthesis. Just as research in economics is severely intellectually limited when not inspired by Darwinism, so is that in political science, anthropology, history, psychology, sociology, medicine, and so on across all the fields that investigate matters of human life. It is our hope that the parasite-stress theory of values will be an important bridge between evolutionary theory and the scholarly traditions that have ignored evolutionary theory.

## 2.4 Other Reasons Why Philosophical Aesthetics Failed

### *2.4.1 Common Sense Is Biased*

> By far the greatest hindrance and aberration of the human understanding proceeds from the dullness, incompetency, and deceptions of the senses; in that things which strike the sense outweigh things which do not immediately strike it, though they be more important. — Francis Bacon (*Novum Organum, Book I*, 1620)

We have given some explanation earlier of why pure philosophical reasoning itself cannot, and without scientific testing, discover the causes of nature. There are additional fatal problems with pure reasoning as a way of knowing. Fundamentally, thinking and deduction are biased, i.e., prejudiced, by intuitions and common sense interpretations that arise from personal values. The psychological machinery of human reasoning and deduction was favored by Darwinian selection because it promoted the reproductive success of those who reasoned and not because it yielded knowledge of the causes of nature's features. Given that human reasoning is biased, as Bacon realized, testing of the empirical implications of thoughts is the only path to discovery of how nature actually is. Because of cognitive biases, to stop at thinking and not test the thinking cannot answer any question about nature.

Common sense, intuition, and emotional validation of ideas actually can be impediments to discovery, because much of the true world is counterintuitive. The Earth is not flat, nor is it the center of the universe, as most thought prior to the relevant scientific discoveries. Life's history on Earth, including that of humans, was an evolutionary history despite the intuitions and common sense that say otherwise to the majority of people. The causes pertaining to the evolution of life, particularly those about the evolution of human behavior and psychology, are often counterintuitive, in part because evolution involves ultimate causes—those that acted prior to the conception and development of the sapient humans who try to contemplate causation (for further discussion, see Goddard 2009).

We have considerable personal experience with the difficulties in teaching university and college undergraduates and university graduate students about evolutionary biology and its application to human behavior and psychology. Thornhill has done so for 40 years. In order to understand themselves, other people, and life in general, students first must be made aware of their fundamental and pervasive reliance on value-based common sense and intuition rather than on scientific evidence. Once that awareness is deeply comprehended and constantly kept in mind, students can become scientifically objective in viewing themselves and others.

### *2.4.2 Intuitions Vary Because Core Values Vary*

People experience and understand the world in terms of their personal moral or value system, which results in a powerful bias in people's thinking. Furthermore, as we show in detail in Chap. 4, the differences are large between conservatives and liberals in intuition and cognition (e.g., Jost et al. 2003, 2009; Carney et al. 2008). For example, collectivists interpret events in the context of the events' relation to in-group togetherness, maintenance, and goals, whereas individualists interpret events as independent of the in-group (Hofstede 1980). Recall, collectivists have an interdependent view of their world, whereas individualists see things as independent. A simplification of a lot of interesting research in educational psychology is the following. A kindergarten teacher holds up an egg and then asks small children to describe the last time they had cooked and eaten eggs. Collectivist children describe family members interacting and cooperating to cook eggs—their thinking is in-group in focus. In contrast, individualist children focus on events and phenomena that are mentally independent of the in-group—they describe the various physical properties of eggs (Trumbull et al. 2001).

Other examples of the distinctly different cognitive styles of collectivists and individualists include collectivists' mental rigidity toward valuing traditional and conformist views—i.e., dislike of change because of its perceived threat to stability and security—in contrast to the open-mindedness and preference for change of individualists (Jost et al. 2003; Carney et al. 2008). This is sometimes described as collectivist neophobia (fear and dislike of new or different ideas and ways) and individualist neophilia (Thornhill et al. 2010). This difference is seen in attitudes about the validity of science and intellectual activity in general, as well as in attitudes about new technology—conservatives are more negative, and liberals are more supportive and open (Jost et al. 2003, 2009; Thornhill and Fincher 2007; Carney et al. 2008; Ferris 2010). What people find funny also depends on their value system. Conservatives enjoy humor that is unambiguous and has closure; liberals respond more positively to humor that has ambiguity and lacks closure (Ruch and Hehl 1983). Also, a difference in thinking between conservatives and liberals is observed when past events come to mind. Conservatives interpret their past in more positive terms than do liberals (Thornhill and Fincher 2007). As a final example, conservatives endorse and support human inequality, whereas liberals endorse and strive for equality, as in all people are equally human and important (Carney et al. 2008; Jost et al. 2009). Both conservatives and liberals are using behavioral displays of values in socially strategic ways—that is, to gain social resources in their respective value-based social niches.

The differences in cognitive styles described in the examples just mentioned reflect those at the two ideological poles. Collectivism–individualism is a continuous variable across individuals (Chap. 4). Hence, any given topic will be seen in a range of ways across people, depending upon where each person falls on the values' continuum. Information processing and deduction are value laden, and values vary among people. Hence, views of reality and priority vary among people. The scientific method is the only known method of knowledge gain that can distinguish how the world actually is versus how people interpret it or want it to be as part of their personal value system.

Aesthetic philosophers tried to solve the question of the nature of a beautiful versus ugly idea or value, using only their biased value systems. "Beauty is truth" was a perpetual and central argument in traditional philosophical aesthetics (Kovach 1974). This arose from the view that beauty, or what makes you feel good to see, hear or think, is therefore morally good, while ugly, or what feels bad, is morally bad. In scientific practice, motional validation of something experienced, in itself, does not count as evidence.

### *2.4.3 Personal Introspection Is Incomplete Analytically*

Another problem with philosophical aesthetics for gaining knowledge is that much of the information processing involved in aesthetic judgment and associated reasoning is not available to analysis by introspection. Even in the imaginary world of a person without any values at all (hence his or her thoughts could not be biased), the causal basis of judgment and action remains unavailable to personal introspection. Most human mental activity is not experienced consciously. For instance, the mental activity that causes routine respiration and blood circulation is continuous, but this information processing is regulated by psychological adaptations that do not produce consciousness. Beauty experiences are felt consciously, but the specific information processed by psychological mechanisms that generates the experiences is not. Given this, thought per se cannot identify the causes of beauty judgments, including the causes of aesthetically pleasant and unpleasant values. Science is required to identify the data processed that result in aesthetic judgments. Conscious beauty experiences upon encountering a beautiful human, habitat, or ideology can be even profound, positive experiences, and so is the case of encountering the opposite of beauty. The conscious profundity of certain aesthetic judgment misled early scholars in philosophical aesthetics to reason that they could identify by reasoning alone the causes of the experiences (Thornhill 1998, 2003).

As Kaplan (1992) pointed out for landscape aesthetic judgments, the preferences are made rapidly and effortlessly, but research subjects' explanations of their choices have no relationship to the differences that are associated empirically with preferred and nonpreferred landscapes. The same is seen in people's judgments of physical attractiveness of faces and bodies: they are made rapidly and effortlessly and without awareness of the features actually assessed (e.g., symmetry) or the features' relationships to health or, more accurately, to phenotypic/bodily quality (for reviews see Thornhill and Gangestad 1993, 1999a, b, 2008).

Moral disgust is a strong emotion felt when a person or group's behavior violates what is considered to be moral by the person disgusted (Tybur et al. 2009). This is felt without knowledge that the behavioral norm violation is an indication of out-group threat and associated novel infectious-disease threat (Chap. 3; Curtis 2007; Schaller and Murray 2008; Oaten et al. 2009; Inbar et al. 2012).

## 2.5 Scientific Aesthetics

For the reasons discussed, the understanding of aesthetic judgments, including those made about morals or values, was not advanced by philosophical aesthetics. In large contrast, scientific aesthetics has greatly advanced knowledge of aesthetic judgments. Scientific aesthetics uses the scientific method to determine the causes of experiences of attractiveness of things, including values, and why beauty assessments exist in the first place.

Scientists studying aesthetics are pursuing knowledge of the two categories of causation in biology, proximate causes and ultimate causes. Proximate causes are those that act during the lifetime of the individual to bring about their effects. These causes include physiology, psychological mechanisms, development (sometimes called "ontogeny"), genes, cues or stimuli arising from the ecological ambience, and information processing. Ultimate causes act in the deep-time past to bring about their effects—they are the evolutionary causes. Proximate and ultimate causes in biology are complementary causes, not competitive/alternative ones. Complete knowledge of a biological effect of interest, such as the mockingbird's song or a person or group's values, requires knowledge of both proximate and ultimate causation of the effect.

## 2.6 Categories of Ultimate Causation

As mentioned earlier in this chapter, ultimate causes are of two categories. One is the evolutionary origin (often called "phylogenetic origin") of a feature on the Tree of Life—the feature's debut in life's history on Earth. Novel traits/phenotypes arise on the tree from developmental processes that modify preexisting traits; developmental causes of phylogenetically new phenotypes are ultimate causes because they brought about effects—new phenotypes—by acting in the deep-time past. When the new phenotype and the preexisting phenotype differ genetically, the potential is created for changes in allele frequencies—that is, for evolution (West-Eberhard 2003). The second category of ultimate causation is the maintenance of the trait after its phylogenetic origin. Maintenance causes are the evolutionary agent's Darwinian selection and drift. Selection for a trait may be the result of the trait itself as the actual target of selection (direct selection). In this case, the trait itself increases reproductive success of individuals that bear it and therefore is favored by selection. Or the selection for the trait may be the result of its correlation with another trait that selection actually favors (indirect selection). Directly selected traits are called evolved adaptations, and indirectly selected traits are called incidental effects/by-products. The two categories of ultimate causation—origin and maintenance—are complementary: comprehensive understanding of ultimate causes of any feature of life requires knowledge of both its evolutionary origin as well as its maintenance after it first appeared on the Tree of Life (Thornhill 2007).

## 2.7 By-products

By-products are traits that have been maintained evolutionarily because they are linked to features that are directly selected. The color of chicken egg yolk is not an evolved adaptation; it is a by-product of a maternal care adaptation: the carotenoid pigment deposition in the yolk giving its color actually functions to nurture the developing embryo in the egg. There was direct selection in the past for carotenoid deposition in the egg because it increased maternal reproductive success, while the color itself is incidental to this functional effect and was indirectly selected. Similarly, in mammals, nipples of males are a by-product of female nipples; the nipples of females were directly selected and function in delivering milk to young offspring.

Many human values and associated behaviors are by-products of adaptation, meaning they are not the effect that gave rise to direct selection. That is, they are not the manifestations of the psychological adaptation involved that caused this feature to be favored by direct selection and to therefore become an evolved adaptation. By-products are common in all categories of phenotypic traits, not just psychological and behavioral traits (Andrews et al. 2002; Thornhill and Gangestad 2008). By-products are as evolutionary as evolved adaptations; the existence of both is caused ultimately by—that is, they exist as a result of—selection's action in the past.

An example of a value-based behavior that is a by-product is the handling of venomous snakes or the drinking of poison as part of religious services in certain Christian groups (Hood and Williamson 2008). There was no direct selection for these behaviors or the very immediate cognitive framework causing them. Indeed, there was past direct selection for avoiding venomous snakes and reducing toxin intake. The psychological adaptation that proximately causes these two religious behaviors, we propose, is functionally designed by past selection to locate and adopt ideas, values, and behaviors that allow the ideologue to honestly signal in-group commitment, allegiance, and boundary (Chap. 9). The same interpretation applies to religious ritual in general. The rituals that are used as in-group commitment displays change periodically in order for members of a group, or members of a budding subgroup, to best exhibit their in-group identity, embeddedness, and boundary. Another example of a by-product value is the preference typical of southerners to drawl the vowel in "dog" as "daaaa … wg." This preference, we hypothesize, is an incidental effect of human-typical linguistic adaptation that functions in signaling in-group affiliation—as is the pronunciation of dog without vowel drawl.

Our point is that there was no direct selection that favored individuals who handled venomous serpents, drank poison, or drawled vowels. These behaviors and related values are incidental effects of directly selected psychological adaptation for acquiring honest signals of in-group affiliation and commitment. These incidental effects are functional, but that in no way makes them evolved adaptations. By-products, which, by definition, are indirectly selected as a result of direct selection for the adaptation that accounts for them, can be functional or not, and currently adaptive, maladaptive, or neutral. Most of the cultural variation in values and associated behavior of people is by-product (also Chap. 3).

Andrews et al. (2002) apply the label "exapted learning mechanisms" to by-product psychological mechanisms that become functional as a result of change or novelty in the cultural environment. As examples, they mention the learned abilities of driving a car and reading. Both of these by-products arise from adaptation for purposes other than driving or reading, because these two tasks are evolutionarily novel and hence there cannot be adaptation designed to accomplish them. Similarly, the handling of venomous snakes, drinking poison, and drawling vowels are also behavioral outputs of exapted learning mechanisms and are socially functional in the local cultural milieu in which they are typical.

## 2.8 Units of Selection and Altruism

A basic finding of evolutionary biology, and the research result most fundamental to understanding all living organisms including people, is that Darwinian selection acts most effectively at the *individual* level to bring about evolution, not at higher levels in the hierarchy of life such as populations, societies, and species. This was the scientific conclusion within evolutionary biology by the 1970s (see Trivers 1985).

We emphasize that this conclusion is not just our opinion or merely a popular opinion in biology unsubstantiated by strong evidence. For example, the discoverers of the basic ideas behind it were given the prestigious Crafoord Prize, a Nobel-like prize that is awarded in the sciences including biology and recognizes science's biggest or most encompassing ideas. William D. Hamilton won this prize in 1993, George Williams and John Maynard Smith in 1999, and Robert Trivers' in 2007.

The ideas behind these awards were published in the 1960s and 1970s. Of course, the masses of biologists who tested and empirically supported these ideas and continue to do so figure centrally, too, in biology's conclusion about the relative effectiveness of selection at different levels of life's hierarchical organization.

Darwinian selection is the nonrandom differential reproduction of entities. Thus, it can act at all levels in life's hierarchy because each level consists of entities with trait variation among them that can result in their differential survival and reproduction. For example, some species multiply (speciate) more frequently than others or others may not persist phylogenetically in the Tree of Life. Also, some human cultures persist and give rise to "offspring" cultures, whereas others do not. When such differentials are caused by trait differences, the differential reproduction is Darwinian selection by definition. When the differential is the result of chance (randomness), rather than trait differences, the process by definition is drift. Individual selection is differential reproduction of individuals resulting from their trait differences.

Individual selection's power in causing evolution, relative to the weak selection processes at higher levels, is derived from two facts. First, individuals are far more common than groups (e.g., populations or human societies), providing more variation among individuals than among groups for selection to act upon. Second, individuals turn over far more quickly—possess a higher rate of reproduction and mortality—than groups. These and other well-established facts—e.g., there is higher intergeneration resemblance in individual traits than in group traits—account for why individual organisms are designed to strive in their own reproductive interests rather than the interests of their group as a whole, and why there is a total absence of evidence for evolved adaptation of organisms that has the evolved function of group benefit at the expense of individual reproductive success.

The view that individual organisms are functionally designed for—i.e., have adaptation for—the good of the group (e.g., population, culture, or species) assumes an evolutionary history of more effective group selection than individual selection. This is because the two levels of selection act on altruism in opposing directions: group selection favors group-benefiting altruism that has a net cost to individual reproductive success, but individual selection always acts against such. Individual selection "wins" because of its much greater power in driving evolution.

The conditions needed in nature for group selection to exceed individual selection in power seem to be outside natural reality and possibility (e.g., Williams 1966; Lewontin 1970; Dawkins 1976; Trivers 1985). The conditions of groups turning over more rapidly than individuals and of group number exceeding individual number are not features of the structure of populations, including those of people. In essence, group selection as a cause of evolution (allele frequency changes) could exceed individual selection as a cause of evolution any time selection on individuals within groups does not exist. This circumstance's occurrence, however, is impossible in nature. It is not surprising, then, that there is no evidence for adaptations of individuals that are functionally organized for group welfare. Of course, group welfare sometimes results from evolved adaptation possessed by individuals, both behavioral and other evolved adaptation, and when this occurs, the group benefit is always a by-product, incidental to the reason the adaptation was favored by individual selection. Differences among groups (cultures) in by-product group welfare may lead to intergroup selection—i.e., differential survival and reproduction among groups due to their trait differences—but it does not follow that this intergroup selection will create group-selected evolved adaptations.

The published literature claiming the inordinate power of group selection and even that the human psychological features that account for culture are group-benefiting adaptations is substantial (reviewed recently by West et al. 2011). Although that literature sometimes gives lip service to the fundamental flaws in the group-selection arguments mentioned earlier, it characteristically lacks serious consideration of these flaws and of the empirical evidence showing the absence of group-selected adaptation. It just proceeds as if human behavior and culture can be productively researched from the viewpoint that effective group selection for group-benefiting altruism created significant features of human social behavior.

We hypothesize that the remarkable tenacity of the group-selection paradigm in the study of human culture reflects a collectivist value in those who advocate it. To the collectivist, things are understood in terms of how they bear on group harmony and well-being. If our hypothesis is supported in future research, it will be a good example of how personal values of people give intuitive, commonsense perspectives that are fundamentally flawed theoretically and empirically. We certainly recognize however that values of group welfare can arise from liberal values because of the importance of the well-being of strangers, even those in future generations. Hence the tenacity of the group-selection paradigm may be mediated by values from either ideological pole alone or by values from both poles.

Some critical commentators may suggest that our opinion that the best theory is differential reproductive success of individuals simply reflects our personal individualistic values. Baschetti (2007a, b) argues this in reply to the widespread negativism among biologists toward any role for group selection in creating phenotypic features of organisms. Baschetti (2007a, b) attributes this negativism simply to Western individualistic ideology and its corresponding anticollectivist ideology. Refuting Baschetti's view of why biologists prioritize individual-level natural selection as ultimate causation are the quality of the theory and the associated vast evidence supporting our favored scientific view of people and other organisms (e.g., see Abbot et al. 2011; West et al. 2011).

## 2.9 Individually Selected Altruism

The absence of evidence of evolved adaptation of individuals that is functionally designed for group benefit contrasts sharply with the universality of altruism adaptations of individuals that have been produced by individual selection. Risks taken in benefit of and resources given to offspring by parents (parental nepotism) are characteristic of parents across all species of plants and animals. Risks taken for and resources given to nondescendant kin, so-called extra-parental nepotism, also are common forms of altruism across many taxa of animals. Parental nepotism and extra-parental nepotism show functional organization. To understand the evolution of these types of altruism requires the realization that individuals reproduce—i.e., propagate their alleles across generations—by producing and taking care of offspring and sometimes other descendant kin (grandchildren), as well as by assisting their nondescendant kin. Assisting the reproduction of full- and half-siblings, nieces/nephews, and first or more distant cousins provides avenues of individual reproductive success as assuredly as having one's own children and investing in them.

That realization, first published by William D. Hamilton (1964), revolutionized biology's concept of fitness. Darwinian or classical fitness is an individual's design for direct reproductive success through production and aid of descendant kin (offspring typically, but in humans grand-offspring commonly as well). Hamiltonian fitness, or so-called inclusive fitness, is an individual's design for direct reproductive success (classical fitness) plus the individual's design for indirect reproductive success arising from the enhanced reproduction of nondescendant kin as a result of its extra-parental nepotism. Hamilton's inclusive-fitness theory is just that—a scientific theory—or more exactly, a major theory of evolutionary biology. Its three major predictions are: kin will be assisted more than nonkin, close kin will be assisted more than distant kin, and kin of high reproductive potential within a category of kin (e.g., full siblings) will receive more nepotism than kin of low reproductive potential in the same category. These three predictions are strongly supported empirically and have led to thousands of supportive studies of the social behavior of nonhuman animals and hundreds of such studies of humans (see Abbot et al. 2011; Bourke 2011; West et al. 2011). Fundamentally, an individual's inclusive fitness determines whether it is favored by Darwinian selection.

Reciprocity among nonrelatives, in which humans in particular engage commonly, is also a product of individual-level selection. Robert Trivers (1971) first proposed this important idea. It is straightforward: if altruists get return goods and/or services from nonrelatives that boost the altruists' reproductive success (inclusive reproductive success) above the costs to the altruists' reproductive success of giving altruism to the nonrelatives, then helping nonrelatives will be favored by individual selection. Trivers' idea about the evolution of social life, like Hamilton's, has been heuristic, generating a huge literature and receiving strong support as a basic part of the design of human social psychology.

After the seminal ideas of Hamilton and Trivers, significant extensions of the evolutionary theory of altruism were recognized: the concepts of indirect reciprocity, competitive altruism, and altruistic punishment. These three topics are related integrally to Trivers' ideas on reciprocity and are often features of nepotistic social networks as well.

Adaptive indirect reciprocity occurs when an altruist receives return altruistic benefits that promote its reproductive success from individuals other than the individual(s) to whom the altruist gives benefit. Onlookers of altruistic acts return the benefits instead of the recipient(s) of the altruism. The onlookers need not actually observe the altruistic behavior; they may selectively direct their altruism based on the reputation of others: giving altruism to those with a reputation for generosity and withholding altruism when reputational information implies an unwillingness to assist others. A reputation for altruism serves as an honest and community-based label of one's worth as a reliable and giving social partner. In human evolutionary history, altruistic reputation affected the quality and quantity of reciprocal alliances one achieved, and thereby affected individual reproductive success. Indirect reciprocity is the basis of the great effort people place in building and maintaining reputations of kindness and helpfulness, as first explained in evolutionary terms by Richard Alexander (1987).

Given the importance of reciprocity in humans, individuals have been designed by evolutionary selection to display their altruism to others. These displays are socially competitive for the altruistic benefits of others that can be received by way of altruistic alliances. Competitive altruistic displays may involve direct acts of altruism by competitors or competition using reputational building.

Altruistic punishment is punitive actions against group members who do not follow normative group behavior. The altruistic punishers suffer costs by engaging in the punishment, while other group members gain when the punishment results in norm conformity. Altruistic punishers obtain their gains through a reputation of being competent social partners, because of their convictions about and knowledge of the norms and goals of their group (see also Pedersen et al. 2013). Indirect reciprocity, competitive altruism, and altruistic punishment are important components of human moral or ideological systems (Alexander 1987).

## 2.10 A Role for Parasites

We have proposed that parasite stress was a major selective force that shaped the condition-dependent human psychological adaptations that regulate nepotistic behavior, both its intensity as well as its extensiveness outside the nuclear family (Fincher and Thornhill 2012). Extended nepotism is important for creating the reliable social network of kin that reduces individuals' morbidity and mortality under high parasite adversity. Under low parasite stress, nepotism is more restricted and may involve only the nuclear family. When the human family organization is based on intensive nepotism of extended family members, it is referred to as "cooperative breeding," because extended family members assist breeding pairs by serving as helpers at the nest, thereby enhancing the helpers' inclusive reproductive success (Hrdy 2009; Kramer 2010; Jones 2011). We hypothesized that parasite-stress variation may be the selective history for the variable family structures seen across nonhuman animals that range from nuclear family only, to small groups of helpers in cooperative breeding efforts, to extreme eusociality with sterile worker castes as in ants and termites (Fincher and Thornhill 2012). Also, variable parasite stress may account for important aspects of the design of human psychological adaptations that regulate reciprocity among nonrelatives (Thornhill et al. 2009). We discuss infectious disease in relation to nepotism, reciprocity, and family organizations more completely in Chap. 5.

## 2.11 The Ontogeny of Culture

Proximate causes produce their effects during an individual's lifetime, including during its development or what biologists call, its ontogeny. An individual's ontogeny is ongoing from the time of conception to old age and death. Oftentimes, organisms' features are caused developmentally during a distinct window—or windows—of ontogeny. An example is the pubertal feminization of women's faces and bodies by estrogen (Thornhill and Gangestad 2008). Other examples are the learning of language and sibling incest avoidance by socialization during childhood. Sibling incest avoidance is caused (proximately) by social interactions of children during childhood (Lieberman et al. 2003, 2007; Lieberman 2009; DeBruine et al. 2011). Hence, language and sibling incest avoidance are socially learned—learned by interaction with conspecifics or members of the same species.

Learning is an ontogenetic event. It is defined typically as the acquisition of cognition and behavior as a result of ontogenetic experiences. All ontogenetic causes of all traits are experienced as the individual develops. Nonlearned phenotypic traits—e.g., the human nose—are caused by ontogenetic experiences that do not bring about subsequent cognition and behavior. Social learning is learning in which a causal ontogenetic experience is conspecific interactions. Such interactions may or may not involve the transmission of information between individuals by teaching or imitation. Sibling incest avoidance is socially learned but does not require imitation and teaching—mere co-socialization during ontogeny can cause it (Lieberman et al. 2003). The acquisition of language and dialect, however, includes social learning by imitation and teaching alongside co-socialization (Nettle 1999).

In people, social learning begins in childhood and continues throughout life. Social learning during ontogeny is a major way humans acquire the values that are optimal for navigation in the local culture. The interactions with conspecifics that proximately cause enculturation give rise to vertical transmission of values/preferences of individuals between generations (by way of genetic relatives or nonrelated individuals) and horizontal transmission between individuals of the same generation (Boyd and Richerson 1985). Both avenues of transmission include psychological adaptations of teaching, differential attention, imitation and anti-imitation, and sometimes of manipulation or coercion. Paying attention to, copying, learning from, and emulation of those people with status, prestige, popularity, skills, knowledge, intelligent action, desirable social networks, health, and physical attractiveness are psychological mechanisms natural selection should have favored directly during human evolutionary history. A reproductive advantage of copying socially successful and high-phenotypic-quality others is that it avoids the costs of time and error inherent in trial-and-error learning. Selection is expected to have favored accurate assessment of models to pay attention to and emulate. Similarly, selection will have favored directly psychological adaptation for anti-imitation of those people without traits and resources that ancestrally promoted reproductive success. Moreover, selection should have favored directly psychological adaptation for identifying and avoiding manipulation and coercion, and for making the best of it, in terms of enhanced reproductive success, when manipulation or coercion occurs in the context of cultural item choice (for more discussion of ancestrally adaptive biases in social learning, see Daly 1982; Flinn and Alexander 1982; Boyd and Richerson 1985, 1996; Henrich and Henrich 2007, 2010; Chudek et al. 2012).

Trial-and-error learning, despite its costs, is a component of the social learning of values that complements copying those with social success and high phenotypic quality. A person's values when manifested in behavior result in responses from others. Modification of one's values based on how one is treated by others is promoted by self-awareness. Alexander (1989, 1990) has hypothesized that self-awareness is the component of human consciousness that is functionally designed to provide information about how others perceive one's actions, thereby allowing adaptive modification based on social feedback. Copying the values of successful models, combined with modifications based on use of those values and resultant social feedback, allow one to develop and maintain a repertoire of values that is socially effective locally. Every social interaction is a scenario for trial-and error learning of more effective values for local social navigation. Moreover, as Alexander (1989, 1990) has explained, people expend great effort in participating as self-aware players in imaginary alternative social scenarios that provide practice of how others will respond.

Another route to cultural learning of values is by way of ontogenetic experiences with environmental features other than people. In this case, because interactions with conspecifics are not causal, this route of cultural acquisition is not social learning. This second route, however, is often influenced by conspecific interactions. Experiences with food types during ontogeny affect the culinary cultural values acquired by people, but which foods taste good versus bad depend, in part, upon their being offered by people and how people feel about and behave toward them locally (Siegal et al. 2011). And fear of spiders and snakes arises ontogenetically in a person, in part, due to social learning: when other people react cautiously or defensively toward these animals (Ohman and Mineka 2001).

The parasite-stress theory of values led us to hypothesize that personally optimal values for avoidance and management of infectious diseases arise ontogenetically, in part, from an individual's experiences with his or her own immune system's reaction to local parasitic adversity (Thornhill et al. 2009). Hence, high and enduring activation of the immune system evokes conservative values, and low activation evokes liberal values. This aspect of value learning is not socially learned—not affected by conspecific interaction—but instead is caused by one's experience with its own immune system.

We expect, however, that various types of socially acquired information will interact with immune-system information to evoke values for dealing with local parasite adversity. This social input will include awareness of local people with debilitations, bodily markers of trauma, developmental irregularities and other bodily deviations from local typicality, as well as folklore and stories about local disease risks.

The term culture is applied to both individual- and group-level features. Learning through social interactions that often includes trial and error as well as learning through nonsocial experiences gives rise to a person's cultural values and behavior. At the group level, culture is the collection of learned values and associated behaviors of a people at a place and time and the related artifacts and institutions such as stories, recipes, schools, agriculture, hand sanitizer, septic tanks, public goods and services, government, law, and so on.

Social learning, ultimately speaking, is phylogenetically ancient, and much older than the ancestral species of the Primates. The scientific study of the positive bias in acquisition of behavior by copying socially successful and high quality individuals and antimimicking socially unsuccessful and low quality individuals is a huge area of cross-species research. Even some fish and birds are cultural strategists in this sense (see useful reviews by Dugatkin 2000; Seppänen et al. 2011). The elaborate human capacities for cultural learning phylogenetically debuted in the branch of the Tree of Life culminating in *Homo sapiens*. Also, ultimately speaking, the evolutionary function of human cultural learning—the reason it was directly favored by selection—is the acquisition of information about the local culture that maximized the learner's survival and, most importantly for the action of the creative Darwinian selection involved, maximized inclusive reproductive success in human evolutionary history.

Enculturation of the individual is an active ontogenetic process of choice of information from the environment that yields reproductive competence in the local culture. Hence, as discussed in Chap. 1, enculturation is not based on passive processes that generate automatic cultural transmission. Although the passive view is assumed widely, its fundamental error is in not recognizing how natural selection will have acted during the evolution of the psychological adaptations involved in the ontogeny of culture. Selection was always against arbitrary learning and automatically accepting into one's cultural repertoire all that others model or tell one to use. Selection was always for psychological features that discriminate among cultural items and acquire those that result in the highest inclusive reproductive success of individuals. The conjecture that people acquire their culture passively sometimes is a part of the empirically unsupported notion that culture is transmitted by a nonmaterial process that works independently of the action of evolved strategic human brains. The view that enculturation is a passive process derives, in part, from human reasoning that fails to recognize or understand that much or most of cultural acquisition is caused entirely by unconscious information processing, cognition, and deduction. If the parasite-stress theory applied to cultural acquisition is correct, the implication is that the cultural repertoires of people are caused importantly by ontogenetic, unconscious experiences with local infectious diseases and their effects on people. That most causation of values is outside consciousness, as we emphasized earlier in this chapter, is one reason why thinking unaccompanied by empirical support cannot provide knowledge of the causes of values, whereas the scientific method does.

We are proposing that the ontogeny of people is designed to choose ideas, ways of thinking, and attitudes—i.e., choose values—that promoted effective and ancestrally adaptive social navigation in the local culture. The recognition that values are chosen by individuals is quite old in the sociological and psychological literature (see Jost et al. 2009). We add to this traditional sociological view by our general theory of values that claims that the choices are by evolved design—that is, they are guided by psychological adaptations dedicated to the function of value acquisition to meet ecological adversity and demands pertaining to infectious-disease stress, and that much of the choice is unconscious by design.

## 2.12 Current Adaptiveness

It is essential to realize that there is no implication here that contemporary culture anywhere or everywhere is currently adaptive, i.e., currently promotes net inclusive reproductive success of individuals. The truism is that culture-acquisition adaptations promoted net inclusive reproductive success of individuals in the environments in which the adaptations were favored by selection. Current circumstances faced by humans (and other organisms) can be evolutionarily novel—meaning that the circumstances are not the same as those in which adaptations were favored by selection—which can lead to maladaptive or adaptively neutral traits, including cultural traits.

Evolutionary novelty is common. Humans have aesthetic adaptations that guide them to prefer social partners with facial and bodily features that corresponded to sound phenotypic condition in evolutionary history (developmental health, hormonal health, and limited senescence) (Thornhill and Gangestad 2008). This does not mean that attractive people now and everywhere are expected to have more children than unattractive people—in much of the world, a multitude of evolutionary novel factors such as hormonal birth control, healthcare, and cosmetic surgery disrupt the ancient, consistent, positive correlation between good looks and reproductive success. Humans, too, have gustatory aesthetic adaptations that motivate them to eat foods with high sugar and fat contents. This does not mean that evolutionary biology predicts a positive correlation between junk food consumption and reproductive success; in this case, a negative correlation is expected because of the novelty of copious sugar and fats in Western modern diets—indeed, negative health effects from overconsuming these food items are significant health concerns. The same goes for values or morals. We argue that people have aesthetic adaptation for acquiring certain values because that acquisition was adaptive historically, whether it is currently.

Evolutionary biologists use what is called adaptationism as a method to discover how organisms are functionally designed—that is, to discover their evolved adaptations. Any given evolved adaptation may be adaptive currently, or not. Current adaptiveness of an evolved adaptation is expected when organisms live in the same environment as that in which the adaptation was created by direct selection in the past. We deal more with the adaptationist method below, but first we treat some additional aspects of ontogeny.

## 2.13 Genes, Environment, and Ontogeny

An evidence-based truth is that each and every feature of individual living things arises during development through gene–environment interactions. This is one of two basic conclusions of developmental biology. Genes are causal in ontogeny and so is the environment. This is the case for each and every trait of the individual, e.g., one's thumbnail, skin color, and values. This does not imply that the specific genetic loci are known for these traits. For most features, the genetic loci are not known; nevertheless, the genes have to be there for each and every trait of the individual to exist. Modern biology is not genetically deterministic; it does not give more developmental power to genes than to environment. Developmental biology has adopted, on the basis of empirical evidence, a position of total democracy or equality for the creative or causal power of genes and environment during ontogeny, because these two categories of causation are partial causes. This means that each is necessary and neither alone is sufficient to yield a phenotypic feature, e.g., one's ear or value system. All proximate causes are partial causes. A cause of a phenotypic feature is that, without which, the feature will not occur. Hence, all proximate causes of the feature are necessary to create it, but individually each cause is inadequate. Similarly, each of the two categories of ultimate causation—phylogenetic origin events and maintenance processes—is a partial cause.

The second basic conclusion of developmental biology is that ontogeny itself is an evolved outcome, just as are the phenotypic products of ontogeny. The developmental pathway that creates the human five-fingered hand is evolved adaptation, just as the hand is. Of course, ontogenetic adaptations give rise to by-products, just as the end-product adaptations of ontogeny do.

As mentioned earlier, the parasite-stress theory of human values proposes that environmental developmental causes in the form of encounters with infectious diseases and related information lead to value systems that are suitable for the local disease ecology. These environmental causes of ideology, whether external disease threat or internal immune activation, are ancestral ontogenetic cues. The value-acquisition psychological adaptations of a human use these developmental cues to adopt the values appropriate for the local disease ecology.

Individual differences in values seem to be due largely to environmental differences experienced—specifically, the individual's experiences with infectious disease and other ancestral cues of contagion risk—rather than genetic differences. We emphasize, however, that individual and group differences in values may be affected by genetic differences—for example, genetic differences in immunity to infectious diseases or in the capability to socially manage the effects of novel infectious diseases. We discuss in more detail the topic of genetic differences in values in Chap. 3. Note that here we are referring to differences *between individuals or groups*. The question of whether genes or environment is more important can be asked scientifically only in reference to individual or group differences in a feature, but certainly not, as we explained, for any trait of the individual. By group differences, we mean differences between human societies or cultures, or even between subcultures such as castes or social strata within a single society.

A person's values are caused by contemporary environmental events as well as by environmental experiences during the person's ontogeny. An example of a relevant contemporary event is when people become immediately more ethnocentric and xenophobic—that is, more conservative—when presented with cues of high infectious-disease salience in their local environment (Faulkner et al. 2004; Navarrete and Fessler 2006). Such cues, in addition to other cues that affect the ontogeny of values as well as the psychological adaptation that reads all these cues and transforms them into values, are proximate causes of values. Proximately, then, a person's values reflect genetic and environmental causes as well as the psychological machinery that interprets ancestral cues of infectious-disease prevalence in the local environment and guides appropriate values' acquisition.

## 2.14 Human Values Are Not an Objectively Delimited Category

An objective definition of values is not possible or necessary. Human values are a category of human preferences, but not an objectively delimited one. Organisms are comprised of a vast multitude of preferences. Indeed, all adaptations are aesthetic or value adaptations, because they interact with the environment, external or internal, and prefer certain outcomes to others. A habitat-aesthetic adaptation designed to bias entry into a productive habitat rather than an unproductive one is no more or less a value or aesthetic adaptation than a physiological adaptation designed to achieve a certain body temperature.

Some scholars have suggested to us that human values correspond specifically to those preferences that really matter to people. But, of course, maintenance of body temperature really matters too, and deviations from the preferred body temperature motivate corrective behavior. There is a great deal of specialized physiological machinery and behavior dedicated to body temperature maintenance. The concept of human values is arbitrarily bounded, too, if one were to claim that human values are the preferences that matter to the human mind. Temperature regulation is physiological, but the mind is part of the brain, a physiological system containing functionally dedicated psychological mechanisms for temperature control.

Living things are those things in nature that are subject to evolution by Darwinian selection. They possess three combined characteristics that make them inevitably and continuously subject to evolution by selection: variation in traits, reproduction, and inheritance (like begets like). All living things, from viruses to petunias and people, have the same fundamental preference or value: they all give the highest valuation to inclusive reproductive success. The preferences and associated goals for which they strive serve that fundamental value. They all owe this homologous similarity in their value systems to direct selection for it in evolutionary history. The value of advancing reproductive success arose with the first life form that possessed all three characteristics—variation, reproduction, and inheritance—and persisted in all branches of the Tree of Life.

Although there is no way to objectively delimit the domain of aesthetics in general or of the sorts of things people refer to as values in particular, there is utility in treating scientifically the topics that have been researched and discussed by scholars of values. Hence, topics included in our book are sexual attitudes and mate choice, political values, personality, religiosity and secularism, parenting and family life in general, prejudice, liberalism, conservatism, egalitarianism, gender relations, authoritarianism, self-concept, provincialism, philopatry, and many others.

There is a debate in the social science literature about what the word "ideology" means (see review in Jost et al. 2009). This literature has been produced primarily by evolutionarily uninformed social and behavioral scientists. Adding Darwinism does not provide an objectively delimited definition of ideology, but does put ideology in the general domain of evolved preferences of people, and hence gives it conceptual unity with the theory of life, evolution. Ideology generally means the composite of ideas reflecting the social needs and aspirations of an individual or group. Thus, a person's ideology is a significant component of her or his social strategy or means of social navigation, including presentation of self, pursuing status, and accessing social groups and family resources, friends, and mates. We use ideology, beliefs, values, morals, and ethos synonymously because they all refer to preferences in the same conceptual evolutionary framework. Synonymy of these preference labels is often followed in the traditional literature dealing with human values (e.g., Jost et al. 2003, 2009).

## 2.15 Adaptationism and Special-Purpose Adaptation

Adaptationism is the method of evolutionary biology that identifies adaptations by documenting functional design, and thereby distinguishing adaptations from by-products of adaptations. Evidence of functional design is obtained by observations, including those resulting from experimentation, of a traits' ability to solve an adaptive problem, and hence a problem that gave rise to differential reproductive success of individuals. Functional design is a property of the individual. It is seen when there is a fit or a correspondence between a phenotypic trait of the individual and an adaptive problem such that the fit solves the problem. Such evidence demonstrates evolved adaptation. This evidence simultaneously provides the necessary and sufficient evidence demonstrating the designer of the adaptation—that is, the type of direct Darwinian selection that acted in the past to create the adaptation. Selection is the only cause of evolution that produces evolved adaptation (meaning that drift, e.g., cannot). The specific type of selection that caused a given adaptation is stamped in the adaptation's functional design.

And importantly, direct selection favors special-purpose adaptations because the adverse ecological problems that give rise to selection are specific problems. For instance, the human digestive system is for digestion of foods that comprise the omnivorous human diet. This is an accurate, but superficial, description of the system's functional design. Actually, it is comprised of many highly specialized adaptations, each dedicated to digest a type of sugar but not another, a type of fat but not another, and so on. This specialization is characteristic of the design of adaptations in general, including psychological ones. The human eye is a psychological adaptation whose function is vision, but the experience of vision is accomplished by a multitude of more specialized psychological adaptations (color assessment, distance of object, object orientation, and so on).

Our hypothesis discussed earlier about how people obtain their culture, and specifically their values, is part of a broader, earlier adaptationist hypothesis (Thornhill 1998, 2003). The more encompassing hypothesis is that aesthetic judgments arise out of numerous, specialized psychological adaptations, each organized to give outputs functionally dedicated to solving domain-specific problems of aesthetic assessment. Accordingly, there are distinct specialized aesthetic psychological adaptations for solving habitat selection, mate selection, values selection, and so on across all domains of aesthetic valuation. A great deal is known about the many psychological aesthetic adaptations that show specialized function in people's long-term and short-term mate choices and romantic lives (Buss 2003; Thornhill and Gangestad 2008). Increasingly, as this book documents, more and more is being discovered about the special-purpose design of human psychological adaptations that assess, adopt, and use values in other categories of human life.

## 2.16 Beauty as Truth

Although the early aesthetic-philosophers' notion that the values that make one feel good—the beautiful ideas—identify objectively moral truth or correctness is naturalistic-fallacy nonsense, in certain ways, beauty is actually truth. The signaling adaptations of animals and plants, from the rooster's comb, the *Habernaria* orchid's flower, to women's facial and bodily estrogenization, are truthful indicators of phenotypic quality (see Thornhill and Gangestad 2008). They are honest signals of the degree of that quality. Beautiful expressions of these signals are truth, and so are the ugly expressions.

Values are truthful in the sense that ideological truth is attributed to a preference when it is understood by the person's evolved brain as providing a path to relatively high reproduction in evolutionarily historical environments. To use the word "understood" in this context does not mean conscious comprehension, but instead a comprehension inherent in the way the values-adopter psychological adaptation works, or from its functional design. The idea that spirits created people is scientifically unsupported and hence is not scientific truth, but it is the ideological truth of most people who live and have lived on Earth. Aesthetic judgments of ideas by the values-adopter adaptation are based importantly on the data one gathers from positive and negative trial-and-error experiences that arise from holding and manifesting beliefs and from observation of the social successes and failures of others with various beliefs. These largely unconsciously processed data of social experience from childhood forward make truthful or self-evident to the theist the spiritual origin and guidance of people. Similarly, the data of social experience makes truthful to the liberal the equality of people, which is displayed in the liberal rhetoric of the Declaration of Independence of the United States: a self-evident truth is that all men are equal. Each value system has its own self-evident truths.

## 2.17 Comparative Methodology

A primary unit of analysis used in this book to study variation in human values across the world is that of a geopolitical region. In most cases, these regions are countries. In some cases, the regions are geographically separate colonies or territories (e.g., Guam), or culturally distinct regions within a country (e.g., Hong Kong). The term "country" or "nation" hereafter will refer to all such units of analysis. Although political borders that divide countries do not always correspond to cultural borders, evidence exists that countries serve as useful proxies for societal cultures. For example, Schwartz (2004) showed that there can be much more variation in values among countries than within countries. Also, country boundaries typically are political boundaries, i.e., ideological boundaries, and hence are the logical units of analysis in our research on the diversity of value systems and related topics. Sample sizes of number of countries vary across the international analyses in our book, but we used all available data for each analysis.

Many of our tests in the empirical chapters use both international analyses and analyses across states of the USA. This allows us to examine patterns in data bearing on our hypotheses at two levels of geographical variation. Even within the single country of the USA, there is considerable interstate variation in variables relevant to testing the parasite-stress theory of values.

Our use of countries and US states as sampling units assumes that they are statistically independent. It is critical to discuss and defend this assumption early in the book because throughout the book we employ comparative analyses that assume statistical independence of cultures across regions. Below we give a condensed defense of the assumption that coincides closely with arguments we present in more detail in our recent paper in *Evolutionary Biology* (Thornhill and Fincher 2013). As we explain, the fundamental reason for the independence of cultures is that psychological adaptations for enculturation are designed to incorporate values and other cultural items that solve local problems of adversity impacting individuals.

In the field of comparative research across cultures, which examines covariation between cultural traits or between cultural traits and ecological variables, the statistical independence of cultures, including that of countries, is debated (e.g., Mace and Pagel 1994; Rogers and Cashdan 1997; Nettle 2009). The debate centers on the transmission of norms, beliefs, ideas, values, i.e., culture, between societies (intercultural diffusion), including retention of cultural traits when a parent culture gives rise to a descendant culture. Some scholars argue that such transmission events make cultures nonindependent (e.g., Murdock and White 1969; Mace and Pagel 1994). The nonindependence among cultures to which this argument refers often is labeled "Galton's problem," because Francis Galton apparently was the first scholar (as early as 1889) to claim that cultures cannot be studied as independent units in cross-cultural research (see Mace and Pagel 1994). For instance, the countries of southern Africa share more historical interactions among themselves through diffusion and common ancestry than shared between cultures of southern Africa and cultures in southeastern Asia. Hence, according to some researchers, southern African countries are not statistically independent, but instead comprise a single observation (Nettle 2009).

This argument goes on to claim that given the nonindependence caused by cultural sharing, researchers should correct for historical cultural sharing by considering as valid units of sampling and analysis only the changes in a cultural feature—the feature's appearance and/or loss or modification—among a historically related group of cultures (e.g., Mace and Pagel 1994). Hence, according to this claim, what counts is how a cultural feature behaves in terms of arising, disappearing, or changing in relationship to a hypothetical causal variable (e.g., a climatic variable or parasite stress). Mace and Pagel (1994) state that "the validity of comparative methods for anthropology depends upon correctly counting [these] independent instances of cultural change" (p. 551). They define such instances as "any instance of the de novo invention or acquisition of a new [cultural] element by copying from another culture or the change or loss of an element." They also emphasize, and certainly correctly, that independent instances of cultural change cannot be identified without a phylogeny of the cultures under study. The phylogeny of a group of cultures identifies their historical interrelationships. Hence, this approach is concerned with identifying instances of cultural change, which then become the units of analysis (the sampling units or sample size) in testing for a relationship between two cultural elements or a relationship between a cultural element and an ecological variable such as parasite stress.

This approach sharply contrasts with one based on the assumption that cultures are independent. With the assumption of cultural independence, each of the cultures under investigation, and whether they share common ancestry or cultural items due to diffusion, is a sample unit, and there is no pseudo-replication (i.e., inflation of sample size) by including in analysis all the cultures as the sample size.

The existence of historical sharing among cultures in a region or across regions (e.g., the USA's cultural sharing with Western Europe), whether by cultural diffusion or common ancestry, is not simultaneously a fact about the nonindependence of the cultures. This is because, as we have explained, cultural transmission processes, whether involving intercultural diffusion, intracultural transmission within or between generations, or common cultural descent, are guided by evolved psychological adaptations that function to selectively invent, adopt, discard, or change cultural features based on local utility of the features for inclusive reproductive success. Moreover, correction for historical cultural influences by determining independent instances of cultural change based on phylogeny, proposed by Mace and Pagel and others, is not a uniformly scientifically superior method for all cross-cultural research and is theoretically inaccurate and empirically misleading for testing cross-cultural hypotheses about the causes of the transmission of cultural elements and the maintenance of cultural differences and similarities.

The argument of cultural nonindependence is simply based on a judgment to give validity, or at least salience, to one scientific question about cultural history, but not another. The question addressed by the nonindependence approach is when and where did a target cultural item(s) change significantly or first arise de novo or become lost from the cultural repertoire. Certainly, these are valid scientific concerns, and ones in which Galton's problem must be considered. As the cultural nonindependence proponents correctly emphasize, to answer questions of the debut, disappearance or change of a cultural item in history requires knowledge of the phylogeny of cultures involved in the analysis. The other scientific question about cultural history deals with the reason the cultural trait was maintained after it arose. For this question, Galton's problem is totally irrelevant. Researchers who propose that the origin causation is the important question are favoring one question over the other.

The nonindependence perspective seems to reflect a view of people's mental design that is quite different from the view of those who assume cultural independence. Is the relevant psychology—that for the assessment, adoption, rejection, and modification of cultural items—the product of past evolution by selection for inclusive reproductive success that acted potently only at the interindividual level, or is it not? One's answer to this question importantly informs one's position on cultural independence or nonindependence. Those for cultural independence understand the relevant psychology as we do: individual humans are highly discriminative cultural strategists, because their enculturation is designed by past selection for maximum inclusive fitness.

Daly (1982), Flinn and Alexander (1982), and Daly and Wilson (2010), in criticizing hypotheses that propose cultural transmission with mere exposure to others' ideas as all that matters—the cultural inertia notion we have criticized and also the similar memes hypothesis—discuss the theoretically, heuristically, and empirically verified strengths of viewing cultural transmission as caused by psychological processes that function to selectively adopt cultural items. Alexander (1979a, b, 1987) and Irons (1979), among others, give overviews of the theory and evidence supporting this view of enculturation. Our book is a testament to the strength of this view as well. Gangestad et al. (2006) focused on evocation of culture—the experiences of people in a cultural ecology that proximately cause their cultural repertoire. Gangestad et al. emphasize also the history of direct Darwinian selection for learning mechanisms that incorporate cultural repertoires that are adaptive (ancestrally) locally. Schaller (2006) explains why this evocation is responsible for the transmission of cultural items. Flinn (1997) applied this view of enculturation to ideas used as competitive tools, as an explanation for the creation or origin of new cultural features and their perpetuation.

Under this way of thinking, cultures are independent because the creation of cultural features and their transmission between individuals within a generation or between generations or between cultures depends on whether the local individuals find attractive the various available values and other cultural items, where attraction is based on evaluation and choice by psychological mechanisms dedicated functionally to this. Hence, transmission, both within and across cultures, is not a given. Instead, it involves very sophisticated preference adaptations dedicated to the assessment and selection of cultural features (just as mate preferences depend on functionally specialized preferences) and to the incorporation of those cultural items that are optimal for that place and generation. In statistics, independence is defined typically as follows: "Two events are independent if the occurrence of one of the events gives no information about whether or not the other event will occur" (Statistics Glossary, www.stats.gla.ac.uk/steps/glossary/probability.html#indepevents). Historically related cultures are statistically independent because knowing the cultural repertoire in one culture does not allow you to conclude automatically that the same repertoire will exist in a closely related culture. If both cultures have the same ecological causes of enculturation then cultural similarities are expected to occur, but if not, they are not. Hence, distinct cultures, as identified in anthropology and cross-cultural psychology, are statistically independent, even if they share a recent history and geographic proximity. With the cultural independence approach to study cultural diversity, the phylogeny of the cultures under investigation is irrelevant.

The ecological variable of primary interest in the parasite-stress theory of culture is parasite stress. Thus, countries that have high parasite prevalence are expected to show similarities in culture (e.g., collectivist values), whether or not the countries are near each other, contiguous, or on separate continents. Likewise, across this range of proximity and distance, countries that have low parasite prevalence are expected to show cultural similarity (e.g., individualist values). According to those with the view of nonindependence among cultures, however, the collectivist countries of the Middle East, for example, share a cultural history and hence are nonindependent, because their culture arose once and then was retained. This claim ignores the cause(s) of the retention and gives importance only to the causation of cultural origin. Understanding retention of values across time and space requires a causal explanation; saying that history accounts for retention is a description of how a culture was in the past but does not address causation. Historical retention of cultural features in a region is never an explanatory substitute for ecological and ultimate causes of temporal consistency in culture. And, as we have emphasized, historical–particularistic explanations based on tradition per se can reduce to a view that quasi-spiritual forces drive cultural retention. The reasoning we advocate provides the causal proximate reason for the retention. Ecologically or proximately, the parasite-stress theory proposes that the retention is due to the continuation of high parasite stress in each of the Middle East countries, and each of these countries is a distinct datum. With this approach, one is spared the difficulties and uncertainties of phylogenetic cultural reconstruction that many have discussed (e.g., Mace and Pagel 1994).

We emphasize, too, that the parasite-stress model of sociality and values specifically offers a hypothesis for the variation in isolation of cultures, or said differently, the variation in intercultural exchange, an important variable and topic in cross-cultural research. For instance, some countries endorse more inflow of immigrants, cultural products, and ideas than others. In the Preface, we mentioned the long-term cultural isolation and parochialism of the Old South (Key 1949; Reed 1972, 1983; Grantham 1994). This cross-cultural variation in isolation or openness is not a variable that confuses independence among regions. Instead, it is the predicted outcome, based on variation in the value systems caused by variation in parasite stress. Openness to new and different people, ideas, and commercial products is prevalent in countries with liberal/individualistic values, whereas collectivist countries are more culturally isolated, parochial, and xenophobic (Chap. 11; Schaller and Murray 2008; Thornhill et al. 2009).

Similarly, the parasite-stress theory of values proposes that promoting innovative thinking or promoting its opposite—adherence to traditional ideas—will be predictable in relation to variation in parasite stress. In collectivist societies innovation is not rewarded and may be discouraged (Chap. 11; Thornhill et al. 2009, 2010). Although this will limit the potential pool of values that arises, it will not affect the values selected locally among those that are available. Also collectivist cultures are more traditional and conformist than individualist cultures (Fincher et al. 2008; Murray et al. 2011). This does not mean, however, that regional history causes the values adopted under collectivism; instead, the values of traditionalism and conformity, which are evoked by high parasite adversity, are the causes, not history.

To attempt to correct for nonindependence due to historical contact, say, by replacing cross-national analyses with ones that examine parasite stress in relation to our focal dependent variables in each of the six world cultural regions identified by Murdock and White (1969), or by giving more credibility to world regions' analysis, is theoretically inappropriate given the independence of countries' value systems. Such analyses are empirically incorrect, too, because they greatly reduce sample size and hence the ability to detect actual patterns across the globe. "Correcting" is the wrong word to call the analysis by world regions, or for the other methods for dealing with alleged historical nonindependence that have been proposed by Mace and Pagel (1994), Nettle (2009), and others. Correction implies a more scientifically valuable procedure and hence the word "correction" does not apply in research on cultural diversity such as ours. The correction procedures address a different question (the origin of a cultural trait in a cultural phylogeny) than what we focus on in this book. Our approach is focused on the causation of cultural diversity arising from psychological adaptation that functions to create, retain, discard, or modify cultural elements based on local utility. Hence, we assume cultural independence across geopolitical regions.

As presented by Mace and Pagel (1994) and others, the correction procedures for nonindependence are statistical adjustments, not evolutionary theoretical adjustments. A statistical argument is not the same as an evolutionary theoretical proposal. We provide here a theoretical rationale for the use of the correction procedures. The procedures are theoretically valid when the research question is about independent instances of cultural change, but invalid when the research question is cultural transmission and maintenance.

## 2.18 A Cultural Phylogenetic Example

An example involving imaginary cultures will illustrate further our view. Figure 2.1 is a phylogeny of nine imaginary cultures showing the relationships between them based on, say, evidence of homology in degree of language similarity. Homology means the degree of similarity among traits compared across kinds (e.g., culture, species, and so on) that is caused by common ancestry, i.e., shared descent from an ancestor with the trait. On the phylogeny, each branch tip is a culture; "yes" refers to camels being used by that culture, and "no" to camels not used. Let's assume that anthropologists recognize each culture as distinct based on language and norm differences among them. The phylogeny shows that "no camel use" was the ancestral cultural state from which "camel use" was derived. The phylogeny also shows that camel use exists in four cultures, but arose independently only three times in the three lineages with an asterisk (*).

Consider the following hypothesis: drought causes people to value camels. Imagine that a study of rainfall data in the regions of the nine cultures shows the following distribution:

| | Cultures with Camels | Cultures without Camels |
|---|---|---|
| Drought | 4 | 0 |
| No drought | 0 | 5 |

The cultural nonindependence scholars (those scholars who insist on a view of cultural interdependence as a mainstay of cross-cultural methodology) say this data distribution cannot be used in testing the hypothesis about cultural history of camel use. They say that the two cultures "yes a" and "yes b" are not two cases, but one, because these cultures are interdependent (nonindependent) historically as a result of camel use arising in their common ancestral culture (existed at →). Certainly, they are correct if their historical question is the independent origins of camel use. Camel use arose three times from no camel use. It always arises with the environmental condition of drought, but you cannot count the four cultures with camel use as four separate data.

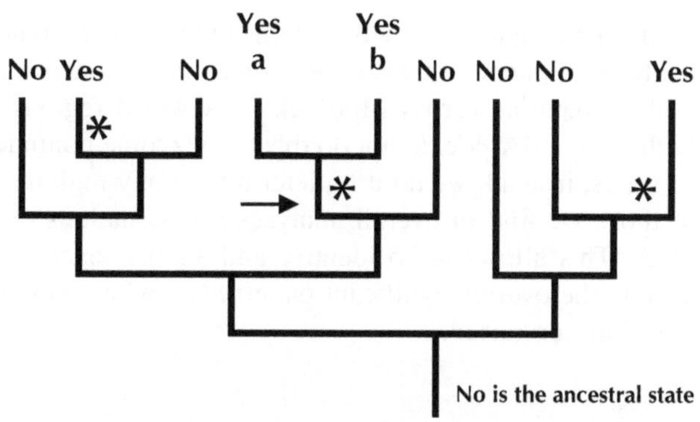


**Fig. 2.1** Phylogeny of nine cultures constructed from the degree of language similarity that is due to common descent, *asterisk* = independent origin of camel use, *yes* = camel use, *no* = no camel use (figure originally published in Thornhill and Fincher 2013)

As discussed earlier, there are, however, two scientific questions/hypotheses about cultural history. One is the origin of a cultural trait (debut on the tree of cultural history); the other is its persistence (or nonpersistence) after its origin. The two questions are complementary, not alternative, and a complete understanding of the history of a cultural trait requires answering both. Because they are not alternatives, though, only one question can be investigated at a time. Furthermore, it is critical and fundamental for researchers to understand the distinction between analyses of cultural trait origin and analyses of cultural trait persistence.

Our point is that, for the question of cultural persistence (not that of independent-origin events), you count all four "yes" cultures as supporting the hypothesis. The reason: people adopt values or other cultural preferences based on local costs and benefits. People are designed by past selection to assess, adopt, reject, or modify values (and other ideas) just as they are designed to make choices of mates, habitats, and so on. All such choices are based on processing local costs and benefits. Even if culture "yes a" got the camel-use idea from "yes b," it was adopted and transmitted by "yes a" people as a result of benefits/costs of the idea to them. "Yes a" people in their place and time independently perceived camel use to be valuable, and it persisted. "Yes a" and "Yes b" have the camel use in common (that particular similarity) for one of two reasons: (1) because it was transmitted through time (through generations) from their common cultural ancestral group, including across generations after "yes a" and "yes b" became distinct cultures; or (2) because one of these two obtained it from the other culture. Either way, the persistence of camel use across time—accounting for its persistence in "yes a" and "yes b"—is not a given, but an interesting problem/question about ecological causation of the persistence of camel use.

To the cultural nonindependence scholars, cultural history is only or primarily about the origin of cultural traits. Their focus is required and correct for their question: find the independent origins, which require a phylogeny of the cultural groups, and then map those origins onto a hypothetical causal ecological factor (drought, in the example). Cultural nonindependence is an issue only for reconstructing the origin of a cultural trait; it is a nonissue for questions such as ours about the persistence of cultural values. This argument applies to any regional analysis of cultures, from ethnographic cultures to average values of people in a country or in a state in the USA.

In later chapters, we will examine both cross-national and more focused regional analyses of parasite stress in relation to value systems. In some of the analyses, we look at patterns across Murdock's six world regions and across census regions within the USA. We do not do this to overcome nonindependence among countries or states; instead, we do it to determine if any regions exhibit exceptional patterns to those we find in overall analyses across nations or across the 50 states of the USA. This allows us to identify and address exceptional regions that may exist despite the overall significant pattern, and when they do exist, additional research questions are raised.

## 2.19 Interspecies Comparative Research

The methods for historical correction in cross-cultural research used and advocated by Mace and Pagel (1994), Nettle (2009), and others were borrowed from the typical phylogenetic analytical approach used in biology for the comparative study of adaptations and other traits across multiple species. This borrowing assumes that cultures in a geographical region are interdependent in ways analogous to the interdependence of closely related species—in both cases, shared history is thought to generate nonindependence. Biologists widely believe that it is necessary to correct for phylogenetic history when comparing traits of species, and, hence, it would seem to follow that cross-cultural analyses must do the same (Ridley 1983; Brooks and McLennan 1991; Harvey and Pagel 1991; Mace and Pagel 1994). Given the similar conception of how phylogeny creates nonindependence in both cross-cultural analyses and interspecies analyses, it is informative to examine critically the popular method of interspecies analyses that is anchored in the assumption that the features of closely related species are not independent.

As in cross-cultural research, the phylogenetic correction procedure is essential when the research question is about trait origin in a phylogeny, but invalid when the question is about trait maintenance after origin. The view that there is a need for phylogenetic correction in the comparative study of diversity in adaptations and other phenotypic traits was criticized by Williams (1992), Westoby et al. (1995a, b, c) and Reeve and Sherman (2001), but their criticisms have not been widely discussed or appreciated. The flaw in the non-independence–phylogenetic-correction opinion is that related species, given that they are distinct species, are historically evolutionarily independent units, and hence the evolutionary processes, including selection, that occurred in one given lineage (a species) are independent of the evolutionary processes that happened in another lineage (a different species), even when species are in the same genus. This independence is in the evolutionary processes that acted historically in each lineage (each species), and it meets the standard definition of statistical independence mentioned earlier.

Consider an example involving two closely related species preyed upon by the same predator. And, in this example, the prey species evolved functionally identical predator-defense adaptations such as camouflage coloration. Even though the "same" selection (differential reproductive success of individuals as a result of their features of defense against a single species of predator) and the "same" adaptation (in terms of functional design) are involved, the evolutionary processes in the two lineages are independent because the two lineages are distinct species—and hence evolved to the status of distinct species through reproductive isolation.

The conclusion of evolutionary independence is equally warranted if you consider a by-product that is similar in each of these species. The by-product in each lineage was maintained by the direct selection that favored the evolved adaptation in each lineage to which the by-product is tied. The direct selection in this case acted independently in each lineage because they are different species. Hence, all the traits, whether adaptations or by-products and whether similar or dissimilar, of closely related species (and distantly related species) are evolutionarily independent. These traits are statistically independent too: a trait's presence (or absence) in one species does not predict its presence (or absence) in a closely related species because of lineage-specific causal processes acting to create (or eliminate) the traits.

Research procedures for distinguishing species in biology, whether based on evidence of morphological differences, evidence from reproductive isolation, and/or evidence from the distinctiveness of branch tips in genetic phylogenetic analysis, are efforts to address historical evolutionary independence of lineages. This is the most widely accepted and encompassing criterion for distinguishing biological species (e.g., Coyne and Orr 2004). Of course, the nature of speciation processes necessitates that the degree of evolutionary isolation among closely related species is a continuum. Closely related species will vary in the length of time they have been evolutionarily independent. Species also will vary in that interspecies' hybridization may or may not occur, and when it does, it varies in degree, persistence, and extent in which it affects traits that are the focus of a particular comparative analysis. This continuum, however, does not cast doubt on our general point that the empirical recognition of evolutionary independent lineages is the criterion for species designation in biology. The concept of species as evolutionarily isolated lineages that have evolved independently and hence have been subjected to lineage-specific evolutionary agents that caused their traits, even their trait similarities, is the virtually universally accepted species' concept in the life sciences. (See Thornhill and Fincher 2013 for an expanded treatment of phylogenetic methodology used in biological research across taxonomic groups.)

Phylogenetic causation and evolutionary maintenance causation are distinct categories of causation. They are complementary categories and not competitive (alternative) ones. To understand fully the historical causation of any trait, cultural or otherwise, requires knowledge of both causal categories. To base a need to correct for cultural history in an investigation of cultural elements of related cultures on the popular practice in biology of phylogenetic correction is replete with misconceptions. In the final chapter of this book, we discuss our hypothesis that the opinion that cultures are not independent reflects the collectivist cognition of interdependence.

## 2.20 Opponents of Evolution Applied to Human Behavior

The non-independence-of-cultures researchers we criticize earlier are not opposed to the study of human behavior and psychology using evolutionary theory. Instead, they have different assumptions than us about how certain methods should be applied in this research. This last section of this chapter briefly addresses the actual opposition to applying evolutionary biology to human social behavior and psychology. Many evolutionary biologists have spent time and effort to address these opponents; a sample of comprehensive responses to those who oppose applying evolutionary biology to human behavior can be found in Thornhill and Palmer

(2000), Alcock (2001), and Kurzban (2010). The opponents are comprised of two ideological camps: (a) certain theists and (b) certain scholars and media commentators, who either self-identify as liberals or are assumed to be liberals. We find both opposing camps scientifically intriguing.

Some theists have ideological issues with the concept of evolution as a whole, but especially to references of humans as evolved. This is fully understood, given that high religiosity coincides with conservative values (Chap. 9). These values include a priority given to supernatural direction of human activity and associated magical thinking, security based on salvation from punishment and pain in afterlife, and conformity and obedience to traditional contra-evidence interpretations. Some theist critics of evolution as applied to humans have proposed that religion is in a moral war with evolution, claiming that their religion endorses a moral worldview, whereas evolution endorses an immoral worldview. Of course, as we emphasized, the study of evolution is science, and science does not endorse any moral system. Thus, evolution does not and cannot threaten one's ideology, religious or secular. The war-of-morals criticism is profoundly misinformed, but keeps appearing in the media and from pulpits. Its basis, in part, we hypothesize, is the ideological nonacceptance of the naturalistic fallacy. As we have explained, collectivists cannot comprehend that facts can be interpreted independently of in-group moral goals and harmony. Given the conservative understanding of self as interdependent, the theist opposition to evolutionary biology will continue as long as there are theists. Furthermore, as mentioned in the previous chapter, some theists are accommodationists and thus believe erroneously that the study of religion or morals is off-limits to and not knowable by scientific investigation.

The liberal critics of evolution applied to human behavior are more challenging to understand scientifically. A core value of liberalism is openness to new ideas and ways, including scientific discoveries (Chap. 7). The biologists Richard Lewontin, Steven Rose, and Steven Gould essentially have claimed in their criticisms that evolutionary theory does not apply to human social behavior. The evidence is strongly against their view, to say the least. How could an open, analytical mind discount evolution applied to any part of human psychology and behavior? The evolutionary theoretical ideas and supporting evidence have gotten richer and more encompassing of human affairs at a fast clip. Actually, the evidence has been overwhelmingly supportive ever since biologists solved the issues of human altruism in the 1960s and 1970s. The biologist critics behave more like accommodationists of a different type. They accept evolution applied to human physiology, biochemistry, genes, and bones, but claim that the human mind and behavior are different. The difference they see is that evolution, the general theory of all life, cannot inform, and should not be used to try to inform, these topics; in their writing they have vehemently opposed the reality that human behavior is evolved.

Some of these critics follow and believe Marxist ideology (see Alcock 2001). Marxism is a collectivist ideology in which the group's goals are all important. People's duty is to the collective or state over their personal interests. Marxism, too, is a holistic ideology in which the collective cannot be understood by the reductionist analysis that is fundamental to science (Gregor 2009). In Chap. 4 we discuss evidence that holistic reasoning is a component of collectivist cognition. Also linking Marxism to collectivism is the fact that Marxist values are the ideological foundation of totalitarian communism, which is characterized by undemocratic values, including elite control of all economic and political matters, and illegality of personal property rights (Gregor 2009). In Chap. 10, we discuss the evidence showing that highly autocratic governments, like ones based on Marxist totalitarianism, are also ones with high collectivism and associated authoritarianism. Marxism is run-of-the-mill collectivism with communist ideology in place of religious dogmatism.

The so-called liberal critics can appear intellectual to naïve audiences when they raise criticisms such as: those people (like us) who discuss human behavior as evolved think that all traits are evolved adaptations. They say, too, that such people do not pay attention to culture. Of course, a huge, published literature documents that they are profoundly mistaken in both claims. One straightforward and reasonable hypothesis for their opposition is that it, like the theist opposition, stems from collectivist values. Historically, the opposition to science has come essentially entirely from conservative ideologues, while liberals have supported science (Ferris 2010). The opposition of critics in both camps is simply a form of ideological opposition to the most significant scientific understanding of humans ever accomplished. Hence, the so-called liberals who oppose evolutionary biology as applied to people's behavior may hold significant conservative values. Certainly, it is not scientifically reasonable to accept the public proclamation of held values by a person or group as the truth of held values. For example, autocratic governments sometimes claim to be democracies, and many prejudiced southerners in the Old South claimed to be democrats. This is a reason why political scientists measure democratization and values rather than taking a person's or a group's word for what is believed. The values, then, of the self-proclaimed liberals who oppose evolution applied to human behavior is an empirical issue and could be studied by obtaining conservatism-scale-questionnaire responses from samples of people who are supportive of science until it is applied to the evolution of human behavior and psychology. Our prediction is that such people will score right of center and thus be more conservative than liberal. Some research has been conducted already to measure the values of graduate students who do research on evolution and human behavior; they are way left of center (Tybur et al. 2007).

Beyond our scientific fascination with those who ideologically oppose evolution being applied to human behavior is our worry that these people hinder the rate of scientific discovery about the causes of human affairs. Such opponents are not rare in academia, and thus sometimes serve as referees for submitted manuscripts to journals or grant proposals, and as consultants to scientific journals and societies. This gives them considerable influence on the research that is conducted and published. Also, some of these opponents actively publish their own papers and books criticizing the concept of evolution applied to human affairs.

As a recent example, in an article in the prestigious scientific journal *Nature*, Bolhuis and Wynne (2009) criticized the application of evolutionary ideas to human psychology with the standard, tired, uninformed comments. They say that the "approach overlooks the importance of culture in shaping the human mind" (p. 832). They go on to say that the approach assumes all traits are the product of the direct action of natural selection. They emphasize, however, that some traits are by-products of selection acting on some other trait. Their final and "most serious" objection is that "cognitive traits of past generations leave little trace in the fossil record" (p. 832). Hence, they claim, the human mind cannot be understood scientifically, or only understood at a trivial level.

All of these criticisms are answered in this chapter and have been answered many times in the earlier scientific literature. Scientists who study the evolution of human behavior and psychology pay a great deal of attention to culture and to by-products. The fossil record is not necessary for robust conclusions about the functional organization of evolved psychological adaptations. The full evidence of the past causal selection process involved in the creation of each psychological adaptation (and all other types of adaptations) is stamped in the functional design of the adaptation.

Evolution and human behavior, like any other scientific field, welcomes criticism and advances because of it; however, the ideological opponents of the research on the evolution of human behavior have not used the validated rules of scientific criticism. Valid scientific criticism is based on some objective examination and understanding of the evidence and ideas in the field of research criticized. Objective criticism is promoted by self-knowledge of one's values and their effects on biasing reasoning. Of course, these critics ignore evidence altogether and/or distort it. They proceed by ideological faith, not by evidence. On scientific grounds, neither of the two groups of critics has had anything to say of scientific interest or relevance. As social commentators, they are maintaining into the twenty-first century the long-standing anti-intellectual tradition that first arose in opposition to the Enlightenment. (For an excellent treatment of the history of the anti-Enlightenment tradition, see Sternhell 2009.)

In the next chapter we give a detailed discussion of the parasite-stress theory of values.

## 2.21 Summary

This chapter discusses the primary methods and assumptions used throughout the remainder of our book.

The scholarly study of aesthetic judgments, including those about the attractiveness/unattractiveness of values, initially arose as a branch of philosophy. This scholarship failed to advance knowledge of the causes of values and other aesthetic judgments and preferences. Its failure resulted, in part, from its use of the philosophical method of evaluating an idea entirely in terms of emotional verification. The essence of the scientific method is its ability to empirically evaluate conjectures. The scientific method replaced the philosophical method and then became and remains the sole way of knowing the causes of natural things, including the causes of morals. Another reason that the philosophical method cannot provide insight into the causes of values is that common sense or intuitive cognitions are values. Hence, human thoughts in themselves lack scientific objectivity because they are biased toward a particular outcome. An additional reason the philosophical method cannot address the causation of values is that human reasoning and rationalization do not access much of the information actually processed in reaching conclusions.

Darwin's method of historical science importantly extended the scientific method to causes in the deep-time past. Darwin's method is the one used for identifying deep-time past causation in all scientific fields that study such causes, including biology.

Biology is the scientific study of the proximate and ultimate causes of all life's features and hence encompasses totally all the scholarly fields that investigate human activity. These fields vary greatly in scientific sophistication and power because of varying awareness and use of biology's general theory: evolutionary theory. The parasite-stress theory of values uses this theory, including its applications to social behavior and culture.

People are enculturated during their ontogeny. We propose that people have values-adopter psychological adaptation that is functionally designed to incorporate during ontogeny the values that provide solutions to local problems that adversely influence the reproductive success of individuals. We emphasize the role of this psychology in adopting values that optimize behavior in relation to the level of local infectious diseases. This psychology, like all evolved adaptation, is the product of evolution by individual-level selection for inclusive fitness maximization. The common, but erroneous, view that culture functions for the good of the group, we hypothesize, reflects collectivist cognition and values about the overriding importance of group prosperity.

Some scholars have proposed that cultures are essentially interdependent because of intercultural flow of cultural items and common cultural phylogenetic descent; some researchers see this as an impediment for cross-cultural analysis. The enculturation process we propose counters this, because people adopt cultural items including values based on assessments of the items' local benefits relative to costs. Hence, it is valid to treat cultures and cultural regions as independent in analyses. The cultural nonindependence view is correct when the question investigated is of the origin of cultural items in a cultural phylogeny, but incorrect when the question is one of the transmission and maintenance of cultural items after their origin.

The scientific study of the evolution of human behavior and psychology continues to advance despite ideological opponents whose striving maintains the anti-Enlightenment tradition that began in the 1700s.

## References

Abbot, P., Abe, J., Alcock, J. et al. (2011). Inclusive fitness theory and eusociality. *Nature* 471: E1-E4.

Alcock, J. (2001). *The Triumph of Sociobiology*. Oxford University Press, New York, NY.

Alexander, R. D. (1979a). Evolution and culture. In *Evolutionary Biology and Human Social Behavior: An Anthropological Perspective* (eds. N. A. Chagnon & W. Irons), pp. 59–78. Duxbury Press, North Scituate, MA.

Alexander, R. D. (1979b). *Darwinism and Human Affairs*. University of Washington Press, Seattle, WA.

Alexander, R. D. (1987). *The Biology of Moral Systems*. Aldine de Gruyter, New York, NY.

Alexander, R. D. (1989). Evolution of the human psyche. In *The Human Revolution* (eds. P. Mellars & C. Stringer), pp. 455–513. University of Edinburgh Press, Edinburgh, U.K.

Alexander, R. D. (1990). *How Did Humans Evolve? Reflections on the Uniquely Unique Species*. Special Publication No. 1. Museum of Zoology, The University of Michigan, Ann Arbor, MI.

Andrews, P. W., Gangestad, S. W., & Matthews, D. (2002). Adaptationism: How to carry out an exadaptationist program. *Behavioral and Brain Sciences* 25: 489–547.

Baschetti, R. (2007a). The dawn of science-based moral reasoning. *Medical Hypotheses* 68: 4–8.

Baschetti, R. (2007b). Evolutionary, neurobiological, gene-based solution of the ideological "puzzle" of human altruism and cooperation. *Medical Hypotheses* 69: 241–249.

Bolhuis, J. J., & Wynne, C. D. L. (2009). Can evolution explain how minds work? *Nature* 458: 832–833.

Bourke, A. F. G. (2011). The validity and value of inclusive fitness theory. *Proceedings of the Royal Society B* 278: 3313–3320.

Boyd, R. & Richerson, P. J. (1985). *Culture and the Evolutionary Process*. University of Chicago Press, Chicago, IL.

Boyd, R., & Richerson, P. J. (1996). Why culture is common, but cultural evolution is rare. *Proceedings of the British Academy* 88: 77–93.

Brooks, D. R., & McLennan, D. A. (1991). *Phylogeny, Ecology, and Behavior: A Research Program in Comparative Biology*. The University of Chicago Press, Chicago, IL.

Buss, D. M. (2003). *The Evolution of Desire: Strategies of Human Mating*. Basic Books, New York, NY.

Carney, D. R., Jost, J. T., Gosling, S. D. et al. (2008). The secret lives of liberals and conservatives: Personality profiles, interaction styles, and the things they leave behind. *Political Psychology* 29: 807–840.

Chudek, M.,Heller, S., Birch, S. et al. (2012). Prestige-biased cultural learning: Bystander's differential attention to potential models influences children's learning. *Evolution and Human Behavior* 33: 46–56.

Coyne, J. A., & Orr, H. A. 2004. *Speciation*. Sinauer Assoc., New York, NY.

Curtis, V. A. (2007). Dirt, disgust and disease: A natural history of hygiene. *Journal of Epidemiology and Community Health* 61: 660–664.

Daly, M. (1982). Some caveats about cultural transmission models. *Human Ecology* 10: 401.

Daly, M., & Wilson, M. (2010). Cultural inertia, economic incentives and the persistence of "Southern violence." In *Evolution, Culture, and the Human Mind* (eds. M. Schaller, A. Norenzayan, S. Heine et al.), pp. 229–241. Psychology Press, New York, NY.

Dawkins, R. (1976). *The Selfish Gene*. Oxford University Press, New York, NY.

DeBruine, L.M., Jones, B.C., Watkins, C.D. et al. (2011). Opposite-sex siblings decrease attraction, but not prosocial attributions, to self-resembling opposite-sex faces. *Proceedings of the National Academy of Sciences* 108: 11710–11714.

Dugatkin, L. A. (2000). *The Imitation Factor*. The Free Press, New York, NY.

Faulkner, J., Schaller, M., Park, J. H. et al. (2004). Evolved disease-avoidance mechanisms and contemporary xenophobic attitudes. *Group Processes and Intergroup Relations* 7: 333–353.

Ferris, T. (2010). *The Science of Liberty: Democracy, Reason, and the Laws of Nature*. Harper, New York, NY.

Fincher, C. L., & Thornhill, R. (2012). Parasite-stress promotes in-group assortative sociality: The cases of strong family ties and heightened religiosity. *Behavioral and Brain Sciences* 35: 61–79.

Fincher, C. L., Thornhill, R., Murray, D. R. et al. (2008). Pathogen prevalence predicts human cross-cultural variability in individualism/collectivism. *Proceedings of the Royal Society of London Biological Sciences* 275: 1279–1285.

Flinn, M. V. (1997). Culture and the evolution of social learning. *Evolution and Human Behavior* 18: 23–67.

Flinn, M. V., & Alexander, R. D. (1982). Cultural theory: The developing synthesis from biology. *Human Ecology* 10: 383–400.

Gangestad, S. W., Haselton, M. G., & Buss, D. M. (2006). Evolutionary foundations of cultural variation: Evoked culture and mate preferences. *Psychological Inquiry* 17: 75–95.

Ghiselin, M. T. (1969). *The Triumph of the Darwinian Method*. University of California Press, Berkeley, CA.

Goddard, M. J. (2009). The impact of human intuition in psychology. *Review of General Psychology* 13: 167–174.

Grantham, D. W. (1994). *The South in Modern America*. HarperCollins, New York, NY.

Gregor, A. J. (2009). *Marxism, Fascism, and Totalitarianism: Chapters in the Intellectual History of Radicalism*. Stanford University Press, Palo Alto, CA.

Hamilton, W. D. (1964). The genetical evolution of social behaviour, I & II. *Journal of Theoretical Biology* 7: 1–52.

Harvey, P. H., & Pagel, M. D. (1991). *The Comparative Method in Evolutionary Biology*. Oxford University Press, Oxford, U.K.

Henrich, N. S., & Henrich, J. (2007). *Why Humans Cooperate: A Cultural and Evolutionary Explanation*. Oxford University Press, Oxford, U.K.

Henrich, J., & Henrich, N. (2010). The evolution of cultural adaptations: Fijian food taboos protect against dangerous marine toxins. *Proceedings of the Royal Society B* 277: 3715–3724.

Hofstede, G. (1980). *Culture's Consequences*. Sage, Beverly Hills, CA.

Hood Jr., R. W., & Williamson, W. P. (2008). *Them That Believe: The Power and Meaning of the Christian Serpent-handling Tradition*. University of California Press, Berkeley, CA.

Hrdy, S. B. (2009). *Mothers and Others*. Belknap, Cambridge, MA.

Inbar, Y., Pizarro, D. A., Iyer, R. et al. (2012). Disgust sensitivity, political conservatism, and voting. *Social Psychological and Personality Science* 5: 537–544.

Irons, W. (1979). Natural selection, adaptation and human social behavior. In *Evolutionary Biology and Human Social Behavior: An Anthropological Perspective* (eds. N. A. Chagnon & W. Irons), pp. 4–38. Duxbury Press, North Scituate, MA.

Jones, C. B. (2011). Are human cooperative breeders? A call for research. *Archive of Sexual Behavior* 40: 479–481.

Jost, J., Glaser, J., Kruglanski, A. et al. (2003). Political conservatism as motivated social cognition. *Psychological Bulletin* 129: 339–375.

Jost, J. T., Federico, C. M., & Napier, J. L. (2009). Political ideology: Its structure, functions, and elective affinities. *Annual Review of Psychology* 60: 307–337.

Kaplan, S. (1992). Environmental preference in a knowledge-seeking, knowledge-using organism. In *The Adapted Mind: Evolutionary Psychology and the Generation of Culture* (eds. J. Barkow, L. Cosmides, & J. Tooby), pp. 581–600. Oxford University Press, Oxford, U.K.

Key, V. O. (1949). *Southern Politics: In State and Nation*. A. F. Knopf, New York, NY.

Kovach, F. J. (1974). *Philosophy of Beauty*. University of Oklahoma Press, Norman, OK.

Kramer, K. L. (2010). Cooperative breeding and its significance to the demographic success of humans. *Annual Review of Anthropology* 39: 417–436.

Kurzban, R. (2010). Grand challenges of evolutionary psychology. *Frontiers in Psychology* 1: 1–3.

Lewontin, R. C. (1970). The units of selection. *Annual Reviews of Ecology and Systematics* 1:1–18.

Lieberman, D. (2009). Rethinking the Taiwanese minor marriage data: Evidence the mind uses multiple kinship cues to regulate inbreeding avoidance. *Evolution and Human Behavior* 30: 153–160.

Lieberman, D., Tooby, J., & Cosmides, L. (2003). Does morality have a biological basis? An empirical test of the factors governing moral sentiments relating to incest. *Proceedings of the Royal Society of London B* 270: 819–826.

Lieberman, D., Tooby, J., & Cosmides, L. (2007). The architecture of human kin detection. *Nature* 445: 727–731.

Mace, R., & Pagel, M. (1994). The comparative method in anthropology. *Current Anthropology* 35: 549–564.

Murdock, G. P., & White, D. R. (1969). Standard cross-cultural sample. *Ethnology* 8: 329–69.

Murray, D. R., Trudeau, R., & Schaller, M. (2011). On the origins of cultural differences in conformity: Four tests of the pathogen prevalence hypothesis. *Personality and Social Psychology Bulletin* 37: 318–329.

Navarrete, C. D., & Fessler, D. M. T. (2006). Disease avoidance and ethnocentrism: The effects of disease vulnerability and disgust sensitivity on intergroup attitudes. *Evolution and Human Behavior* 27: 270–282.

Nettle, D. (1999). *Linguistic Diversity*. Oxford University Press, Oxford, NY.

Nettle, D. (2009). Ecological influences on human behavioural diversity: A review of recent findings. *Trends in Ecology and Evolution* 24: 618–624.

Oaten, M., Stevenson, R. J. & Case, T. I. (2009). Disgust as a disease-avoidance mechanism. *Psychological Bulletin* 135: 303–321.

Ohman, A. & Mineka, S. (2001). Fears, phobias, and preparedness: Toward an evolved module of fear and fear learning. *Psychological Reviews* 108: 483–522.

Pedersen, E. J., Kurzban, R., & McCullough, M. E. (2013). Do humans really punish altruistically? A closer look. *Proceedings of the Royal Society B* 280: 20122723.

Pelikan, J. J., & Lehmann, H. T. Eds. (1955–1986). *Luther's Works* (LW), 55 vols. Fortress Press, St. Louis, MO: Concordia, and Philadelphia, PA.

Reed, J. S. (1972). *The Enduring South: Subcultural Persistence in Mass Society*. Lexington Books, Lexington, MA.

Reed, J. S. (1983). *Southerners: The Social Psychology of Sectionalism*. University of North Carolina Press, Chapel Hill, NC.

Reeve, H. K., & Sherman, P. W. (2001). Optimality and phylogeny: A critique of current thought. In *Adaptationism and Optimality* (eds. S. Orzack & E. Sober), pp. 64–113. Oxford University Press, Oxford, U.K.

Ridley, M. (1983). *The Explanation of Organic Diversity: The Comparative Method and Adaptations for Mating*. Clarendon Press, Oxford, U.K.

Rogers, A. R., & Cashdan, E. (1997). The phylogenetic approach to comparing human populations. *Evolution and Human Behavior* 1: 353–358.

Ruch, W., & Hehl, F. J. (1983). Intolerance of ambiguity as a factor in the appreciation of humor. *Personality and Individual Differences* 4: 443–449.

Schaller, M. (2006). Parasites, behavioral defenses, and the social psychological mechanisms through which cultures are evoked. *Psychological Inquiry* 17: 96–101.

Schaller, M., & Murray, D. (2008). Pathogens, personality, and culture: Disease prevalence predicts worldwide variability in sociosexuality, extraversion, and openness to experience. *Journal of Personality and Social Psychology* 95: 212–221.

Schwartz, S. H. (2004). Mapping and interpreting cultural differences around the world. In *Comparing Cultures: Dimensions of Culture in a Comparative Perspective* (eds. H. Vinkin, J. Soeters, & P. Ester). Brill, Leiden, The Netherlands.

Seppänen, J. T., Forsman, J. T., Mönkkönen, M. et al. (2011). New behavioural trait adopted or rejected by observing heterospecific tutor fitness. *Proceedings of the Royal Society B: Biological Sciences* 278: 1736–1741.

Siegal, M., Fadda, R., & Overton, P. G. (2011). Contamination sensitivity and the development of disease-avoidant behavior. *Philosophical Transactions of the Royal Society B: Biological Sciences* 366: 3427–3432.

Sternhell, Z. (2009). *The Anti-Enlightenment Tradition*. Yale University Press, New Haven, CT.

Thornhill, R. (1998). Darwinian aesthetics. In *Handbook of Evolutionary Psychology: Ideas, Issues and Applications* (eds. C. Crawford & D. Krebs), pp. 543–572. Lawrence Erlbaum Assoc., Mahwah, NJ.

Thornhill, R. (2003). Darwinian aesthetics informs traditional aesthetics. In *Evolutionary Aesthetics* (eds. K. Grammer & E. Voland), pp. 9–38. Springer-Verlag, Berlin, Germany.

Thornhill, R. (2007). Comprehensive knowledge of human evolutionary history requires both adaptationism and phylogenetics. In *The Evolution of Mind: Fundamental Questions and Controversies* (eds. S. W. Gangestad & J. A. Simpson), pp. 31–37. Guilford Press, New York, NY.

Thornhill, R., & Fincher, C. L. (2007). What is the relevance of attachment and life history to political values? *Evolution and Human Behavior* 28: 215–222.
Thornhill, R. & Fincher, C. L. (2013). The comparative method in cross-cultural and cross-species research. *Evolutionary Biology* 40: 480–493.
Thornhill, R., & Gangestad, S. W. (1993). Human facial beauty: Averageness, symmetry, and parasite resistance. *Human Nature: An Interdisciplinary Biosocial Perspective* 4: 237–269.
Thornhill, R., & Gangestad, S. W. (1999a). Facial attractiveness. *Trends in Cognitive Sciences* 3: 452–460.
Thornhill, R., & Gangestad, S. W. (1999b). The scent of symmetry: A human sex pheromone that signals fitness? *Evolution and Human Behavior* 20: 175–201.
Thornhill, R. & Gangestad, S. W. (2008). *The Evolutionary Biology of Human Female Sexuality*. Oxford University Press, New York, NY.
Thornhill, R. & Palmer, C. (2000). A Natural History of Rape: Biological Bases of Sexual Coercion. MIT Press, Cambridge, MA.
Thornhill, R., Fincher, C. L., & Aran, D. (2009). Parasites, democratization, and the liberalization of values across contemporary countries. *Biological Reviews* 84: 113–131.
Thornhill, R., Fincher, C. L., Murray, D. R. et al. (2010). Zoonotic and non-zoonotic diseases in relation to human personality and societal values: Support for the parasite-stress model. *Evolutionary Psychology* 8: 151–169.
Trivers, R. L. (1971). The evolution of reciprocal altruism. *Quarterly Review of Biology* 46: 35–57.
Trivers, R. L. (1985). *Social Evolution*. Benjamin/Cummings, Menlo Park, CA.
Trumbull, E., Rothstein-Fish, C., Greenfield, P. M. et al. (2001). *Bridging Cultures between Home and School: A Guide for Teachers*. WestEd, New York, NY.
Tybur, J. M., Miller, G. F., & Gangestad, S. W. (2007). Testing the controversy: An empirical examination of adaptationists' attitudes toward politics and science. *Human Nature* 18: 313–328.
Tybur, J. M., Lieberman, D., & Griskevicius, V. (2009). Microbes, mating, and morality: Individual differences in three functional domains of disgust. *Journal of Personality and Social Psychology* 97: 103–122.
West, S. A., El Mouden, C., & Gardner, A. (2011). Sixteen common misconceptions about the evolution of cooperation in humans. *Evolution and Human Behavior* 32: 231–262.
West-Eberhard, M. J. (2003). *Developmental Plasticity and Evolution*. Oxford University Press, Oxford, U.K.
Westoby, M., Leishman, M. R., & Lord, J. M. (1995a). On misinterpreting the 'phylogenetic correction.' *Journal of Ecology* 83: 531–534.
Westoby, M., Leishman, M., & Lord, J. (1995b). Issues of interpretation after relating comparative datasets to phylogeny. *Journal of Ecology* 83: 892–893.
Westoby, M., Leishman, M., & Lord, J. (1995c). Further remarks on phylogenetic correction. *Journal of Ecology* 83: 727–729.
Williams, G. C. (1966). *Adaptation and Natural Selection*. Princeton University Press, Princeton, NJ.
Williams, G. C. (1992). *Natural Selection: Domains, Levels, and Challenges*. Oxford University Press, New York, NY.

# Chapter 3
# The Parasite-Stress Theory of Values

## 3.1 Introduction

This chapter elaborates on the brief sketch of the parasite-stress theory of values in Chap. 1. It also treats briefly the earliest research findings inspired by that theory.

## 3.2 Immunity

Established knowledge of the ecology and evolution of parasitic diseases (= infectious diseases = pathogenic diseases) provides part of the foundation for the parasite-stress theory of values. (We also refer to this theory as the parasite-stress theory of sociality.) Infectious diseases are significant causes of Darwinian selection acting on all life. For modern humans, parasites appear to be the number one cause of evolutionary change. Geneticists who study evolutionary change in genes of the human genome (all the genes of the human species) report that parasites account for more evolutionary action across the genome than other environmental factors that are also sources of selection. Recently, Fumagalli et al. (2011) reviewed much of the published evidence of recent evolution in the human genome in response to infectious diseases. Moreover, their extensive study across 55 contemporary human populations shows that, compared to genes involved in dealing with 13 other environmental challenges (climatic and geographic factors, metabolic traits, diet, subsistence strategies), genes related to immunity exhibit significantly more change across geographic regions. Immunity genes as evolutionary hot spots means that selection is acting more strongly on these genes than other genes so far studied. These findings are consistent with the fact that a large portion of the current human morbidity and mortality across the world and even across the USA states is attributable to parasitic diseases (Chap. 8).

In addition, infectious diseases were a major source of morbidity and mortality, and hence of natural selection, in deep-time human evolutionary history (Anderson

R. Thornhill and C.L. Fincher, *The Parasite-Stress Theory of Values and Sociality: Infectious Disease, History and Human Values Worldwide*,
DOI 10.1007/978-3-319-08040-6_3, © Springer International Publishing Switzerland 2014

and May 1991; Ewald 1994; Dobson and Carper 1996; McNeill 1998; Wolfe et al. 2007; Volk and Atkinson 2013). Volk and Atkinson (2013) published an important review of rates and causes of human juvenile mortality in three ethnographic samples representative of ecological conditions in human evolutionary history: hunter–gatherer societies without contact with modern technology that can affect mortality (e.g., medicine, sanitation, education, birth control), agriculturalist indigenous societies with limited access to modern technology, and ancient historical populations, extending in some cases as far back as several hundred years B.C. The data on rates of infant and child mortality reflected the probability of mortality by age one year and by approximate sexual maturity at age 15 years, respectively. Across the three samples combined, infant mortality showed an average of about 23 % and child mortality about 48 %, and similar patterns were seen across all three samples. Volk and Atkinson limited their study to samples with relatively large sample sizes from reliable sources and emphasize that their estimates are probably considerably below the actual mortality rates. The two largest mortality factors were infectious disease (especially gastrointestinal and respiratory illnesses) and infanticide, with the former greatly predominating. As discussed in Chap. 8 in our treatment of child maltreatment by parents, sick children suffer higher rates of maltreatment than well children. Thus, part of the mortality attributed to infanticide by Volk and Atkinson is likely mediated by infectious disease. In sum, infectious disease was the chief cause of juvenile mortality in the evolutionary historical settings comprising the juvenile mortality data reviewed by Volk and Atkinson. Finally, the existence of complex, evolved human adaptations that are designed functionally to defend against parasites document that natural selection in the deep-time past directly favored individuals with defenses against infectious diseases.

Humans have two immune systems. One is the classical immune system: the physiological, cellular, and tissue-based mechanisms of defense against parasites. The second is the behavioral immune system, comprised of the psychology and behavior for infectious-disease avoidance (Schaller and Duncan 2007) as well as for managing the effects of diseases when they strike (Fincher and Thornhill 2008a).

The behavioral immune system—the focus of this book—includes ancestrally adaptive feelings (e.g., disgust), cognition (e.g., worry about contagion), values about and behavior toward out-group and in-group members, caution about or unwillingness to interact with out-group people, and prejudice against people perceived as unhealthy, contaminated, or unclean. The design of the prejudicial aspects of the human behavioral immune system extends prejudice to people with symptoms of noncontagious diseases, physically and mentally handicapped people, extra-thin and obese people, homosexuals, and the elderly, because human psychological and behavioral adaptation for avoidance of contagion is designed for an adaptive (ancestrally) oversensitivity to stimuli that may even remotely suggest contagion risk (Park et al. 2003, 2007, 2013; Curtis et al. 2004, 2011; Faulkner et al. 2004; Navarrete and Fessler 2006; Curtis 2007; Schaller and Duncan 2007; Fincher et al. 2008; Ackerman et al. 2009; Duncan and Schaller 2009; Oaten et al. 2009; Terrizzi et al. 2010; Kouznestsova et al. 2012; Miller and Maner 2012; Ryan et al. 2012). The behavioral immune system also includes avoidance and regulation of contact with nonhuman animals that pose human infectious-disease threats (Prokop et al. 2010a, b).

## 3.3 Host–Parasite Coevolution

Hosts and their parasites coevolve in antagonistic and perpetual races with adaptation, counteradaptation, counter-counteradaptation and so on for both host and parasite; there is no lasting adaptive solution that can be mounted by either side against the other (Haldane 1949; Van Valen 1973; Tooby 1982; Ridley 1993; Ewald 1994; Thompson 2005). In the human case, this dynamic, antagonistic interaction is illustrated by the fact that, despite the large somatic allocation made to the classical immune system, people still get sick and even small reductions in immunocompetence increase vulnerability to infectious disease.

Furthermore, host–parasite arms races are localized geographically across the range of a host species and its parasite, creating a coevolutionary mosaic involving genetic and phenotypic differences in host immune adaptation and corresponding parasite counteradaptation (Thompson 2005). An important outcome of the geographical localization of parasite–host coevolutionary races is that host defense works most effectively, or only, against the local parasite species, strains, or genotypes, but not against those evolving in nearby host groups. Hence, out-groups may often harbor novel parasites that cannot be defended against very well or at all by an individual or his or her immunologically similar in-group members (Fincher and Thornhill 2008a, b). Out-group individuals pose the additional infectious-disease threat of lacking knowledge of local customs, manners, and norms in general, many of which (e.g., methods of hygiene or food acquisition and preparation) may prevent infection from local parasites (Fincher et al. 2008; Schaller and Neuberg 2008). Norms of many types—culinary, linguistic, moral, sexual, nepotistic, religious, dress, and so on—are used by people both to portray in-group affiliation and associated values and to distinguish in-group from out-group members. Norm differences between groups are often the basis of intergroup prejudice and hostility (i.e., xenophobia). Likewise, norm similarity is the basis of positive valuation and altruism among people (Park and Schaller 2005; Norenzayan and Shariff 2008; Murray et al. 2011).

Evidence for geographically localized host–parasite coevolutionary races is convincing. On the parasite side of the race, parasite geographical mosaics were found, for example, in recent research on the human protozoan parasite *Leishmania braziliensis*. Rougeron et al. (2009) described the high genetic diversity and subdivided population structure of this parasite across both Peru and Bolivia. They found high levels of microgeographic variation identifiable by at least 124, highly localized, physiologically, and genetically distinct strains. The strains showed strong evidence of high degrees of close inbreeding and thus resembled genetic clones.

This extremely fine-grained geographic mosaic in *L. braziliensis* implies a similar microgeographic immunological genetic mosaic in human hosts. This type of spatial variation in host adaptation against local parasites, or said differently, in host immune maladaptation against out-group-typical parasites, is a general pattern in the animal and plant infectious disease literature (e.g., Kaltz et al. 1999; Thompson 2005; Tinsley et al. 2006; Dionne et al. 2007; Corby-Harris and Promislow 2008).

Specific human cases showing this include the caste-specific infectious diseases and associated caste-specific immunity among sympatric Indian castes (Pitchappan 2002). Indeed, McNeill (1998) suggested that the castes of India initially formed, in part, from differential localized cultural responses to local parasite stress. Other cases include the village-specific immune defenses against leishmania parasites in adjacent Sudanese villages (Miller et al. 2007) and the variation in virulence of human African trypanosomiasis in northern versus southern human populations in East Africa (MacLean et al. 2004). In particular regions, the localization of host immunity to local parasites is so fine-grained that people inbreed, risking the potential costs of inbreeding depression, in order to maintain coadapted gene complexes important for coping with parasite infection in their offspring, as Denic and colleagues have shown for malaria across regions (Denic and Nicholls 2007; Denic et al. 2008a, b) and we and colleagues have proposed and empirically supported for parasite stress in general across countries (Hoben et al. 2010; Chap. 6). On a broad scale, the localization of host–parasite coevolutionary races in humans is seen dramatically in the findings of the human genetic research noted just above: there is more regional variation in genes affecting classical immunity than in many other human genes affecting fitness (Fumagalli et al. 2011).

There are other bodies of evidence of localized host immunity. One familiar type of evidence involves events where individuals from isolated groups interact with novel groups by conquest or trade and infectious disease transmission ensues, sometimes with drastic effects. This has occurred after the intra- and intercontinental movement of individuals brought about intergroup contact (Good 1972; Dubos 1980; Jenkins et al. 1989; Diamond 1998; McNeill 1998). Some other human examples of localized immunity are discussed in Fincher and Thornhill (2008a) and Tibayrenc (2007). The final evidence we will mention for local host adaptation to parasites is found in the literature showing that the hybridization between adjacent, closely related conspecific populations results in hybrid individuals with reduced immunocompetence and thus reduced fitness (e.g., house mice: Sage et al. 1986, cottonwood trees: Floate et al. 1993; also see Thompson 2005 for other examples).

## 3.4 Assortative Sociality: An Aspect of the Behavioral Immune System

Due to localized host immune adaptation, in an ecological setting of high disease stress, reduced dispersal, xenophobia, and ethnocentrism are adaptive preferences/values and behaviors for avoiding novel parasites contained in out-groups and for managing local infectious disease (Fincher and Thornhill 2008a, b). Philopatry—the absence of dispersal away from the natal range for reproduction—reduces contact with out-groups and their habitats that may contain new parasites. Likewise, xenophobia—the avoidance and dislike of out-group members—discourages contact with out-groups and their likely different parasites. Neophobia—the dislike of new ideas and ways of doing—is a component of xenophobia; according to the parasite-stress theory of values, neophobia functions like xenophobia. Ethnocentrism is in-group favoritism entailing nepotism toward both nuclear and extended family, as well as altruism toward other, yet immunologically similar, in-group members. This support and loyalty toward in-group members is a defense against the morbidity and mortality effects of parasites (Sugiyama 2004; Sugiyama and Sugiyama 2003; Navarrete and Fessler 2006). Sugiyama (2004) reported that in the Shiwiar, an Amazonian society without ready access to modern medicine, healthcare in the forms of food and other assistance from in-group members to persons suffering from infectious disease is a major factor lowering mortality. This parasite-management benefit of local embeddedness in in-groups seems to characterize numerous traditional human societies in the ethnographic record (Gurven et al. 2000; Sugiyama 2004; Sugiyama and Sugiyama 2003). To paraphrase Navarrete and Fessler (2006), in human evolutionary history, under high parasite stress, in-group members were the only health insurance one had, and it was adaptive to have always paid your premiums—in terms of social investment and loyalty toward in-group allies that buffer one and one's family against the morbidity and mortality of infectious disease.

Hence, philopatry, xenophobia (including neophobia), and ethnocentrism—the basic features of assortative sociality and simultaneously of the behavioral immune system—are expected to be values and normative behaviors predominantly in areas of high parasite stress (Fincher et al. 2008; Thornhill et al. 2009). This prediction is strongly empirically supported, as we document in subsequent chapters.

Parasite stress is not the same across the globe nor has it been the same across time. Humans have experienced parasite gradients throughout history and continue to do so today (McNeill 1980, 1998; Low 1990; Dobson and Carper 1996; Guernier et al. 2004; Lopez et al. 2006; Crawford 2007; Smith et al. 2007; Wolfe et al. 2007; Smith and Guégan 2010). Thus, we expect that the benefits and costs of assortative sociality will shift along the parasite-stress gradient such that in some circumstances (elevated parasite stress) high levels of assortative sociality will be more beneficial than in other circumstances (low parasite stress). As parasite stress declines, the infectious-disease risks to individuals of dispersal and interaction with out-groups decrease. Consequently, for individuals in ecological settings that are relatively low in parasite stress, out-group contacts and alliances may provide greater benefits than costs. The benefits of out-group interactions can be many and include gains through intergroup exchange of goods and ideas, and diversified and sometimes larger social networks for marriage and other social alliances (Fincher et al. 2008; Thornhill et al. 2009). We show in subsequent chapters that the components of assortative sociality/behavioral immunity respond quantitatively to parasite stress across regions as predicted by this reasoning.

## 3.5 The Genesis of Cultures and Species

The parasite-stress theory of values is a general theory of human sociality and of cultural diversity. As we document in this book, it seems to explain many features of people's social behaviors and their variation across the globe and across time.

As a theory of cultural diversity, the parasite-stress theory informs the processes causing new cultures to originate. McElreath et al. (2003) and Nettle (1999) argued that social preference for in-group members could cause cultural isolation and hence cultural divergence and emergent new cultures in the absence of geographic barriers such as mountains or rivers that fractionate a culture's original range into isolated segments. Building on this, we argued that, given the ecological localization of host defenses against parasites, the components of assortative sociality—limited dispersal, ethnocentrism, and xenophobia—by functioning in parasite avoidance and management, fractionate or segment an original culture's range and thereby contribute to the independence of the resulting segments (Fincher and Thornhill 2008a, b). Thus, the parasite-stress theory includes a theory about the genesis of cultural or ethnic diversity, and some of the predictions related to this aspect of the theory have been supported empirically. We have shown that endemic religion diversity (both major religions and ethnoreligions) and indigenous language diversity, across contemporary countries worldwide, are related strongly and positively to parasite stress (Chap. 13). Also consistent with this aspect of the parasite-stress theory is Cashdan's (2001) finding that high parasite-stress regions have more ethnic groups than low parasite-stress regions.

The parasite-stress theory has also been applied to species formation. Fincher and Thornhill (2008a; Thornhill and Fincher 2013) developed the parasite-driven-wedge model of speciation, a new speciation hypothesis in which parasite stress and the behaviors it selects for—limited dispersal and in-group social preference including local mating—segment an original species' range and cause divergence of the segments to the status of reproductive isolation and hence distinct species. This speciation model is supported by a variety of evidence presented in Chap. 13.

Our earlier published ideas about the role of parasite stress in leading to new kinds of cultures and species are expanded considerably in Chap. 13.

## 3.6 Conditional Behavioral Immunity

The parasite-stress theory of sociality posits an adaptive (ancestrally), condition-dependent adoption of in-group and out-group values and related social tactics by individuals dependent on variable local parasite stress. This condition-dependent adaptation, like other condition-dependent adaptation, requires for its evolution, local variation on a short time scale in the selection pressures responsible for it. Hence, evolutionary historical selection due to morbidity and mortality from pathogens varied locally in individual lifetimes and thereby favored contingent behavioral and psychological adaptations for assortative sociality.

The evolution of conditionality as an important feature of assortative sociality's design, rather than region-specific genetically distinct adaptation, is consistent with knowledge about infectious diseases. The dynamics of an infectious disease can generate high variation in prevalence, transmissibility, and pathogenicity of the disease agent across the range of its host species, as well as on a fine-grained, local scale within an individual's lifetime. Important factors affecting this variability at a single locale and in a single generation are temporal changes in host group size, weather, disease-vector abundance and behavior, and the number, virulence, and dynamics of the different infectious diseases infecting hosts (Anderson and May 1991; Ewald 1994; Guernier et al. 2004; Prugnolle et al. 2005; Corby-Harris and Promislow 2008; Loker 2012). Thus, in-group assortative sociality is an example of adaptive phenotypic plasticity within individuals. That is, the individual possesses a conditional strategy with multiple contingent tactics (Fincher et al. 2008; Schaller and Murray 2008; Thornhill et al. 2009). Such plasticity in traits is favored by Darwinian selection when phenotypic change allows the individual to modify its phenotypic expression in directions that give greater net inclusive fitness benefit than that achieved by a single phenotype. Conditional strategies in behavior, psychology, development, and physiology are very common across animal taxa (West-Eberhard 2003). Cultural behavior in humans is a category of behavioral and psychological plasticity that evolved, at least in part, as a solution to the evolutionary historical, fitness-relevant problem of complex short-term change in the social environment (Alexander 1979; Flinn 1997; Flinn and Coe 2007). We have argued that a major part of this social change was adjustments by individuals in their in-group and out-group values and behavior in order to deal adaptively with temporally varying local parasite problems (Thornhill et al. 2009; Chap. 5).

A considerable body of research supports the hypothesis of an evolved contingent assortative sociality in people that functions against contagion. For example, Faulkner et al. (2004) and Navarrete and Fessler (2006) provide evidence, based on numerous and diverse Western samples, that scores among individuals on scales that measure the degree of xenophobia and ethnocentrism correspond to chronic individual differences in worry about catching infectious disease (measured by scores on the perceived-vulnerability-to-disease scale [Duncan et al. 2009]); those who perceive high infectious disease risk are more xenophobic and ethnocentric than those who perceive low disease risk. Importantly, this research also showed that xenophobia and ethnocentrism within individuals increase under experimental primes of greater pathogen salience in the current environment. Other research that we discuss later in this chapter documents within-individual shifts in personality—toward greater introversion and avoidance of novelty—and in heightened classical immune responses as well as behavioral avoidance of strangers immediately after research participants view cues of infectious-disease salience.

Hence, there is considerable evidence of both interindividual stable differences as well as within-individual conditionality in xenophobic and ethnocentric values and related personality features and behaviors, and that both the interindividual consistency and within-individual contingency are caused by infectious-disease problems in the local environment.

As mentioned in Chap. 2, the proximate means by which individuals assess local parasite stress—and thereby ontogenetically and contingently express the locally adaptive degree of assortative sociality—may include activation of the classical immune system (such as, the frequency of infection; Stevenson et al. 2009) and social learning of local disease risks (Fincher et al. 2008). Both of these causes may act and account for the interindividual and within-individual variation in values affecting in- and out-group behavioral preferences.

Our emphasis on adaptive contingency in the expression/adoption of assortative sociality does not imply that we expect no variation across human groups in genetic adaptation for assortative sociality. Durham (1991) and Blute (2010) treated culture–gene coevolution in detail. It involves allelic frequency changes (i.e., evolution) that correspond to changes in cultural traits. Culture–gene coevolution may produce genetically differentiated cross-cultural variation in the values and behaviors of assortative sociality. For example, in areas of high parasite prevalence, cultural practices of xenophobia, philopatry, and ethnocentrism may effectively select for alleles affecting psychological features that promote the learning and effective use of these values (Fincher et al. 2008). Our argument is that infectious disease problems are locally variable on a short time scale as a result of temporal changes mentioned earlier, and hence significant conditionality will be favored and maintained by selection even in the presence of localized genetic adaptation functioning in adoption and use of local values and behaviors. There is some evidence, although to date quite limited, that culture–gene coevolution may play a role in cross-national variation in the value dimension collectivism–individualism (Chiao and Blizinsky 2010; Way and Lieberman 2010). That genetically distinct adaptation for coping with an ecological problem and condition-dependent adaptation for the same problem domain can co-occur is well established in the literature of alternative reproductive tactics (recent review in Oliveira et al. 2008).

## 3.7 Behavioral Immunity Adaptively Manages False Positives

Natural selection in all species favors individuals with indirect knowledge of infectious disease risk and the avoidance of such risks. Hence, there are directly selected human psychological features that pay attention to, and process information about, environmental cues that, across generations of human evolutionary history, corresponded with the presence of contagion. Moreover, given that an error in judging a contagion risk can be literally grave, selection has built behavioral immunity to accept adaptively many false positives—i.e., deduce contagion risk when it is actually absent (Curtis 2007; Duncan and Schaller 2009; Oaten et al. 2009; Miller and Maner 2012). As a result, people's behavioral immune system sometimes overreacts to even the hint of contagion danger in our environment, including our social environment. This is why a person's encounter with a stranger who speaks a different dialect or believes in a different mythical system may evoke strong xenophobia toward the stranger. This, too, is the cause of prejudice of many people toward sexual minorities (homosexuals and bisexuals), obese or very thin people, the elderly, people with noncontagious diseases, or people in wheelchairs or who show behavior that deviates from normative patterns. (See Duncan and Schaller 2009; Ryan et al. 2012; Kouznestsova et al. 2012, and Terrizzi et al. 2012 for reviews of these studies.)

## 3.8 Brief History of Research Inspired by the Parasite-Stress Theory of Sociality

Research began fairly recently on human values and associated emotions, cognition, and behavior as adaptations for dealing with variable parasite stress, and on how these adaptations may regulate enculturation and determine cultural diversity. Here we briefly sketch some of the contributions of this research, which we treat more fully in later chapters that connect these research findings to broader areas of research related to the parasite-stress theory of values.

Bobbi Low (1988, 1990) did very early work on human social life inspired by the parasite-stress theory of sociality; she proposed that across indigenous societies polygynous marriage systems will be concentrated in regions of high parasite stress and monogamous systems will be concentrated in low parasite-stress regions. Her thinking was influenced by the parasite theory of sexual selection proposed by Hamilton and Zuk (1982). Sexual selection is the component of variance in reproductive success of individuals resulting from their traits that affect the number and quality of mates obtained. Hamilton and Zuk's idea was that hosts' variation in genetic resistance to parasites causes sexual selection on hosts, both intrasexual (within-sex competition) and intersexual (between-sex choice) sexual selection. According to this idea, the winners of within-sex competition for mates and of contests to impress mates possess relatively high parasite resistance. Moreover, parasite-driven sexual selection varies in intensity in direct relation to parasite adversity faced by hosts. High parasite stress in a region, according to Low's hypothesis, generates high variance among men in genetic and phenotypic quality related to parasite resistance, which makes polygynous unions with the men of highest genetic quality adaptive for women and thus valued by them. Hence, human polygynous mating systems and the values of people promoting them are predicted to be more common in high than in low parasite-stress areas. Low's empirical work across indigenous societies in the ethnographic record of anthropology supported this hypothesis.

In related research about the same time, Gangestad and Buss (1993) reported that, across many contemporary countries, people of both sexes in high parasite regions more strongly value attractive others as long-term mates than do people living in low parasite-stress areas. This pattern, like that Low discovered, was predicted on the basis of the parasite theory of sexual selection, because physical attractiveness is a marker of phenotypic and genetic quality, including health and disease resistance (Thornhill and Gangestad 1993, 1999a, b, 2008). Subsequently, the Gangestad and Buss (1993) study was expanded and showed the same result (Gangestad et al. 2006).

Several years after the first study by Gangestad and Buss, Billing and Sherman (1998) and Sherman and Billing (1999) hypothesized that the value people place on the use of spices in cooking is a defense against food-borne human parasites. To test this, they investigated the types and numbers of spices used in recipes across many regions of the world. They found that temperature positively correlates with antipathogen spice use across regions. Temperature is a useful surrogate for parasite stress, with warmer equating with more parasite adversity (Billing and Sherman 1998). Later research by Murray and Schaller (2010) reported a robust positive relationship across countries between spice use and parasite stress per se.

About the same time, Flaxman and Sherman (2000) published their ideas about the function of "morning sickness," a condition common in women during the first 16 weeks of pregnancy. They encouraged the label of "nausea and vomiting in pregnancy" (NVP) for the condition rather than sickness because of evidence they gathered for NVP's evolved function as defense against parasites of mother and fetus. The adaptive immunosuppression of mothers during early pregnancy apparently functions to prevent the rejection by pregnant mothers of the half-foreign fetus, but with the cost of greater susceptibility to infections. They present a range of evidence that NVP, in part, is a component of behavioral immunity and guides diet choice of pregnant mothers away from ingestion of foods with contagion risk.

Fessler (2001) recognized the adaptive challenge presented to women by the immunosuppression during both pregnancy and the luteal phase of the menstrual cycle, which is characterized by a similar immunosuppression, and proposed that females would compensate for this immunosuppression through behavioral means such as increased disgust sensitivity and other forms of behavioral disease avoidance—this was later described as the "compensatory behavioral prophylaxis hypothesis." While an initial test failed to support the hypothesis (Fessler and Navarrete 2003), later studies provided support for the hypothesis (Fessler et al. 2005; Fleischman and Fessler 2011).

Curtis and colleagues' research on the emotion of disgust as an antipathogen adaptation was published after that on food spicing and NVP (Curtis and Biran 2001; Curtis et al. 2004; Curtis 2007). Their approach to disgust has inspired considerable research on the topic. Consequently, it is now known that disgust is not only evoked in the context of perception of disease-laden cues, such as contaminated foods, sick people, parasites (e.g., worms), or parasite reservoirs (e.g., cockroaches), but also is commonly generalized to include (a) groups of people who are perceived as harboring infectious disease and (b) cultural behaviors that are different or unfamiliar. Thus, disgust directed toward out-group people, ideas, beliefs, and behavior appears to be a pathogen defense. This includes so-called moral disgust toward others in which others' beliefs, norms, values, manners, or behavior are deemed morally undesirable or repugnant (Oaten et al. 2009; Schnall et al. 2008; Curtis et al. 2011; Inbar et al. 2012, but see Tybur et al. 2010). Furthermore, young children use expressions of disgust in the faces of parents and others to discriminate contaminated objects from safe objects and even to judge normative or moral behaviors (Stevenson et al. 2010).

About the same time as the research on disgust began to appear, other research relating prejudicial attitudes to parasite stress was published in the scientific literature. Prejudice against out-groups (xenophobia) and against physically abnormal or disabled and obese or very thin people as contagion-avoidance adaptations was first proposed and documented empirically by Kurzban and Leary (2001), Park et al. (2003), Faulkner et al. (2004), and Park et al. (2007). Navarrete and Fessler (2006) also provided evidence that xenophobia is for contagion avoidance, and that ethnocentrism is a strategy for in-group alliances to manage the debilitating effects of infectious diseases. In related research, Navarrete et al. (2007) documented that women in their first trimester of pregnancy show a facultative shift to greater ethnocentrism, probably as an adaptation to build and maintain in-group alliances that promote in-group investment in the woman and in her offspring after birth, as well as to avoid novel parasites of out-groups during the trimester of pregnancy when a woman's immune system is adaptively suppressed. Also, Quinlan (2007) reported that, across indigenous societies, there is more extended parental care (measured as increased nursing duration) in high-parasite-stress ecological settings than under low-infectious-disease levels. Extensive parental care is an aspect of in-group investment or ethnocentrism (Chap. 5).

We began research on parasites and social life in 2004. The published work of Buss, Curtis, Faulkner, Fessler, Flaxman, Gangestad, Kurzban, Low, Navarrete, Park, Sherman, and their coauthors especially influenced our thinking about human psychology and behavior in relation to infectious disease. Thornhill also acknowledges the large influence that his interactions with Bill Hamilton and Marlene Zuk in the 1980s had on his thinking about the connection between infectious diseases and sexually selected social behavior. Also influential for Thornhill was hearing a lecture by Bill Freeland in the summer of 1973. At the time, Freeland and Thornhill were fellow doctoral students in biology at the University of Michigan. Freeland's lecture presented his new ideas on nonhuman primate xenophobia and ostracism in relation to contagion avoidance. These ideas were published later (Freeland 1976, 1979). Ben Hart (Hart et al. 1987; Hart 1988, 1990) also was a pioneer of research in behavioral immunity of nonhuman animals; he continues to make discoveries of infectious-disease defenses (for a recent review, see Hart 2011). As well, Kim Lewis (1998) published an interesting paper suggesting that association with and altruism toward kin were favored by selection because such behaviors reduce contact with nonrelatives who carry diseases to which one's kin group is not resistant.

Hamilton and Zuk's (1982) ideas about parasites driving sexual selection had immediate impact in the scientific community and began to be tested soon after their publication in a diversity of animal species (See the papers in the journal edited by Hausfater and Thornhill 1990.) Freeland's ideas were generally ignored until very recently, despite an important paper by Loehle (1995) that sought to create interest in them. We credit Freeland with the first ideas about how parasites affect the evolution of sociality, including the role of parasites in the sexual aspect of social behavior (see Freeland 1976). Hamilton and Zuk's and Hart's work followed soon after.

In 2004, when we began research on parasites and sociality, there was no evolutionary theory about the major human value dimension collectivism–individualism. A great deal of descriptive research had been published already about these values. We thought that regional variation in this value dimension might be caused by regionally variable parasite stress, since it was known that collectivism–individualism corresponds to differences in norm following and in in-group versus out-group interaction preferences and behavior. Steve Gangestad mentioned to us that Mark Schaller was thinking similarly. Steve had heard Mark give a talk at a conference and said it was theoretically similar to the framework we were developing and testing. We contacted Mark and his collaborating graduate student, Damian Murray, which began a research collaboration that resulted in Fincher, Thornhill, Murray and Schaller (2008), which showed, as we had predicted, that infectious-disease stress positively correlates with collectivism, or said differently, negatively correlates with individualism, across many nations of the world.

Fincher and Thornhill went on to develop the hypothesis that the parasite-evoked values of philopatry, ethnocentrism, and xenophobia create new cultures in a region. We supported this perspective with cross-national studies of two major types of cultural diversity, language, and religion diversity (Fincher and Thornhill 2008a, b; Chap. 13).

We also proposed a hypothesis of parasite-driven parapatric speciation (Fincher and Thornhill 2008a). Parapatric speciation involves a "parent" species' range becoming regionally subdivided into contiguous segments as a consequence of localized adaptation leading to maladaptive mating between individuals across different segments. The selection for local adaptation and the associated maladaptive hybridization can result in evolutionary divergence and independence of the segments without segmentation by the various geographic barriers that bring about allopatric speciation. We suggested that localized host–parasite arms races generate Darwinian selection for limited dispersal and in-group mating and other localized social preferences and that evolutionary divergence of contiguous segments of an original species to the status of new species with reproductive isolation can be the outcome. The parasite-stress theory of parapatric speciation was supported by various data already published in the literature. For example, compared to higher latitudes, at low latitudes and hence areas of higher parasite stress, there are more species, ranges of species are smaller, and nearby populations are more genetically dissimilar (Fincher and Thornhill 2008a; Thornhill and Fincher 2013; Chap. 13).

Subsequently, with colleagues, Fincher and Thornhill extended the parasite-stress theory to explain regional variation in democratization, conservatism–liberalism, gender inequality, sexual restrictiveness, property rights, personality, family values, and religiosity (Thornhill et al. 2009; Thornhill et al. 2010; Fincher and Thornhill 2012; also see Gangestad et al. 2006 and Schaller and Murray 2008); civil and other intrastate warfare, revolutions, and coups (Letendre et al. 2010; Letendre et al. 2012; also see Schaller and Neuberg 2008); cognitive ability (Eppig et al. 2010, 2011); and marital inbreeding and outbreeding (Hoben et al. 2010). In later chapters, we treat in detail each of these topics.

Hence, the parasite-stress theory of sociality has been quite heuristic; that is, it has produced a range of new ideas and newly discovered patterns and offered new interpretations of some previously described patterns. Moreover, it continues to generate new findings and research directions. We give here examples of this heuristic continuation.

A recent study by Schaller et al. (2010) reported that research subjects who observed slides of people with infectious disease symptoms (e.g., pox, skin lesions, sneezing) immediately mounted a classical immune response. Their white blood cells produced elevated amounts of inflammatory cytokine-interleukin-6 when exposed to bacterial antigens. This immune response was not seen in research subjects who viewed control slides, including subjects who saw slides depicting a person pointing a gun directly at the viewer. Hence, the classical immune response was not a general reaction to danger or threat, but was specific to cues of other people with symptoms of parasitic infection.

Research by Stevenson et al. (2011) compared salivary immune markers between research participants in whom disgust was induced by disease-relevant pictorial cues documented to be disgust elicitors (e.g., a dirty toilet, an eye infection) and other participants who were exposed to either negative, but disease-irrelevant, pictures or neutral pictures. The disgust-primed group showed an oral immune response, but the other groups did not.

The recent study by Mortensen et al. (2010) reported that subjects viewing slides with disease-salient cues immediately exhibited greater feelings promoting between-person avoidance (extraversion, openness to experiences, and agreeableness were reduced) in comparison to these subjects' feelings upon viewing control slides. These researchers also found that subjects with high scores on the scale of perceived vulnerability to disease, which measures a person's concern about contagion in the environment, reacted more strongly—showed greater feelings of interpersonal avoidance—than did subjects with low scores on the same scale. Finally, this same paper reported that viewing parasite-salient slides resulted in increased avoidant arm movements when subjects viewed facial photos of strangers, especially for subjects high in perceived vulnerability to disease.

When considered together, these studies by Mortensen et al. (2010), Schaller et al. (2010), and Stevenson et al. (2011) reveal that visually perceiving cues pertinent to risk of parasitic infection generate immediate cellular and biochemical immune responses, a change in perceptions of one's own personality, and behavioral actions that defend against or avoid infectious people. Hence, such cues activate markedly the classical immune system as well as the behavioral immune system, and the dual activation is functionally coordinated to defend against infectious-disease threat.

Further evidence of dual activation and functional coordination of the classical and behavioral immune systems was reported by Miller and Maner (2011). In this case the activation of the classical immune system leads to activation of the behavioral immune system, a coordination that is the converse of that reported by Schaller et al. (2010) and Mortensen et al. (2010). Miller and Maner found that recently ill people were more attentive to and avoidant of disfigured human faces than were people who were not recently ill.

Olsson et al. (2014) reported a complementary result to that of Miller and Maner (2011), but involving experimental activation of the classical immune system of research participants using hypodermic injection of an antigen. Each participant also received a saline injection. The antigen injection and control injection were given about one month apart. Other research participants smelled and rated the collected body odor of these participants a few hours after each injection. Raters showed an aversive response to the body odor of participants whose classical immunity had been activated, but not to the body odor of controls.

Another example of the continuing heuristic nature of the parasite-stress theory is the recent research by Stevenson et al. (2009) who reported that people with high contamination sensitivity and disgust sensitivity had fewer recent infectious diseases than people with low sensitivities, providing evidence of a protective function of these emotions against these diseases. These researchers also found that high contamination sensitivity, in particular, was associated positively with a history of contracting infectious diseases (but not with recency of infections), implying that an ontogeny of repeated activation of the classical immune system may underlie the adoption of conservative values and associated behavior.

A related study by de Barra et al. (2013) found that adults with a childhood history of more infectious diseases, in comparison to adults with a childhood of fewer parasitic diseases, placed greater importance on facial attractiveness in a mate. As discussed earlier in this chapter, increased salience given to physical attractiveness is a manifestation of the behavioral immune system because physical attractiveness is a certification of health (Chap. 6).

Recent research on conformity in relation to infectious-disease risk is another example of the continuing heuristics of the parasite-stress theory of values. Conforming to the beliefs and values of the majority has benefits and costs. Benefits of socially navigating in a conformist group include the predictability of the way people think and behave. Moreover, when conformity is coupled with aversion toward those who do not conform to the majority behavior, as it typically is, conformity will be protective against novel parasites in out-groups to which the conforming in-group is not immune (Fincher et al. 2008; Murray et al. 2011; Murray and Schaller 2012; Wu and Chang 2012). Costs of conformity include the low rate of generating and of adopting ideas, especially ideas that are unfamiliar locally. However, preferring traditional ways of thinking and avoiding foreign ideas can be defenses against novel parasites in out-groups. In line with this reasoning and supporting the parasite-stress theory of values, Murray et al. (2011) showed that cross-national variation in extent of conformity correlates positively with parasite adversity. Also, Murray and Schaller (2012) in research in Canada and Wu and Chang (2012) in research in China examined individual differences in conformity values and found that perceptions of personal vulnerability to infectious disease correlated positively with conformity. Each of these two studies also included experiments that made infectious-disease risk salient to research participants. The participants immediately became more conformist, but this change in values was not observed in control groups of participants, including controls presented with disease-irrelevant threat cues. In the Murray and Schaller study, the participants exposed to parasite-salient cues showed increased positivism toward conforming others. Murray and Schaller's and Wu and Chang's findings indicate that an individual's perception of threat of infectious disease, either arising from the individual-difference measure on the perceived vulnerability to disease questionnaire or due to immediate stimuli of parasite presence, causes her or him to adopt conformist values. The Murray and Schaller study also showed that individuals presented with cues of parasite presence in their immediate environment became prejudiced in favor of others with conformist values.

Other recent research also reveals the heuristic impact of the parasite-stress theory of values. Terrizzi et al. (2010, 2012) recently investigated individual differences in the relationship of disgust sensitivity with the conservative values of religiosity and prejudice against sexual minorities (homosexuals and bisexuals). They reported that disgust sensitivity positively predicts these values and argued that disgust, religiosity, and prejudice against sexual out-groups are components of the human behavioral immune system. In complementary work, Clay et al. (2012) showed that individual differences in disgust sensitivity and perceived vulnerability to disease positively correlate with collectivism and several other variables that reflect conservatism (e.g., traditionalism, conformity, and importance of societal stability and security). Terrizzi et al. (2013) conducted a synthetic meta-analysis of 22 studies of individual differences in various components of collectivism or of conservatism in relation to perceived vulnerability to disease or disgust sensitivity. They reported robust positive relationships among the variables. They concluded that conservative values are defenses that reduce contact with infectious diseases.

Inspired by the parasite-stress theory of values, Scott Reid and colleagues (2012) made an important discovery for the scholarly discipline of linguistics. Reid and colleagues researched disgust sensitivity in relation to sound perception of dissimilarity to self's accent of foreign-accented English. Americans of high disgust sensitivity rated foreign-accented English as more dissimilar to their own accent than did Americans of low disgust sensitivity. The study also showed that research participants who viewed parasite-salient stimuli perceived a greater difference in foreign-accented English compared to their own accent, but participants viewing other threat stimuli (unrelated to parasite threat) did not. Given the positive relationship between conservatism and disgust, these results imply that conservatives perceive greater differences between in-group and out-group spoken language than do liberals. We hypothesize from the parasite-stress theory of values that the greater sensitivity of high-disgust people to differences between us and them, which was documented by Reid et al. for accents, may extend to many differences outside of language, such as the perception of value differences, skin color, and behavior. Consistent with this suggestion, as it pertains to values, are the research findings by Reid and colleagues (personal communication) on people's perception of differences between one's own and others' religions. Research participants of high disgust sensitivity rated unfamiliar religions (but not familiar religions) as more dissimilar to their own religion than did participants of low disgust sensitivity. The study showed, too, that research participants who viewed parasite-salient stimuli perceived a greater difference between their own religious beliefs and unfamiliar religious beliefs, but participants viewing other threat stimuli (unrelated to parasite threat) did not. Reid and colleagues conclude that many people perceive unfamiliar religions as a threat of infectious disease. The kinds of perception biases discovered by Reid and colleagues may underlie the xenophobia sensitivity of conservatives.

Prokop and colleagues' recent research extended the parasite-stress theory of values to human interactions with nonhuman animals. These scholars showed that the human behavioral immune system includes avoidance and regulation of contact with pets that pose human infectious-disease threats (Prokop et al. 2010a, b).

Domesticated dogs harbor many diseases that can be transmitted to people (Alcock 2001; Prokop et al. 2010b). In regions in which parasite stress is high, dogs will harbor more of these diseases. Although in such regions dogs may be valued for their utility, they may be treated differently than in regions of low parasite stress. Specifically, in high-parasite-stress regions, dogs may be segregated to reduce human contact with them. This was the finding of Prokop et al. (2010b) when they compared rates of keeping dogs in homes between two countries with dogs present, but with differing parasite stress. Research subjects in Turkey reported lower rates of pets in home (which included dogs) than did Slovakians; parasite stress is higher in Turkey than in Slovakia. Furthermore, in the same study, in each of the two countries, having pets in the home was related negatively to individual differences in the perception of vulnerability to infectious disease.

Earlier in this chapter we mentioned the research that reported the positive association across nations between antimicrobial spice use in cooking recipes and parasite stress. Additional evidence that spicing foods is a form of behavioral immunity has been found by Prokop and Fančovičová (2011). They showed that individual differences in preference for and use of spiced food corresponded with concern about infectious diseases. Individuals who were high in worry about contagion had stronger preference for, and consumption of, spicy foods than individuals who were low on such concern.

In other research, Huang et al. (2011) document that individuals shift to lower out-group prejudice when they perceive cues indicating reduced contagion risk. They conducted three related studies. Study 1 was done during the H1N1 swine-flu epidemic. The disease-primed research participants read a story about the potential widespread health problems with the epidemic and the value of vaccination in defense against H1N1. Control participants read a story unrelated to disease. All participants then completed a questionnaire that measures attitudes toward immigrants; also they indicated whether or not they had recently received an H1N1 flu vaccination. Results indicated that the disease-primed people reported more prejudice against immigrants than control people and that vaccinated people showed reduced prejudice compared to unvaccinated people. Moreover, among disease-primed people, the prejudice reduction related to being vaccinated was attributable to their own perception that the vaccine offered protection from H1N1. Huang et al. included a second study that eliminated certain confounds or alternative conclusions from study 1. Overall, the results of the two studies provide evidence that vaccination reduces prejudice against out-groups and it does so by causing people to feel protected from contagion. In the third study, Huang et al. (2011) researched the effect of hand washing with a commercial sanitizing hand wipe on attitude toward out-groups. Huang et al. (2011) discuss earlier research showing that hand washing reduces a person's contact with gastrointestinal and respiratory pathogens. Evidence from the third study indicated that hand washing reduced prejudice against out-groups.

Huang et al.'s (2011) research indicating that negative attitudes toward out-group people can be reduced by perception that disease risk is lowered provides additional evidence of people's ability to adjust conditionally social behavior in relation to immediate change in risk of contact with infectious disease. The studies discussed earlier in this section reveal that cues of parasite salience in the current environment cause people to immediately adopt personality features, values, and behavioral avoidance that reduce contact with strangers and new experiences. Huang et al.'s (2011) research shows that when people perceive lower risk of contagion they shift in the opposite direction—toward more openness to out-groups.

The continuing importance of the parasite-stress theory of values in guiding researchers to discoveries about human and other animal behavior and psychology is documented also in the chapters that follow.

The research discussed in this section was inspired by and derived from the parasite-stress theory of social behavior. What about earlier research on human values conducted without awareness of this theory? Is the earlier research consistent with the parasite-stress theory of values? The next chapter discusses social-science research findings on values published either prior to development of the parasite-stress theory of values or subsequent to the theory, but not inspired by it. The rest of this book builds on these and other findings and documents that the parasite-stress theory offers a coherent and encompassing interpretation of research on values, both before and after its origin.

## 3.9 Summary

The parasite-stress theory of values/sociality is presented in detail. Humans have two immune systems: the classical physiological, cellular, and tissue-based defense system and the behavioral immune system. Only recently has the latter been researched in any detail; it is comprised of two parts: (a) psychology and behavior for infectious-disease avoidance and (b) psychology and behavior for managing the fitness-reducing effects of parasitic infection. Our focus in this book is on the behavioral immune system. This immune system is comprised of the adaptations of philopatry, xenophobia, neophobia, and ethnocentrism, which are the basic features of assortative sociality. These values and norms of the behavioral immune system are predicted to be most strongly held in areas of high parasite-stress because parasite–host coevolutionary arms races are geographically localized. Philopatry, xenophobia, and neophobia reduce contact with novel parasites inhabiting out-groups. Ethnocentrism builds dutiful and embedded social ties with in-group members who have similar immunity. Such ties provide reliable social investments and network that protect one's inclusive fitness interests from adverse effects of present parasites. The adversity of infectious disease varies geographically. As parasite-stress declines, the infectious-disease risks to individuals of interaction with out-groups decrease. As a result, in areas that are relatively low in parasite stress, out-group contacts and alliances provide greater benefits than costs to individuals. The benefits of out-group interactions and transactions include access to other groups' resources and ideas and more diverse social networks for social alliances. In subsequent chapters, we show that the components of assortative sociality/behavioral immunity increase with increasing parasite stress across regions, as predicted by the parasite-stress theory of values.

The three components of assortative sociality—limited dispersal, ethnocentrism, and xenophobia—also fractionate cultures and thereby contribute to the genesis of new cultures. Thus, the parasite-stress theory includes a hypothesis about the origin of cultural or ethnic diversity. The parasite-stress theory of sociality also may be an important engine of speciation. Later in our book, we present empirical support for the ethnogenesis and speciation aspects of the parasite-stress theory of sociality.

Assortative sociality, like human cultural behavior in general, is a conditional strategy of the individual. Conditional strategies are favored by selection when phenotypic change allows the individual to diversify its phenotypic expression adaptively. Temporal variation in infectious-disease stress over a short time frame on a local scale has favored the phenotypic contingency in valuation and use of tactics reflecting degree of assortative sociality.

Some of the research findings inspired by the parasite-stress theory of sociality are briefly reviewed. This theory has produced numerous new discoveries and new interpretations of previously described findings.

## References


Ackerman, J. M., Becker, D. V., Mortensen, C. R. et al. (2009). A pox on the mind: Disjunction of attention and memory in the processing of facial disfigurement. *Journal of Experimental Social Psychology* 45: 478–485.

Alcock, J. (2001). *The Triumph of Sociobiology*. Oxford University Press, New York, NY.

Alexander, R. D. (1979). *Darwinism and Human Affairs*. University of Washington Press, Seattle, WA.

Anderson, R. M., & May, R. M. (1991). *Infectious Disease of Humans: Dynamics and Control*. Oxford University Press, Oxford, U.K.

Billing, J., & Sherman, P. W. (1998). Antimicrobial functions of spices: Why some like it hot. *Quarterly Review of Biology* 73: 3–49.

Blute, M. (2010). *Darwinian Sociocultural Evolution: Solutions to Dilemmas in Cultural and Social Theory.* Cambridge University Press, Cambridge, U.K.

Cashdan, E. (2001). Ethnic diversity and its environmental determinants: Effects on climate, pathogens, and habitat diversity. *American Anthropology* 103: 968–991.

Chiao, J. Y., & Blizinsky, K. D. (2010). Culture–gene coevolution of individualism–collectivism and the serotonin transporter gene. *Proceedings of the Royal Society B* 277: 529–537.

Clay, R., Terrizzi Jr., J. A., & Shook, N. J. (2012). Individual differences in the behavioral immune system and the emergence of cultural systems. *Journal of Social Psychology* 43: 174–184.

Corby-Harris, V., & Promislow, D. E. L. (2008). Host ecology shapes geographical variation for resistance to bacterial infection in *Drosophila melanogaster. Journal of Animal Ecology* 77: 768–776.

Crawford, D. H. (2007). *Deadly Companions: How Microbes Shaped our History*. Oxford University Press, New York, NY.

Curtis, V., & Biran, A. (2001). Dirt, disgust, and disease: Is hygiene in our genes? *Perspectives in Biology and Medicine* 44: 17–31.

Curtis, V. A. (2007). Dirt, disgust and disease: A natural history of hygiene. *Journal of Epidemiology and Community Health* 61: 660–664.

Curtis, V., Aunger, R., & Rabie, T. (2004). Evidence that disgust evolved to protect from risk of disease. *Proceedings of the Royal Society B* (Supplement) 271: 17–31.

Curtis, V., de Barra, M., & Aunger, H. (2011). Disgust as an adaptive system for disease avoidance behaviour. *Philosophical Transactions of the Royal Society B* 366: 389–401.

de Barra, M., DeBruine, L., Jones, B. et al. (2013). Illness in childhood predicts face preferences in adulthood. *Evolution and Human Behavior* 43: 384–389.

Denic, S., & Nicholls, M. G. (2007). Genetic benefits of consanguinity through selection of genotypes protective against malaria. *Human Biology* 79: 145–158.

Denic, S., Nagelkerke, N., & Agarwal, M. M. (2008a). Consanguineous marriages and endemic malaria: Can inbreeding increase population fitness? *Malaria Journal* 7: 150.

Denic, S., Nagelkerke, N., & Agarwal, M. M. (2008b). Consanguineous marriages: Do genetic benefits outweigh its costs in populations with alpha(+)-thalassemia, hemoglobin s, and malaria? *Evolution and Human Behavior* 29: 364–369.

Diamond, J. (1998). *Guns, Germs and Steel: The Fates of Human Societies*. W. W. Norton and Co., New York, NY.

Dionne, M., Miller, K. M., Dodson, J. J. et al. (2007). Clinical variation in MHC diversity with temperature: Evidence for the role of host–pathogen interaction on local adaptation in Atlantic salmon. *Evolution* 61: 2154–2164.

Dobson, A. P., & Carper, E. R. (1996). Infectious diseases and human population history. *BioScience* 46: 115–126.

Dubos, R. (1980). The evolution of microbial diseases. In *Man Adapting* (ed. R. Dubos), pp. 163–195. Yale University Press, New Haven, CT.

Duncan, L. A., & Schaller, M. (2009). Prejudicial attitudes toward older adults may be exaggerated when people feel vulnerable to infectious disease: Evidence and implications. *Analyses of Social Issues and Public Policy* 9: 97–115.

Duncan, L. A., Schaller, M., & Park, J. H. (2009). Perceived vulnerability to disease: Development and validation of a 15-item self-report instrument. *Personality and Individual Differences* 47: 541–546.

Durham, W. H. (1991). *Coevolution: Genes, Culture and Human Diversity*. Stanford University Press, Stanford, CA.

Eppig, C., Fincher, C. L., & Thornhill, R. (2010). Parasite prevalence and the worldwide distribution of cognitive ability. *Proceedings of the Royal Society of London B* 277: 3801–3808.

Eppig, C., Fincher, C. L., & Thornhill, R. (2011). Parasite prevalence and the distribution of intelligence among the states of the USA. *Intelligence* 39: 155–160.

Ewald, P. W. (1994). *Evolution of Infectious Disease*. Oxford University Press, New York, NY.

Faulkner, J., Schaller, M., Park, J. H. et al. (2004). Evolved disease-avoidance mechanisms and contemporary xenophobic attitudes. *Group Processes and Intergroup Relations* 7: 333–353.

Fessler, D.M.T. (2001). Luteal phase immunosuppression and meat eating. *Rivista Di Biologia* 94: 403–426

Fessler, D.M.T., & Navarrete, C.D. (2003). Domain-specific variation in disgust sensitivity across the menstrual cycle. *Evolution and Human Behavior* 24: 406–417.

Fessler, D.M.T., Eng, S.J., & Navarrete, C.D. (2005). Elevated disgust sensitivity in the first trimester of pregnancy: evidence supporting the compensatory prophylaxis hypothesis. *Evolution and Human Behavior* 26: 344–351.

Fincher, C. L., & Thornhill, R. (2008a). A parasite-driven wedge: Infectious diseases may explain language and other biodiversity. *Oikos* 117: 1289–1297.

Fincher, C. L., & Thornhill, R. (2008b). Assortative sociality, limited dispersal, infectious disease and the genesis of the global pattern of religion diversity. *Proceedings of the Royal Society of London, Biological Sciences* 275: 2587–2594.

Fincher, C. L., & Thornhill, R. (2012). Parasite-stress promotes in-group assortative sociality: The cases of strong family ties and heightened religiosity. *Behavioral and Brain Sciences* 35: 61–79.

Fincher, C. L., Thornhill, R., Murray, D. R. et al. (2008). Pathogen prevalence predicts human cross-cultural variability in individualism/collectivism. *Proceedings of the Royal Society of London Biological Sciences* 275: 1279–1285.

Flaxman, S. M., & Sherman, P. W. (2000). Morning sickness: a mechanism for protecting mother and embryo. *Quarterly Review of Biology* 75: 113–148.

Fleischman, D.S., & Fessler, D.M.T. (2011). Progesterone's effects on the psychology of disease avoidance: support for the compensatory behavioral prophylaxis hypothesis. *Hormones and Behavior* 59: 271–275.

Flinn, M. V. (1997). Culture and the evolution of social learning. *Evolution and Human Behavior* 18: 23–67.

Flinn, M. V., & Coe, K. (2007). The linked red queens of human cognition, reciprocity, and culture. In *The Evolution of Mind* (eds. S. W. Gangestad & J. A. Simpson), pp. 339–347. Guilford Press, New York, NY

Floate, K. D., Kearsley, M. J. C., & Whitham, T. G. (1993). Elevated herbivory in plant hybrid zones: *Chrysomela confluens*, *Populus* and phenological sinks. *Ecology* 74: 2056–2065.

Freeland, W. J. (1976). Pathogens and the evolution of primate sociality. *Biotropica* 8: 12–24.

Freeland, W. J. (1979). Primate social groups as biological islands. *Ecology* 60: 719–728.

Fumagalli, M., Sironi, M., Pozzoli, U. et al. (2011). Signatures of environmental genetic adaptation pinpoint pathogens as the main selective pressure through human evolution. *PLoS Genetics* 7: e1002355.

Gangestad, S. W., & Buss, D. M. (1993). Pathogen prevalence and human mate preference. *Ethology and Sociobiology* 14: 89–96.

Gangestad, S. W., Haselton, M. G., & Buss, D. M. (2006). Evolutionary foundations of cultural variation: Evoked culture and mate preferences. *Psychological Inquiry* 17: 75–95.

Good, C. M. (1972). Salt, trade, and disease: Aspects of development in Africa's northern Great Lakes region. *International Journal of African Historical Studies* 5: 543–586.

Guernier, V., Hochberg, M. E., & Guégan, J. -F. (2004). Ecology drives the worldwide distribution of human diseases. *PLoS Biology* 2: 740–746.

Gurven, M., Allen-Arave, W., Hill, K. et al. (2000). "It's a Wonderful Life": Signaling generosity among the Ache of Paraguay. *Evolution and Human Behavior* 21: 263–282.

Haldane, J. B. S. (1949). Disease and evolution. *Ricerca Scientifica Suppl. A* 19: 68–76.

Hamilton, W. D., & Zuk, M. (1982). Heritable true fitness and bright birds: A role for parasites? *Science* 218: 284–387.

Hart, B. L. (1988). Biological basis of the behavior of sick animals. *Neuroscience and Biòbehavioral Reviews* 12: 123–137.

Hart, B. L. (1990). Behavioral adaptations to pathogens and parasites: Five strategies. *Neuroscience and Biobehavioral Reviews* 14: 273–294.

Hart, B. L. (2011). Behavioural defences in animals against pathogens and parasites: Parallels with the pillars of medicine in humans. *Philosophical Transactions of the Royal Society B* 366: 3406–3417.

Hart, B. L., Korinek, E., & Brennan, P. (1987). Postcopulatory genital grooming in male rats: Prevention of sexually transmitted infections. *Physiological Behavior* 41: 321–325.

Hausfater, G., & Thornhill, R. (1990). Organized and edited the collection of 12 papers by 25 authors on parasites and sexual selection. *American Zoologist* 30: 2.

Hoben, A. D., Buunk, A. P., Fincher, C. L. et al. (2010). On the adaptive origins and maladaptive consequences of human inbreeding: Parasite prevalence, immune functioning, and consanguineous marriage. *Evolutionary Psychology* 8: 658–676.

Huang, J. Y., Sedlovskaya, A., Ackerman, J. M. et al. (2011). Immunizing against prejudice: Effects of Disease Protection on Attitudes Toward Out-groups. *Psychological Science* 22: 1550–1556.

Inbar, Y., Pizarro, D. A., Iyer, R. et al. (2012). Disgust sensitivity, political conservatism, and voting. *Social Psychological and Personality Science* 5: 537–544.

Jenkins, C., Dimitrakakis, M., Cook, I. et al. (1989). Culture change and epidemiological patterns among the Hagahai, Papua New Guinea. *Human Ecology* 17: 27–57.

Kaltz, O., Gandon, S., Michalakis, Y. et al. (1999). Local maladaptation in the anther-smut fungus *Microbotryum violaceum* to its host plant *Silene latifolia*: Evidence from a cross-inoculation experiment. *Evolution* 53: 395–407.

Kouznestsova, D., Stevenson, R. J., Oaten, M. J. et al. (2012). Disease-avoidant behaviour and its consequences. *Psychology and Health* 27: 491–506.

Kurzban, R., & Leary, M. R. (2001). Evolutionary origins of stigmatization: The functions of social exclusion. *Psychological Bulletin* 127: 187–208.

Letendre, K., Fincher, C. L., & Thornhill, R. (2010). Does infectious disease cause global variation in the frequency of intrastate armed conflict and civil war? *Biological Reviews* 85: 669–683.

Letendre, K., Fincher, C. L., & Thornhill, R. (2012). Infectious disease, collectivism, and warfare. In *The Oxford Handbook on Evolutionary Perspectives on Violence, Homicide, and Warfare* (eds. T. Shackelford & V. Weekes-Shackelford), pp. 351–371. Oxford University Press, New York, NY.

Lewis, K. (1998). Pathogen resistance as the origin of kin altruism. *Journal of Theoretical Biology* 193: 359–363.

Loehle, C. (1995). Social barriers to pathogen transmission in wild animal populations. *Ecology* 76: 326–335.

Loker, E. S. (2012). Macroevolutionary immunology: A role for immunity in the diversification of animal life. *Frontiers in Immunology* 3: 25.

Lopez, A. D., Mathers, C. D., Ezzati, M. et al. (2006). Global and regional burden of disease and risk factors, 2001: Systematic analysis of population health data. *Lancet* 367: 1747–1757.

Low, B. S. (1988). Pathogen stress and polygyny in humans. In *Human Reproductive Behavior: A Darwinian Perspective* (eds. L. Betzig, M. Borgerhoff Mulder & P. Turke), pp. 115–128. Cambridge University Press.

Low, B. S. (1990). Marriage systems and pathogen stress in human societies. *American Zoologist* 30: 325–339.

McElreath, R., Boyd, R., & Richerson, P. J. (2003). Shared norms and the evolution of ethnic markers. *Current Anthropology* 44: 122–129.

McNeill, W. H. (1980). Migration patterns and infection in traditional societies. In *Changing Disease Patterns and Human Behaviour* (eds. N. F. Stanley & R. A. Joske), pp. 27–36. Academic Press, New York, NY.

McNeill, W. H. (1998). *Plagues and Peoples*. Anchor, Harpswell, ME.

Miller, E. N., Fadl, M., Mohamed, H. S. et al. (2007). Y chromosome lineage- and village-specific genes on chromosomes 1p22 and 6q27 control visceral leishmaniasis in Sudan. *PLoS Genetics* 3: 679–688.

Miller, S. L., & Maner, J. K. (2011). Sick body, vigilant mind: The biological immune system activates the behavioral immune system. *Psychological Science* 22; 1467–1471.

Miller, S.L., & Maner, J.K. (2012). Overperceiving Disease Cues: The Basic Cognition of the Behavioral Immune System. *Journal of Personality and Social Psychology* 102: 1198–1213.

Mortensen, C. R., Becker, D. V., Ackerman, J.M. et al. (2010). Infection breeds reticence: The effects of disease salience on self-perceptions of personality and behavioral avoidance tendencies. *Psychological Science* 21: 440–447.

Murray, D. R., & Schaller, M. (2010). Historical prevalence of infectious diseases within 230 geopolitical regions: A tool for investigating origins of culture. *Journal of Cross-Cultural Psychology* 41: 99–108.

Murray, D. R., & Schaller, M. (2012). Threat(s) and conformity deconstructed: Perceived threat of infectious disease and its implications for conformist attitudes and behavior. *European Journal of Social Psychology* 42: 180–188.

Murray, D. R., Trudeau, R., & Schaller, M. (2011). On the origins of cultural differences in conformity: Four tests of the pathogen prevalence hypothesis. *Personality and Social Psychology Bulletin* 37: 318–329.

Navarrete, C. D., & Fessler, D. M. T. (2006). Disease avoidance and ethnocentrism: The effects of disease vulnerability and disgust sensitivity on intergroup attitudes. *Evolution and Human Behavior* 27: 270–282.

Navarrete, C. D., Fessler, D. M. T., & Eng, S. J. (2007). Elevated ethnocentrism in the first trimester of pregnancy. *Evolution and Human Behavior* 28: 60–65.

MacLean, L., Chisi, J. E., Odiit, M. et al. (2004). Severity of human African trypanosomiasis in East Africa is associated with geographic location, parasite genotype, and host inflammatory cytokine response profile. *Infection and Immunity* 72: 7040–7044.
Nettle, D. (1999). *Linguistic Diversity*. Oxford University Press, Oxford, New York, NY.
Norenzayan, A., & Shariff, A. F. (2008). The origin and evolution of religious prosociality. *Science* 322: 58–62.
Oaten, M., Stevenson, R. J., & Case, T. I. (2009). Disgust as a disease-avoidance mechanism. *Psychological Bulletin* 135: 303–321.
Oliveira, R. F., Taborski, M., & Brockman, H. J. (2008). *Alternative Reproductive Tactics: An Integrated Approach*. Cambridge University Press, Cambridge, U.K.
Olsson, M. J., Lundstrom, J. N. Kimball, B. A. et al. (2014). The scent of disease: Human body odor contains an early chemosensory cue of sickness. *Psychological Science* 25: 817–823.
Park, J. H. & Schaller, M. (2005). Does attitude similarity serve as a heuristic cue for kinship? Evidence of an implicit cognitive association. *Evolution and Human Behavior* 26: 158–170.
Park, J. H., Faulkner, J., & Schaller, M. (2003). Evolved disease avoidance processes and contemporary anti-social behavior: Prejudicial attitudes and avoidance of people with disabilities. *Journal of Nonverbal Behavior* 27: 65–87.
Park, J. H., Schaller, M., & Crandall, C. S. (2007). Pathogen-avoidance mechanisms and the stigmatization of obese people. *Evolution and Human Behavior* 28: 410–414.
Park, J. H., Van Leeuwen, F., & Ypapanti, C. (2013). Disease-avoidance processes and stigmatization: Cues of substandard health arouse heightened discomfort with physical contact. *Journal of Social Psychology* 153: 212–228.
Pitchappan, R. M. (2002). Castes, migration, immunogenetics and infectious diseases in south India. *Community Genetics* 5: 157–161.
Prokop, P. & Fačovičová, J. (2011). Preferences for spicy foods and disgust of ectoparasites are associated with reported health in humans. *Psihologija* 44: 281–293.
Prokop, P., Usak, M. & Fačovičová, J. (2010a). Health and the avoidance of macroparasites: A preliminary cross-cultural study. *Journal of Ethology* 28: 345–351.
Prokop, P., Usak, M. & Fačovičová, J. (2010b). Risk of parasite transmission influences perceived vulnerability to disease and perceived danger of disease-relevant animals. *Behavioural Processes* 85: 52–57.
Prugnolle, F., Manica, A., Charpentier, M. et al. (2005). Pathogen-driven selection and worldwide HLA Class I diversity. *Current Biology* 15: 1022–1027.
Quinlan, R. J. (2007). Human parental effort and environmental risk. *Proceedings of the Royal Society B* 274: 121–125.
Reid, S. A., Zhang, J., Anderson, G. L. et al. (2012). Parasite primes make foreign-accented English sound more distant to people who are disgusted by pathogens (but not by sex or morality). *Evolution and Human Behavior* 33: 471–478.
Ridley, M. (1993). *The Red Queen: Sex and the Evolution of Human Nature*. Macmillan Publishing Company, New York, NY.
Rougeron, V., De Meeus, T., Hide, M. et al. (2009). Extreme inbreeding in *Leishmania braziliensis*. *Proceedings of the National Academy of Sciences USA* 106: 10224–10229.
Ryan, S., Oaten, M. Stevenson, R. J. et al. (2012). Facial disfigurement is treated like an infectious disease. *Evolution and Human Behavior* 33: 639–646.
Sage, R. D., Heyneman, D., Lim, K. C. et al. (1986). Wormy mice in a hybrid zone. *Nature* 324: 60–63.
Schaller, M., & Duncan, L. (2007). The behavioral immune system: Its evolution and social psychological implications. In *Evolution and the Social Mind: Evolutionary Psychology and Social Cognition* (eds. J. P. Forges, M. G. Haselton & W. Von Hippel), pp. 293–307. Psychology Press, New York, NY.
Schaller, M., & Murray, D. (2008). Pathogens, personality, and culture: Disease prevalence predicts worldwide variability in sociosexuality, extraversion, and openness to experience. *Journal of Personality and Social Psychology* 95: 212–221.

Schaller, M., & Neuberg, S. L. (2008). Intergroup prejudices and intergroup conflicts. In *Foundations of Evolutionary Psychology* (eds. C. Crawford & D. L. Krebs), pp. 399–412. Lawrence Erlbaum Associates, New York, NY.

Schaller, M., Miller, G. E., Gervais, W. M. et al. (2010). Mere visual perception of other people's disease symptoms facilitates a more aggressive immune response. *Psychological Science* 21: 649–652.

Schnall, S., Haidt, J., Clore, G. L. et al. (2008). Disgust as embodied moral judgment. *Personality and Social Psychology Bulletin* 34: 1096–1109.

Sherman, P. W., & Billing, J. (1999). Darwinian gastronomy: Why we use spices. *BioScience* 49: 453–463.

Smith, K. F., & Guégan, J. -F. (2010). Changing geographic distributions of human pathogens. *Annual Review of Ecology, Evolution & Systematics*, 41, 231–250.

Smith, K. F., Sax, D. F., Gaines, S. D. et al. (2007). Globalization of human infectious disease. *Ecology* 88: 1903–1910.

Stevenson, R. J., Case, T. I. & Oaten, M. J. (2009). Frequency and recency of infection and their relationship with disgust and contamination sensitivity. *Evolution and Human Behavior* 30: 363–368.

Stevenson, R. J., Oaten, M. J., Case, T. I. et al. (2010). Children's response to adult disgust elicitors: Development and acquisition. *Developmental Psychology* 46: 165–177.

Stevenson, R. J., Hodgson, D., Oaten, M. J. et al. (2011). The effect of disgust on oral immune function. *Psychophysiology* 48: 900–907.

Sugiyama, L. S. (2004). Illness, injury, and disability among Shiwiar forager–horticulturalists: Implications of human life history. *American Journal of Physical Anthropology* 123: 371–389.

Sugiyama, L. S., & Sugiyama, M. S. (2003). Social roles, prestige, and health risk: Social niche specialization as a risk-buffering strategy. *Human Nature* 14: 165–190.

Terrizzi Jr., J. A., Shook, N. J., & Ventis, W. L. (2010). Disgust: A predictor of social conservatism and prejudicial attitudes toward homosexuals. *Personality and Individual Differences* 49: 587–592.

Terrizzi Jr., J. A., Shook, N. J., & Ventis, W. L. (2012). Religious conservatism: an evolutionarily evoked disease-avoidance strategy. *Religion, Brain and Behavior* 2: 105–120.

Terrizzi Jr., J. A., Shook, N. J., & McDaniel, M. A. (2013). The behavioral immune system and social conservatism: A meta-analysis. *Evolution and Human Behavior* 34: 99–108.

Thompson, J. N. (2005). *The Geographic Mosaic of Coevolution*. University of Chicago Press, Chicago, IL.

Thornhill, R., & Fincher, C. L. (2013). The parasite-driven-wedge model of parapatric speciation. *Journal of Zoology* 291: 23–33.

Thornhill, R., & Gangestad, S. W. (1993). Human facial beauty: Averageness, symmetry, and parasite resistance. *Human Nature: An Interdisciplinary Biosocial Perspective* 4: 237–269.

Thornhill, R., & Gangestad, S. W. (1999a). Facial attractiveness. *Trends in Cognitive Sciences* 3: 452–460.

Thornhill, R., & Gangestad, S. W. (1999b). The scent of symmetry: A human sex pheromone that signals fitness? *Evolution and Human Behavior* 20: 175–201.

Thornhill, R., & Gangestad, S. W. (2008). *The Evolutionary Biology of Human Female Sexuality*. Oxford University Press, New York, NY.

Thornhill, R., Fincher, C. L., & Aran, D. (2009). Parasites, democratization, and the liberalization of values across contemporary countries. *Biological Reviews* 84: 113–131.

Thornhill, R., Fincher, C. L., Murray, D. R. et al. (2010). Zoonotic and non-zoonotic diseases in relation to human personality and societal values: Support for the parasite-stress model. *Evolutionary Psychology* 8: 151–169.

Tibayrenc, M. (2007). Human genetic diversity and the spread of infectious diseases. In *Encyclopedia of Infectious Diseases: Modern Methodologies* (ed. M. Tibayrenc), pp. 321–335. John Wiley and Sons, Inc., Hoboken, New Jersey.

Tinsley, M.C., Blanford, S., & Jiggins, F.M. (2006). Genetic variation in *Drosophila melanogaster* pathogen susceptibility. *Parasitology* 132: 767–773.

Tooby, J. (1982). Pathogens, polymorphism, and the evolution of sex. *Journal of Theoretical Biology* 97: 557–576.

Tybur, J. M., Merriman, L. A., Caldwell Hooper, A. E. et al. (2010). Extending the behavioral immune system to political psychology: Are political conservatism and disgust sensitivity really related? *Evolutionary Psychology* 8: 599–616.

Van Valen, L. (1973). A new evolutionary law. *Evolutionary Theory* 1: 1–30.

Volk, A. A., & Atkinson, J. A. (2013). Infant and child death in the human environment of evolutionary adaptation. *Evolution and Human Behavior* 34: 182–192.

Way, B. M., & Lieberman, M. D. (2010). Is there a genetic contribution to cultural differences? Collectivism, individualism and genetic markers of social sensitivity. *Social Cognitive and Affective Neuroscience* 5: 203–211.

West-Eberhard, M. J. (2003). *Developmental Plasticity and Evolution*. Oxford University Press, Oxford, U.K.

Wolfe, N. D., Dunavan, C. P., & Diamond, J. (2007). Origins of major human infectious diseases. *Nature* 447: 279–283.

Wu, B., & Chang, L. (2012). The social impact of pathogen threat: How disease salience influences conformity. *Personality and Individual Differences* 53: 50–54.

# Chapter 4
# Human Values Research Prior to the Parasite-Stress Theory

## 4.1 Introduction

Prior to the recent research on human values in relation to parasite stress (briefly introduced in Chap. 3), political scientists, sociologists, and psychologists produced a huge and important descriptive literature about variation in people's values across countries and the USA states, as well as across individuals in certain regions. Largely, this literature was not generated using hypotheses inspired by evolutionary theory. All of the literature, however, is scientific; thus, the scholars producing it were pursuing an understanding of cause and effect, specifically the causes of values and associated behavior.

Oftentimes in this research tradition, wealth and economic development are assumed to be the most encompassing or fundamental causes of variation in values (e.g., Lipset 1959; Triandis 1995; Hofstede 2001). Temperature, rainfall, and related climatic variables also are seen as important causes of cross-cultural value systems (Van de Vliert 2009). This research indicates that economic and climatic factors do covary systematically with values. The limitation of the traditional scientific literature on values is that it does not consider ultimate causation through evolutionary processes and its product of evolved values-adopter psychological adaptation. As a result, this research was limited to identifying some proximate causes of ideology that lack a coherent and unifying theoretical foundation. The parasite-stress theory of values is not an alternative to this traditional scientific approach; instead, it is complementary and more causally synthetic and encompassing.

The parasite-stress theory can explain why ecological factors such as temperature and rainfall, as well as economic factors, affect values. Parasites thrive in hot and moist ecological settings, but are reduced in cold or dry regions (Low 1990; Cashdan 2001; Guernier et al. 2004; Dunn et al. 2010). Consequently, these climatic factors are proximate causes of the optimal values in a region by way of their influence on parasite stress in the region (Fincher and Thornhill 2008a, b). That is, climatic conditions in a region are part of the causal chain leading to the region's

R. Thornhill and C.L. Fincher, *The Parasite-Stress Theory of Values and Sociality: Infectious Disease, History and Human Values Worldwide*,
DOI 10.1007/978-3-319-08040-6_4, © Springer International Publishing Switzerland 2014

value system. We treat in more detail the interrelationship between climatic variables, values, and parasite stress in Chap. 14.

We have proposed that human parasitic diseases and the values they evoke are causes of cross-national economic variables such as Gross Domestic Product through three general mechanisms (Chap. 11, Fincher et al. 2008; Thornhill et al. 2009). First, parasites cause lethargy and morbidity that limit people's ability to work and produce (e.g., Landes 1998; Price-Smith 2002; McGuire and Coelho 2011; Bonds et al. 2012). Compared to a healthy person, a person with schistosomiasis, hookworm, malaria, amoebic dysentery, flu, or any other kind of the roughly 1,400 human infectious diseases will more often lack the energy and stamina to be on the job. (On the number of kinds of human parasites, see Taylor et al. 2001.) Also, many parasites lower the work capability of hosts by reducing visual, auditory, and other sensory competence. Moreover, they damage additional physiological systems, tissues, and organs and thereby cause permanent negative effects on personal productivity throughout the life of hosts.

Second, parasites cause people to adopt conservative values that cause low economic productivity. Conservative values are preferences for the local community and thus foster in-group production, even only by family or at most by close ethnic group, rather than production by larger realms and markets. As importantly, conservative values are preferences for traditional and conformist ideas and ways with a concomitant dislike and avoidance of new ideas, technologies, and means. As shown later in the book (Chap. 11), the neophobia of conservatism reduces the flow and adoption of new ideas, including innovations that promote health, scientific progress, technological advance, and economic productivity. Liberalism, however, promotes those innovations and their diffusion. The parasite-stress theory of sociality is a general theory of human culture and affairs, because parasite levels in the environment of humans proximately cause people's core values, and values impact so many, if not all, realms of human activity.

The third way in which parasites affect economics is through the lowering of cognitive ability, which limits innovation and understanding of new ideas required for economic, scientific, and technological advances. Recent research reveals that parasite stress is correlated negatively with cognitive ability, measured as IQ, across nations and states of the USA. This may result from an adaptive ontogenetic tradeoff in increased allocation to classical immunity at the expense of the brain as parasite stress increases (Eppig et al. 2010, 2011; Chap. 11).

Thus, according to the parasite-stress theory of values, parasite adversity and associated values are important causes of the economic conditions in a region. In addition, the causation in the parasite-stress theory's application to economic productivity is bidirectional—the values evoked by a region's level of parasite adversity feedback and affect parasite stress. Chapter 11 deals in detail with economics in relation to the parasite-stress theory of values. In that chapter, we argue that the huge variation in the wealth of nations can be illuminated importantly by the parasite-stress theory.

The earlier scientific literature on values provided much of the data that has been used for testing the parasite-stress theory of sociality as it applies to diversity across regions: data on cultural diversity in collectivism, personality, religiosity, democratization, gender equality, civil conflicts, property rights, and so on. Other data sources that have been used include public data archived at websites made by scholars of economics, religion, linguistics, political science, and related disciplines. As we document in subsequent chapters, the application of the parasite-stress theory of values to these two types of data sources has shown their consistency with that theory. The parasite-stress theory of sociality has successfully predicted numerous new patterns in values and their interrelationships that were not known to exist prior to the emergence of the theory.

Below, we review various traditional scientific findings on values. Subsequent chapters reveal what these findings mean—that is, how they all can be put together into a unified intellectual framework based on the parasite-stress theory of values. Thereby, the parasite-stress theory provides a general scientific theory comprised of (a) a fact-based set of conceptual research principles that unifies previously unconnected findings about values discovered by political scientists, historians, psychologists, anthropologists, sociologists, and other scholars, and (b) a framework for future research in ideology. It is from the synthetic understanding of values allowed by the parasite-stress theory that we can say something new and meaningful about cultures across the world.

## 4.2 Collectivism–Individualism Is Conservatism–Liberalism

Traditional research effort in the investigation of values, especially cross-nationally, has focused on collectivism–individualism. Many cross-cultural psychologists feel collectivism–individualism is the best way to characterize the general value system of a country. Collectivism–individualism is typically considered a unidimensional variable (Gelfand et al. 2004), as is conservatism–liberalism (Carney et al. 2008; Jost et al. 2009). Below, we show that these two value dimensions are very similar. Hence, high collectivism is high conservatism, and high individualism is high liberalism. Correspondingly, low collectivism equates with low conservatism, and low individualism with low liberalism. Before discussing the correspondence of collectivism–individualism with conservatism–liberalism, we briefly discuss traditional research on conservatism–liberalism.

The labels "conservatives" and "liberals" are used widely across cultures and identify distinctly different clumps of values (see meta-analysis by Jost et al. 2003 for 12 countries, 88 samples, and 23,000 people; also see Feather 1979; Laponce 1981; Knight 1993, 1999; Forabosco and Ruch 1994; Carney et al. 2008; Graham et al. 2009; Jost et al. 2009). The labels "rightist" for conservative and "leftist" for liberal are similarly common across cultures (Laponce 1981; Jost et al. 2009). The labels "right" and "left" arose during the French Revolution (1789–1799), which was a time period of increased democratization in France. The monarchy that had ruled France for centuries collapsed quickly, and French society underwent a rapid transformation from conservative values of traditionalism, authoritarianism, and religiosity to liberal values based on the Enlightenment principles of citizenship and inalienable rights for all. The left-minded were seated on the left side of the French General Assembly and the right-minded on the right side. These two ideologies of the Assembly were divided on the grounds of conservation in maintaining tradition/status quo, rule by and respect for authorities (religious, masculine, and elites) and inequality of people versus liberation from tradition/status quo with priority on social change, freedom from rule by authorities, and all people as equal and deserving of opportunity, dignity, respect, and participation in societal matters (see Laponce 1981; Jost et al. 2009). Today, these remain core ideological differences between liberals and conservatives. We return to the French Revolution in Chap. 10, where we discuss the relationship between infectious-disease reduction and democratization.

### *4.2.1 Psychometric Studies*

Western political scientists typically measure individual differences in conservatism–liberalism in questionnaire-based research. Numerous questionnaires have been developed to measure these values (Knight 1993, 1999). Many of these questionnaires have validity as seen in both the intercorrelation of the questions within a questionnaire—i.e., the items or questions of a questionnaire measure the same psychological dimension—and in people's behavior—e.g., scores predict people's political involvement, voting activity, and other behavioral differences across the right–left ideological continuum (Feather 1979; Knight 1993, 1999; Altemeyer 1996; Carney et al. 2008). As examples, we mention two similar questionnaires based on Wilson and Patterson's (1968) earlier questionnaire on conservatism. One of these, the 28-item C-scale, assesses numerous conservative (C)–liberal (L) values: attitude about the death penalty (C for, L against), abortion (C against, L for), minorities (C against, L for), immigration (C against, L for), racial segregation (C for, L against), censorship (C for, L against), gay's and women's rights (C against, L for), X-rated movies (C against, L for), military draft (C for, L against), modern art (C against, L for), pacifism (C against, L for), and so on across 28 value domains that separate the two ideological poles according to prior research. The measure of a person's values is calculated such that a high score is high conservatism and thus low liberalism, and a low score the reverse (see Thornhill and Fincher 2007). A second scale is a reduced version (18 items) of the 28-item scale with wording modifications to make it more relevant to contemporary Western people (Oxley et al. 2008).

As we mentioned, although some researchers disagree, the bulk of the evidence indicates that there is a single right–left dimension. Jost et al. (2009) review evidence for this single dimension, as well as evidence identifying many of the interrelated components of the ideology of each of the two wings (see also Jost et al. 2003; Carney et al. 2008). Conservatives and liberals differ reliably in the following ways. Conservatives place salience on salvation and religious participation, social stability or maintenance of status quo, inequality of people and out-group inferiority, out-group prejudice, traditional hierarchy and status, norm conformity and obedience, management of threat and uncertainty, need for closure and intolerance of ambiguity, conventional wisdom, simplicity and internal consistency, and need for social order and order in general. Liberals place salience on social change; openness to other groups, ideas, and new experiences; analytical and rational ways of knowing (as opposed to contra-evidence, traditional and conformist opinion); cognitive complexity; and tolerance of ambiguity and uncertainty. Liberals also are low in authoritarianism, whereas conservatives are high in authoritarianism.

Right-wing Authoritarianism (RWA) and Social Dominance Orientation (SDO) are two dimensions of values related to conservatism–liberalism. Extensive prior research has established that RWA, typically measured by a 30-item validated scale, and conservatism, measured by the C-scale or similar conservatism scales, are strongly and positively correlated, but that RWA and conservatism are not completely identical measures (e.g., Altemeyer 1996; Thornhill and Fincher 2007). RWA measures the conservative values of high regard for and obedience to authority and its associated traditional hierarchy, rules, and norms. People who score high on RWA are highly authoritarian: they hold authority figures in high esteem and want (perhaps need) to be dominated by them. Those high in authoritarianism additionally overlap with conservatives in being conventional, rigid with regard to moral absolutes, and distrusting and dehumanizing of out-groups (Jost et al. 2003; Hodson and Costello 2007; Carney et al. 2008; Napier and Jost 2008). In contrast, people low on RWA are norm and rule violators and are disrespectful of traditional hierarchies. They are more independent and free—liberated from traditional values and authority figures. The liberals who comprised the Western hippie movement of the 1960s and 1970s are an example of low RWA people. They were insurrectionists who opposed many major traditional values and power asymmetries and thus were against war, imperialism, sexual restrictions, racism, male domination, female subordination, religious authority, authority of parents and other elders, and legalized control of behavior by a conservative government.

SDO scores across individuals, measured on the 14-item SDO questionnaire, correlate positively, but moderately, with scores on RWA and conservatism. People high on SDO want to become the dominating authorities themselves (Pratto et al. 1994; Altemeyer 1996; Thornhill and Fincher 2007). High SDO scorers support traditional power asymmetries and hence are racist, ethnically intolerant, sexist, and nonequalitarian (Pratto and Hegarty 2000; Hodson and Costello 2007).

Some researchers have suggested that conservatism, at least in principle, is divisible into two components, economic conservatism and social conservatism. If this were true, there would exist multiple dimensions of conservatism–liberalism rather than a single dimension. For instance, in principle, one could be economically conservative, but socially liberal. We have noticed that people commonly express this distinction in describing their personal values. However, the body of evidence reviewed by Jost et al. (2009) indicates that economic conservatism and social conservatism are positively correlated overall, not negatively correlated, as they would be if the two types were opposed. The positive relationship between the two types of conservatism certainly seems accurate because the value of human inequality characterizes conservatism. Social conservatism is a prejudice against out-groups and low-status people. Economic conservatism places importance on maintaining unequal resource distribution across a society. Both of these aspects of conservatism arise from viewing some humans as better or more human than others. It has been found, however, that SDO scores correlate more strongly with economic conservatism than with social conservatism, but vice versa for RWA scores (see Jost et al. 2009 for a review of relevant studies.)

In sum, on the basis of evidence from research, it is reasonable to treat conservatism–liberalism, RWA, and SDO as closely related ways to characterize human values. Of course, more research is needed to explore the differences and similarities of these value types.

### *4.2.2 Correlates of Collectivism–Individualism*

Table 4.1 summarizes the published findings about collectivism–individualism from numerous cross-national studies (Table 4.1a) and studies across the USA states (Table 4.1b). A number of the findings reported in Table 4.1a are overlapping, which is because we have described in Table 4.1 each study's findings to reflect the study's own conclusions. Redundancies across entries in Table 4.1 are equivalent to replications of findings. For clarity, the studies' results are presented in Table 4.1 as the value poles of the unidimension of collectivism–individualism. The methods used by scholars to measure collectivism–individualism are described in the next chapter. According to the parasite-stress theory of sociality, all the differences between the two ideological poles listed in Table 4.1 are caused proximately by the greater parasite prevalence in collectivist regions than in individualist locales. Although the majority of findings in Table 4.1 are derived from research conducted independently of the parasite-stress theory of values, to be more comprehensive, we include in the table several findings that were discovered from the application of the parasite-stress theory of values.

### *4.2.3 Cross-National Findings*

#### 4.2.3.1 Conservatism–Liberalism

As seen in the first entry in the cross-national portion of Table 4.1, collectivist countries have conservative values and individualist countries have liberal values. For example, this is apparent in the differences between collectivist countries and individualist countries in people's preferences about individuals' rights, freedom, and equality. Also, it is seen in the differences between the two types of countries in people's willingness to socialize with in-group and out-group members. Collectivists

**Table 4.1** Comparison of values and parasite stress of collectivist versus individualist cultures, based on cross-national (part a) and interstate USA (part b) published studies

| Collectivists | Individualists | Reference(s) |
|---|---|---|
| *(a) Cross-national findings* | | |
| Conservatism; restriction of individuals' rights and freedoms | Liberalism; individuals' rights and freedoms paramount | Gelfand et al. (2004), Thornhill et al. (2009, 2010) |
| Property rights limited to elites | Property rights widespread across citizenry | Thornhill et al. (2009) |
| Low interest in wealth redistribution and welfare outside of the dominant in-group | High interest in well-being of entire populace | Thornhill et al (2009) |
| Inequality of people | Equality of people | Hofstede (1980) |
| Trust and social-capital network restricted to in-group | Trust and social-capital network extensive outside in-group | Allik and Realo (2004) |
| Strangers distrusted; in- and out-group members are fixed | Strangers may become friends or allies | Oyserman and Uskul (2008), Gheorghiu et al. (2009) |
| Less helpful toward strangers | More helpful toward strangers | Knafo et al. (2009) |
| More wary of contact with foreigners and other out-group members | Less wary of contact with foreigners and other out-group members | Schwartz and Sagiv (1995) |
| Harsh and unsympathetic treatment of out-groups | Seek out-group contact and alliance | Triandis (1995) |
| Tight social network | Loose social network | Triandis (1995), Gelfand et al. (2011) |
| In-group goals paramount | Personal autonomy and self-fulfillment paramount | Gelfand et al. (2004) |
| Relationships and group memberships are ascribed and fixed, to which people must accommodate | Relationships and group membership are impermanent and nonintensive | Oyserman and Uskul (2008) |
| Group-identity and in- and out-group distinctions | Self-identity and dynamic group affiliation | Gelfand et al. (2004) |
| Prefer to engage in group activities | Often engage in activities alone | Gelfand et al. (2004) |
| More cohesive friendship groups | Less cohesive friendship groups | Gelfand et al. (2004) |
| Fewer, but more durable and intimate social interactions | More, but briefer and less intimate social interactions | Gelfand et al. (2004) |
| Greater distinctions between in- and out-groups | Fewer distinctions between in- and out-groups | Gelfand et al. (2004) |
| High cooperation within in-group | Less cooperation within in-group | Gelfand et al. (2004) |
| Motivation: fulfill duties and obligations that contribute to the group welfare | Motivation: fulfill personal interests, needs, and success | Gelfand et al. (2004) |
| High in-group embeddedness | Low in-group embeddedness | Gelfand et al. (2004) |

(continued)

**Table 4.1** (continued)

| Collectivists | Individualists | Reference(s) |
|---|---|---|
| Low self-expression | High self-expression | Inglehart and Carballo (1997) |
| Self is malleable, based on context | Permanent self, separate from context, trait-like | Oyserman and Uskul (2008) |
| Interdependent agency and self | Independent agency and self | Markus and Kitayama (1991), Kashima et al. (2004), Kitayama and Uchida (2005) |
| Self-esteem a weak predictor of life satisfaction | Self-esteem a strong predictor of life satisfaction | Diener and Diener (1995) |
| Cultural norms and emotions similarly important for making life satisfaction judgments | Emotions most important for making life satisfaction judgments | Suh et al. (1998) |
| Duty and obligations to in-group | Individuality | Inglehart and Carballo (1997), Hofstede (1980), Gelfand et al. (2004) |
| High respect for family and other in-group members | Less respect for family and other in-group members | Gelfand et al. (2004) |
| Extended family embeddedness | Self and nuclear family investment | Hofstede (1980) |
| Extended family focus | Nuclear family focus | Triandis (1989) |
| Strong family ties | Weak family ties | Gelfand et al. (2004), Fincher and Thornhill (2012) |
| More parental influence in marriage decisions of children | Less parental influence in marriages | Buunk et al. (2010) |
| High family harmony, respect and loyalty | Low family harmony, respect, and loyalty | Gelfand et al. (2004) |
| Live closer to extended family relatives | Live farther from family | Georgas et al. (2001) |
| Philopatry | Dispersal, emigration, the frontier spirit | Kitayama et al. (2006), Alesina and Giuliano (2010) |
| Visit and telephone extended family relatives more frequently | Visit and telephone family relatives less frequently | Georgas et al. (2001) |
| Honor and modesty paramount | Honor and modesty less important | Oyserman and Uskul (2008), Vandello et al. (2009) |
| Reasoning: a tool to make sense of whole rather than its parts (holistic cognition) | Reasoning: a tool for separating out main causes from background (analytical cognition) | Oyserman and Uskul (2008) |
| Low divorce rate | High divorce rate | Vandello and Cohen (1999), Gelfand et al. (2004) |

(continued)

**Table 4.1** (continued)

| Collectivists | Individualists | Reference(s) |
|---|---|---|
| High rate of male-against-female aggression in mateships | Lower rate of male-against-female aggression in mateships | Archer (2006) |
| Tolerance of male-against-female aggression in mateships | Intolerance of domestic abuse | Vandello et al. (2009) |
| Highly value female mateship fidelity | Female mateship fidelity valued less | Vandello et al. (2009) |
| Restricted/conservative female sexuality | Unrestricted/liberated female sexuality | Schaller and Murray (2008), Thornhill et al. (2009, 2010), Fong and Goetz (2010) |
| Gender inequality | Gender equality | Hofstede (1980), Gelfand et al. (2004), Archer (2006), Thornhill et al. (2009, 2010) |
| High elder respect | Low elder respect | Gelfand et al. (2004) |
| Autocratic governance | Democratic governance | Gelfand et al. (2004), Thornhill et al. (2009, 2010) |
| Traditionalist political culture emphasizing hierarchy and elite rule | Moralistic political culture emphasizing participatory egalitarianism | Hofstede (1980) |
| More legal restrictions of people's behavior | Less legal restrictions on people's behavior | Conway et al. (2006), study 3 |
| Slow pace of life | Fast pace of life | Levine and Norenzayan (1990) |
| Rural | Urban | Gelfand et al. (2004) |
| Low socioeconomic status | High socioeconomic status | Gelfand et al. (2004) |
| Developing countries | Developed countries | Hofstede (1980), Gelfand et al. (2004) |
| Indirect in communication | Direct, forthright, and literal in communication | Holtgraves (1997) |
| Attend more to the status of people | Attend less to the status of people | Gelfand et al. (2004) |
| High respect for high status (high authoritarianism) | Low respect for high status (low authoritarianism) | Gelfand et al. (2004) |
| Personal pronoun drop | No pronoun drop | Kashima and Kashima (1998) |
| Verbal abuse of in-group | Verbal abuse of the individual | Semin and Rubini (1990) |
| Emotional content of language paramount | Words themselves paramount | Ishii et al. (2003) |
| High conformity to tradition and norms | Low conformity to tradition and norms | Hofstede (1980), Bond and Smith (1996), Gelfand et al. (2004), Murray et al. (2011) |
| Knowledge transmitted from elders | Knowledge sought by the individual | Hofstede (1980), Gelfand et al. (2004) |

(continued)

**Table 4.1** (continued)

| Collectivists | Individualists | Reference(s) |
|---|---|---|
| Low federal monetary investment in quality education | High federal investment in quality education | Cheung and Chan (2008) |
| Reward conformity and normative behavior | Reward deviation from status quo toward creative ends | Cukur et al. (2004), Murray et al. (2011) |
| Low rate of innovation | High rate of innovation | Thornhill et al. (2009), Gorodnichenko and Roland (2011), Taylor and Wilson (2012) |
| High frequency of civil war | Low frequency of civil war | Letendre et al. (2010) |
| High frequency of clan and tribal (nonstate) wars | Low frequency of clan and tribal (nonstate) wars | Letendre et al. (2012) |
| High frequency of coups and revolutions | Low frequency of coups and revolutions | Letendre et al. (2012) |
| Low openness to experiences | High openness to experiences | Schaller and Murray (2008) |
| High avoidance of uncertainty | Low avoidance of uncertainty | Gelfand et al. (2004 |
| Low intellectual autonomy | High intellectual autonomy | Gelfand et al. (2004) |
| Low economic productivity | High economic productivity | Triandis (1995), Ball (2001), Hofstede (2001), Gelfand et al. (2004) |
| Low success in science and technology | High success in science and technology | Gelfand et al. (2004), Taylor and Wilson (2012) |
| High religious participation and commitment | Low religious participation and commitment | Fincher and Thornhill (2012) |
| High religious devotion and dogmatism | Low religious devotion and dogmatism | Gelfand et al. (2004), Fincher and Thornhill (2012) |
| Low human condition index (societal health, life expectancy, Human Development Index) | High human condition index | Gelfand et al. (2004) |
| High rates of violent crime | Low rates of violent crime | Karstedt (2006) |
| More homicide | Less homicide | Thornhill and Fincher (2011) |
| High infectious-disease severity | Low infectious-disease severity | Fincher et al. (2008) |
| High nonzoonotic disease prevalence | Low nonzoonotic disease prevalence | Thornhill et al. (2010) |
| *(b) Interstate USA findings* | | |
| Trust and social-capital network restricted to in-group | Trust and social-capital network extensive outside in-group | Allik and Realo (2004) |
| More legal restrictions on people's behavior | Less legal restrictions on people's behavior | Conway et al. (2006) |
| High religious participation and commitment | Low religious participation and commitment | Fincher and Thornhill (2012) |

(continued)

**Table 4.1** (continued)

| Collectivists | Individualists | Reference(s) |
|---|---|---|
| Few elderly living alone | More elderly living alone | Vandello and Cohen (1999), Fincher and Thornhill (2012) |
| High percentage of homes with grandparents and grandchildren coresident | Low percentage of homes with grandparents and grandchildren coresident | Vandello and Cohen (1999), Fincher and Thornhill (2012) |
| More carpooling | Less carpooling | Vandello and Cohen (1999) |
| Stable, enduring marriage | High divorce rate | Vandello and Cohen (1999) |
| More homicide | Less homicide | Thornhill and Fincher (2011) |
| More domestic-partner violence | Less domestic-partner violence | Archer (2006), Thornhill and Fincher (2011) |
| More infectious disease | Less infectious disease | Fincher and Thornhill (2012) |
| High nonzoonotic disease prevalence | Low nonzoonotic disease prevalence | Chap. 5 |
| More frequent naming of sons after male forebears | Less frequent use of patronyms | Brown et al. (2013) |

are wary, untrusting, and avoiding of contact with foreigners and other out-group people, and support harsh and unsympathetic treatment of out-groups. This xenophobia of collectivists contrasts sharply with the xenophilic values of individualists. As a final example—one also stemming from the greater xenophobia of collectivists—people in collectivist countries (measured by what Knafo et al. 2009 label "embeddedness," a part of the ethnocentrism of collectivism), compared to people in individualistic countries, exhibit less willingness to help strangers.

#### 4.2.3.2 Self-concept

Collectivist and individualist countries differ in how the self is understood—the meaning of the individual person (Table 4.1a). Is a person inseparably and interdependently part of a collective (an in-group) or is one an independent/autonomous agent with personal rights and freedoms? Collectivists understand the person in terms of the former, while individualists comprehend the person according to the latter. In collectivist regions, a person is relatively indivisible within his/her in-group. The collectivist self is expressed in relation to in-group goals—the goals defined by one's extended family and other like-minded in-group members. Collectivists virtually blend into a background of in-group social striving. Collectivist people are somewhat like worker ants that strive selflessly for the goals and harmony of their collective, the colony-family as a whole, and have no goals as independent agents. Individualists, in contrast, possess a self-concept that reflects the individual's own aspirations, not the in-group's, but individualists support their nuclear family's goals and harmony to the extent that the goals overlap with the individualist's personal goals. Not surprisingly, then, collectivists have low intellectual autonomy. The in-group authorities think for their collectivist membership and set the normative path to follow and obey. The individualist thinks for him- or herself. The analogy with the ant worker applies here too: the worker ant obeys the colony's rules and goals in a seemingly mindless manner, an automaton.

Note that the comparison here between worker ants and collectivists is descriptive, not derogatory. Indeed, the authors' values are that ants are noble creatures that have fascinated us since early childhood. For those with knowledge of insects, we add that the appropriate comparison for individualist humans is the common burying beetle with a nuclear family life and biparental care and nothing more. In Chap. 5, we treat in detail comparative family life across animal species in relation to parasite stress.

The difference in the meaning of self between collectivists and individualists manifests in many aspects of human everyday behavior, including language (Table 4.1a). Verbally abusive language differs in content between the two value categories in a way consistent with the difference in the self-concept. The target of collectivist verbal abuse is both the person and his or her in-group. Individualists, however, restrict such abuse to the individual target. Also, collectivists tend to drop from their languages the pronouns "I" and "you"; individualists retain them. Collectivists replace "I" with "we," which expresses the in-group. The pronoun "ya'll" is an example of pronoun drop. Ya'll is a word that is commonly used in the southeastern US "Ya'll come" or "How are ya'll doing?" is sometimes spoken to a person, but refers to that person's collective, not the individual. Ya'll is not a contraction of you and all, but a new word that lacks recognition of you the person altogether. "You" as a person is inconsistent with collectivist values, specifically with the collectivist self-concept, just as is "I." The individualist is you—and I and me—focused because the individual understands self's and others' autonomy that way. The "I" focus of individualists is seen as well in their personal expression and desire to stand out as an individual and in the salience they give to personal self-esteem. Collectivists are less motivated to stand out personally, and their self-esteem emphasizes family and other in-group esteem, respect, and honor. The "we" focus of collectivists reflects their embeddedness in their in-group, and the extent that you and I hardly exist.

Twenge et al. (2013) studied the presence of pronouns in the text of three-quarters of a million American English books published 1960–2008 and digitized as part of the Google ngram database. A range of evidence indicates that the USA has become increasingly individualistic over the last three generations and correspondingly less collectivistic. (A human generation equals about 20 years.) Twenge et al.'s (2013) study was inspired by their hypothesis that this pattern of increasing individualism would extend to frequencies of value-based pronouns used in books that reflected either a priority of autonomous self or of interdependence on and embeddedness in an in-group. The results strongly supported their hypothesis. Across the three generations, the second-person pronouns "you" and "your" quadrupled in use, first-person singular pronouns "I" and "me" increased in use by 42%, and first-person plural pronouns (e.g., "us" and "we") decreased 10%. In a separate study of the same database, Twenge et al. (2012) show that the pattern for pronoun use also holds for the frequency of use of individualistic words and phrases over the period 1960–2008. Examples of the analyzed individualistic linguistic items are: unique, self, all about me, I am special. As Twenge et al. point out in their papers we have cited, their linguistic findings likely stem from widespread linguistic changes and preferences since the 1960s that are associated with increasing individualism in the USA. In Chap. 10 we discuss the pattern of increased individualism in the USA (and the West generally) in recent generations in relation to reduced parasite stress.

Above, we compared collectivist people to worker ants. Other analogies are seen in military groups and athletic teams. Their success depends upon obedience to authority, self-sacrifice for the group's prosperity, group unity, and avoidance of thinking of oneself as autonomous. Collectivists make cooperative soldiers and team players. Military and team-sports training promotes the importance of collectivist values and discourages individualist values. The independence of individualists makes their within-group dynamics more conflictual, with each group member engaged in self-promotion. These conflicts, however, are reconciled by compromise and diplomacy, which take time to achieve because of the many personal opinions aired and respected. Obviously, groups of liberals do get things done—advances in democratization and economic productivity are testament to that (see Chaps. 10 and 11). Conservatives settle in-group disagreement with an appeal to tradition, omniscient gods or authority figures' opinions or rulings, or with aggression (on interpersonal violence, see Chap. 8). Obviously, this also works, but primarily benefits those in roles of authority (men, elders, elites), with the most extreme exclusive benefits going to the autocratic leadership of highly collectivist societies.

#### 4.2.3.3 Reasoning Styles

The reasoning styles of the two ideologies differ in ways consistent with their differences in in-group versus autonomous-self-conceptions (Table 4.1a). The holistic reasoning style of collectivists interprets events and other things in terms of the whole system. This is why we suggested in Chap. 2 that researchers who feel that cultural behavior reflects human adaptation that functions to benefit the culture as a whole are using collectivist cognition.

The antireductionist cognition of collectivists can make science a strange and difficult topic to comprehend and endorse, given science's focus on causes as partial determinants, which can be separated from a whole system, and then analyzed and understood independent of the whole. The whole-system cognition of the collectivist includes his or her in-group and can be extended as patriotism and support to the region or to society or nation, specifically the part of society or nation sharing collectivist values. To the collectivist, nothing has meaning independent of the in-group's goals and harmony. The individualist, in contrast, sees the whole as comprised of separable parts that are important separately and can be understood by dissecting them out from the whole. This analytical-reasoning style is an essence of scientific investigation. Science illuminates the whole through synthesis of the causal components of the whole into a single concept or a few basic concepts.

In later chapters, we do this type of synthesis repeatedly. For example, we provide evidence that geographically variable parasite stress causes geographically variable value systems, which, in turn, cause geographically variable political and economic systems, which, in turn, feedback and affect parasite stress, which, in turn, affects values, and so on. Other causes at work here include the developmental events that proximately cause people's values and include as well the ultimate cause of evolution by selection for the psychological adaptations that function during discriminative adoption of values. All these tiers of causation and their interactions are based fundamentally on how parasites build us proximately and have built us ultimately to develop, think, and behave. Hence, we call this synthetic way of analyzing human affairs the parasite-stress theory of values or of sociality. Parasite stress is the fundamental causal concept at both proximate and ultimate levels.

We hypothesize that the collectivist holistic style of reasoning explains why the naturalistic fallacy remains such a widespread way of thinking. (This fallacy was introduced in Chap. 1.) Despite the efforts of many scientists to emphasize that "is" does not equal "ought," the naturalistic fallacy persists as a common criticism of the study of human behavior and psychology in evolutionary terms. To most scientists, "is" just is and the facts of nature's causes exist independently of societal moral goals and are precious in their own right. To the collectivist, these facts cannot be considered as independent of in-group goals and well-being. That rape by men is ultimately the product of evolution by sexual selection is simply a fact to the biologist—and a fact in itself without any moral implications (Thornhill and Palmer 2000). To the collectivist, however, nothing is independent of its impact on group well-being. Thus, for the collectivist, the statement that rape is evolved by selection simultaneously and necessarily makes a value judgment pertinent to the in-group.

Analytical reasoning is required for achievement of societal moral goals. As we discussed in Chap. 1, scientific knowledge cannot identify moral goals, but provides the basis for their achievement after their identification by moralizing humans. The more that is known about the causes of rape, the more effective could be policies to reduce it. Any problem facing humanity, social or otherwise, can be solved only through knowledge of its causation. Suggested solutions to humanity's problems that are not based on understood causation are impotent (see Thornhill and Palmer 2000 for further discussion). Thus, whatever a society's moral goals are—be they conservative or liberal—requires a scientific community of analytical thinkers to achieve the facts that are needed to attain the goals.

#### 4.2.3.4 Social Network

Table 4.1a lists other documented differences between collectivist and individualist cultures. The social network of the collectivist is intensive, thick or viscous, exclusive, local and permanent, with restriction of membership to value-similar others.

That of the individualist is loose, nonintensive, nonlocal, impermanent, and diverse with regard to inclusion of other members. The collectivist distrusts and avoids strangers and hence in- and out-group memberships are fixed. Individualists' groups change membership and are open to outsiders. Collectivists accommodate to their expected role in the in-group network, while individualists are less role-bound and roles are impermanent. In addition, collectivists' social activities are predominantly with other in-group members, while individualists enjoy solitude and engaging in activities alone. This manifests in higher rates of contact (e.g., by telephone) with family by collectivists than by individualists, and in the greater geographic separation of family members in individualistic families than in collectivist families. Collectivists are more philopatric, compared to individualists.

#### 4.2.3.5 Intersexual Relationships

The marriage relationships of the two ideologies differ (Table 4.1a). Collectivists have relatively permanent marriages (as well as other alliances), whereas individualists divorce more commonly. This is the case in spite of the greater importance of romantic love in marital decisions in individualist cultures than in collectivist cultures (Gelfand et al. 2004). In collectivist cultures, marriage is more in accordance with in-group membership and goals and less in terms of personal romantic feelings. Also, under collectivism, compared to individualism, parents have more influence on marriage partners of their children (Buunk et al. 2010). These patterns are consistent with the greater inbreeding (e.g., cousin marriage) in collectivist cultures than in individualist societies (Chap. 6).

There are other differences between collectivists and individualists in intersexual relationships (Table 4.1a). Individualists value gender equality, whereas collectivists value gender inequality. The collectivist value of traditional sex roles, with assumed male superiority over females, likely is a cause of the higher rate of male-perpetuated spousal abuse in collectivist cultures than in individualist cultures. Gender inequality of collectivist cultures contributes to the moral endorsement and justification of men's abusive treatment of women to whom they are pair-bonded. Women's acceptance of the ideology of gender inequality, which is manifested in norms of female honor, is part of that endorsement and justification (Chap. 8 treats in detail interpersonal violence).

Collectivist countries show more sexual restrictiveness in both sexes, but especially in women, than do individualist countries. This topic is treated in detail in later chapters.

#### 4.2.3.6 Hierarchy, Honor, and Norms

Compared to individualist societies, collectivist societies more strongly respect all traditional hierarchies, not just that of the superiority of masculinity above femininity (Table 4.1a). The difference between collectivist and individualist in respect for traditional hierarchy affects the cultural difference in knowledge acquisition. Tradition is the most important source of knowledge under collectivism, and knowledge is transmitted from elders and elites. In contrast, individualists seek their own knowledge and rely less on traditional wisdom. This cultural difference accounts, we propose, for the reduced value placed on modern education and educational support by collectivist cultures compared to individualist cultures (Table 4.1a). Collectivists show more elder respect than individualists. In the South, people of all ages commonly address elders by "sir" and "ma'am" as long as the person addressed is of the appropriate race and social class. Hierarchy under collectivism also involves rigid family roles, with men at the top, women next, and children at the bottom. Boys are taught to value and portray traditional masculine behavior of toughness, bravery, and willingness to protect the reputation of family and other in-group members by aggression and violence. This is an aspect of the "culture of honor," as it is oftentimes called in reference to the culture of the South (Nisbett and Cohen 1996), and is actually a basic aspect of collectivist culture wherever it is found (Chap. 8).

As just mentioned, compared to individualists, collectivists have greater respect for the status hierarchy. Not surprisingly, then, collectivists are more attentive of the status of the individuals with whom they interact, which is a manifestation of the higher authoritarianism of collectivists (Table 4.1a). In Western academics, a generally liberal culture, now it is not unusual for undergraduate students to address their professors by their first name.

Under collectivism, females are trained in modesty and continence pertaining to sexual matters, as well as behavior in general, and deference to and obedience of the significant men in their lives (Table 4.1a). This is the female culture of honor in collectivist societies. Women are expected to control themselves sexually and otherwise, and accept the value that men are superior humans and only men have the wisdom to direct social affairs, including women's and children's lives.

The role of children in collectivist cultures is to learn and obey all the collectivist norms and strictly obey and honor parents and other elders within the in-group; this is the children's culture of honor in collectivist societies. The fixity of family roles and acceptance of one's family role in collectivist cultures promotes family harmony, respect, cooperativeness, and loyalty and thereby reduces aspects of within family conflict. In collectivist cultures, the concept of family honor cements a family together, and deviations from expectations of the family values, e.g., a daughter not following father's preferences for her sexual behavior and marriage, may lead to her disownment or disinheritance, or even her death by suicide or homicide (Chap. 8).

The published literature strongly supports the finding that, in collectivist cultures, rules abound and traditional norms are taken very seriously; there is high surveillance for norm violations, and punitive consequences are culturally justified and expected. Even recently, across the USA, Republican-party states endorse corporal punishment of children, whereas Democratic-party states prefer time-outs as punishment (Hetherington and Weiler 2009). This same difference is seen when USA southern states are compared with the other states: in the South, relative to other regions, there is much less sparing of the rod in controlling children's behavior at home by parents and at school (Nisbett and Cohen 1996).

In-group norms function as in-group markers that unify, stabilize, and simplify the actions of all participants in the in-group. If all behave according to established and understood norms, social life is highly predictable—the navigation of social life is relatively easy for the collectivist. In-group norms are fundamentally protective of in-group members. Violators of the norms create social uncertainty and generate social complexity for the nonviolators, but most importantly, violators may be from an out-group, or have been contaminated by out-group contact and ideas, and hence perhaps infected with a novel parasite. It is better to be safe than sorry regarding catching their parasite—watch for norm violators and punish or ostracize them when found.

This is true for collectivist norms in general (Murray et al. 2011), but hygienic norm violations are particularly transparent as parasite-avoidance values. Bullying in schools has been related to hostility, aggression, and ostracism toward hygiene violators (Turagabeci et al. 2008). For conservatives, malodorous body scent from the persistent absence of bathing is likely especially conducive to disgust reactions. Conservatives are more easily disgusted than liberals (Inbar et al. 2009, 2012; Terrizzi et al. 2010, 2012, 2013; Clay et al. 2012), as expected given conservatives' greater concern about contagion.

In general, collectivist countries show higher rates of violent crime, including homicide, than do individualist countries (Table 4.1a). This difference appears to stem importantly from the greater degree of culture-of-honor ideology and unequal resource distribution in collectivist cultures. We treat some major categories of violent crime in relation to collectivism–individualism in Chap. 8.

#### 4.2.3.7 The Pace of Life

The pace of life differs between the two ideological poles (Table 4.1a). Individualists engage in a more fast-paced life than do collectivists. Collectivism involves striving after ascertaining that moving forward is consistent with in-group harmony and goals. Individualists act more impulsively and autonomously with less concern about violating norms or what other group members would want them to do. An essence of risk-taking is the willingness to accept uncertainty and proceed in thinking or acting. Individualists are more risk prone—that is, more accepting of uncertainty—than conservatives (Table 4.1a).

#### 4.2.3.8 Language

Communication styles differ between individualistic and collectivistic cultures (Table 4.1a). We mentioned above the cultural variation in verbal abuse, use of personal pronouns, and word and phrase-use patterns. Cultures also differ in whether they express their meanings directly or indirectly, and whether they look for indirect meanings in spoken words. Collectivists engage in more indirect communication than do individualists. In collectivist talk, the wants or preferences of the speaker are less often stated literally, but are understood by in-group members because the members are emotionally and ideologically connected/embedded. Collectivists assume there is unarticulated meaning in remarks made by others and look for and find it. Individualist speech is more forthright and literal; individualists say what they mean and listen to others based on what others literally say. In order to communicate, individualists must be literal because they address a variety of types of people in terms of backgrounds, values, and opinions. Collectivists communicate more with expressed feelings and with familiar others or at least people with familiar ideology.

Related to this aspect of linguistic style, collectivists prioritize the emotional content of language, whereas individualists prioritize the words themselves (Table 4.1a). For collectivists, meaning is reflected more in the emotions accompanying the words than in the words. Collectivists exhibit less emotional autonomy than individualists; that is, how collectivists emote in speech reflects the feelings and values of the in-group. In contrast, the feelings in individualist speech reflect more of the speaker's own personalized opinions.

We hypothesize that collectivist speech is a signal of emotional connectedness to, and embeddedness in, the in-group. It is an honest signal of in-group commitment because of the large amount of time associated with its acquisition. Localized emotional nuances in speech are reflected in dialects and even finer grained speech patterns or word use. The ability to use local language in the way that the locals understand to reflect true local feelings and values is obtained only through a long ontogeny of hearing and using the local language and its associated emotional nuances and assessing the effects of one's language on the locals. The language of fundamentalist Christian groups is an obvious example of this. An outsider cannot walk into a church that prescribes speaking in tongues as a mechanism for in-group identification and commitment and begin speaking the tongue in the normative manner. Without the local-in-group upbringing, the ability to use this language signal is impossible to display accurately. Similarly, the emotional expressions, including those accompanying language, needed to convincingly display commitment to any local ideology are difficult, if not impossible, to fake by an outsider of the in-group.

We suggest that the parasite-stress theory of values has much to offer the scholarly study of linguistics. This is implied by our comments just above, as well as by the research on pronoun drop and recent linguistic changes in the USA and Reid et al.'s (2012) research on accent perception (discussed in Chap. 3). As well, in Chap. 13 we show that the number of languages across the world is predictable from the parasite-stress theory of values—specifically from the parasite-driven diversification aspect of the theory.

#### 4.2.3.9 Governmental Systems, Resource Distribution, and Economics

Collectivist countries are more autocratic than individualist countries (Table 4.1a). We propose that the high respect for authority and associated low intellectual autonomy, as well as the high value placed on human inequality, yield collectivists' need for and willingness to accept elite rule and autocracy in government. Such need and acceptance justifies the ruling class's disproportionate control of social power and other resources, leading to high wealth disparity, restriction of opportunity and property rights to elites, and widespread poverty and reduced health-related infrastructure and longevity among the general populace. The dehumanization of those low in societal rank justifies the differences across social strata in social and economic access and opportunity. In contrast, individualistic societies are antiauthoritarian, democratic/equalitarian, and value and provide widespread health infrastructure that increases longevity. Chapter 10 treats fully cross-national variation in political systems.

Collectivist societies show a disinterest in social welfare for the populace as a whole; specifically, there is less distribution of wealth and other goods and services and education outside the socially dominant and privileged in-group. In contrast, individualist cultures show an interest in welfare and other resources and opportunities being distributed across the populace (Table 4.1a).

As mentioned earlier, collectivists respect, conform to, and reward traditional norms in behavior; individualists instead reward deviations from the status quo including ones that enhance individual achievement (Table 4.1a). This difference stagnates collectivist cultures, but promotes intellectual, technical and scientific innovation, and advances in individualistic cultures. Given the egalitarian values of individualists, scientific and technological advances are transformed into public goods and services and humanitarian advances in individualistic societies (Chap. 11).

#### 4.2.3.10 Civil Conflict

Collectivist societies exhibit more civil (within country) conflicts of all forms than do individualist societies (Table 4.1a). Collectivist countries have higher rates of civil wars (intrastate wars involving the federal government versus a group or allied groups within the same country), tribal and clan wars, and revolutions and coups. According to the parasite-stress theory, the difference in intranation conflict frequencies between the two ideological poles arises from the greater in-group embeddedness, in-group boundary recognition and defense, and xenophobia of collectivists compared to individualists. We examine fully the topic of the relationship between the collectivism–individualism value dimension and civil conflicts in Chap. 12.

#### 4.2.3.11 Religiosity

The people of collectivist countries differ in religiosity from those of individualist countries (Table 4.1a). Collectivist countries show higher religious devotion, dogmatism, and participation than do individualist countries. The topic of religiosity in relation in collectivism–individualism is treated in detail in Chap. 9.

#### 4.2.3.12 Infectious Diseases

Human infectious diseases are more prevalent in collectivist countries than in individualist countries. This pattern is seen, too, for infectious diseases that are transmitted among humans (i.e., nonzoonotic diseases) (Table 4.1a). These topics are treated fully in Chap. 5.

### *4.2.4 Interstate USA Findings*

Research findings related to collectivism–individualism across the USA states are presented in Table 4.1b. Compared to individualist states, collectivist states have more infectious diseases; stable or enduring marriages (lower divorce rate); car-pooling, presumably, in part, reflective of in-group trust; multigeneration family residences, reflective of extended family embeddedness, loyalty, and philopatry; religious commitment and participation; and legal restrictions on personal behavior. Homicide, including domestic-partner slayings, is more frequent in collectivist states than in individualist states (Chap. 8 treats interpersonal violence). Collectivist states also have fewer elderly people living alone than individualist states, reflective of greater familial in-group support under collectivism, and more desire for solitude in individualists. The social networking and trust of collectivists is restricted to in-group others, whereas the social network and trust of individualists extend outside the in-group to encompass more variable others and groups.

The naming of children, in terms of their personal names (as opposed to their surnames), across the states of the USA was studied by Brown et al. (2013). They separated personal names into patronyms (names of male relatives in previous generations) and matronyms (names of female forebears). Across the 50 states, the use of patronyms in naming boys (but not matronyms) across recent generations correlated highly with collectivism scores of states. In their paper, Brown et al. (2013) emphasize the use of patronyms and male-lineage identity in general, as well as the greater value placed on male babies than on female babies, in states of high honor ideology, compared to low-honor-ideology states.

Hence, overall, where comparable differences between collectivists and individualists across USA states have been examined, there is similarity to the differences between countries in these two value systems.

## 4.3 Overview of Patterns in Table 4.1

There is variation in the strength of conclusions about, or said differently, the scientific confidence in, the patterns depicted in Table 4.1. They are all statistically robust patterns. There is, however, variation in sample size across the studies referenced. The international differences between collectivists and individualists listed in

Table 4.1a derive from studies of a few countries to studies of virtually all of the countries of the world. The value differences in Table 4.1a refer to contemporary countries around but preceding the time of the publication date of the studies listed as references. Many of the patterns listed in Table 4.1a are replicated across multiple research investigations, whereas some of the patterns have not been replicated yet. The USA interstate differences between collectivists and individualists listed in Table 4.1b are based on all or in some cases most of the USA states.

The extensive and intensive research background depicted in Table 4.1 indicates that collectivism equates with conservatism, and individualism with liberalism. Each of the value differences between collectivism and individualism depicted in the table has been shown to correspond to related or identical differences between conservatives and liberals. Hence, we will use interchangeably the terms conservative and collectivist and liberal and individualist.

## 4.4 Additional Differences Between Conservatives and Liberals

Table 4.2 lists some differences between conservatives and liberals reported by researchers. (A more exhaustive and detailed list of differences between conservatives and liberals can be found in Carney et al. 2008.) The patterns in Table 4.1 are for measured collectivism and individualism, whereas those in Table 4.2 address measured conservatism–liberalism. As we have shown just earlier, however, conservatism corresponds to collectivism, and liberalism with individualism. The contents of Table 4.2 bolster this claim and add some background studies of correlates of conservatism–liberalism. Most of the differences in Table 4.2, along with those in Table 4.1, will be empirically explored further in subsequent chapters and tied to the parasite-stress theory of values. We have discussed already the differences between conservatives and liberals in out-group attitudes, prejudice, and tolerance. The differences in sexual behavior of conservatives and liberals are discussed in Chap. 6; as mentioned, collectivists are more sexually restricted, especially among women. Several research studies using various valid metrics have measured the Intelligence Quotient (IQ) differences between conservatives and liberals, with similar results: conservatives have lower cognitive abilities than do liberals. In Chap. 11, we look in detail at the relationship between cognitive ability and collectivism–individualism across countries and USA states. Earlier in the book we mentioned the greater disgust sensitivity and past positiveness of conservatives compared to liberals. Prior research also has documented the greater openness to new experience of liberals compared to conservatives. The relationship between openness to new experience (a personality variable) and collectivism–liberalism is considered further in Chap. 7; not surprisingly, individualists are more open-minded, creative, and curious than collectivists. The value differences in openness to new experiences are manifested in choices of supermarket items—conservatives buy traditional brands, whereas

**Table 4.2** Some differences between conservatives and liberals reported in the scientific literature. Conservatism corresponds to collectivism, and liberalism to individualism (see Table 4.1)

| Conservatives | Liberals | Reference(s) |
|---|---|---|
| More out-group avoidance and racism (xenophobia) | Less out-group avoidance and racism | Sibley and Duckitt (2008), Hodson and Busseri (2012) |
| Sexually reserved | Sexually adventurous | Feather (1979) |
| Low cognitive ability (IQ) | High cognitive ability (IQ) | Deary et al. (2008a, b), Kanazawa (2010), Fraley et al. (2012), Hodson and Busseri (2012), also Woodley (2010) and Hodson and Busseri (2012) for references to other studies |
| High disgust sensitivity | Low disgust sensitivity | Inbar et al. (2009, 2012), Terrizzi et al. (2010, 2012, 2013) |
| Low openness to new experiences | High openness to new experiences | Jost et al. (2003), Carney et al. (2008), also see Woodley (2010) |
| Closed-minded and unimaginative | Creative and curious | Carney et al. (2008) |
| Buy traditional supermarket items | Buy new products at supermarkets | Khan et al. (2013) |
| Interpret past experiences positively | Interpret past events negatively | Thornhill and Fincher (2007) |
| Prefer simple paintings | Prefer complex paintings | Wilson et al. (1973) |
| Prefer conventional music | Prefer complex and rebellious music | Rentfrow and Gosling (2003) |
| Greater volume of amygdala of the brain | Greater volume cingulated cortex of the brain | Kanal et al. (2011) |
| Perceptions of a threatening and dangerous world | Perceptions of a more secure world | Rokeach and Fruchter (1956), Jost (2006), Van Leeuwen and Park (2009) |
| Intolerant of ambiguity | Tolerant of ambiguity | Jost et al. (2003), Carney et al. (2008) |
| High contagion concern (PVD) | Low contagion concern | Park and Isherwood (2011), Terrizzi et al. (2013) |
| Highly authoritarian | Less authoritarian | Carney et al. (2008), Fraley et al. (2012) |
| Raised by authoritarian parents | Raised by egalitarian parents | Block and Block (2006), Fraley et al. (2012) |
| Less travel from homebase | More travel from homebase | Carney et al. (2008) |
| Binding moral intuitions | Individualizing moral intuitions | Haidt (2007), Graham et al. (2009), Park and Isherwood (2011), Kidwell et al. (2013) |

liberals buy generics and new products on the market. Given the value differences of conservatives and liberals, it is not surprising that the two poles differ in art preferences. Conservatives value traditional styles of music and simplistic paintings more than liberals, whereas liberals value more complex music and paintings.

Given the difference between conservatives and liberals in thinking and behavior, there have to be brain differences between the two value systems. Recent research is finding some of the differences. There are volumetric differences between certain brain parts between the two ideological poles as listed in Table 4.2. Also, conservatives and liberals differ in the activity levels in certain brain parts (Amodio et al. 2007; Rule et al. 2010). These size and activity differences have been related to the consistent differences in cognitive styles of conservative versus liberals by the researchers cited.

Continuing on with differences between conservatives and liberals listed in Table 4.2, conservatives highly value a secure and stable world to live in and are fearful of and vigilant to threats to such a world. Liberals, on the other hand, view their world as a safer and more secure place, and welcome change. A threat of more concern to conservatives than to liberals is vulnerability to infectious disease. This is measured often as individual differences in scores on the psychometric scale referred to as the perceived-vulnerability-to-disease (PVD) scale. This scale is what Park and Isherwood (2011) and Terrizzi et al. (2013) used to show the greater worry of conservatives than of liberals about contagion. PVD and disgust sensitivity are highly, positively related variables (Terrizzi et al. 2013). Given the greater concern about contagion of conservatives, in comparison to liberals, it is not surprising that multiple studies indicate a relatively greater priority placed on cleanliness and hygiene by conservatives (Carney et al. 2008).

As we have emphasized, conservatives are more authoritarian than liberals. In part, this difference in authoritarianism reflects a difference in parental enculturation of children. Conservatives grow up in homes with parental authoritarianism, whereas liberals grow up in homes in which parents value input from their children in making decisions.

Conservatives are more philopatric than liberals. This is seen in research examining the residential contents of people across the values continuum. Compared to conservatives' homes, liberals' homes contain more travel paraphernalia (travel tickets, receipts, and memorabilia) and more travel books (Carney et al. 2008).

Conservatives and liberals have different moral intuitions. Certain recent research has cast this difference as one of "moral foundations" (Haidt 2007; Graham et al. 2009). This research showed that liberal values prioritize what is referred to as an "individualizing moral foundation" that emphasizes individual autonomy and well-being. In contrast, conservatives moralize more from a foundation that prioritizes collective or in-group integrity, honor, and well-being (Table 4.2). This research and the other research summarized in Table 4.2 is consistent with the general point we are making—that individualism corresponds with liberalism and collectivism with conservatism.

In Chap. 5 we turn to an empirical analysis of the parasite-stress theory of sociality in relation to collectivism–individualism and related variables.

## 4.5 Summary

The large scientific literature on human values produced prior to the recent publication of the parasite-stress theory of values is reviewed and discussed. This traditional literature on values has provided some of the data used to test the parasite-stress theory of values. The major causal frameworks in that literature, notably climate and wealth, are not alternatives to the parasite-stress of values—they are complementary proximate causes of values. The parasite-stress theory of values is a synthetic theory of values encompassing both proximate and ultimate causation of values.

Traditional research effort in the investigation of values, especially cross-nationally, has focused on the unidimensional value system referred to as collectivism–individualism because of its ability to capture differences in values across regions. The values that correlate with collectivism versus individualism are reviewed. The great similarity of the values dimension of collectivism–individualism to the values dimension of conservatism–liberalism is documented by examining cross-national studies as well as studies across the states of the USA. Additional studies are discussed that reported differences between conservatives and liberals, but did not measure collectivism–individualism per se. These additional studies also support the correspondence between collectivism–individualism and conservatism–liberalism.

Conservative/collectivist values and liberal/individualist values differ in many ways that correspond to differences in authoritarianism, social prejudices, equalitarianism, social hierarchy, self-concept, reasoning style, linguistic behavior, personality, religiosity, the structure of social networks, in-group and out-group transactions, economics, governmental systems, dispersal, family relationships, violence, warfare, adherence to tradition, norm adherence, honor ideology, sexual behavior, and marriage. According to the parasite-stress theory of values, the form that each of these features takes in a region is caused proximately by the region's level of parasite adversity and associated evoked values.

## References

Alesina, A., & Giuliano, P. (2010). The power of the family. *Journal of Economic Growth* 15: 93–125.

Allik, J., & Realo, A. (2004). Individualism–collectivism and social capital. *Journal of Cross-Cultural Psychology* 35: 29–49.

Altemeyer, B. (1996). *The Authoritarian Spector*. Harvard University Press, Cambridge, MA.

Amodio, D. M., Jost, J. T., Master, S.L. et al. (2007). Neurocognitive correlates of liberalism and conservatism. *Nature Neuroscience* 10: 1246–1247.

Archer, J. (2006). Cross-cultural differences in physical aggression between partners: A social-role analysis. *Personality and Social Psychology* 10: 133–153.

Ball, R. (2001). Individualism, collectivism, and economic development. *Annals of the American Academy of Political and Social Science* 573: 57–84.

Block, J., & Block, J. H. (2006). Nursery school personality and political orientation two decades later. *Journal of Research in Personality* 40: 734–749.

Bond, R., & Smith, P. B. (1996). Culture and conformity: A meta-analysis of studies using Asch's (1952, 1956) line judgment task. *Psychological Bulletin* 119: 111–137.

Bonds, M. H., Dobson, A. P., & Keenan, D. C. (2012). Disease ecology, biodiversity, and the latitudinal gradient in income. *PLoS Biology* 10: e1001456.

Brown, R. P., Carvallo, M., & Imura, M. (2013). Naming patterns reveal cultural values: Patronyms, matronyms, and the U.S. culture of honor. *Personality and Social Psychology Bulletin*. doi: 10.1177/0146167213509840.

Buunk, A. P., Park, J. H., & Duncan, L. A. (2010). Cultural variation in parental influence on mate choice. *Cross-Cultural Research* 44: 23–40.

Carney, D. R., Jost, J. T., Gosling, S. D. et al. (2008). The secret lives of liberals and conservatives: Personality profiles, interaction styles, and the things they leave behind. *Political Psychology* 29: 807–840.

Cashdan, E. (2001). Ethnic diversity and its environmental determinants: Effects on climate, pathogens, and habitat diversity. *American Anthropology* 103: 968–991.

Cheung, H.Y. & Chan, A.W.H. (2008). Relationships amongst cultural dimensions, educational expenditures and class size of different nations. *International Journal of Educational Development* 28: 698–707.

Clay, R., Terrizzi Jr., J. A., & Shook, N. J. (2012). Individual differences in the behavioral immune system and the emergence of cultural systems. *Journal of Social Psychology* 43: 174–184.

Conway III, L. G., Sexton, S. M., & Tweed, R. G. (2006). Collectivism and governmentally initiated restrictions: A cross-sectional and longitudinal analysis across nations and within a nation. *Journal of Cross-Cultural Psychology* 37: 20–41.

Cukur, C. S., De Guzman, M. R. T., & Carlo, G. (2004). Religiosity, values, and horizontal and vertical individualism–collectivism: A study of Turkey, the United States, and the Philippines. *Journal of Social Psychology* 144: 613–634.

Deary, I. J., Batty, G. D., & Gale, C. R. (2008a). Bright children become enlightened adults. *Psychological Science* 19: 1–6.

Deary, I. J., Batty, G. D., & Gale, C. R. (2008b). Childhood intelligence predicts voter turnout, voting preferences, and political involvement in adulthood: The 1970 British Cohort Study. *Intelligence* 36: 548–555.

Diener, E., & Diener, M. (1995). Cross cultural correlates of life satisfaction and self-esteem. *Journal of Personality and Social Psychology* 68: 653–663.

Dunn, R. R., Davies, T. J., Harris, N. C. et al. (2010). Global drivers of human pathogen richness and prevalence. *Proceedings of the Royal Society B* 27: 2587–2595.

Eppig, C., Fincher, C. L., & Thornhill, R. (2010). Parasite prevalence and the worldwide distribution of cognitive ability. *Proceedings of the Royal Society of London B* 277: 3801–3808.

Eppig, C., Fincher, C. L., & Thornhill, R. (2011). Parasite prevalence and the distribution of intelligence among the states of the USA. *Intelligence* 39: 155–160.

Feather, N. T. (1979). Value correlates of conservatism. *Journal of Personality and Social Psychology* 37: 1617–1630.

Fincher, C. L., & Thornhill, R. (2008a). A parasite-driven wedge: Infectious diseases may explain language and other biodiversity. *Oikos* 117: 1289–1297.

Fincher, C. L., & Thornhill, R. (2008b). Assortative sociality, limited dispersal, infectious disease and the genesis of the global pattern of religion diversity. *Proceedings of the Royal Society of London Biological Sciences* 275: 2587–2594.

Fincher, C. L., & Thornhill, R. (2012). Parasite-stress promotes in-group assortative sociality: The cases of strong family ties and heightened religiosity. *Behavioral and Brain Sciences* 35: 61–79.

Fincher, C. L., Thornhill, R., Murray, D. R. et al. (2008). Pathogen prevalence predicts human cross-cultural variability in individualism/collectivism. *Proceedings of the Royal Society of London Biological Sciences* 275: 1279–1285.

Fong, K.E., & Goetz, A.T. (2010). Mating strategies along narrowing definitions of individualism and collectivism. *Journal of Social, Evolutionary, and Cultural Psychology* 4: 128–141.

Forabosco, G., & Ruch, W. (1994). Sensation seeking, social attitudes and humor appreciation in Italy. *Personality and Individual Differences* 16: 515–528.

Fraley, R. C., Griffin, B. N., Belsky, J. et al. (2012). Developmental antecedents of political ideology: A longitudinal investigation from birth to age 18 years. *Psychological Science* 20: 1–7.

Gelfand, M. J., Bhawuk, D. P. S., Nishii, L. H. et al. (2004). Individualism and collectivism. In *Culture, Leadership, and Organizations: The GLOBE Study of 62 Societies* (eds. R. J. House, P. J. Hanges, M. Javidan et al.), pp. 437–512. Sage Publications, Thousand Oaks, CA.

Gelfand, M. J., Raver, J.L., Nishii, L. et al. (2011). Differences between tight and loose cultures: A 33-nation study. *Science* 332: 1100–1104.

Georgas, J., Mylonas, K., Bafiti, T. et al. (2001). Functional relationships in the nuclear and extended family: A 16-culture study. *International Journal of Psychology* 36: 289–300.

Gheorghiu, M. A., Vignoles, V.L., & Smith, P.B. (2009). Beyond the United States and Japan: Testing Yamagishi's Emancipation Theory of Trust across 31 Nations. *Social Psychology Quarterly* 72: 365–383.

Gorodnichenko, Y., & Roland, G. (2011). Individualism, innovation, and long-run growth. *Proceedings of the National Academy of Sciences* 108: 21316–21319.

Graham, J., Haidt, J., & Nosek, B. A. (2009). Liberals and conservatives rely on different sets of moral foundations. *Journal of Personality and Social Psychology* 96: 1029–1046.

Guernier, V., Hochberg, M. E., & Guégan, J. -F. (2004). Ecology drives the worldwide distribution of human diseases. *PLoS Biology* 2: 740–746.

Haidt, J. (2007). The new synthesis in moral psychology. *Science* 316: 998–1002.

Hetherington, M. J., & Weiler, J. D. (2009). *Authoritarianism and Polarization in American Politics*. Cambridge University Press, Cambridge, U.K.

Hodson, G., & Busseri, M. A. (2012). Bright minds and dark attitudes: Lower cognitive ability predicts greater prejudice through right-wing ideology and low intergroup contact. *Psychological Science* 20: 1–9.

Hodson, G., & Costello, K. (2007). Interpersonal disgust, ideological orientations, and dehumanization as predictors of intergroup attitudes. *Psychological Science* 18: 691–197.

Hofstede, G. (1980). *Culture's Consequences.* Sage, Beverly Hills, CA.

Hofstede, G. (2001). *Culture's Consequences: Comparing Values, Behaviors, Institutions, and Organizations Across Nations*, 2nd ed. Sage Publications, Thousand Oaks, CA.

Holtgraves T. (1997). Styles of language use: Individual and cultural variability in conversational indirectness. *Journal of Personality and Social Psychology* 73: 624–637.

Inbar, Y., Pizarro, D. A., & Bloom, P. (2009). Conservatives are more easily disgusted than liberals. *Cognition and Emotion* 23: 714–725.

Inbar, Y., Pizarro, D. A., Iyer, R. et al. (2012). Disgust sensitivity, political conservatism, and voting. *Social Psychological and Personality Science* 5: 537–544.

Inglehart, R., & Carballo, M. (1997). Does Latin America exist? (And is there Confucian culture?): A global analysis of cross-cultural differences. *PS: Political Science and Politics* 30: 34–47.

Ishii, K., Reyes, J. A., & Kitayama, S. (2003). Spontaneous attention to word content versus emotional tone: Differences among three cultures. *Collectivism* 14: 39–46.

Jost, J. T. (2006). The end of the end of ideology. *American Psychologist* 61: 651–670.

Jost, J., Glaser, J., Kruglanski, A. et al. (2003). Political conservatism as motivated social cognition. *Psychological Bulletin* 129: 339–375.

Jost, J. T., Federico, C. M., & Napier, J. L. (2009). Political ideology: Its structure, functions, and elective affinities. *Annual Review of Psychology* 60: 307–337.

Kanal, R., Fellden, T., Firth, C. et al. (2011). Political orientations are correlated with brain structure in young adults. *Current Biology* 21: 677–680.

Kanazawa, S. (2010). Why liberals and atheists are more intelligent. *Social Psychology Quarterly* 73: 33–57.

Karstedt, S. (2006). Democracy, values and violence: Paradoxes, tensions, and comparative advantages of liberal inclusion. *Annals of the American Academy of Political and Social Science* 605: 50–81.

Kashima, E. S., & Kashima, Y. (1998). Culture and language: The case of cultural dimensions and person al pronoun use. *Journal of Cross-Cultural Psychology* 29: 461–486.

Kashima, Y., Kokubo, T., Kashima, E. S. et al. (2004). Culture and self: Are there within-culture differences in self between metropolitan areas and regional cities? *Personality and Social Psychology Bulletin* 30: 816–823.

Khan, R., Misra, K., & Singh, V. (2013). Ideology and brand consumption. *Psychological Science* 24: 326–333.

Kidwell, B., Farmer, A., & Hardesty, D. M. (2013). Getting liberals and conservatives to go green: Political ideology and congruent appeals. *Journal of Consumer Research* 40: 350–367.

Kitayama, S., & Uchida, Y. (2005). Interdependent agency: An alternative system for action. In *Culture and Social Behavior: The Ontario Symposium*, Vol. 10 (eds. R. Sorrentino, D. Cohen, J. M. Olsen et al.), pp. 165–198. Erlbaum, Mahwah, NJ.

Kitayama, S., Ishii, K., Tmada, T. et al. (2006). Voluntary settlement and the spirit of independence: Evidence from Japan's "Northern Frontier." *Journal of Personality and Social Psychology* 91: 369–384.

Knafo, A., Schwatrz, S. H., & Levine, R. V. (2009). Helping strangers is lower in embedded cultures. *Journal of Cross-Cultural Psychology* 40: 875–879.

Knight, K. (1993). Liberalism and conservatism. In *Measures of Social Psychological Attitudes. Measures of Political Attitudes*, Vol. 2 (eds. J. P. Robinson, P. R. Shaver & L. S. Wrightsman), pp. 59–158. Sage Publishing, San Francisco, CA.

Knight, K. (1999). Liberalism and conservatism. In *Measures of Political Attitudes* (eds. J. P. Robinson, P. R. Shaver & L. S. Wrightsman), pp. 59–158. Academic Press, San Diego, CA.

Landes, D. S. (1998). *The Wealth and Poverty of Nations*. W.W. Norton and Co., New York, NY.

Laponce, J. A. (1981). *Left and right: The topography of political perceptions*. University of Toronto Press, Toronto, Ontario, Canada.

Letendre, K., Fincher, C. L., & Thornhill, R. (2010). Does infectious disease cause global variation in the frequency of intrastate armed conflict and civil war? *Biological Reviews* 85: 669–683.

Letendre, K., Fincher, C. L., & Thornhill, R. (2012). Infectious disease, collectivism, and warfare. In *The Oxford Handbook on Evolutionary Perspectives on Violence, Homicide, and Warfare* (eds. T. Shackelford & V. Weekes-Shackelford), pp. 351–371. Oxford University Press, New York, NY.

Levine, R. V., & Norenzayan, A. (1990). The pace of life in 31 countries. *Journal of Cross-Cultural Psychology* 30: 178–205.

Lipset, S. M. (1959). Some social requisites of democracy: Economic development and political legitimacy. *American Political Science Review* 53: 69–105.

Low, B. S. (1990). Marriage systems and pathogen stress in human societies. *American Zoologist* 30: 325–339.

Markus, H. R., & Kitayama, S. (1991). Cultures and the self: Implications for cognition, emotion and motivation. *Psychological Review* 98: 224–253.

McGuire, R. A., & Coelho, P. R. P. (2011). *Parasites, Pathogens, and Progress: Diseases and Economic Development*. MIT Press, Cambridge, MA.

Murray, D. R., Trudeau, R., & Schaller, M. (2011). On the origins of cultural differences in conformity: Four tests of the pathogen prevalence hypothesis. *Personality and Social Psychology Bulletin* 37: 318–329.

Napier, J. L., & Jost, J. T. (2008). The "antidemocratic personality" revisited: A cross-national investigation of working-class authoritarianism. *Journal of Social Issues* 64: 595–617.

Nisbett, R. E., & Cohen, D. (1996). *Culture of Honor: The Psychology of Violence in the South*. Westview, Boulder, CO.

Oxley, D. R., Smith, K. B., Alford, J. R. et al. (2008). Political attitudes vary with physiological traits. *Science* 321: 1667–1670.

Oyserman, D., & Uskul, A. (2008). Individualism and collectivism: Societal-level processes with implications for individual-level and society-level outcomes. In *Multilevel Analysis of Individuals and Cultures* (eds. F. J. R. van de Vijver, D. A. van Hemert & Y. H. Poortinga), pp. 145–173. Lawrence Erlbaum Associates, New York, NY.

Park, J. H. & Isherwood, E. (2011). Effects of concerns about pathogens on conservatism and anti-fat prejudice: Are they mediated by moral intuitions? *The Journal of Social Psychology* 15: 391–394.

Pratto, F., & Hegarty, P. (2000). The political psychology of reproductive strategies. *Psychological Science* 11: 57–62.

Pratto, F., Sidanius, J., Stallworth, L. M. et al. (1994). Social-dominance orientation: A personality variable predicting social and political-attitudes. *Journal of Personality and Social Psychology* 67: 741–763.

Price-Smith, A. T. (2002). *The Health of Nations: Infectious Disease, Environmental Change, and Their Effects on National Security and Development.* MIT Press, Cambridge, MA.

Reid, S. A., Zhang, J., Anderson, G. L. et al. (2012). Parasite primes make foreign-accented English sound more distant to people who are disgusted by pathogens (but not by sex or morality). *Evolution and Human Behavior* 33: 471–478.

Rentfrow, P. J., & Gosling, S. D. (2003). The do re mi's of everyday life. The structure and personality correlates of music preference. *Journal of Personality and Social Psychology* 84: 1236–1256.

Rokeach, M., & Fruchter, B. (1956). A factorial study of dogmatism and related concepts. *The Journal of Abnormal and Social Psychology* 53: 356–360.

Rule, N. O., Freeman, J. B., Moran, J. M. et al. (2010). Voting behavior is reflected in amygdala response across cultures. *Social Cognitive and Affective Neuroscience* 5: 349–355.

Schaller, M., & Murray, D. (2008). Pathogens, personality, and culture: Disease prevalence predicts worldwide variability in sociosexuality, extraversion, and openness to experience. *Journal of Personality and Social Psychology* 95: 212–221.

Schwartz, S. H., & Sagiv, L. (1995). Identifying culture specifics in the context and structure of values. *Journal of Cross-Cultural Psychology* 26: 92–116.

Semin, G. R., & Rubini, M. (1990). Unfolding the concept of person by verbal abuse. *European Journal of Social Psychology* 20: 463–474.

Sibley, C. G., & Duckitt, J. (2008). Personality and prejudice: A meta-analysis and theoretical review. *Personality and Social Psychology Review* 12: 248–279.

Suh, E., Diener, E., Oishi, S. et al. (1998). The shifting basis of life satisfaction judgments across cultures: Emotions versus norms. *Journal of Personality and Social Psychology* 74: 482–493

Taylor, L. H., Latham, S. M., & Woolhouse, M. E. (2001). Risk factors for human disease emergence. *Philosophical Transactions of the Royal Society of London Biological Sciences* 356: 983–989.

Taylor, M. Z & Wilson, S. (2012). Does culture still matter? The effects of individualism on national innovation rates. *Journal of Business Venturing* 27: 234–247.

Terrizzi Jr., J. A., Shook, N. J., & Ventis, W. L. (2010). Disgust: A predictor of social conservatism and prejudicial attitudes toward homosexuals. *Personality and Individual Differences* 49: 587–592.

Terrizzi Jr., J. A., Shook, N. J., & Ventis, W. L. (2012). Religious conservatism: an evolutionarily evoked disease-avoidance strategy. *Religion, Brain and Behavior* 2: 105–120.

Terrizzi Jr., J. A., Shook, N. J., & McDaniel, M. A. (2013). The behavioral immune system and social conservatism: A meta-analysis. *Evolution and Human Behavior* 34: 99–108.

Thornhill, R., & Fincher, C. L. (2007). What is the relevance of attachment and life history to political values? *Evolution and Human Behavior* 28: 215–222.

Thornhill, R., & Fincher, C. L. (2011). Parasite stress promotes homicide and child maltreatment. *Philosophical Transactions of the Royal Society: Biological Sciences* 366: 3466–3477.

Thornhill, R. & Palmer, C. (2000). *A Natural History of Rape: Biological Bases of Sexual Coercion.* MIT Press, Cambridge, MA.

Thornhill, R., Fincher, C. L., & Aran, D. (2009). Parasites, democratization, and the liberalization of values across contemporary countries. *Biological Reviews* 84: 113–131.

Thornhill, R., Fincher, C. L., Murray, D. R. et al. (2010). Zoonotic and non-zoonotic diseases in relation to human personality and societal values: Support for the parasite-stress model. *Evolutionary Psychology* 8: 151–169.

Triandis, H. C. (1989). The self and social behavior in differing cultural contexts. *Psychological Review* 96: 506–520.

Triandis, H. C. (1995). *Individualism and Collectivism.* Westview Press, Boulder, CO.

Turagabeci, A. R., Nakamura, K., & Takano, T. (2008). Healthy lifestyle behaviour decreasing risks of being bullied, violence, and injury. *PLoS ONE* 3: e1585. doi: 10.1371/journal.pone.0001585.

Twenge, J.M., Campbell, W.K., & Gentile, B. (2012). Increases in Individualistic Words and Phrases in American Books, 1960–2008. *PLoS ONE* 7: e40181.doi:10.1371/journal.pone.0040181.

Twenge, J.M., Campbell, W.K., & Gentile, B. (2013). Changes in Pronoun Use in American Books and the Rise of Individualism, 1960–2008. *Journal of Cross-Cultural Psychology* 44: 406–415.

Vandello, J. A., & Cohen, D. (1999). Patterns of individualism and collectivism across the United States. *Journal of Personality and Social Psychology* 77: 279–292.

Vandello, J. A., Cohen, D., Grandon, R. et al. (2009). Stand by your man: Indirect prescriptions for honorable violence and feminine loyalty in Canada, Chile, and the United States. *Journal of Cross-Cultural Psychology* 40: 81–104.

Van de Vliert, E. (2009). *Climate, Affluence, and Culture.* Cambridge University Press, New York, NY.

Van Leeuwen, F., & Park, J. H. (2009). Perceptions of social dangers, moral foundations, and political orientation. *Personality and Individual Differences* 47: 169–173.

Wilson, G. D., & Patterson, J. R. (1968). A new measure of conservatism. *British Journal of Social and Clinical Psychology* 7: 264–269.

Wilson, G.D., Ausman, J., & Mathews, T.R. (1973). Conservatism and art preferences. *Journal of Personality and Social Psychology* 25: 286–288.

Woodley, M. A. (2010). Are high-IQ individuals deficient in common sense? A critical examination of the 'clever sillies' hypothesis. *Intelligence* 38: 471–480.

# Chapter 5
# Collectivism–Individualism, Family Ties, and Philopatry

## 5.1 Introduction

This chapter's three main topics, which are stated in the chapter title, are empirically analyzed in sequence in relation to the parasite-stress theory of values. All three are examined cross-nationally and across states of the USA. Also collectivism–individualism and philopatry are investigated across indigenous societies in the ethnographic record. Then we deal with the topic of family organization across species in relation to parasite stress. We discuss also reciprocal altruism of people and human-specific cognitive ability in relation to the parasite-stress theory. In this chapter, too, we address the ecological fallacy as well as some misconceptions about the scientific validity of the comparative method.

## 5.2 Collectivism–Individualism Across Countries

### *5.2.1 The Origin of the Hypothesis*

Individualism and collectivism are fundamental to social scientists' descriptions of culture and cross-cultural differences (Triandis 1995; Hofstede 2001; Heine 2008). Until recently, however, a gap in the large research effort directed at cross-cultural variation in collectivism–individualism was the absence of a theoretical framework that can explain the variation.

With our colleagues Damian Murray and Mark Schaller, we hypothesized that collectivism (in contrast to individualism) functions as a defense against infectious diseases, and thus is more likely to be evoked in cultures that have a greater prevalence of parasites. The logical basis of this hypothesis is evident in all the major defining features of collectivist (versus individualist) value systems tabulated and discussed in the previous chapter (see Tables 4.1 and 4.2). Here we mention only a few of these.

R. Thornhill and C.L. Fincher, *The Parasite-Stress Theory of Values and Sociality: Infectious Disease, History and Human Values Worldwide*,
DOI 10.1007/978-3-319-08040-6_5, © Springer International Publishing Switzerland 2014

First, collectivists, in comparison to individualists, are embedded in their in-group and form durable in-group relations. This provides the health "insurance" of nepotism and other in-group altruism that manages the negative effects of contagion when it occurs in the in-group.

Second, collectivists make strong distinctions between in-groups and out-groups, whereas among individualists the in-group–out-group distinction is weak. Consequently, collectivists are more distrustful and avoidant of contact with out-group people. This xenophobic attitude can serve an effective anticontagion function by reducing exposure to novel infectious diseases that may be harbored in out-groups.

A third distinction between collectivism and individualism lies in their different emphases on conformity versus tolerance for deviance from the norms. Collectivism is characterized by a strong value placed on tradition and conformity, whereas individualism is characterized by a greater tolerance, and even encouragement of, deviation from the traditional status quo (Oishi et al. 1998; Cukur et al. 2004; Murray et al. 2011). Given that many specific customs and norms—such as those pertaining to food preparation (Sherman and Billing 1999) and hygiene—can serve as defenses against pathogen transmission, deviation from the customary status quo may pose a contagion risk to self and others, whereas conformity helps to maintain the integrity and effectiveness of these defenses against infectious disease. In sum, the behaviors arising from collectivist values (compared with the behaviors of individualism) are more likely to provide defenses against infectious diseases.

In contrast, individualistic values promote different important benefits to individuals. For instance, consider the individualistic values of independent self, intellectual autonomy, and openness to new social contacts and new experiences in general discussed in the previous chapter. Both the discovery and the spread of beneficial new ideas and technologies will occur more frequently when individuals are encouraged to think independently, deviate from existing traditions, and engage in interactions with out-group people. In geographical regions characterized by relatively low parasite stress, the benefits of collectivism (in terms of antiparasite defense) will be minimal, compared with the benefits associated with individualism. Under these ecological circumstances, individualistic values are more optimal or socially effective. However, within geographical regions characterized by a greater prevalence of parasites, the benefits of collectivism to individuals also will be greater, and outweigh benefits conferred by individualistic values. Under these circumstances, collectivistic values are likely to have more utility in terms of individual's inclusive fitness.

From this reasoning, it follows that worldwide variation in infectious diseases is expected to correspond to cultural norms toward individualistic versus collectivistic values. Specifically, our colleagues and we hypothesized that countries with low parasite stress will be individualistic, and that countries with high parasite stress will be collectivistic. Prior to our study, it had been shown that equatorial countries are more collectivistic than societies at higher latitudes (Hofstede 2001). This was consistent with our hypothesis because infectious diseases are more prevalent in equatorial regions (e.g., Guernier et al. 2004). Our hypothesis was tested directly first in the paper by Fincher et al. (2008). We discuss that study next.

### 5.2.2 Measures of Collectivism–Individualism

In our cross-national research with colleagues on collectivism–individualism in relationship to parasite stress, reported in Fincher et al. 2008, we used published data from four studies that provided somewhat different, but conceptually overlapping, highly correlated, measures of collectivism–individualism at the country level. Gelfand et al. (2004) reported on the "Global Leadership and Organizational Behavior Effectiveness Research Program" (GLOBE) measures of in-group collectivism across 62 contemporary societies. The GLOBE research program sampled 17,370 middle managers from 951 organizations in three industries (food processing, financial services, and telecommunications services). Samples were gathered during the years 1994–1997. Our analyses were of GLOBE's "in-group collectivism practices" measure. Gelfand et al. (2004) reported that these particular scores showed the greatest convergent validity with other independent measures of collectivism–individualism. Moreover, compared with the alternative measures summarized by Gelfand et al., this particular measure is the one most clearly based on actual behavior.

This scale consists of adding together responses to four questions that comprise a unidimension measuring the "degree to which individuals express pride, loyalty and interdependence in their families" (Gelfand et al. 2004, p. 463). Respondents indicated agreement on a 1–7 scale for each collectivism question. The largest values reflected greater collectivism and the smallest values were the highest individualism (Gelfand et al. 2004). We refer to this scale as *Gelfand In-group Collectivism.*

Hofstede (2001) presented data that was collected from 1967 to 1973 on an IBM employee attitude survey comprising 116,000 questionnaires from 72 countries. From these data, Hofstede created an individualism index, which we refer to as *Hofstede Individualism.* Hofstede (2001, p. 225) describes individualism–collectivism as: "Individualism stands for a society in which the ties between individuals are loose: everyone is expected to look after him/herself and her/his immediate family only. Collectivism stands for a society in which people from birth onwards are integrated into strong, cohesive in-groups, which throughout people's lifetime continue to protect them in exchange for unquestioning loyalty."

Suh et al. (1998) studied the role of following one's emotions (doing what one wants to do) versus that of following cultural norms (doing what one "should" do) for determining life satisfaction or well-being among members of nations. For this study, they developed a new measure of collectivism–individualism that combined Hofstede's values and the rankings of individualism for 41 nations by Triandis (a pioneering researcher in the field of cross-cultural psychology) in 1996. The inter-rater correlation between Hofstede's empirical values and Triandis' ratings for the 26 overlapping nations was 0.78 ($p<0.001$). Thus, Suh et al. combined Hofstede's empirical data from the 1960s and 1970s and Triandis' more recent rankings to develop a measure of individualism presented on a scale from 1 (most collectivist) to ten (most individualist). We refer to this scale as *Suh Individualism.*

A fourth measure of collectivism–individualism was used in Fincher et al. (2008). Kashima and Kashima (1998) studied the phenomenon of pronoun-drop across nations (71 countries and 39 languages). Pronoun-drop was defined as the acceptable dropping of first- and second-person pronouns from the dominant language. They demonstrated that pronoun-drop was found primarily in collectivist cultures, while the persistent use of the pronouns was found in individualist cultures. The correlation between first-person pronoun use and *Hofstede Individualism* across cultures was 0.75 ($p<0.01$). Kashima and Kashima (1998) argued that the dropping of first-person pronouns is predominant in collectivist cultures, because in such cultures the individual is de-emphasized (e.g., the de-emphasis of "I") to such an extent that the personal self is unimportant relative to the in-group or collective (see also Chap. 4). Pronoun-drop, then, provides an independent measure of collectivism–individualism among cultures, and one based on linguistic norms. We recoded their data such that pronoun-drop cultures scored a 1; those that used first-person pronouns were scored a 0. We refer to this scale as *Kashima Collectivism.*

### *5.2.3 Cross-National Parasite Severity*

In part with our colleagues Murray and Schaller, we developed multiple measures of parasite severity across countries that were used in Fincher et al. (2008). By parasite severity, we mean the number of infectious disease cases, not the number of infectious diseases. In Fincher et al. (2008), we used two highly intercorrelated parasite severity measures to study cross-national variation in values. One focuses on historical parasite severity, and the other on contemporary parasite severity.

The measure of *Historical Parasite Severity* estimates the severity of nine pathogens detrimental to human survival and reproductive success (leishmanias, trypanosomes, malaria, schistosomes, filariae, leprosy, dengue, typhus, and tuberculosis) within each of 93 countries worldwide. (Murray and Schaller later updated this measure to include more of the world's countries (Murray and Schaller 2010). Throughout the book, when we analyze *Historical Parasite Severity*, we are referring to our earlier measure based on 93 countries.) The severity of the nine parasites was estimated on the basis of old atlases of infectious diseases and other historical epidemiological information dating back to the early 1900s. The severity estimates (coded on either three- or four-point scales) were standardized (transformed into *z*-scores), and the mean of these nine standardized scores served as the estimate of the historical severity of pathogens in each country (for details, see Fincher et al. 2008). The validity of this measure is shown for example by its very high correlation with an index of similar parasites used by Gangestad and Buss (1993) to assess historical parasite severity within a small sample of countries (see Fincher et al. 2008). Their index was based on one derived independently by Low (1990, 1994) from historical records of parasite severity from disease atlases extending back to the early 1900s.

Our measure of *Contemporary Parasitic Severity* used data we obtained in June–August 2007 from the Global Infectious Disease and Epidemiology Network (GIDEON) (described below). For each country, we calculated a parasite severity index for a subset of infectious diseases, using a set of parasites similar to those used by others in previous cross-cultural research on parasite severity and human social life (e.g., Low 1990, 1994; Gangestad and Buss 1993; Gangestad et al. 2006; Quinlan 2007). We used the same seven classes of infectious diseases, but expanded the parasites included in the classes to all entries in GIDEON in each class (a total of 22 parasites). We recorded the country-wide disease level of the seven groups of parasites: leishmaniasis, trypanosomes, malaria, schistosomes, filariae, spirochetes, and leprosy. We used GIDEON's three-point scale of parasite severity (3 = endemic, 2 = sporadic, 1 = not endemic) depicted in the geographical distribution maps of the diseases in GIDEON.

The validity of our *Contemporary Parasite Severity* index is shown by two analyses (Fincher et al. 2008). First, it is highly correlated with an index mentioned earlier of similar parasites used by Gangestad and Buss (1993) to assess historical parasite severity within a sample of countries. Second, our index also shows a high correlation with a separate measure of contemporary parasite severity known as DALY (Disability Adjusted Life Years). DALY is a measure of morbidity and mortality across the globe from many sources, as reported by the World Health Organization (WHO). We used the DALY for infectious diseases per 100,000 population, as reported by WHO for the year 2002 (WHO 2004). This DALY combines into one measure the time lived with disability and the time lost due to premature mortality; one DALY equals one lost year of healthy life and the associated burden of infectious disease. It is a measurement of the gap between current health status and an ideal situation where everyone lives into old age free of infectious disease and disability from contagious agents (www.who.int). For the 192 countries for which we had data on both this DALY corrected for population size (ln transformed due to skew) and our contemporary parasite-severity index, the correlation is high: $r=0.74$, $p<0.0001$. Hence, our contemporary parasite severity index correlates highly with historical and contemporary parasite stress as measured by others. *Infectious Disease DALY* is our label throughout the book for the DALY for infectious diseases corrected for population size and log transformed.

## 5.3 Findings

The results supported the hypothesis that, across countries, pathogen severity will correlate negatively with measures of individualism and positively with measures of collectivism (see Table 5.1). Across two measures of pathogen severity, and the four measures of individualism–collectivism, the results are consistent with the hypothesis. As presented in Table 5.1, *Historical Parasite Severity* was an especially strong predictor of both individualism and collectivism (absolute magnitude of the $r$'s ranged from 0.63 to 0.73; all $p$'s$<0.001$). *Contemporary Parasite Severity* showed the identical pattern of results, although the magnitude of the correlations was somewhat less strong.

**Table 5.1** Pearson zero-order correlations between two measures of parasite severity, and four measures of individualism–collectivism

| | Parasite-severity measure | |
|---|---|---|
| Values measure | Historical | Contemporary |
| Hofstede Individualism | −0.69 (68) | −0.59 (68) |
| Suh Individualism | −0.71 (58) | −0.58 (58) |
| Gelfand Collectivism | 0.73 (52) | 0.56 (57) |
| Kashima Collectivism | 0.63 (70) | 0.44 (70) |

All $p$'s $<0.001$; the number of countries in each analysis is in parentheses (results were originally reported in Fincher et al. 2008; reprinted with permission)

To address potential alternative causal explanations, in Fincher et al. (2008), we assessed the relation between *Historical Parasite Severity* and collectivism–individualism while statistically controlling for other variables that earlier published research mentioned as possible causes of collectivism–individualism. Four additional variables were assessed: Gross Domestic Product (GDP) per capita (a measure of wealth of the average person in a country), Gini (the wealth disparity among the people of a country), population density, and residual life expectancy. High values of GDP per capita and of Gini mean high average wealth per person and high wealth disparity among people, respectively. Residual life expectancy means the deviations from the general statistically fitted regression line when life expectancy of both sexes combined is regressed on *Historical Parasite Severity*. This residual is the variation in life expectancy not accounted for by *Historical Parasite Severity*. Of the four additional variables, only GDP per capita and Gini were correlated reliably with collectivism–individualism. Residual life expectancy correlated significantly with only one of the four collectivism–individualism measures, and population density failed to correlate significantly with any of the four measures. By contrast, GDP per capita and Gini were substantially and significantly correlated with all four individualism–collectivism measures (all $p$'s $<0.05$). Consequently, we conducted four multiple regression analyses, in which *Historical Parasite Severity*, GDP per capita and Gini were entered simultaneously as predictors of each of the four individualism–collectivism measures. An identical pattern of results emerged across all four analyses. There were no unique effects of Gini (all $p$'s $>0.05$). By contrast, GDP per capita did exert unique predictive effects (all $p$'s $<0.05$); greater GDP per capita was associated with greater individualism and less collectivism. Of primary interest, parasite severity also uniquely predicted all four measures of individualism–collectivism (all $p$'s $<0.05$). Thus, while other variables (like economic development) also may predict cultural differences in collectivism–individualism, these other variables cannot account for the predictive effects of pathogen severity (also see Chap. 11 on economics and parasite adversity).

Furthermore, Fincher et al. (2008) reported that the pattern of results above relating collectivism–individualism to parasite stress was replicated when cultural regions (rather than countries) were treated in analysis. Regardless of whether the world is divided up according to Murdock's (1949) six world regions, or Gupta and

Hanges' (2004) ten cultural regions of the world, composite scores on pathogen severity correlated negatively with composite scores on individualism, and positively with composite scores on collectivism. These correlations were very strong in some cases. For example, when Murdock's six world regions were used in analysis, the correlation between *Historical Parasite Severity* and the Gelfand et al. collectivism measure was 0.93 ($p=0.004$, $n=6$); and when Gupta and Hanges' (2004) ten cultural regions were used in analysis, the correlation was 0.80 ($p=0.003$, $n=10$). Hence, the variation in the two key variables of parasite stress and collectivism within world or cultural regions is consistent with the same pattern worldwide. (Murdock's world regions, a typical division of the world by anthropologists into cultural regions, are described more fully later in the book.)

In sum, across multiple measures, we found that worldwide variation in parasite severity substantially predicted societal values of individualism–collectivism. Within ecological regions characterized by higher severity of infectious diseases, human cultures are characterized by greater collectivism. The size of this effect was substantial and remained significant even when controlling statistically for potential confounding variables. The effect also remained strong when broader cultural regions (rather than individual countries) were used in analysis. These findings are consistent with the hypothesis that individualism confers benefits upon individuals, but the behaviors that define individualism also enhance the likelihood of pathogen transmission, and thus are maladaptive under ecological conditions in which pathogens are highly prevalent. In contrast, the behaviors of collectivism function in antipathogen defense, and thus are adaptive under conditions of high pathogen prevalence.

These findings help to explain additional variables that were known prior to Fincher et al. (2008) to be correlated with individualism–collectivism. A positive correlation between individualism and latitude has frequently been noted, but never explained (Cohen 2001; Hofstede 2001). Our results imply that this correlation is substantially accounted for by parasite adversity: the meteorological and ecological conditions associated with lower latitudes provide the ideal circumstances for the proliferation of parasites (Guernier et al. 2004), which, in turn, evoke collectivist cultural values as a defense against the high parasite adversity.

Multiple researchers have observed a strong, positive correlation between GDP per capita and individualism and have suggested ways through which economic affluence might lead to individualism (Triandis 1995; Hofstede 2001). Our results indicate that the sizeable correlation between economic productivity and individualism results, in part, from each of these two variables sharing variance with parasite severity. Even the apparently unique effect of GDP per capita may indirectly reflect some causal role of pathogens, given that infectious diseases are potent inhibitors of economic development (Chap. 11). Thus, the traditional literature on collectivism–individualism over-estimates economic influences, while underestimating the causal influences of parasites, a topic we return to later in the book when we treat fully parasite stress and the wealth of nations (Chap. 11).

## 5.4 Nonzoonosis and Collectivism–Individualism Cross-Nationally

One limitation of the study reported in Fincher et al. (2008) is that it employed relatively crude measures of pathogen severity. One measure estimated overall parasite severity (number of cases of disease) on the basis of data pertaining to a diverse set of nine human infectious diseases represented in epidemiological atlases that refer back to the early 1900s. A second measure estimated overall parasite severity (number of disease cases) on the basis of data pertaining to a diverse set of 22 human infectious diseases, obtained in 2007 from an online database of contemporary human infectious diseases (GIDEON). Statistical analyses attest to the reliability and validity of these measures (e.g., Fincher et al. 2008; Thornhill et al. 2009; Murray and Schaller 2010), but these measures are indicators only of overall parasite adversity.

Most importantly, these measures fail to discriminate between conceptually distinct categories of human parasites defined by different modes of transmission to humans. Parasitologists and epidemiologists classify human diseases into three distinct categories based on their modes of transmission: zoonotic, multihost, and human-specific (Smith et al. 2007). Zoonotic parasites develop and reproduce entirely in nonhuman hosts (livestock and wildlife) and can infect humans as well, but are not transmitted directly from human to human. Multihost parasites can use both nonhuman and human hosts to complete their life cycle and may be transmitted directly from human to human as well as to humans through between-species transmission. Human-specific parasites are transmitted only from human to human (although ancestrally they often had a zoonotic transmission origin; see Pearce-Duvet 2006).

These categorical distinctions are important in the parasite-stress theory of values. The cross-national differences in collectivism–individualism discussed earlier are predicted by a parasite-stress theory of *human sociality* that especially emphasizes the potential infection risks associated with interactions with other humans. Hence, according to the parasite-stress theory of human sociality, worldwide differences in the domains of human values are unlikely to correlate substantially with the presence of zoonotic parasites (which cannot be transmitted from human to human), but should correlate strongly with the presence of nonzoonotic parasites (which have the capacity for human-to-human transmission). Empirical evidence consistent with this prediction would provide unique and novel support for the parasite-stress theory of human sociality. Thornhill, Fincher, Murray et al. (2010) reported such evidence, which is summarized below. (See the original publication for further details about methodology and results.)

### 5.4.1 *Methods*

For each of 227 countries, we computed three indices of human *Disease Richness* (number of kinds of human parasitic diseases) based on the presence or absence of every human infectious disease cataloged in the GIDEON database. GIDEON is a frequently updated, subscription-based online database of human infectious diseases available to the medical community and researchers. GIDEON data have been used extensively in research on the global distribution of infectious diseases (e.g., Guernier et al. 2004; Smith et al. 2007; Thornhill et al. 2009; Fumagalli et al. 2011). Our indices were generated from data obtained from GIDEON in 2008. We classified each human infectious disease as either zoonotic, multihost, or human-specific, according to Smith et al.'s (2007) classification scheme, with updates based on more recent epidemiological information in GIDEON and in other sources (see Thornhill et al. 2010 for details of updating). This classification has 154 diseases as zoonotic (e.g., rabies, hantavirus), 40 diseases as multihost (e.g., leishmaniasis, Chagas disease, Dengue fever), and 117 diseases as human-specific (e.g., measles, cholera, filariasis). For each country, we computed separately the sums of all zoonotic diseases, multihost diseases, and human-specific diseases that GIDEON listed as having a presence within that country. These three sums represented three distinct indices of *Disease Richness*. Across all countries combined, the mean parasite richness scores were as follows: zoonotic: mean (M) ± standard deviation = 53.92 ± 10.40 (range = 38–87); multihost: M ± SD = 23.59 ± 2.81 (range = 20–32); human-specific: M ± SD = 102.33 ± 2.96 (range = 98–110). In Thornhill et al. (2010) we did not publish the data for the numbers of the three disease categories per country; these are given in this chapter's Appendix 1, and the methods used in distinguishing the three disease categories are described in Thornhill et al. (2010).

The three parasite indices do not distinguish between certain aspects of disease transmission—e.g., vector-borne versus those requiring direct contact—nor need they. Whether a disease transmitted between people is carried through the air by way of a mosquito or by expelled mucus droplets is not relevant to our main hypothesis about differences between nonzoonotic and zoonotic infectious diseases and values. Similarly, the taxon of the disease—e.g., fungi, viral, helminth ("worm"), etc.—is not relevant to this hypothesis. An argument could be made for the importance of investigating the impact of differences among infectious disease virulence in understanding the evolution of values, but that research has not been done yet.

Note that these measures of parasite *richness* are only indirect measures of the *severity* (the number of infectious-disease cases) that parasites impose on people. Nevertheless, when we did the research using the richness measures, there was evidence that human parasite richness covaries substantially with human parasite severity (Fincher et al. 2008; Fincher and Thornhill 2008a, b); consequently, these measures of parasite richness were used to test hypotheses derived from the parasite-stress theory of human sociality. After we present the results from our study of richness of the types of human infectious diseases, we then discuss new research on severity of the types in relation to collectivism–individualism. The results are similar across the two infectious-disease measures.

**Table 5.2** Pearson zero-order correlations and *p*-values between each measure of parasite richness and each measure of collectivism–individualism

| Values measure | Parasite richness measure | | | | | | |
|---|---|---|---|---|---|---|---|
| | Human-specific | *p* | Multihost | *p* | Zoonotic | *p* | n |
| Hofstede Individualism | −0.60 | <0.001 | −0.70 | <0.001 | −0.17 | >0.10 | 67 |
| Suh Individualism | −0.58 | <0.001 | −0.61 | <0.001 | −0.20 | >0.10 | 57 |
| Gelfand Collectivism | 0.51 | <0.001 | 0.51 | <0.001 | 0.27 | 0.04 | 57 |
| Kashima Collectivism | 0.35 | 0.003 | 0.45 | <0.001 | 0.19 | >0.10 | 70 |

*n* = the number of countries in each analysis (results were originally reported in Thornhill et al. 2010)

## 5.4.2 Findings

The findings of the study based on *Disease Richness* are presented in Table 5.2. Each of the two individualism measures correlated substantially negatively with both human-specific and multihost parasite richness; in contrast, they correlated only weakly with zoonotic parasite richness. Analogously, each of the two collectivism measures showed moderate to strong positive correlations with both human-specific and multihost parasite richness, and weaker correlations with zoonotic parasite richness.

Follow-up regression analyses included all three parasite-richness indices as predictors of the values. The results revealed that the predictive effects associated with human-specific parasite richness remained significant with *Suh Individualism* ($p < 0.001$), and marginally significant on two other measures (*Hofstede Individualism and Gelfand In-group Collectivism*; $p$'s = 0.09 and 0.12, respectively). The predictive effects of multihost parasite richness remained significant on three of the measures (both individualism measures, as well as the *Kashima Collectivism*; all $p$'s < 0.001) and marginally significant on the additional measure (the *Gelfand In-group Collectivism*; $p = 0.10$). In contrast, the modest relations with zoonotic parasite richness actually *reversed* in sign when controlling for shared variance with the other parasite-richness measures. For the two individualism measures, the reversal in sign actually resulted in significant *positive* relations with zoonotic parasite richness (both $p$'s < 0.002), in direct contrast with the significant *negative* relations with human-specific and multihost parasite richness.

Additional regression analyses that included the zoonotic index and the nonzoonotic composite index (multihost and human-specific combined) as predictors revealed a clear distinction: nonzoonotic parasite richness was a unique negative predictor of individualism (both $p$'s < 0.001), and a unique positive predictor of collectivism (all $p$'s < 0.001); zoonotic parasite richness had no consistent unique effect, and any such effect at all (on the two individualism measures) was exactly opposite to that indicated by the correlations in Table 5.2.

In sum, although the Fincher et al. (2008) study (discussed earlier in this chapter) provided a substantial body of evidence linking the prevalence of human infectious diseases to the human value dimension of collectivism–individualism, that study was limited by the fact that (a) the previous indices of human parasite adversity represented only a small fraction of the many infectious diseases that affect human health, and (b) these indices failed to distinguish between different human disease categories defined by their mode of transmission to humans. To address these limitations, Thornhill et al. (2010) used data bearing on more than 300 different human infectious diseases, computed separate indices assessing the prevalence of three functionally distinct categories of these diseases (human-specific, multihost, zoonotic), and examined the extent to which each index uniquely predicted cross-national differences in societal values. The results were convincing.

Both human-specific and multihost parasite richness predicted uniquely cross-national differences in collectivist–individualist values. Zoonotic parasite richness contributed little, if at all, to previously documented cross-national relationships between parasite prevalence and these values. Thus, worldwide variation in the values predicted by parasite prevalence appears to be attributable almost entirely to the prevalence of nonzoonotic diseases. The richness of human-specific parasites and the richness of multihost parasites were approximately equally predictive of collectivistic–individualistic values. These findings conform to the functional logic of the parasite-stress model, because collectivism–individualism has consequences for a broad range of behaviors, including behaviors with implications for interpersonal contact (e.g., approach versus avoidance of unfamiliar peoples), as well as behaviors with implications for interspecies transmission of pathogens (e.g., violation versus conformity to cultural norms pertaining to hygiene).

The preceding analyses found that relationships linking parasite prevalence to collectivism–individualism are attributable primarily to the prevalence of nonzoonotic parasites (human-specific and multihost parasites). Compared to the effects of nonzoonotic parasite richness, any effects associated with zoonotic parasite richness were negligible. Before conclusively ruling out an important contribution of zoonotic parasites to these worldwide values, it is critical to consider an alternative explanation, based on differential measurement error. It is possible that epidemiologists and health agencies are especially attentive to diseases that are transmitted from human to human, whereas the presence of human zoonotic parasites may be relatively poorly recorded. If so, then simply for reasons of differential measurement error, zoonotic parasite richness would be expected to correlate less strongly than nonzoonotics with any outcome variable. The plausibility of this alternative explanation is undermined by evidence that many zoonotic diseases are monitored by the Centers for Disease Control and Prevention and other relevant agencies worldwide as zoonotics are thought to be an important source of emerging human infectious diseases (Greger 2007; Jones et al. 2008). Some zoonotics, however, may escape surveillance by these agencies (e.g., Maudlin, Eisler and Welburn 2009). One way to address this alternative explanation empirically is to show that the zoonotic parasite-richness index is measured with sufficient fidelity to predict additional outcome variables to which it is conceptually related—such as the presence of livestock within a country. Many zoonotic diseases are contracted from livestock, and so we should observe an especially strong relationship between livestock and zoonotic parasite richness—but only if the index of zoonotic parasite richness is measured with a high degree of validity.

For 205 countries, we computed the total number of avian and mammalian livestock over the period from 2000 to 2004 (data obtained from the Global Livestock Atlas of the World Agricultural Information Center; http://www.fao.org/index_en.htm). To correct for skew, this value was log-transformed prior to analyses. Correlations with the three parasite richness indices were as follows: human-specific, $r=0.31$; multihost, $r=0.44$; zoonotic, $r=0.78$ (all $p$'s $<0.001$). In a follow-up regression analysis with all three parasite-richness indices simultaneously entered as predictors, only zoonotic parasite richness remained significantly, positively related to the total number of livestock ($p<0.001$). These results reveal that the zoonotic parasite-richness index is measured with sufficient accuracy to be a uniquely powerful predictor in domains of theoretical relevance. Differential measurement error, therefore, is unlikely to account for the fact that nonzoonotic parasite richness predicted cross-national variability in collectivism–individualism to a much greater extent than did zoonotic disease richness.

### 5.4.3 *Severity of Disease Types*

Subsequent to the cross-national research on collectivism–individualism published in Thornhill et al. (2010) (just described) based on parasite-richness measures, we tallied the parasite severity (number of cases) for all the diseases in GIDEON in 2009 separated by zoonotic and nonzoonotic categories, and then obtained severity indices for each category for each of the countries of the world; our methods are described fully in Fincher and Thornhill (2012) and the parasite severity data are published as electronic supplements to Fincher and Thornhill (2012). The above results for parasite richness in relation to collectivism–individualism were repeated with the new parasite severity measures. All analyses showed that nonzoonotic severity related more strongly to collectivism–individualism than did zoonotic severity. For example, with *Suh Individualism*, nonzoonotics, $r=-0.62$, $p<0.0001$, and zoonotics, $r=-0.23$, $p=0.09$; $n=57$ for both. Also we found that the measures of parasite richness were correlated nearly perfectly with measures of parasite severity. For example, nonzoonotic severity and nonzoonotic richness showed an $r=0.96$ across 222 countries, and zoonotic severity and zoonotic richness correlated $r=0.98$ across the same countries.

## 5.5 Scientific Strengths of the Research

We emphasize that the research studies reported earlier on collectivism–individualism in relation to parasite richness and severity across countries have two major scientific strengths. First, they offer general and synthetic comparative perspectives on the important value dimension collectivism–individualism that may explain it in terms of parasite stress as both proximate and ultimate causation. Another strength of the studies pertains to the data sets involved. Data collection was not biased in favor of the hypothesis that parasite adversity causes values. Unconscious biases can enter at the data-collection stage in scientific studies to bias results toward a hypothesis held by researchers, but this cannot be the case here because the data sets on values and parasite adversity were assembled by researchers who were unaware of the parasite-stress theory of values.

## 5.6 A Caution About Interpreting Correlational Results

At this point in this book, it is important to emphasize that sometimes the scientific significance of the patterns of results presented in this chapter is misunderstood. People sometimes stumble intellectually when shown these kinds of results, and make statements such as, "[t]he researchers have some correlations only and correlation can not prove causation," or "[t]o be convincing, they must conduct experiments." Or, as one researcher put this, "[o]nly experiments can truly test theory." (Stearns 1976, p. 42)

The lines of thinking behind these statements, which are articulated often in the form of criticisms of comparative research findings like those presented earlier, are scientifically erroneous. They fail to understand the concept of correlation in scientific research. More specifically, they fail to recognize that each method of applying the scientific method—lab experiments, field experiments, observation, and the comparative method—has strengths and weaknesses, uses correlational data, and can demonstrate causation. Our discussion below of common misunderstandings and controversy about comparative analyses is based on the more extensive treatment in Thornhill and Fincher (2013).

First, note that *all* scientific findings are correlational—that is, they are *co-relationships* between dependent and independent variables. Alexander (1978) pointed this out in response to creationist critics' claims that evolutionary biology is fatally flawed because its core evidence provided by Darwin is based on correlations obtained by the comparative method. (Also see Thornhill and Palmer (2000) for a related response to certain secularist critics of evolutionary theory as applied to human behavior.) So, to say that correlational data do not count or are less convincing than data generated from experimentation is to misunderstand fundamentally scientific methodology and evidence; the findings from experiments are just as correlational as those from statistical analysis called correlational or regression analysis. In all cases, the scientific value of a finding—its ability to address causation—depends solely upon the control of confounders, not the type of method itself. This is true for studies conducted at individual or group levels.

In discussing scientific methodology with students and colleagues, we have observed that the equivalence of all scientific findings as correlated variables can be understood easily by using the simple example of an imaginary botanical greenhouse experiment to test the hypothesis, say, that potassium causes the growth of hibiscus plants. The researcher grows hibiscus under three different levels of potassium supplementation to the soil. The same basic soil is used in each of three treatments of ten plants per treatment, and all plants have the same watering schedule and genetic background. The researcher's hypothesis will be supported if the data show that the average hibiscus growth across treatments corresponds positively to the amount of experimentally added potassium; that is, by data of a positive correlation between the level of potassium and hibiscus growth. In this case, the actual analysis used to see if the correlation is present is not a statistical correlation analysis per se, but the test is indeed for a corelationship between variables. Note that if the experimenter uses only two treatments and tests the hypothesis with a *t*-test for a difference in average growth between treatments, the procedure is still directed at establishing whether the predicted corelationship exists between the two variables, potassium level and plant growth. Statistics, whether *t*-statistics, correlation coefficients, or other statistics are interchangeable because they are all for determining if there is a correlation between variables. Note, too, that this experiment is based on comparative data, comparing plant growth under different levels of a nutrient. All scientific results are fundamentally both correlational and comparative.

Say the researcher finds the positive correlation predicted by the hypothesis and submits a paper describing the research to a scientific journal or presents the work at a scientific conference. Critics would point to the potential confounds unaddressed; they cannot accept the results as proof of causation, because another factor (potential confound) may be responsible for the correlation of the two variables. For example, it may be that a fungal parasite of hibiscus was present in part of the greenhouse, but not in other parts. Presumably, hibiscus has hundreds of types of parasites, so there is much room for potential confounds, just in the domain of parasites alone. Or perhaps there was a slight, but significant lighting or temperature or humidity variation across parts of the greenhouse, and so on. This example illustrates the general rule in interpreting scientific results, all of which are correlational: the confidence one can have in a study's finding depends entirely upon the ability of the procedure used to control potential confounds. Hence, this confidence is independent entirely of what kind of test method is conducted (experimentation, observations, or a study employing the comparative method).

The comparative method we have used in this chapter to produce findings uses statistical controls, which are routine, straightforward, and scientifically respected analytical procedures for controlling potential confounds. Recently, Minkov (2011, p. 35) criticized our research with colleagues on the cross-national relationship between parasite stress and collectivism–individualism (Fincher et al. 2008) by saying that, although the patterns found in the research are very strong ones, even strong correlations may not identify causation. Of course, this is a correct point. As discussed earlier in this chapter, it is the reason we used control procedures in our cross-national analysis, which allowed us to conclude that parasite stress significantly and positively predicts collectivism when various potential confounds are controlled (Fincher et al. 2008, and see earlier). And this is why we also draw on many lines of evidence to evaluate hypotheses. For example, our predictions about collectivism and parasite stress at the cultural level are bolstered by findings of studies of individuals and their xenophobic attitudes in relation to pathogen sensitivity (Faulkner et al. 2004).

The comparative method also employs the control procedure of randomization. This is used commonly also, for example, in field experimentation in biological research to randomize the locations of treatment replicates in order to control site effects. The randomization procedure in such field experimentation attempts to control unknown potential confounds by making any influence these confounds may have independent of the treatments used. In comparative studies, the kind and diversity of comparison groups allow randomization of the influence of many potential confounders on an effect of interest (Alexander 1978). The parasite-stress theory of values predicts a positive correlation between the degree of collectivism and the degree of parasite stress. In our cross-national analyses, parasite stress corresponds to the "treatment" variable. Consider nations with high parasite stress: They vary in many ways such as language, religion, cooking and clothing styles, diet, subsistence type, social and other ecological features, and so on. The same can be said for nations with low parasite stress. Indeed, at any point across the treatment variable there is much variation in many cultural traits. Thus, across the comparisons of values systems in different countries, many potential confounds are randomized with regard to the treatment variable of degree of parasite adversity.

Now suppose the hibiscus researcher repeats the work, but with careful attention to the criticisms of the first study such that numerous potential confounds are controlled: lighting, temperature, humidity, and the use of fungicide and other pesticides are equal across all treatments, and also the locations of the plants within the greenhouse are randomized in regard to potassium treatment. Again, the correlation is found and the results from the study are presented to scientists. Critics could agree that there is now good evidence that potassium causes hibiscus growth; that is, there is evidence of a correlation—as predicted by theory—between potassium and hibiscus growth unconfounded by other factors so far examined. This confidence that the results show causation stems from the new procedures used to control for confounders, which always constitutes credibility of causation, regardless of the method of testing (experiment or otherwise).

The confidence these critics have in the study, however, has important limits: the research was conducted in a greenhouse, a very unnatural environment. Critics would reasonably still want to see results from nature showing the same positive correlation that was found in the "laboratory" setting. Say the researcher obtains data that show that, within a uniform geographical region, high-potassium-soil locations contain hibiscus plants that grow faster than those in low-potassium-soil locations, which provides evidence that the lab results have meaning in nature. Relevant here, also, would be a field experiment in which potassium levels are manipulated by the researcher in a more natural ecological setting than the lab. It is nature, after all, that science is charged with discovering; unnatural or laboratory results are supportive of a hypothesis about nature only when they are shown to address the natural environment. We emphasize that our studies using the comparative method inherently contain the naturalness of the groups compared, which, as in our analyses earlier, are people living in their cultural ecology.

All the standard methods of testing hypotheses are valid scientific procedures and each can address causal conjectures (hypotheses). Each method has advantages and disadvantages, and no method is superior to all others. Experiments typically involve manipulation of presumed causes, which can yield manipulation anomalies that render the results useless or misleading. The strengths and weaknesses of each of the methods of testing are discussed further in Thornhill (1984). Given that no single method is superior to another, confidence in a hypothesis is increased by the use of multiple methods of testing a prediction of a hypothesis: lab experiment, field experiment, comparative method, and field observations without any manipulations of variable(s). As we show in this book, the parasite-stress theory of values is supported by the full range of scientific-testing methods: experimental, comparative, and observational in both contemporary and ethnographic societies.

We emphasize again that the ability of findings to address causation (to test any hypothesis) depends only on the quality of controls for confounding variables. There is no qualification or exception to this basic feature of the scientific method. Thus, the specific method of testing is always, in itself, irrelevant. When thorough controls are in place, correlation documents causation.

In each of the methods of scientific testing, researchers must decide which variables to control, whether by manipulation or statistically. Because there are always many possible confounds to be considered, in practice, scientists choose appropriate controls based on the hypothesis under investigation; yet sometimes, in order to get their paper published, they must control for the favorite candidate confounds of particular reviewers of the submitted papers; this applies equally across methods.

Critics' demands for controls, however, can lead to an analysis that commits the "partialling fallacy" (Gordon 1968). This fallacy is the use of control variables that are not based on a specified theoretical model. It is called the partialling fallacy in reference to the statistical procedure of multiple regression and partialling (controlling) many different variables that are claimed by critics to be confounds according to their intuitions only. Yet the partialling fallacy must be considered in any application of the scientific method, regardless of whether it tests with experimentation, observation, or comparative analysis. This fallacy is committed widely and is criticized appropriately because it is not a test of a causal conjecture (scientific hypothesis). The only thing it tests is the purely statistical notion that an observed result can survive the addition of any conceivable control variable(s). In proper hypothesis testing, a control is included in a study only when a specified theoretical context demands it. Without this basic understanding of hypothesis testing, the partialling fallacy may be committed, and if so, results are useless for testing causal ideas (hypotheses).

In Chap. 2, we mentioned the great importance of the scientific method—it is *the* method for understanding causes of natural things. Some scholars have pointed out that Francis Bacon, one of the method's founders, was adamant about the requirement of "experimentation" to test ideas and thereby learn the truth of our universe. This is certainly true, but to Bacon "experiment" meant simply searching for data that a hypothesis says must exist (for further discussion, see Wilson 1998). The word "experiment" now has a different meaning in science in that it is typically applied when a method of testing a hypothesis manipulates a presumed cause. The erroneous opinion that only this form of experiment provides accurate scientific results has led to inappropriate conclusions and research directions in biology. Mayr (1982) discusses some well-documented cases of this in the history of biology.

Science is the study of the causes of the effects making up the natural world. Most fundamentally, the process of discovering such causes is by construction and refinement of correlations. Science progresses toward deeper and deeper knowledge by improving the understanding of correlation between hypothesized cause and effect. In Thornhill and Fincher (2013), we discuss further the fundamental role of correlation in scientific research.

## 5.7 Ecological Correlations, the Ecological Fallacy, and Multilevel Modeling

Our correlational comparative results presented earlier, as well as other such results throughout the rest of the book, most directly apply to aggregated data on values and parasite stress. We warned against making the naturalistic fallacy in Chap. 1: the logical error of concluding that fact arising from scientific evidence identifies moral direction. Just above we warned against making the partialling fallacy. A third fallacy we want to address is using aggregate results obtained from a group to which individuals belong to infer the features of the individual; this is the so-called ecological fallacy (Robinson 1950), which is discussed widely in the literature and appropriately cautioned against.

There are multiple forms of ecological-fallacy thinking. One has to do with inferring an individual's score on a variable from a regional aggregate score of the same variable. In aggregate or on average, people in some regions experience more infectious disease than in other regions. This is not to say that everyone in high parasite-stress regions experiences a lot of infectious diseases. Some people in such regions will have encountered few, some more, and some many infectious diseases. Some people will have more effective immune systems and thereby not become "infected," even when exposed to the same parasite adversity. A randomly selected individual from a high parasite-stress region could fall in any of these categories. Similarly, we are not saying that all individuals in high parasite-stress regions are extreme collectivists or that all individuals in low parasite-stress regions are ultra-liberal. There will be variation in values among individuals in any such region. We are saying that the measures of values discussed earlier identify the values of individuals in a region on average. We are saying also that the aggregate scores we use have meaning in terms of the general pattern of cultural behavior or of parasite stress experienced in regions and therefore can be used to test predictions of the parasite-stress theory of values.

Robinson (1950) has been criticized for taking an overly individual-centric view because it ignores contextual variation (Subramanian et al. 2009). Many researchers have suggested multilevel modeling is what is needed to account simultaneously for individual level processes and contextual differences (Subramanian et al. 2009; Pollet et al. in press). When these researchers suggest multilevel modeling, what they typically mean is conducting studies that incorporate simultaneously within a single statistical model both individual-level and group-level data. Multilevel modeling also takes place when researchers synthesize studies conducted at cross-cultural levels and individual levels, even when conducted by different researchers. We have incorporated individual-level and group-level research throughout this book in order to evaluate the basic hypotheses of the parasites-stress theory of sociality.

Another form of ecological-fallacy thinking is when one automatically assumes that two variables generally co-occur within the individual (such as collectivism and parasite stress or individualism and gender equity) in a region. This may or may not be the case for any given set of variables. There is much evidence that our assumption of co-occurrence within the individual of our central variables is the case in general. Vandello and Cohen (1999) showed that their measure of USA state-level collectivism–individualism does not make this form of the ecological fallacy. Also, a range of studies has documented the association between individual-level values and individual-level strength of the behavioral immune response. Clay et al. (2012) have provided evidence that disgust sensitivity and various other measures of concern about infectious diseases are associated positively with collectivism at the level of individual people. Inbar et al. (2012) and Terrizzi et al. (2013) reported this same association, but with conservatism rather than collectivism. This association is an assumption in the cross-cultural research that shows the positive relationship between collectivism and parasite stress. Hence, evidence indicates that it is actually the case that individuals with high pathogen-related disgust and cognitions of contagion concern are more collectivist/conservative, whereas people with low disgust and contagion concern are more individualistic/liberal. Throughout this book we marshal both aggregate and individual-level analyses to make claims about pathogen stress as a causal force.

Although the ecological fallacy warning is typically applied in interpreting comparative correlation results such as those presented earlier, it applies to any finding of difference between two groups or among more groups. This is the case with any method of testing a hypothesis, whether it involves experimentation or other methods, and regardless of the statistical methodology used.

## 5.8 Family Ties Cross-Nationally

As an additional measure of collectivist and individualist values in our cross-national research, we conducted analyses on the strength of "family ties" within each country—measured as a numerical composite variable of multiple self-report items included in the World Values Survey ($n=78$ countries). These items quantify allegiance to the extended family, which, as we have emphasized, is a defining feature of collectivistic value systems. The family-ties variable correlates highly with the measures of collectivism–individualism discussed earlier (see Thornhill et al. 2010). Thornhill et al. (2010) reported that parasite stress was positively associated with a measure of family ties across modern countries. Subsequently, we explored a new and better measure of the strength of family ties at the cross-national level using updated World Values Survey files from a recently produced publicly available dataset unavailable at the time of Thornhill et al. (2010). We now turn to a summary of our research using the updated measure of family ties (for details about methods and results, see Fincher and Thornhill 2012).

### 5.8.1 Methods

Our new measure of the strength of family ties indexed the importance of family loyalty and interdependence, and was similar to the measure of collectivism–individualism used by the GLOBE project (see above; House et al. 2004) and by Alesina and Giuliano (2010). Our new data came from the 1981–2007 pooled dataset of the World Values Survey (www.worldvaluessurvey.org) for the following five items: (1) how important is family in your life? (We used the proportion of those who chose "very important."); (2) the respondent had to endorse one of two statements: (a) regardless of what the qualities and faults of one's parents are, one must always love and respect them, (b) one does not have the duty to respect and love parents who have not earned it [we used the proportion of those who chose "a"]; (3) the respondent was asked to endorse one of two statements: (a) it is the parents' duty to do their best for their children, even at the expense of their own well-being, (b) parents have a life of their own and should not be asked to sacrifice their own well-being for the sake of their children [we used the proportion of participants that chose "a"]; (4) respondents were asked whether they lived with their parents [we used the proportion who indicated they did live with their parents]; (5) respondents were asked about their goals in life [we used the proportion of respondents who said one of their goals in life was to make their parents proud]. All proportions were arcsine-square-root transformed and then standardized prior to analysis. All five items were highly interrelated (Cronbach's $\alpha = 0.86$, $n = 72$ countries). All five components were summed to become our measure, *Strength of Family Ties*. Larger values indicate stronger family ties, while smaller values indicate weaker family ties. National values are published in the supplementary material to Fincher and Thornhill (2012).

Using multiple measures of parasite stress, we tested the prediction that, across countries, parasite stress would be correlated positively with the strength of family ties. One such measure we used was the WHO variable *Infectious Disease DALY* (mentioned earlier in this chapter), which is a cross-national measure of morbidity and mortality attributed to 28 different "infectious and parasitic diseases" for the year 2002 (e.g., tuberculosis, measles, leprosy, dengue; WHO 2004). As explained earlier, an important element of the parasite-stress theory is the costs associated with acquiring diseases from out-group humans. Thus, infectious diseases that are transmissible between humans are predicted to be more important in evoking collectivism than human infectious diseases that are not transmitted between humans. We summed the number of cases of human-specific and multihost infectious diseases per country (called nonzoonotic) and the number of cases of zoonotic diseases, based on data from GIDEON for each disease's severity in each country (see Sect. 5.4). This provided measures of parasite severity for the two disease categories. Nonzoonotic disease was correlated positively with zoonotic disease ($r=0.61$, $p<0.0001$, $n=226$). Nonzoonotic disease was correlated positively with *Infectious Disease DALY* ($r=0.76$, $p<0.0001$, $n=192$). Zoonotic disease was correlated positively with *Infectious Disease DALY* ($r=0.16$, $p=0.03$, $n=192$).

Because there is considerable overlap and covariation in our infectious disease measures, we standardized each of the two variables—*Infectious Disease DALY* and nonzoonotic disease severity—and then summed these standardized scores for each country. Zoonotic disease was not included because of its minimal relationship with collectivism–individualism (see Sect. 5.4 and below). This sum we refer to as *Combined Parasite Stress* (Cronbach's $\alpha=0.76$, $n=192$); national values are published in the supplementary material to Fincher and Thornhill (2012).

### 5.8.2 Findings

The results were as follows. *Strength of Family Ties* was correlated positively and significantly with each of the two separate parasite-stress variables mentioned just earlier; the correlation coefficients were 0.64 for *Infectious Disease DALY* ($n=69$ countries) and 0.57 for nonzoonotic disease ($n=72$; both $p$'s$<0.0001$). *Combined Parasite Stress* was correlated similarly with the *Strength of Family Ties* ($r=0.63$, $p<0.0001$, $n=69$). *Strength of Family Ties* and zoonotic disease showed a small and insignificant correlation ($r=0.15$, $p>0.22$, $n=72$).

There are variables other than parasite stress that have been proposed in the literature as explanations of the strength of family ties (e.g., economic development is often assumed to reduce family ties). We examined correlations between the possible confounding factors (described below) and our dependent variable *Strength of Family Ties*. Potentially confounding variables that were correlated significantly ($p\leq0.05$) were then entered into multiple regressions with *Combined Parasite Stress* and *Strength of Family Ties* to determine whether the predicted correlation remained after removing the effect of the potentially confounding variables.

We examined the effects of national wealth (GDP per capita over the years 1960–2008; data obtained from data.worldbank.org) and the equitability of resource distribution within a nation. For the equitability of resource distribution, we used the measure produced by Vanhanen (2003). It combines GDP per capita, percentages of university students and literates, the degree to which land ownership is widespread, and the degree of decentralization of nonagricultural economic resources in a single value. We called this variable *Resource Distribution*. Too, we examined the effects of human freedom (e.g., the freedoms of expression and belief) using the average of cross-national scores of civil liberties from Freedom House for the years 1972–2008 (www.freedomhouse.org). This became our variable *Civil Liberty*. In our regression analyses, we used two model specifications. The most general model contained

*Combined Parasite Stress*, *Civil Liberty*, and *Resource Distribution* as the predictor variables of *Strength of Family Ties*. *Resource Distribution* includes GDP per capita; however, because of the large amount of cross-cultural research that focuses on GDP per capita, we tested a second model that used GDP per capita and *Combined Parasite Stress* as the predictor variables. All of these potential confounders have been validated and used widely in prior research (see Thornhill et al. 2009).

While we have identified some potentially confounding factors, there may be others that we have not identified. Because we have proposed that parasite stress is an encompassing causal factor, we determined the residual lifespan score for each country by regressing the average life expectancy at birth (for the year 2008) for both sexes combined (data from data.worldbank.org) on nonzoonotic disease ($r^2=0.51$, $p<0.0001$, $n=190$). *Infectious Disease DALY* was not included because its calculation by WHO incorporates life expectancy. These residual lifespan scores represent the variation in lifespan expectancy that cannot be explained by parasite stress (i.e., potentially, this variation represents other causal factors besides parasite stress). We then used these residual lifespan scores in correlations with the *Strength of Family Ties* to address the potential of causal factors besides parasite stress to account for international variation in strength of family ties. This analysis of residuals is similar in logic to that mentioned earlier.

Among the focal, potentially confounding variables, all had significant correlations with the *Strength of Family Ties*. All correlations between *Combined Parasite Stress* and the *Strength of Family Ties* remained positive and significant after removing the confounding effects of the potential confounders (std. $\beta$ coefficient for the effect of *Combined Parasite Stress* on *Strength of Family Ties* was 0.36 when controlling *Civil Liberty* and *Resource Distribution*; the std. $\beta$ coefficient was 0.34 when controlling GDP per capita; $p$'s$<0.01$). Thus, the correlation between parasite-stress and family ties was robust to the effects of freedom or economic development and equivalence as captured by Vanhanen's resource distribution, Freedom House ratings, and GDP per capita.

The correlation between residual lifespan and *Strength of Family Ties* was small and insignificant ($r=0.06$, $p=0.64$, $n=71$). Thus, the variation in life expectancy explained by family ties that was independent of that explained by parasite stress was minimal.

To look for patterns of regional variation that might be inconsistent with the overall pattern across the globe, we used the following approach: we divided the countries into six world regions according to the method devised by Murdock (1949), which is based on geographical proximity and cultural historical contact. Then, we conducted correlations using the mean values for each of the variables for each world region. This six-world-region approach allowed us to characterize a region comprised of multiple countries into a single value (thus deflating sample size). The small sample size of six makes the $p$-values unreliable, but it does allow us to test whether the correlations remain in the predicted direction after reducing the sample size. We also conducted a nested-effect linear regression model that accounts for the nested design of our analysis. In the cross-national case, *Combined Parasite Stress* was nested within each world region as the independent variable and used to predict the *Strength of Family Ties*. Our findings are repeated in the world regions. At the world region level, *Combined Parasite Stress* and *Strength of Family Ties* correlated highly and positively, and thus in the predicted direction ($r=0.94$, $n=6$). Also, the *Strength of Family Ties* was predicted significantly by *Combined Parasite Stress* when nested within world region ($r^2=0.47$, $p<0.0001$, $n=69$). Hence, there is no good evidence that the overall global pattern of positive relationship between parasite stress and strength of family ties is not repeated across each world region.

Again, the details of methods and results used in our research on family ties across countries are in Fincher and Thornhill (2012).

## 5.9 United States: Collectivism and Family Ties

This section gives a brief presentation of the study of collectivism in relation to parasite stress across USA states by Fincher and Thornhill (2012); for details about methods and results, consult that paper.

### *5.9.1 Methods*

In order to investigate family ties in the United States, we used a measure of USA state-level collectivism compiled by Vandello and Cohen (1999), because collectivism includes strong family ties; it also includes preferential assortment with in-group members outside the extended family (Triandis 1995; Hofstede 2001; Gelfand et al. 2004; Fincher et al. 2008; Thornhill et al. 2009). Vandello and Cohen (1999) provided a validated measure of collectivism across the USA states by standardizing and summing the following items obtained from USA state data archives: percentage of people living alone (reversed), percentage of elderly people (65+) living alone (reversed), percentage of households with grandchildren in them, divorce to marriage ratio (reversed), percentage of people with no religious affiliation (reversed), average percentage voting Libertarian over the four presidential elections during 1980–1992 (reversed), ratio of people carpooling to work to people driving alone, and percentage of self-employed workers (reversed). Values ranged from 31 for Montana (highest individualism) to 91 for Hawaii (highest collectivism). We predicted a positive correlation between parasite stress and Vandello and Cohen's (1999) measure, which we called *Collectivism*.

Furthermore, from the original state data sources, we extracted the components that address most directly family ties (as described by Vandello and Cohen 1999). These were the items percentage of people living alone (reversed), percentage of elderly people (65+) living alone (reversed), and the percentage of households with grandchildren in them. All three items were interrelated (Cronbach's $\alpha=0.73$, $n=50$). The three items were summed to become the variable *Strength of Family Ties USA*.

We obtained the annual *Morbidity and Mortality Weekly Report*'s "Summary of Notifiable Diseases, United States" from the Centers for Disease Control (CDC) for the years 1993–2006 (available at http://www.cdc.gov). For each year, we adjusted the number of cases of all infectious diseases tracked by the CDC for that year by the CDC-reported population size for each state. We only included infectious diseases that the CDC had information on from each state. Thus, some diseases that were unreported in some states (meaning that, for some states, there was no information on the disease's presence or absence, not just that there were zero cases reported) were not included in the index. For each state, we determined the average $z$-score of this population adjusted disease incidence score for all years. This approach was necessary because the infectious diseases tracked by the CDC can vary between years, though there was often great similarity between years. The standardization allowed us to pinpoint a state's position along a parasite gradient relative to the other states. This index of parasite severity, *Parasite Stress USA*, is ecologically validated by the fact that it shows a negative correlation with latitude (−0.45, $n=50$, $p=0.001$; or after removing the latitudinal outliers Alaska and Hawaii, −0.71, $n=48$, $p<0.0001$), as do global measures of parasite stress (Cashdan 2001a; Guernier et al. 2004). Further ecological validation is demonstrated by the negative correlation of *Parasite Stress USA* with the average lifespan expectancy at birth for both sexes in the year 2000 (data collected from www.census.gov): $r=-0.67$, $p<0.0001$, $n=50$). (A similar pattern between infectious disease stress and lifespan expectancy is found in cross-national analyses; see Chap. 8). The list of diseases across years that comprise *Parasite Stress USA* as well as the actual values of the variable for each state is in the supplementary materials published with Fincher and Thornhill (2012).

### 5.9.2 Results

The results were as follows. Across US states, collectivism and family ties are predicted positively by parasite severity. *Parasite Stress USA* was correlated positively with *Collectivism* ($r=0.60$, $p<0.0001$, $n=50$) and the *Strength of Family Ties USA* ($r=0.46$, $p=0.0007$, $n=50$).

For addressing potentially confounding variables in the USA analysis, we followed a similar approach as in our cross-national analysis described earlier. We examined correlations between *Collectivism* and *Strength of Family Ties USA*, and potentially confounding factors. Variables that were significantly correlated ($p\leq 0.05$) then were entered into partial correlations with *Parasite Stress USA* to examine whether the predicted correlations between parasite stress and the dependent variables remained after partialling the potentially confounding factors. The potentially confounding factors we considered were GDP per capita and Gini. GDP per capita is an average of the values for years 1999–2007 obtained from the Bureau of Economic Analysis (http://www.bea.gov). Gini was measured at the family level for 1999, the last year available at the time for the variable at the US Census Bureau (http://www.census.gov). Of the two potentially confounding variables, only Gini correlated significantly with *Collectivism* and *Strength of Family Ties USA*. Given these significant zero-order correlations, the effect of Gini was partialled from the correlations between *Parasite Stress USA* and *Collectivism* and between *Parasite Stress USA* and *Strength of Family Ties USA*. After statistically controlling the effect of Gini, *Parasite Stress USA* remained positively correlated with *Collectivism* ($r=0.46$, $p=0.0009$, $n=50$) and *Strength of Family Ties USA* ($r=0.35$, $p=0.014$, $n=50$). Thus, the correlations between parasite stress and family ties or collectivism were not confounded with the effects of economic inequality and development as captured by the Gini index and GDP per capita.

As in the cross-national analysis, we regressed average life expectancy at birth for both sexes combined for the year 2000 (obtained from www.census.gov) on parasite stress. This regression was significant ($r^2=0.45$, $p<0.0001$, $n=50$). The residuals of the regression represent the variation in life expectancy that is not explained by our measure of parasite stress—that is, residual lifespan. The finding of statistically significant covariation between residual lifespan and any of the dependent variables would imply causation other than parasite stress. No such covariation was detectable. Residual lifespan was not correlated significantly with *Collectivism* ($r=0.11$, $p=0.44$, $n=50$) or *Strength of Family Ties USA* ($r=0.22$, $p=0.13$, $n=50$). Therefore, parasite stress accounts for much of the state-level variation in collectivism and family ties as they relate to life expectancy.

For the USA regional analysis, we divided the states into the nine geographic regions used by the Census Bureau and used both approaches as we did for the cross-national analysis described earlier. The correlation between the dependent variables and *Parasite Stress USA* at the regional level all were in the predicted direction (*Collectivism*: $r=0.83$; *Strength of Family Ties USA*: $r=0.51$, $n=9$ for both). The *Strength of Family Ties USA* was predicted significantly by *Parasite Stress USA* when nested within USA region ($r^2=0.34$, $p=0.033$, $n=50$). *Collectivism* was predicted significantly by *Parasite Stress USA* when nested within USA region ($r^2=0.45$, $p=0.0021$, $n=50$). Hence, the overall pattern across the USA states is a positive relationship between parasite severity and each of the two variables, collectivism and family ties, and this relationship is not specific only to certain regions of the USA.

Again, see Fincher and Thornhill (2012) for the details about the methods and results discussed earlier.

Following the publication of Fincher and Thornhill (2012), we explored the potentially different effects of zoonotic and nonzoonotic infectious-disease severities for the USA. Appendix 2 contains the list of USA infectious diseases and their classification into nonzoonotic (multihost and human-specific comprise nonzoonotic) or zoonotic. The classification scheme was based on Smith et al. 2007 and Fincher and Thornhill 2012, or on our additional research when a disease was not listed in either of those sources. Appendix 3 contains the severity scores for each USA state. Nonzoonotic infectious diseases were correlated positively with *Collectivism* ($r=0.53$, $p<0.0001$, $n=50$), and so were zoonotic infectious diseases, but much less and with marginal significance ($r=0.32$, $p=0.03$, $n=50$). Nonzoonotic infectious diseases were correlated positively with the *Strength of Family Ties USA* ($r=0.47, p=0.0005, n=50$), and so were zoonotic infectious diseases, but much less so and not significantly ($r=0.23, p=0.11, n=50$). Additional regression analyses showed that nonzoonotic diseases were strongly predictive of *Collectivism* and *Strength of Family Ties USA* in comparison to zoonotic diseases: when predicting *Collectivism* (nonzoonotic std. $\beta=0.50, p=0.0001$; zoonotic std. $\beta=0.24, p=0.05$), when predicting *Strength of Family Ties USA* (nonzoonotic std. $\beta=0.45, p=0.0009$; zoonotic std. $\beta=0.16, p=0.21$). Thus, just as we found when comparing nations, across the USA states nonzoonotic infectious diseases are more paramount for explaining collectivism and the strength of family ties than are zoonotic infectious diseases.

We also conducted regression analyses to examine the unique predictive effects of each of the indices of transmission types for the USA states (zoonotic, multihost, and human-specific). For *Collectivism*, only human-specific infectious diseases were significantly associated (zoonotic std. $\beta=0.20, p=0.10$; multihost std. $\beta=0.04$, $p=0.75$; human-specific std. $\beta=0.53$, $p=0.0001$). For the *Strength of Family Ties USA*, only human-specific infectious diseases were significantly associated (zoonotic std. $\beta=0.16$, $p=0.23$; multihost std. $\beta=0.21$, $p=0.13$; human-specific std. $\beta=0.34, p=0.02$). Thus, for both dependent variables, human-specific diseases have a greater effect than either of the other two disease transmission categories. Unlike the cross-national comparisons, multihost diseases, considered alone, were not predictive of collectivism and the strength of family ties within the USA states.

## 5.10 Summary: Cross-National and USA States

In summary, our cross-national analysis showed that the strength of family ties was correlated positively with parasite stress. As predicted also, the cross-national analysis showed that the strength of family ties was correlated more strongly with nonzoonotic infectious diseases than with zoonotic infectious diseases. Within the United States, the strength of family ties and collectivism were correlated positively with parasite stress and more strongly with nonzoonotic infectious diseases than with zoonotic infectious diseases. The potential confounds examined did not change these conclusions. The patterns appear robust at regional levels both cross-nationally and across the USA.

## 5.11 Minkov (2011)

Minkov (2011) recently proposed a new measure of collectivism–individualism across countries that is called exclusionism–universalism, with high values equaling high exclusionism (collectivism) and low universalism (individualism). He offered it as a new collectivism–individualism variable to be added to three traditional measures we discussed at the beginning of this chapter (*Hofstede Individualism, Suh Individualism, Gelfand In-group Collectivism*). Exclusionism is characterized by strong social ties of relatives between generations and within groups of relatives, as well as a low value placed on the interests and well-being of strangers. In contrast, universalism is weak ties of relatives between generations and within groups of relatives, as well as a high degree of interest in the needs and welfare of strangers. Given the composition of Minkov's variable, it is a standard collectivism–individualism metric. Indeed, it is highly correlated with the three traditional measures mentioned from an $r$ of −0.71 with *Hofstede*'s *Individualism* to an $r$ of 0.81 with *Gelfand In-group Collectivism.* Importantly, exclusionism–universalism correlates strongly with *Combined Parasite Stress* ($r=0.72$, $p<0.0001$, $n=86$ countries).

## 5.12 Cashdan and Steele (2013): The Standard Cross-Cultural Sample

Cashdan and Steele (2013) have conducted an important first study of collectivism–individualism in relation to parasite severity in the 186 indigenous societies comprising the Standard Cross-Cultural Sample. They used a measure of these values based on how children are inculcated across the societies as reported by the ethnographers working in the societies. In this sample, child training ranges from high obedience to parents and other authorities (collectivist inculcation) to high self-reliance (individualist inculcation). They used a parasite severity measure similar to that used by Low (1988) for the same societies. Their results supported the parasite-stress theory of values: in societies experiencing high parasite stress, children are taught obedience whereas in low parasite stress conditions children are taught self-reliance.

## 5.13 Convergence of Evidence

So far in this chapter, we have provided convergent evidence that parasite stress directly predicts collectivist values across countries, USA states, and societies in the Standard Cross-Cultural Sample, or, said differently, parasite stress negatively predicts individualism across all three of these levels of analysis. The findings cross-nationally and across the USA that strong family ties are correlated with parasite stress complements our earlier work, discussed earlier, that linked collectivism–individualism with parasite stress. Our finding of the strong positive correlation between Vandello and Cohen's measure of collectivism and parasite stress within the USA is also an important complement to the cross-national findings of the same relationship. Finally, the findings of Cashdan and Steele (2013) showing that collectivism is positively related to parasite stress across indigenous societies complements all these other sources of evidence for the relationship between collectivism–individualism and parasite adversity.

## 5.14 Collectivism and Family Ties: Other Implications

### 5.14.1 Life History

All the findings presented earlier are major cross-cultural extensions to the ethnographic research on indigenous societies that has documented the important role of nepotistic and other in-group altruistic connections and support for surviving parasitic infections (discussed in Chap. 3) and conducted by Gurven et al. (2000), Sugiyama and Sugiyama (2003), Sugiyama (2004), and Hill and Hurtado (2009). However, whenever parasite stress is extremely high, collectivistic, including nepotistic, investment may not be optimal because the extreme parasite stress yields extrinsic mortality (Quinlan 2007 and references therein). In the area of evolutionary theory dealing with life history patterns, extrinsic mortality is mortality independent of individuals' efforts to combat it. Because extrinsic mortality, by definition, cannot be reduced by collectivist social investment, comparatively low levels of investment are expected from life history theory when extrinsic mortality is high. Therefore, we expect reduced collectivist investment in offspring, other kin, and the in-group in general in the face of extremely high parasite stress. In this setting, as in other contexts of high extrinsic mortality, early reproduction with minimum nepotistic investment per family member (e.g., offspring) is expected from life history theory (Charnov 1993; Kaplan and Gangestad 2005). Consistent with this thinking, Quinlan (2007) found in a sample of traditional societies that maternal investment in the form of nursing duration increased along with pathogen stress, but then began to decrease after pathogen stress became extreme. We reasoned that this same pattern would be seen in human value systems as well.

Supporting this reasoning, when focusing on Murdock's six world regions, we reported in Fincher and Thornhill (2012) that the correlation in Africa between *Combined Parasite Stress* and a variable we called *In-group Assortativeness* was negative (–0.31), instead of positive as in the other five world regions. *In-group Assortativeness* is a combination variable that we have constructed and explored cross-nationally in Fincher and Thornhill (2012). It combines strength of family ties and religiosity and is a validated measure of embeddedness in the in-group and in-group favoritism. We will discuss this variable further when we treat the relationship between religiosity and parasite stress (Chap. 9). The exceptional case of Africa mentioned might be explained by the fact that parasite stress is exceptionally high in Africa—and therefore generally yields extrinsic mortality—as compared to the five other world regions. (A *post-hoc* Tukey HSD means test showed Africa to be distinctly high in parasite stress: Africa, mean ($M$) = 3.36, A (world areas not followed by the same letter are significantly different); South America, $M = 0.85$, B; East Eurasia, $M = 0.53$, BC; North America, $M = -0.51$, BC; Insular Pacific, $M = -0.65$, C; West Eurasia, $M = -2.28$, D.) Finding the different pattern in Africa shows the importance of regional analysis to identify patterns that may be contrary to worldwide relationships of variables and masked without regional analytical follow-up.

In the final chapter of this book we return to the idea of extremely high parasite stress as extrinsic mortality and provide additional evidence that under very high parasite stress collectivism shows the predicted decline.

### 5.14.2 Implications for Future Regional Analyses

The history of infectious-disease reduction in the USA and other Western regions is reviewed in Thornhill et al. (2009) and involved a multitude of factors varying from antibiotic availability, vaccination programs, chlorinated and fluoridated public drinking water availability, vector control, food handling regulations, and other sanitation changes and technology. As we discuss more fully in Chap. 10, we have argued that the reduction in parasite stress, beginning in the early to mid-1900s (depending on the particular health intervention), accounts for the cultural revolution in the huge increase in liberalization of values in the 1960s and 1970s in the West—the sexual revolution, antiauthoritarianism, women's rights, gay rights, changes in divorce law, civil rights, and so on (Thornhill et al. 2009). Although these health improvements had the overall effect of reducing parasitic infections across the USA (and the West in general) and associated morbidity and mortality, the degree of their effect remains variable across the USA, as seen in the data we presented earlier on the power of parasite-stress variation for predicting variation in values across the states of the USA.

It would be relevant to explore regional analysis within other countries that contain significant parasite gradients. For example, Japan's northern island of Hokkaido rivals the high individualism in the United States (Kitayama et al. 2006); likely, Hokkaido has a much lower level of parasite stress than does southern Japan, given the negative covariation of parasite stress and latitude (Guernier et al. 2004). Also, in China, historically, much of China's innovation came from the northern side, which was much lower in parasite adversity than the southern portion below the Yellow River (McNeill 1998). Innovation—both its generation and the willingness of people to adopt it—corresponds to individualistic values (Thornhill et al. 2009, Chap. 11). The regional development of innovation in China and elsewhere could be studied more thoroughly in its relation to pathogen stress and associated evoked value systems. In the USA, we found significant regional variation in values in spite of generally low parasite adversity in the country. We expect this pattern to be repeated within nations across the world in cases in which a nation contains a parasite gradient.

### 5.14.3 The Demographic Transition

One aspect of family life that has been studied often is the demographic transition from large families to smaller families. One of the more convincing explanations for this phenomenon comes from Newson et al. (2005), who argue that the demographic transition arose from an increase over time in the ratio of nonkin to kin in individuals' social networks. We agree, but offer a reinterpretation of the meaning of the demographic transition. Based on our studies on collectivism and family ties discussed earlier, it is reasonable to assume that this increase in the non-kin-to-kin ratio is related to a decrease in parasite stress over time and corresponding increase in individualism and associated nuclear family focus. Moreover, as predicted by the parasite-stress theory of values, those countries where the demographic transition has occurred are the same ones that have experienced a relatively recent emancipation from infectious disease (Thornhill et al. 2009).

## 5.15 Philopatry

Philopatry—the preference for remaining in the natal location for reproduction (absence of dispersal)—reduces contact with out-groups and their habitats that may contain new parasites to which the in-group has no or limited immunity. Or, said differently, dispersal has the potential cost of contacting infectious diseases that could be avoided by remaining philopatric. In areas of high pathogen stress, compared with those of low pathogen stress, limited dispersal will be favored by natural selection and attractive for cultural adoption by individuals owing to the corresponding advantage of increased association with immunologically similar individuals and decreased contact with more distant, and differently parasitized, other individuals. Freeland (1976, 1979), Møller et al. (1993), and Loehle (1995) all discuss how limited dispersal in nonhuman animals reduces exposure to a diversity of types of infectious diseases and argue for the importance of territoriality and restricted home ranges, forms of limited dispersal, as adaptive means for reducing contact with dissimilar conspecifics that may carry novel diseases.

In humans, philopatry keeps people near to their natal locale and social community, and hence contributes to collectivism, ethnocentrism, and in-group assortative sociality in general. Alesina and Giuliano (2010) demonstrated that, across countries, adults who are more embedded in their extended family (and demonstrably more collectivist) disperse for shorter distances from their natal locale and are, hence, more philopatric than are less embedded adults (individualists). They reported that the relatively limited dispersal of collectivists was apparent both in their adult presence in the natal region and in adult residence in their natal house. The research outside of humans, mentioned earlier, as well as that of Alesina and Giuliano on people, supports the hypothesis that infectious diseases cause values or preferences pertaining to dispersal behavior—where infectious disease is more stressful, animals, including humans, disperse over shorter distances than where infectious disease is less stressful. The human research indicates that high philopatry is a core value of collectivists and low philopatry is a core value of individualists.

Below we present research on human movement patterns in relation to parasite stress. First, we summarize our research findings on movement patterns in ethnographic societies reported in Fincher and Thornhill (2008b). We then examine the pattern of interstate residential movement across states of the USA.

### *5.15.1 Ethnographic Societies*

#### 5.15.1.1 Methods

According to the parasite-stress theory of values, the area of land that individuals inhabit should correlate negatively with pathogen stress, reflecting limited dispersal in the face of high contagion risk associated with contact with out-groups and their habitats. In order to further test this proposition's application to human movement, we examined the relationship between Binford's (2001) measure of societal range size, reflecting the aggregate use of space by individuals within a society, for his large sample of 339 traditional societies across the globe, and an estimate of pathogen stress for each of these societies (Fincher and Thornhill 2008b). Binford's measures are based on his estimates taken from ethnographic sources. We estimated pathogen stress by first establishing the linear equation that best predicted parasite severity (the number of parasitic disease cases), as measured by Low (1994), for the 186 societies in the Standard Cross-Cultural Sample (SCCS, Murdock and White 1969). We examined absolute latitude, mean annual temperature, and mean yearly rainfall as predictor variables from data provided by Binford (2001) for each of the 339 societies in his sample. The data for the SCCS were collected from the *World Cultures* journal website (www.worldcultures.org). The best single predictor of parasite severity was absolute latitude ($r = -0.51$, $p < 0.0001$). Both temperature and rainfall significantly predicted parasite severity, but to lower degrees than absolute latitude; therefore, we generated a linear equation to predict parasite severity based on absolute latitude. Then, we used this equation to estimate parasite severity for each of the 339 societies from data on the latitudinal location of each society in Binford (2001).

Societies with more individuals also might have larger home ranges; thus, we examined the potentially confounding effect of population size using data provided by Binford (2001). Also, societal range size might be related positively to reliance on hunting of terrestrial animals for subsistence; thus, we examined the potentially confounding effect of the proportion of subsistence derived from hunting of terrestrial animals, as provided by Binford (2001).

Additionally, we considered the patterns of mobility of each society with respect to estimated parasite severity because high levels of infection and associated lethargy and incapacitation may reduce mobility of people, generating a negative relationship between societal range size and parasite adversity irrespective of philopatric values serving to minimize dispersal. We did this by analyzing the average distance moved annually and the average annual number of moves (1 was added to the number of moves prior to ln-transformation to eliminate negative values).

Lastly, we used an analysis of variance with country in which a traditional society was located, as provided in Binford (2001), as the independent variable predicting societal range size to examine the possibility of regional dissimilarity in results. This method is similar in logic to our use of world regions, cultural regions, and USA census regions in other analyses we have discussed earlier in this chapter. (All values analyzed for each of the 339 societies are contained in Fincher and Thornhill 2008b.)

#### 5.15.1.2 Findings

We found that societal range size was negatively and significantly correlated with parasite severity ($r=-0.48$, $p<0.0001$, $n=339$). After partialling the effect of population size or the proportion of subsistence from hunting, the correlation between range size and parasite severity remained negative and significant (partialling population size: $r=-0.42$; partialling proportion hunting: $r=-0.38$; $p<0.0001$, $n=339$ for each correlation). Removing the effects of population size and the proportion of subsistence from hunting simultaneously, the correlation between parasite severity and home range size was negative and significant ($r=-0.27$, $p<0.0001$, $n=339$). Thus, as predicted by the parasite-stress theory, there was a robust reduction in range size and associated limitation of dispersal in areas with greater pathogen severity compared to societies residing under reduced parasite severity.

In general, a society's range size was predicted by the country in which the society occurred ($p<0.0001$). Thus, we computed the average value of home range size, parasite severity, population size, and the proportion of subsistence from hunting for the cultures of each country and conducted correlations with these composite values. At the country level, the correlation between home range size and parasite severity was $r=-0.60$ ($p<0.0001$, $n=67$ countries). After partialling the effects of population size and proportion of subsistence from hunting, the correlation between home range size and parasite severity at the country level was $r=-0.43$ ($p<0.0004$, $n=67$). Considered separately, partialling the effect of proportion of subsistence from hunting and the effect of population size, the correlation between home range size and parasite severity was $r=-0.51$ ($p<0.0001$; $n=67$) and $r=-0.52$ ($p<0.0001$; $n=67$), respectively.

Considering mobility patterns within home ranges, the number of moves (ln) annually was positively correlated ($r=0.12$, $p<0.03$), while the distance moved annually was negatively correlated ($r=-0.21$, $p<0.0001$) with parasite severity ($n=339$ for both). Moreover, partialling the effect of distance moved increases the correlation between the number of moves and parasite severity to 0.45 ($p<0.0001$), while partialling the effect of number of moves increases the correlation between parasite severity and distance moved to $r=-0.47$ ($p=0.0001$). Thus, people in traditional societies in high parasite-stress areas move more often, but over shorter distances. Conversely, in areas with lower parasite stress, they move less often but over longer distances.

#### 5.15.1.3 Discussion and Conclusions

The main finding in Fincher and Thornhill (2008b) was that range sizes for ethnographic societies are smaller in areas of the world where parasite severity was more intense. Moreover, this relationship is not confounded by population size or the proportion of subsistence from hunting. This finding is consistent with the notion that human societal range sizes, as well as species' range sizes, are generally reduced in the tropics. Ecologists call this pattern for species' ranges "Rapoport's rule" (Stevens 1989). Our research gives this ecological rule a causal explanation, which it has lacked, that is supported by the increasing evidence that parasite adversity affects dispersal behavior. We have hypothesized that this general pattern results from the evolved response of limited dispersal in tropical regions and other high pathogen areas owing to strong selection against out-group contact (Fincher and Thornhill 2008a). We discuss Rapoport's rule further in Chap. 13.

According to the data, individuals in ethnographic societies in areas with high pathogen stress move often, but over shorter distances. These findings certainly negate the common-sense notion that limited societal range size is due primarily to incapacitation, lethargy, and physical inability to move due to a heavy infectious-disease burden. We have hypothesized that this pattern of short, but frequent movement of peoples in high parasite areas is also an aspect of evolved antipathogen behavior (Fincher and Thornhill 2008b). Individuals within these societies may move strategically often within a restricted territory to optimally distance themselves from parasites, especially intestinal parasites, that persist in the soil (McNeill 1981; Loehle 1995). Freeland (1976) discusses similar localized movements of ungulates that seem to function this way.

Cashdan and Steele's (2013) study on collectivism–individualism in the 186 societies in the Standard Cross-Cultural Sample (discussed above) included an analysis of a variable measuring adult movement of residence between communities. They reported that, in indigenous peoples, parasite stress negatively predicts this movement. Hence, in comparison to people living in low parasite-stress locales, people in high parasite-stress locales have smaller home ranges and lower intercommunity residential movement, as expected from the parasite-stress theory of values.

### *5.15.2 Residential Movement Between USA States*

Based on the hypothesis that higher levels of parasite stress will evoke philopatry, we predicted that people in the USA will move from one state to another to establish a new residence less often in areas of high parasite stress than in areas of low parasite stress. We tested this prediction in a study first reported here. We collected migration data from the Census Bureau for the year 2005 (source: http://www.census.gov/hhes/migration/data/acs/state-to-state.html). We chose 2005 as a point prior to the economic upheaval in 2008. Our measure is the number of people leaving a given state (emigrants) within the year 2005 to establish residence in any other state within the USA. Because of outlying data in the samples, we used Spearman's correlations to examine the relationships between variables. Unsurprisingly, the number of emigrants was correlated positively with the state's population size for year 2005 ($\rho = 0.97$, $p < 0.0001$, $n = 50$) (source for population size: http://www.census.gov/popest/data/historical/2000s/vintage_2005/index.html). Therefore, we calculated an *Emigration* variable for each state that is the number of emigrants from the state divided by the state's population size. As predicted, the correlation between *Parasite Stress USA* (introduced in Sect. 5.9.1) and *Emigration* was significantly negative ($\rho$=−0.31, $p$=0.03, $n$=50 states). This means that there were relatively fewer residential emigration events from states that have high parasite stress; conversely, there were more emigrations from states that have lower parasite stress. There are other potential explanations of emigration rate. One particularly cognizant one is the average wealth within a state. We collected data on GDP per capita for the year 2005 from the Bureau of Economic Analysis (source: www.bea.gov). State-level GDP per capita was unimportant for explaining *Emigration* ($\rho$=−0.01, $p$=0.95, $n$=50). Thus, we have here concordant evidence across USA states that philopatry is more prevalent in areas with high parasite adversity than in areas of low parasite adversity. We add, however, that the relationship between Vandello and Cohen's (1999) measure of collectivism and *Emigration* is in the predicted direction, but statistically insignificant ($\rho$=−0.19, $p$=0.18).

### *5.15.3 Overview: Philopatry*

In sum, we hypothesized that reduced dispersal (high philopatry) reflects adaptation to reduce contact with novel parasites, and that increased dispersal is the optimal preference under low parasite stress. Various lines of evidence support this hypothesis. Across nations, collectivist people are more philopatric than individualists, and correspondingly collectivists more often live in areas of higher parasite severity than do individualists. Across indigenous societies, range size is related negatively to parasite stress. People in indigenous societies move a lot under high parasite stress, but not very far. Furthermore, interstate movement of residence by people in the USA is reduced in states with high parasite adversity compared to states with low parasite adversity. A similar pattern is seen in the small-scale societies in the Standard Cross-Cultural sample, but involved reduced intercommunity residential movement under high parasite stress. A pattern mentioned in Chap. 4 is also relevant to the difference in philopatry of collectivists versus individualists. Liberals are more interested in and engage in more travel from their home region. This is seen from scores of degree of conservatism/liberalism of individuals (Carney et al. 2008). Finally, the parasite-stress theory of movement gives new meaning to an important general rule of ecological science, Rapoport's rule.

## 5.16 Collectivism, Family Ties, and Cooperative Breeding

Collectivism and associated strong family ties are not restricted to humans; indeed, they appear to be widespread across animal taxa. We have argued that parasite adversity was one of the main forces of selection responsible for adaptation that functions in extended family investment. Thus, we proposed that variation in parasite prevalence is a cause of the large variation across animal species in the degree of extended nepotism exhibited outside the social unit of parent(s) and offspring (Fincher and Thornhill 2012).

The literature on the evolution of family life is voluminous, with important reviews provided by Andersson (1984), Brockmann (1984), and Emlen (1994, 1995, 1997). The study of family life first became fully encompassed by evolutionary biology with Hamilton's realization that an individual's fitness is more than its phenotypic design for production of descendant relatives—that is, more than the individual's classical or Darwinian fitness (Chap. 2). Indeed, one's inclusive fitness is the classical component plus design for assisting nondescendant kin by nepotism and thereby promoting one's reproduction. However, current inclusive fitness theory, the fundamental component of modern evolutionary social theory, does not account for why nepotism is variable across social systems. Why is nepotism limited to the nuclear family in many systems, but extended beyond the nuclear family in others to entail varying degrees of cooperative breeding? In this chapter, we provide considerable evidence that parasite stress explains this variation across human social systems. Family ties, or collectivism, measures the investment in the extended family and hence measures cooperative breeding. We proposed in Fincher and Thornhill (2012) that a general theory of family life across taxa is accomplished by coupling the parasite-stress theory of sociality with Hamilton's theory.

The social organization of animal species varies along a cooperative breeding continuum, or said differently, a continuum of eusociality (Andersson 1984; Sherman et al. 1995). A mother alone investing in her offspring or, much less commonly across species, a father alone investing in offspring, is on the highest asociality end of the continuum. Species in which both mother and father nepotistically assist the offspring (so-called biparental species) are more social in degree on the continuum. This is followed by varying degrees of nepotism extended outside parental care (i.e., extra-parental nepotism). Sterility, shown by adult members of the group who serve as helpers and thereby assist relatives of varying degrees or in some cases nonrelatives instead of producing their own offspring, occurs in certain taxa of vertebrates, including the human species, and certain taxa of invertebrates. Depending on the species, this sterility ranges from temporary to permanent. The temporary case is delayed striving to produce descendent kin while, instead, engaging in extra-parental nepotism and other in-group altruism (e.g., certain human groups (Hill and Hurtado 2009) and certain species of wasps, birds, and carnivores). The permanent case is lifelong exclusive extra-parental nepotism (as is characteristic of worker and soldier castes in ants and termites). Both temporary and permanent cases are cooperative breeding, a feature of in-group assortative sociality. Also, the relatively eusocial species on the continuum—i.e., the more cooperative in terms of group breeding—exhibit, in general, strong sedentism, delayed or no dispersal from the natal location, and territory defense by the family group or in some cases by the larger in-group (e.g., Arnold and Owens 1998). According to the parasite-stress theory, the sedentism and limited dispersal are analogs (similarity resulting from independent evolution by selection, i.e., convergent evolution) or in some cases homologs (similarity resulting from common ancestry) of human philopatry. The territoriality is the analog or homolog, depending on the comparison, of human xenophobia.

At a minimum, our conjecture is supported by the fact that cooperative breeding in birds and eusocial systems in insects are more common, or in the case of eusociality, more eusocial, in tropical regions for many different taxa (e.g., birds (Brown 1987; Ekman 2006; Blumstein and Møller 2008), wasps (Wilson 1971)). The incidence of cooperative breeding in birds has been shown to correlate positively with temporal variation in certain climatic factors, especially rainfall (Jetz and Rubenstein 2011). It is not known, however, whether parasite stress in birds also corresponds to temporal variation in the factors Jetz and Rubenstein studied, but it is established that climatic factors correlate strongly with human infectious disease adversity (e.g., Dunn et al. 2010). Data that would allow a comparison of parasite stress in cooperative and noncooperative breeding nonhuman vertebrates are unavailable. A test of our hypothesis that cooperative breeding is favored by selection under high parasite adversity would include measures of allocation to the immune system between the two types of social systems. Møller (1998) reported that tropical bird species show greater immune-system allocation than do temperate bird species, which is expected on the basis of greater parasite adversity in the tropics. The prediction that cooperative breeders will invest more in immune defense than closely related species that breed in pairs (noncooperative breeders) was supported by a study of 66 species of African birds, of which 18 were cooperative breeders (Spottiswoode 2008). Similar comparisons could be conducted on cooperative-breeding species versus non-cooperative-breeding species in other taxa containing cooperative breeders.

## 5.17 Reciprocity

In this and the previous chapter, we have seen that, for humans, evidence strongly supports the following: in comparison to individualists, collectivists are less widely egalitarian, favor in-group over out-group, dislike and avoid out-groups, define sharp and permanent in- versus out-group boundaries, and are less motivated to help strangers. As degree of collectivism increases across human cultural groups, so do each of these features of in-group assortative sociality. The ethnocentrism component of in-group assortative sociality involves more than just investment in the nuclear and extended family. It extends to nonrelated others with like values and hence, in human evolutionary history, to other in-group people with similar immunity. As explained in Chap. 3's discussion of the basic components of the parasite-stress theory of values, the ethnocentrism functionally is for both avoidance and management of infectious diseases. The management part of ethnocentrism is designed to produce and maintain socially supportive networks, based on nepotism and pure reciprocity (reciprocity without a kinship component), with other people of similar values and norms, which are therefore safe for social interaction from the standpoint of reduced risk of infection by a novel parasite. This network was the only insurance against the morbidity and mortality resulting from infectious disease in evolutionary ancestral times of the hominin lineage. Hence, the quality and reliability of this support network affected differential inclusive reproductive success of individuals. Those individuals with high quality and reliable networks out-reproduced those without such networks, and hence became modern human's evolutionary ancestors.

In altruistic social interactions, as the coefficient of relatedness between benefactor and recipient declines, nepotistic behavior becomes increasingly similar to pure reciprocity—the altruist's return benefits affecting the adaptiveness of altruism increasingly depend upon resources returned, rather than the return arising from the recipient's enhanced reproductive success that results from the altruism. Hence, nepotism grades into pure reciprocity; this graded social life was part of the social evolutionary legacy of people, even in the kin-based groups of humans' deep-time background.

We have emphasized that human values are conditionally adopted and used in social navigation. In comparison to other species, *Homo sapiens* is adapted to unique degrees to conditionally interact and ally with people in distinct out-groups with dissimilar phenotypes, including dissimilarity in values, behavior, and appearance. The important condition making this xenophilia adaptive, according to the parasite-stress theory, is low parasite stress.

As explained in Chap. 3, parasite adversity is variable on the localized spatial and temporal scale. Hence, in any given locale, within their lifetime, individuals face varying amounts of parasite stress. We suggest that, in human evolutionary history, it was the regularity of relatively low parasite-stress conditions, in which out-group interactions, transactions, and alliances were adaptive, that crafted human psychological adaptation functioning in altruistic reciprocation among nonrelatives, both in direct and indirect reciprocity (also see Thornhill et al. 2009). As explained in Chap. 2, altruism among individuals unrelated by recent common descent is favored by selection when the altruist receives return benefits from another or others that exceed the costs of the altruism. We propose that, in human evolutionary history, this condition was met consistently when contagion risk associated with intergroup contact was relatively low.

The structure seen in indigenous foraging societies/hunter–gatherers—the social organization that characterized human evolutionary history—is that individuals are surrounded primarily by close and distant genetic relatives. In addition, a minority of the group members comes from the outside through marriage, capture, or other sources of immigration (e.g., Van den Berghe 1981; Low 2000). Although the composition of the group in human evolutionary history included very distant relatives and some nonrelatives, giving context for natural selection favoring reciprocity in the broader social network, we propose that an important context for the evolution of reciprocity was in gaining benefits from out-group interactions during periods of relatively low disease threat. Thus, the parasite-stress theory of values offers a novel perspective to explain the evolution of human reciprocal altruism. Accordingly, natural selection is expected to have designed our reciprocity activities to be conditionally sensitive to the variable risk of contagion in the local ecological setting. Under low such risk, reciprocity is more beneficial than under high risk.

Generally consistent with this reasoning is the favorable attitude of liberals versus the unfavorable attitude of conservatives about out-group transactions. Also consistent is the experimental evidence we have discussed of a within-individual shift to values of interpersonal avoidance upon encountering cues of parasite threat in the immediate environment. Finally, in Chap. 11 we take up topics central to understanding the variation in patterns of diffusion of innovations. Such diffusion relies on willingness to interact with out-groups and hence, we argue, is reflective of reciprocity psychological adaptation designed by an evolutionary history of variable parasite stress locally.

## 5.18 Human-Specific Cognitive Ability

The parasite-stress theory of sociality, moreover, provides a new hypothesis for the uniquely sophisticated cognitive ability of humans. The hypothesis that natural selection, in the context of interactions with conspecifics, was a major evolutionary force responsible for increased brain size and the concomitant increased intelligence and associated sophistication of social behavior in various bird and mammalian taxa, including primates—the "social brain hypothesis"—is well supported (see recent overviews by Dunbar and Shultz 2007; Shultz and Dunbar 2007). A version of this idea applied to the cognitive abilities of humans was discussed by Alexander (1987, 1990) and Flinn et al. (2005). Alexander argued that, as we became ecologically dominant during human evolution as a result of the evolved psychological capacity for inventing weapons and other technology, the most important selective agents were not "… climate, weather, food shortages, or parasites—not even predators" (Alexander 1990, p. 4). Rather, he proposed that, as we gained relative freedom (compared to other species) from these forms of mortality, other humans became the greatest force of Darwinian selection. This led to runaway social selection in the human lineage, generating many aspects of human mental uniqueness, such as consciousness, theory of mind, creativity, exquisite linguistic and deceptive abilities, reputation building, and many others (Flinn et al. 2005).

We specifically take issue with Alexander's (1990) notion that our ecological dominance freed us from the importance of infectious diseases as agents of selection. We suggest a different scenario: as our physical environmental problems and predators became less important as mortality agents, parasites became more important as agents of selection. Indeed, only since the inventions of vaccines, antibiotics and modern sanitation have humans achieved any significant dominance over parasites. Also, such dominance is seen primarily only in the West. As discussed in Chap. 3, in many geographic areas infectious disease appears to be the leading factor bringing about natural selection on contemporary humans, and infectious disease is likely the leading cause of juvenile mortality in indigenous peoples in the ethnographic record. Even now in the USA, a relatively low parasite-stress country, much of the variation in human lifespan may result from parasitic disease (see above and Chap. 8). Moreover, we suggest that it was the salience of the variability of local parasite stress that accounts for the runaway social selection in the human lineage and thus for important aspects of human uniqueness in cognitive ability.

This relative increase in the importance of parasites versus other mortality and morbidity factors for a species may mark a novel event in the entire evolution of the diversity of life. If we are correct about collectivism–individualism and related values being causally related to pathogen stress, then human ecological dominance (relative freedom from predators and the physical environment, but of increased importance of parasites) would lead to the same social features considered important in Alexander's (1990) hypothesis: intensive and extensive nepotism, male kin and nonkin coalitions, male philopatry, pervasive intergroup and intragroup conflict, raiding, war, complex reciprocity, and the like. Thus, much of human uniqueness may arise from our ecological dominance generating runaway social selection in concert with the avoidance and management of parasites. Flinn et al. (2005) provided evidence that the parts of the brain evolutionarily enlarged in humans, and which account for the brain-size differences between chimpanzees and humans, function in human social life. We agree with Flinn et al.'s (2005) emphasis on coevolutionary, antagonistic social races in the human lineage. We add that these races were fueled by variable parasite stresses that gave rise to adaptive variation through time and space in the use of collectivist and individualist tactics in defining and delimiting social networks and in-group and out-group interactions. It is the variation in parasite stress in a locale that gives rise to the social complexity and difficulty involved in adaptive use of these tactics in social interactions. In essence, the unique cognitive abilities of *Homo sapiens* may reflect, in large part, mental and associated behavioral adaptations that evolved in the context of the relatively greater selection from pathogen stresses in humans compared to other species, and it was these abilities, we argue, that provided adaptive solutions to the complex and contingently optimal social decisions arising from variable pathogen stresses.

Thus, we disagree, in part, with a major, prior theory for the evolution of unique features of human cognition. We agree that social coevolutionary races were salient; therefore, what accounts for the comparatively lofty mental capabilities of humans is past selection in dealing with conspecifics. Parasite-stress variation, however, is central to adopting adaptive behavioral contingencies for social behavior. We propose that the context of spatially and temporally variable parasite stress and associated in-group and out-group behavioral solutions to the variation was a chief adaptive problem that ultimately created human abilities in intensive and extensive nepotism, complex reciprocity involving distant relatives and unrelated others, and the psychology involved in intergroup interactions, antagonisms, and alliances.

Anders Møller, whose research contributions extend across so many areas of biology, including parasitology, proposed recently a hypothesis that is relevant to our hypothesis of the relatively greater impact of parasites as selection agents in human evolution than in other lineages. After summarizing numerous research findings that predators differentially kill and eat parasitized prey, compared to their depredation of healthy prey items, he proposed that as predator abundance declines, parasites evolve higher virulence (pathogenicity) (Møller 2008). Møller's reasoning was that, under elevated natural selection on a host species from predators, the greater predation on parasitized prey would select for reduced pathogenicity in the parasites. This is because parasites with reduced virulence would be less likely to die before transmission to a new host as a result of their host being depredated. If this is correct, then the relative emancipation of human evolutionary ancestors from predators resulting from their weaponry and other technology contributed to the relatively greater impact of parasite adversity in human evolution through increased parasite virulence (A. Møller, personal communication, August 5, 2010).

Hence, the great reduction in impact of many typical sources of Darwinian selection, including predation, on human ancestral populations makes the human lineage unique, compared to other lineages of living things, in terms of the elevated impact of parasites as agents of selection. Also, the decline in the relative importance of predation in human ancestral generations specifically may have made the parasites of humans more virulent, compared to parasites of other species with higher rates of predation. These factors combined lend theoretical support to our hypothesis that parasites were fundamentally salient as causes in evolutionary history of human sociality and may account for many of the unique aspects of human social life, intellectual capacity, and behavioral immunity.

Parasites, besides perhaps providing the Darwinian selection that created human-unique mental capacity, appear to be important in another aspect of human mental life. The large brains and human-specific lofty mental capabilities of *Homo sapiens* are very energetically costly to produce and maintain during their ontogeny. These costs, we have argued, result in a negative ontogenetic interaction between cognitive ability and classical immunity. Parasitic infections result in a greater allocation to the classical immune system, and hence limit the energy available for cognitive development. These infections also reduce energy availability more directly. In research with Chris Eppig, we have shown that, both cross-nationally and across the states of the USA, parasite stress is strongly, negatively related to cognitive ability (IQ) (Eppig et al. 2010, 2011; Chap. 11). This, we argue, helps to explain why conservatives have lower cognitive ability than liberals and why collectivist cultures, relative to individualist ones, are more economically depressed (see Chaps. 4 and 11). Although parasites ultimately made humankind's large brains, these brains are susceptible to indirect degradation by parasites on a proximate developmental timescale.

## 5.19 Patriotism

We compiled a cross-national measure we call *Strength of National Ties* from the World Values Survey. (See Fincher and Thornhill 2012 for its composition and data within the supplementary materials.) This measure taps into the value placed on an individual for adopting the customs, being born in, and having ancestors from a particular country in order to make a claim of citizenship in that country. The higher the *Strength of National Ties*, the more importance placed on a parochial background and knowledge of local customs for granting citizenship. As we reported in Fincher and Thornhill (2012), the *Strength of National Ties* was correlated positively and strongly with the *Strength of Family Ties* ($r=0.74$, $p<0.0001$, $n=30$

countries) and *Combined Parasite Stress* ($r=0.71$, $p<0.0001$, $n=40$ countries). This relationship could be studied more thoroughly to explore the role of parasite adversity and associated collectivism in nationalism and other similar patriotic cultural features.

## 5.20 Xenophobia

We have stressed that xenophobic attitudes cross-nationally are correlated positively with parasite adversity (as expected from the parasite-stress theory of sociality). Here we mention briefly analyses, reported first in Fincher and Thornhill (2012), based on this relationship that used *Combined Parasite Stress* and *Strength of Family Ties*. Participants in the World Values Survey were asked about different types of people that they would not want as a neighbor. The proportion of those who said they did not want to live next to someone of a different race was associated positively with *Combined Parasite Stress* ($r=0.35$, $p=0.0009$, $n=88$ countries; see also Schaller and Murray 2011) and *Strength of Family Ties* ($r=0.45$, $p<0.0001$, $n=71$ countries). Similar questions are posed in the World Values Survey with similar relationships to *Combined Parasite Stress* and *Strength of Family Ties* (e.g., *Combined Parasite Stress* in relation to the proportion not wanting to live next to someone who speaks a different language: $r=0.42$, $p=0.004$, $n=44$ countries).

Throughout the book, we treat ethnocentrism and xenophobia as though they are positively associated. However, xenophobia and ethnocentrism can arise from separate causes, leading to cases where they may be uncorrelated or potentially even negatively correlated (Brewer 1999; Cashdan 2001b). Cashdan (2001b) demonstrated that ethnocentrism was high in traditional societies that experienced catastrophic food shortage, while xenophobia was high where the threat of intergroup violence was great. Further extension of the parasite-stress model of sociality can provide a basis for making more refined predictions about the patterns of xenophobia and ethnocentrism. For example, in a given area, zoonotic diseases may generate high morbidity and mortality, but nonzoonotics low morbidity and mortality; in this setting, ethnocentrism is predicted to be high, but xenophobia low, because zoonotic infections are not transmitted between human hosts. We discuss these issues in more detail in Chap. 14.

## 5.21 Moral Foundations Theory

We conclude this chapter, and before summarizing it, with a discussion of how we interpret the connection between the parasite-stress theory of values and the recently proposed Moral Foundations Theory of Haidt and Graham (2007). Where a person stands in terms of moral foundations is measured by questionnaires. We mentioned in the previous chapter that liberals moralize in relation to an "individualizing" moral foundation that prioritizes individual autonomy and success, whereas conservatives moralize more from a "binding" foundation of morals that prioritizes the well-being, loyalty, and integrity of the collective. There, too, we tied these two factors of moral foundations theory to individualism and collectivism. Van Leeuwen et al. (2012) have connected aspects of moral foundations theory to the parasite-stress theory in showing across a large sample of nations that parasite stress is related positively and significantly to the three subfactors of the binding moral foundation (specifically, endorsement of in-group loyalty, respect of authority, and purity and holiness). This is as expected by the parasite-stress theory, because collectivism includes high value given to in-group embeddedness and loyalty, authoritarianism, and religiosity. No significant relationship was found by Van Leeuwen et al. (2012) between parasite stress and either of the two individualizing subfactors: harm or care, fairness/reciprocity. These two subfactors address how people ought (morally) to treat other people. We suggest that there is a methodological problem with the measures of both of the individualizing subfactors. From the parasite-stress-theory perspective, it is essential to distinguish between in-group and out-group harm or care or fairness/reciprocity. Hence, the binded people (collectivists) primarily care for in-group members, whereas individualists show more care toward a broad network of people. If the distinction between in-group and out-group altruism were included in a future questionnaire, we predict that high parasite stress would be associated primarily with in-group care, fairness and reciprocity, and low parasite stress predominantly with out-group care, fairness, and reciprocity.

## 5.22 Summary

The cross-national cultural variable collectivism–individualism is a major dimension for describing cross-cultural differences. This value dimension has been studied in some detail and measured in multiple highly correlated ways. Fincher et al. (2008) hypothesized that regional differences in parasite adversity cause this variable, with the following reasoning. Individualism confers benefits upon individuals such as independent thinking, openness to new and nontraditional ideas and ways, and willingness to interact with a diversity of people. These same traits, however, have the cost of an enhanced likelihood of contracting infectious disease. Thus, the lower the parasite stress, the greater the benefits of individualism relative to its costs. In contrast, the behaviors that define collectivism, such as ethnocentrism, xenophobia, and adherence to traditional ideas and ways, function in antipathogen defense, and thus are optimal under conditions of high parasite stress.

Across multiple measures of collectivism–individualism, Fincher et al. (2008) found that worldwide variation in parasite stress robustly predicted cross-national values of collectivism–individualism. Within regions with high severity of infectious diseases, human cultures are characterized by high collectivism whereas in regions of low parasite stress cultures are highly individualistic. This pattern remained significant when controlling statistically for potential confounding variables. Moreover, the pattern was strong when broad cultural regions (rather than individual countries) were used in analysis.

Subsequently, Thornhill et al. (2010) computed separate indices assessing the richness (number) of three functionally distinct categories of human parasitic diseases (human-specific, multihost, zoonotic), and examined the extent to which each index uniquely predicted cross-national differences in collectivism–individualism. The parasite-stress theory of values proposes that infectious disease transmissible among humans (human-specific and multihost parasites) will be more important in predicting collectivism–individualism than those that humans can contract only from nonhuman animals (zoonotics). As predicted, both human-specific and multihost parasite richness predicted uniquely cross-national differences in collectivist–individualist values. Zoonotic parasite richness contributed little, if at all, to cross-national relationships between parasite adversity and these values. Thus, worldwide variation in these values predicted by parasite adversity appears to be attributable almost entirely to the prevalence of nonzoonotic diseases.

These cross-national results for richness of diseases in the transmission categories in relation to collectivism–individualism were repeated with parasite-severity measures (measures of number of infectious-disease cases). Nonzoonotic severity related more strongly to collectivism–individualism than did zoonotic severity. Also the measures of parasite richness were correlated nearly perfectly with measures of parasite severity.

Also, across states of the USA and societies in the Standard Cross-Cultural Sample, collectivism correlated positively with parasite stress. Furthermore, as with the cross-national results, collectivism across the USA states correlated more strongly with nonzoonotic than zoonotic human diseases.

The strength of family ties, a measure of collectivism we compiled, was correlated positively with parasite stress. This was found in cross-national analysis and analysis across states of the USA. And, as predicted, the cross-national analysis and the analysis across USA states showed that the strength of family ties was correlated more strongly with nonzoonotic infectious diseases than with zoonotic infectious diseases.

The potential confounds examined did not change these conclusions. Also, the basic relationships of values and parasite stress are robust at regional levels both cross-nationally and across the USA.

We hypothesized that reduced dispersal (high philopatry) is a defense against contact with novel parasites in out-groups and their habitats, and that reduced dispersal is the optimal preference under high parasite stress. Evidence in support of this hypothesis is seen in movement patterns across nations, states of the USA, and indigenous societies. Moreover, the parasite-stress theory of sociality provides an explanation for a general rule of ecological science, Rapoport's rule: the positive relationship between latitude and species' range size.

The social organization of animal species varies along a cooperative-breeding continuum. Cultures with high degrees of family ties have high degrees of cooperative breeding. Evidence is provided that parasite stress accounts for this variation across human social systems. We propose that the parasite-stress theory of sociality offers a general theory of family life across humans as well as nonhuman animal taxa.

We propose that a major context for the evolution of reciprocity was in gaining benefits from out-group interactions during periods of relatively low disease threat. Thus, the parasite-stress theory of values offers a novel perspective to explain the evolution of reciprocity. Accordingly, natural selection is expected to have designed our reciprocity activities to be conditionally sensitive to risk of contagion in the local ecological setting. Under low contagion risk, reciprocity is more beneficial than under high risk.

We suggest that, in human evolutionary history, as the physical environmental problems and predators became less important as mortality agents, parasites became more important as agents of natural selection. The human lineage may be unique among branches in the Tree of Life in the relatively great importance of parasite stress as a source of Darwinian selection. Also, the reduction in predator-based natural selection in human evolutionary history may have selected for relatively high virulence in human parasites. Moreover, we suggest that it was local change in parasite adversity, and challenges for adoption of values to cope with this change, that accounts for the runaway social selection in the human lineage that produced important aspects of human uniqueness in cognitive ability.

The parasite-stress theory of values suggests useful new research directions for the study of the demographic transition, patriotism, xenophobia, ethnocentrism, and moral foundations theory. The demographic transition's association with increases in non-kin-to-kin ratio in people's social networks may simply reflect individualism and corresponding reduction of parasite adversity. Patriotism may be a manifestation of collectivist values and concomitant high parasite stress. Xenophobia and ethnocentrism often covary positively, but there are circumstances identified by the parasite-stress theory in which they should not do so. So-called moral foundations theory could be improved by distinguishing the in-group and out-group components of altruism and how each relates to parasite stress.

Certain misunderstandings of correlational findings generated by comparative methodology are discussed. It is explained that all scientific results are correlational, including those from experiments. Many of the results supporting the parasite-stress theory of sociality are from application of the comparative method and associated statistical correlation with statistical controls; other research supporting it uses experimental and/or observational methods. The scientific value of any finding depends upon the control of confounders, not the type of method itself. Thus, the method of testing is always, in itself, irrelevant. When thorough controls are in place, correlation documents causation whether the correlation arises from experimentation, comparative methodology, or observational analysis.

We discuss ecological correlations and the ecological fallacy and how they relate to testing of the parasite stress theory of sociality.

## 5.23 Appendix 1

Cross-national scores of human infectious-disease richness (number of diseases) by type of transmission to humans. The type "Nonzoonotic" is the sum of "Human-specific" and "Multihost"

| Country | Human-specific | Multihost | Nonzoonotic | Zoonotic |
|---|---|---|---|---|
| Afghanistan | 105 | 23 | 128 | 53 |
| Albania | 100 | 21 | 121 | 53 |
| Algeria | 104 | 22 | 126 | 55 |
| American Samoa | 101 | 21 | 122 | 42 |
| Andorra | 100 | 20 | 120 | 44 |
| Angola | 107 | 25 | 132 | 61 |
| Anguilla | 100 | 21 | 121 | 40 |
| Antigua and Barbuda | 101 | 23 | 124 | 41 |
| Argentina | 102 | 25 | 127 | 68 |
| Armenia | 102 | 21 | 123 | 59 |
| Aruba | 100 | 21 | 121 | 41 |
| Australia | 101 | 22 | 123 | 64 |
| Austria | 99 | 22 | 121 | 52 |
| Azerbaijan | 100 | 24 | 124 | 57 |
| Azores | 98 | 21 | 119 | 41 |
| Bahamas, The | 99 | 22 | 121 | 41 |
| Bahrain | 100 | 21 | 121 | 42 |
| Bangladesh | 103 | 27 | 130 | 52 |
| Barbados | 100 | 22 | 122 | 40 |
| Belarus | 99 | 23 | 122 | 60 |
| Belgium | 99 | 22 | 121 | 51 |
| Belize | 101 | 25 | 126 | 53 |
| Benin | 107 | 27 | 134 | 53 |
| Bermuda | 99 | 22 | 121 | 39 |
| Bhutan | 101 | 23 | 124 | 51 |
| Bolivia | 104 | 29 | 133 | 59 |
| Bosnia and Herzegovina | 99 | 24 | 123 | 54 |
| Botswana | 104 | 22 | 126 | 53 |
| Brazil | 107 | 30 | 137 | 86 |
| British Virgin Islands | 101 | 21 | 122 | 40 |
| Brunei | 102 | 21 | 123 | 46 |
| Bulgaria | 100 | 22 | 122 | 60 |
| Burkina Faso | 108 | 27 | 135 | 58 |
| Burundi | 106 | 24 | 130 | 56 |
| Cambodia | 105 | 26 | 131 | 58 |
| Cameroon | 109 | 27 | 136 | 63 |
| Canada | 100 | 21 | 121 | 70 |
| Canary Islands | 100 | 21 | 121 | 48 |
| Cape Verde | 105 | 21 | 126 | 46 |
| Cayman Islands | 100 | 21 | 121 | 41 |

(continued)

(continued)

| Country | Human-specific | Multihost | Nonzoonotic | Zoonotic |
|---|---|---|---|---|
| Central African Republic | 109 | 26 | 135 | 71 |
| Chad | 110 | 25 | 135 | 58 |
| Chile | 102 | 23 | 125 | 57 |
| China | 105 | 27 | 132 | 87 |
| Christmas Island | 100 | 20 | 120 | 38 |
| Colombia | 107 | 29 | 136 | 71 |
| Comoros | 104 | 21 | 125 | 50 |
| Congo, Democratic Republic of the | 108 | 25 | 133 | 71 |
| Congo, Republic of the | 107 | 24 | 131 | 69 |
| Cook Islands | 101 | 22 | 123 | 41 |
| Costa Rica | 103 | 27 | 130 | 58 |
| Cote d'Ivoire | 110 | 28 | 138 | 63 |
| Croatia | 100 | 21 | 121 | 60 |
| Cuba | 101 | 21 | 122 | 52 |
| Cyprus | 100 | 21 | 121 | 45 |
| Czech Republic | 100 | 21 | 121 | 62 |
| Democratic People's Republic of Korea (North) | 101 | 28 | 129 | 59 |
| Denmark | 99 | 20 | 119 | 51 |
| Djibouti | 105 | 23 | 128 | 52 |
| Dominica | 101 | 23 | 124 | 43 |
| Dominican Republic | 103 | 23 | 126 | 51 |
| East Timor | 107 | 27 | 134 | 75 |
| Ecuador | 105 | 30 | 135 | 63 |
| Egypt | 104 | 23 | 127 | 69 |
| El Salvador | 102 | 24 | 126 | 56 |
| Eritrea | 107 | 25 | 132 | 63 |
| Estonia | 100 | 21 | 121 | 56 |
| Ethiopia | 108 | 26 | 134 | 65 |
| Falkland Islands | 100 | 20 | 120 | 43 |
| Fiji | 101 | 22 | 123 | 44 |
| Finland | 100 | 20 | 120 | 53 |
| France | 100 | 22 | 122 | 66 |
| French Guiana | 103 | 27 | 130 | 54 |
| French Polynesia | 101 | 21 | 122 | 44 |
| Gabon | 107 | 24 | 131 | 67 |
| Gambia, The | 105 | 26 | 131 | 60 |
| Georgia | 101 | 21 | 122 | 56 |
| Germany | 100 | 20 | 120 | 60 |
| Ghana | 109 | 26 | 135 | 61 |
| Gibraltar | 100 | 21 | 121 | 44 |
| Greece | 100 | 22 | 122 | 59 |
| Greenland | 100 | 20 | 120 | 42 |
| Grenada | 100 | 21 | 121 | 42 |
| Guadeloupe | 102 | 21 | 123 | 49 |
| Guam | 100 | 21 | 121 | 44 |

(continued)

(continued)

| Country | Human-specific | Multihost | Nonzoonotic | Zoonotic |
|---|---|---|---|---|
| Guatemala | 104 | 28 | 132 | 58 |
| Guinea | 107 | 27 | 134 | 57 |
| Guinea-Bissau | 108 | 24 | 132 | 51 |
| Guyana | 104 | 26 | 130 | 63 |
| Haiti | 104 | 22 | 126 | 49 |
| Honduras | 103 | 26 | 129 | 56 |
| Hong Kong | 103 | 22 | 125 | 50 |
| Hungary | 100 | 21 | 121 | 58 |
| Iceland | 100 | 20 | 120 | 42 |
| India | 105 | 27 | 132 | 86 |
| Indonesia | 107 | 27 | 134 | 79 |
| Iran | 103 | 23 | 126 | 68 |
| Iraq | 104 | 22 | 126 | 55 |
| Ireland | 100 | 20 | 120 | 48 |
| Israel | 100 | 22 | 122 | 55 |
| Italy | 100 | 21 | 121 | 64 |
| Jamaica | 100 | 21 | 121 | 49 |
| Japan | 101 | 24 | 125 | 69 |
| Jordan | 102 | 22 | 124 | 51 |
| Kazakhstan | 100 | 21 | 121 | 64 |
| Kenya | 107 | 27 | 134 | 72 |
| Kiribati | 101 | 22 | 123 | 40 |
| Kuwait | 100 | 22 | 122 | 48 |
| Kyrgyzstan | 101 | 21 | 122 | 60 |
| Laos | 104 | 25 | 129 | 58 |
| Latvia | 100 | 21 | 121 | 53 |
| Lebanon | 100 | 22 | 122 | 50 |
| Lesotho | 103 | 21 | 124 | 51 |
| Liberia | 106 | 25 | 131 | 60 |
| Libya | 102 | 22 | 124 | 54 |
| Liechtenstein | 100 | 20 | 120 | 46 |
| Lithuania | 100 | 21 | 121 | 53 |
| Luxembourg | 100 | 20 | 120 | 45 |
| Macau | 100 | 23 | 123 | 42 |
| Macedonia | 100 | 21 | 121 | 58 |
| Madagascar | 106 | 22 | 128 | 62 |
| Malawi | 107 | 23 | 130 | 57 |
| Malaysia | 106 | 27 | 133 | 72 |
| Maldives | 104 | 21 | 125 | 45 |
| Mali | 109 | 25 | 134 | 61 |
| Malta | 100 | 21 | 121 | 45 |
| Marshall Islands | 101 | 21 | 122 | 43 |
| Martinique | 101 | 22 | 123 | 47 |
| Mauritania | 107 | 23 | 130 | 55 |
| Mauritius | 103 | 21 | 124 | 47 |

(continued)

(continued)

| Country | Human-specific | Multihost | Nonzoonotic | Zoonotic |
|---|---|---|---|---|
| Mexico | 104 | 28 | 132 | 74 |
| Micronesia, Federated States of | 102 | 22 | 124 | 43 |
| Moldova | 100 | 21 | 121 | 56 |
| Monaco | 100 | 20 | 120 | 43 |
| Mongolia | 101 | 24 | 125 | 53 |
| Montserrat | 100 | 22 | 122 | 40 |
| Morocco | 103 | 24 | 127 | 53 |
| Mozambique | 107 | 27 | 134 | 57 |
| Myanmar | 104 | 30 | 134 | 59 |
| Namibia | 104 | 25 | 129 | 54 |
| Nauru | 100 | 22 | 122 | 39 |
| Nepal | 103 | 26 | 129 | 51 |
| Netherlands | 99 | 21 | 120 | 54 |
| Netherlands Antilles | 99 | 22 | 121 | 41 |
| New Caledonia | 100 | 22 | 122 | 45 |
| New Zealand | 99 | 21 | 120 | 45 |
| Nicaragua | 102 | 28 | 130 | 51 |
| Niger | 108 | 27 | 135 | 57 |
| Nigeria | 110 | 31 | 141 | 71 |
| Niue | 100 | 22 | 122 | 38 |
| Norfolk Island | 99 | 21 | 120 | 38 |
| Northern Mariana Islands | 100 | 22 | 122 | 42 |
| Norway | 99 | 21 | 120 | 52 |
| Oman | 103 | 23 | 126 | 48 |
| Pakistan | 104 | 29 | 133 | 58 |
| Palau | 100 | 22 | 122 | 42 |
| Panama | 104 | 31 | 135 | 61 |
| Papua New Guinea | 103 | 25 | 128 | 54 |
| Paraguay | 100 | 29 | 129 | 53 |
| Peru | 104 | 32 | 136 | 64 |
| Philippines | 104 | 27 | 131 | 65 |
| Pitcairn Islands | 99 | 21 | 120 | 38 |
| Poland | 99 | 22 | 121 | 56 |
| Portugal | 99 | 24 | 123 | 57 |
| Puerto Rico | 101 | 22 | 123 | 50 |
| Qatar | 101 | 23 | 124 | 43 |
| Republic of Korea (South) | 103 | 28 | 131 | 63 |
| Reunion | 101 | 22 | 123 | 47 |
| Romania | 100 | 24 | 124 | 57 |
| Russia | 103 | 27 | 130 | 81 |
| Rwanda | 106 | 27 | 133 | 57 |
| Saint Helena | 99 | 21 | 120 | 40 |
| Saint Kitts and Nevis | 102 | 22 | 124 | 41 |
| Saint Lucia | 102 | 23 | 125 | 41 |
| Saint Vincent and Grenadines | 101 | 23 | 124 | 39 |

(continued)

(continued)

| Country | Human-specific | Multihost | Nonzoonotic | Zoonotic |
|---|---|---|---|---|
| Samoa | 100 | 22 | 122 | 40 |
| San Marino | 99 | 21 | 120 | 41 |
| Sao Tome and Principe | 104 | 22 | 126 | 49 |
| Saudi Arabia | 105 | 24 | 129 | 55 |
| Scotland | 99 | 22 | 121 | 53 |
| Senegal | 108 | 29 | 137 | 62 |
| Seychelles | 100 | 22 | 122 | 42 |
| Sierra Leone | 107 | 26 | 133 | 55 |
| Singapore | 101 | 24 | 125 | 50 |
| Slovakia | 99 | 22 | 121 | 58 |
| Slovenia | 99 | 24 | 123 | 58 |
| Solomon Islands | 101 | 24 | 125 | 42 |
| Somalia | 105 | 27 | 132 | 56 |
| South Africa | 103 | 25 | 128 | 70 |
| Spain | 99 | 24 | 123 | 59 |
| Sri Lanka | 103 | 26 | 129 | 57 |
| Sudan | 109 | 30 | 139 | 68 |
| Suriname | 105 | 28 | 133 | 57 |
| Swaziland | 103 | 22 | 125 | 50 |
| Sweden | 99 | 21 | 120 | 51 |
| Switzerland | 99 | 21 | 120 | 55 |
| Syria | 100 | 23 | 123 | 47 |
| Taiwan | 100 | 27 | 127 | 61 |
| Tajikistan | 101 | 23 | 124 | 56 |
| Tanzania | 107 | 27 | 134 | 64 |
| Thailand | 104 | 32 | 136 | 73 |
| Togo | 107 | 27 | 134 | 53 |
| Tokelau | 100 | 22 | 122 | 38 |
| Tonga | 100 | 22 | 122 | 41 |
| Trinidad and Tobago | 102 | 25 | 127 | 51 |
| Tunisia | 102 | 25 | 127 | 54 |
| Turkey | 102 | 24 | 126 | 62 |
| Turkmenistan | 100 | 24 | 124 | 55 |
| Turks and Caicos Islands | 100 | 22 | 122 | 39 |
| Tuvalu | 100 | 23 | 123 | 39 |
| Uganda | 109 | 29 | 138 | 72 |
| Ukraine | 100 | 23 | 123 | 60 |
| United Arab Emirates | 101 | 22 | 123 | 47 |
| United Kingdom | 99 | 22 | 121 | 55 |
| United States | 103 | 26 | 129 | 86 |
| Uruguay | 99 | 22 | 121 | 57 |
| Uzbekistan | 101 | 24 | 125 | 55 |
| Vanuatu | 101 | 23 | 124 | 41 |
| Venezuela | 107 | 30 | 137 | 65 |
| Vietnam | 103 | 30 | 133 | 64 |

(continued)

(continued)

| Country | Human-specific | Multihost | Nonzoonotic | Zoonotic |
|---|---|---|---|---|
| Virgin Islands, US | 100 | 23 | 123 | 40 |
| Wake Island | 99 | 21 | 120 | 38 |
| Wallis and Futuna Islands | 100 | 22 | 122 | 40 |
| Western Sahara | 99 | 21 | 120 | 43 |
| Yemen | 104 | 25 | 129 | 51 |
| Zambia | 106 | 25 | 131 | 58 |
| Zimbabwe | 105 | 23 | 128 | 65 |

## 5.24 Appendix 2

Notifiable human infectious diseases within the United States tracked by the Centers for Disease Control and Prevention (CDC) from the years 1993 to 2007 classified by transmission type (Human-specific, Multihost, or Zoonotic) or if the disease was not included in this classification (labeled here as "Removed"). There is some redundancy in disease names because in some cases a disease was variably named by the CDC across years

| Disease name | Transmission type |
|---|---|
| AIDS | Human-specific |
| Chancroid | Human-specific |
| Chlamydia | Human-specific |
| Cholera | Human-specific |
| Diphtheria | Human-specific |
| Gonorrhea | Human-specific |
| Granuloma inguinale | Human-specific |
| *Haemophilus influenzae* | Human-specific |
| *Haemophilus influenzae*—Age <5 years—Nonserotype b | Human-specific |
| *Haemophilus influenzae*—Age <5 years—Serotype b | Human-specific |
| *Haemophilus influenzae*—Age <5 years—Unknown serotype | Human-specific |
| *Haemophilus influenzae*—All ages all serotypes | Human-specific |
| *Haemophilus influenzae*, invasive disease (age <5 years) unknown serotype | Human-specific |
| *Haemophilus influenzae*, invasive disease (all ages, serotypes) | Human-specific |
| *Haemophilus influenzae*, invasive disease (age <5 years) Nonserotype b | Human-specific |
| *Haemophilus influenzae*, invasive disease (age <5 years) Serotype b | Human-specific |
| Hepatitis B | Human-specific |
| Hepatitis C/Non-A, Non-B | Human-specific |
| Hepatitis Non-A, Non-B | Human-specific |
| Hepatitis unsp. | Human-specific |
| *Lymphogranuloma venereum* | Human-specific |
| Malaria | Human-specific |
| Measles—Imported | Human-specific |

(continued)

(continued)

| Disease name | Transmission type |
|---|---|
| Measles—Indigenous | Human-specific |
| Meningococcal disease (all serogroups) | Human-specific |
| Meningococcal disease (other serogroup) | Human-specific |
| Meningococcal disease (Serogroup A, C, Y, and W-135) | Human-specific |
| Meningococcal disease (Serogroup B) | Human-specific |
| Meningococcal disease (Serogroup unknown) | Human-specific |
| Meningococcal disease, invasive—All serogroups | Human-specific |
| Meningococcal disease, invasive—Other serogroup | Human-specific |
| Meningococcal disease, invasive—Serogroup A, C, Y, and W-135 | Human-specific |
| Meningococcal disease, invasive—Serogroup B | Human-specific |
| Meningococcal disease, invasive—Serogroup unknown | Human-specific |
| Meningococcal infections | Human-specific |
| Mumps | Human-specific |
| Pertussis | Human-specific |
| Poliomyelitis, paralytic | Human-specific |
| Rubella | Human-specific |
| Rubella—Congenital syndrome | Human-specific |
| SARS-CoV[a] | Human-specific |
| *Streptococcus pneumoniae*, invasive disease, drug-resistant (age <5years) | Human-specific |
| Syphilis—All stages | Human-specific |
| Syphilis—Congenital (<1 year) | Human-specific |
| Syphilis—Primary and secondary | Human-specific |
| Typhoid fever | Human-specific |
| Varicella deaths | Human-specific |
| Cryptosporidiosis[b] | Multihost |
| *Escherichia coli* O157:H7 | Multihost |
| Hansen disease | Multihost |
| Hepatitis A | Multihost |
| Influenza-associated pediatric mortality | Multihost |
| Listeriosis | Multihost |
| Novel influenza A virus infections | Multihost |
| Shigatoxin-producing *E. coli* (STEC) | Multihost |
| Shigellosis | Multihost |
| Tuberculosis | Multihost |
| Yellow fever | Multihost |
| Anthrax | Zoonotic |
| Botulism—Infant | Zoonotic |
| Botulism—Foodborne | Zoonotic |
| Botulism—Other | Zoonotic |
| Brucellosis | Zoonotic |
| California serogroup virus disease (neuro-invasive) | Zoonotic |
| Domestic arboviral diseases—California serogroup—Neuro-invasive | Zoonotic |
| Domestic arboviral diseases—California serogroup—Non-neuro-invasive | Zoonotic |
| Domestic arboviral diseases—Eastern Equine—Neuro-invasive | Zoonotic |
| Domestic arboviral diseases—Eastern Equine—Non-neuro-invasive | Zoonotic |

(continued)

(continued)

| Disease name | Transmission type |
|---|---|
| Domestic arboviral diseases—Powassan—Neuro-invasive | Zoonotic |
| Domestic arboviral diseases—Powassan—Non-neuro-invasive | Zoonotic |
| Domestic arboviral diseases—St. Louis—Neuro-invasive | Zoonotic |
| Domestic arboviral diseases—St. Louis—Non-neuro-invasive | Zoonotic |
| Domestic arboviral diseases—West Nile—Neuro-invasive | Zoonotic |
| Domestic arboviral diseases—West Nile—Non-neuro-invasive | Zoonotic |
| Eastern equine encephalitis (neuro-invasive) | Zoonotic |
| Eastern equine encephalitis (non-neuro-invasive) | Zoonotic |
| Encephalitis—California serogroup viral | Zoonotic |
| Encephalitis—Eastern Equine | Zoonotic |
| Encephalitis—Postinfectious | Zoonotic |
| Encephalitis—Powassan | Zoonotic |
| Encephalitis—Primary infections | Zoonotic |
| Encephalitis—St. Louis | Zoonotic |
| Encephalitis—West Nile | Zoonotic |
| Leptospirosis | Zoonotic |
| Lyme disease | Zoonotic |
| Murine typhus fever | Zoonotic |
| Plague | Zoonotic |
| Powassan virus disease (neuro-invasive) | Zoonotic |
| Powassan virus disease (non-neuro-invasive) | Zoonotic |
| Psittacosis | Zoonotic |
| Rabies—Animal | Zoonotic |
| Rabies—Human | Zoonotic |
| Rocky Mountain spotted fever (RMSF) | Zoonotic |
| Salmonellosis | Zoonotic |
| St. Louis encephalitis (neuro-invasive) | Zoonotic |
| St. Louis encephalitis (non-neuro-invasive) | Zoonotic |
| Trichinosis | Zoonotic |
| Tularemia | Zoonotic |
| West Nile virus disease (neuro-invasive) | Zoonotic |
| West Nile virus disease (non-neuro-invasive) | Zoonotic |
| Legionellosis | Removed |
| Tetanus | Removed |
| Toxic-shock syndrome | Removed |

[a]SARS-CoV was classified as multihost in Fincher and Thornhill (2012)
[b]Cryptosporidiosis was classified as zoonotic in Fincher and Thornhill (2012)

## 5.25 Appendix 3

Standardized pathogen severity scores for USA states for different transmission types with or without the District of Columbia (DC). The transmission type "Nonzoonotic" is the sum of "Human-Specific" and "Multihost"

| State | Human-specific with DC | Multihost with DC | Nonzoonotic with DC | Zoonotic with DC | Human-specific without DC | Multihost without DC | Nonzoonotic without DC | Zoonotic without DC |
|---|---|---|---|---|---|---|---|---|
| Alabama | 0.90 | –0.10 | 0.79 | –0.19 | 1.40 | –0.08 | 1.33 | –0.20 |
| Alaska | 0.43 | 0.28 | 0.71 | –0.65 | 0.60 | 0.31 | 0.91 | –0.65 |
| Arizona | –0.15 | 1.61 | 1.46 | –0.15 | –0.10 | 1.68 | 1.58 | –0.16 |
| Arkansas | 0.24 | –0.01 | 0.23 | 0.86 | 0.51 | 0.02 | 0.53 | 0.85 |
| California | –0.11 | 0.88 | 0.77 | –0.43 | –0.05 | 0.94 | 0.90 | –0.44 |
| Colorado | –0.27 | –0.16 | –0.43 | –0.17 | –0.27 | –0.13 | –0.40 | –0.18 |
| Connecticut | –0.23 | –0.71 | –0.94 | 1.91 | –0.17 | –0.69 | –0.86 | 1.88 |
| Delaware | 0.55 | –0.07 | 0.48 | 0.88 | 0.90 | –0.03 | 0.87 | 0.86 |
| District of Columbia | 4.82 | 1.52 | 6.33 | –0.44 | | | | |
| Florida | 0.09 | 0.58 | 0.67 | 0.45 | 0.30 | 0.64 | 0.94 | 0.44 |
| Georgia | 1.01 | 0.76 | 1.77 | 0.66 | 1.56 | 0.82 | 2.38 | 0.65 |
| Hawaii | –0.16 | 0.36 | 0.21 | 0.88 | –0.15 | 0.40 | 0.25 | 0.87 |
| Idaho | –0.93 | –0.14 | –1.07 | –0.29 | –1.13 | –0.13 | –1.26 | –0.29 |
| Illinois | 0.46 | 0.09 | 0.56 | –0.43 | 0.76 | 0.13 | 0.89 | –0.43 |
| Indiana | –0.15 | –0.47 | –0.62 | –0.87 | –0.07 | –0.44 | –0.52 | –0.87 |
| Iowa | –0.63 | –0.12 | –0.75 | –0.43 | –0.75 | –0.09 | –0.84 | –0.43 |
| Kansas | –0.37 | –0.41 | –0.78 | –0.35 | –0.38 | –0.39 | –0.77 | –0.35 |
| Kentucky | –0.53 | –0.21 | –0.74 | –0.90 | –0.57 | –0.18 | –0.75 | –0.91 |
| Louisiana | 1.28 | 0.00 | 1.29 | 0.19 | 1.93 | 0.04 | 1.97 | 0.18 |
| Maine | –1.11 | –1.09 | –2.20 | –0.33 | –1.38 | –1.09 | –2.47 | –0.34 |
| Maryland | 0.59 | –0.13 | 0.46 | 0.49 | 1.01 | –0.09 | 0.92 | 0.48 |
| Massachusetts | –0.64 | –0.31 | –0.95 | 0.87 | –0.73 | –0.28 | –1.01 | 0.86 |
| Michigan | 0.21 | –0.45 | –0.23 | –0.95 | 0.41 | –0.42 | –0.01 | –0.95 |
| Minnesota | –0.64 | –0.15 | –0.79 | –0.28 | –0.75 | –0.12 | –0.87 | –0.29 |
| Mississippi | 1.87 | 0.05 | 1.92 | 0.98 | 2.74 | 0.07 | 2.81 | 0.97 |
| Missouri | 0.29 | 0.41 | 0.70 | –0.52 | 0.53 | 0.44 | 0.98 | –0.53 |
| Montana | –0.83 | –0.61 | –1.44 | –0.37 | –1.04 | –0.60 | –1.64 | –0.38 |
| Nebraska | –0.47 | –0.05 | –0.53 | –0.44 | –0.54 | –0.03 | –0.57 | –0.45 |

| | | | | | | | | |
|---|---|---|---|---|---|---|---|---|
| Nevada | −0.20 | −0.03 | −0.23 | −0.66 | −0.16 | 0.00 | −0.16 | −0.67 |
| New Hampshire | −1.11 | −0.95 | −2.05 | −0.11 | −1.36 | −0.94 | −2.31 | −0.12 |
| New Jersey | −0.37 | −0.23 | −0.60 | 0.32 | −0.33 | −0.20 | −0.53 | 0.30 |
| New Mexico | 0.01 | 0.89 | 0.90 | −0.07 | 0.08 | 0.93 | 1.01 | −0.08 |
| New York | 0.24 | 0.33 | 0.57 | 0.64 | 0.50 | 0.38 | 0.88 | 0.63 |
| North Carolina | 0.63 | −0.17 | 0.46 | 0.24 | 1.06 | −0.14 | 0.91 | 0.23 |
| North Dakota | −0.86 | −0.47 | −1.33 | 0.12 | −1.09 | −0.46 | −1.54 | 0.12 |
| Ohio | 0.27 | −0.45 | −0.18 | −0.64 | 0.48 | −0.44 | 0.04 | −0.64 |
| Oklahoma | 0.06 | 1.44 | 1.50 | −0.29 | 0.21 | 1.53 | 1.74 | −0.29 |
| Oregon | −0.71 | 0.15 | −0.56 | −0.83 | −0.85 | 0.18 | −0.67 | −0.83 |
| Pennsylvania | −0.19 | −0.59 | −0.78 | 0.17 | −0.14 | −0.56 | −0.70 | 0.16 |
| Rhode Island | −0.41 | −0.50 | −0.91 | 1.10 | −0.45 | −0.48 | −0.93 | 1.08 |
| South Carolina | 1.12 | −0.18 | 0.94 | 0.50 | 1.71 | −0.14 | 1.57 | 0.49 |
| South Dakota | −0.58 | 0.90 | 0.32 | 0.87 | −0.72 | 0.92 | 0.21 | 0.86 |
| Tennessee | 0.62 | 0.21 | 0.83 | −0.59 | 1.03 | 0.25 | 1.28 | −0.59 |
| Texas | 0.26 | 1.27 | 1.53 | −0.36 | 0.50 | 1.35 | 1.85 | −0.37 |
| Utah | −0.87 | 0.34 | −0.53 | −0.25 | −1.08 | 0.36 | −0.72 | −0.26 |
| Vermont | −1.00 | −0.86 | −1.86 | 0.51 | −1.22 | −0.85 | −2.07 | 0.50 |
| Virginia | −0.04 | −0.40 | −0.44 | 0.17 | 0.09 | −0.37 | −0.28 | 0.16 |
| Washington | −0.52 | 0.00 | −0.52 | −0.69 | −0.60 | 0.03 | −0.57 | −0.69 |
| West Virginia | −0.95 | −1.18 | −2.14 | −0.60 | −1.16 | −1.19 | −2.34 | −0.60 |
| Wisconsin | −0.06 | −0.10 | −0.16 | 0.23 | 0.02 | −0.07 | −0.05 | 0.21 |
| Wyoming | −0.90 | −0.79 | −1.69 | 0.24 | −1.11 | −0.80 | −1.90 | 0.23 |

## References


Alesina, A., & Giuliano, P. (2010). The power of the family. *Journal of Economic Growth* 15: 93–125.

Alexander, R. D. (1978). Evolution, creation and biology teaching. *American Biology Teacher* 40: 91–107.

Alexander, R. D. (1987). *The Biology of Moral Systems*. Aldine de Gruyter, New York, NY.

Alexander, R. D. (1990). *How Did Humans Evolve? Reflections on the Uniquely Unique Species*. Special Publication No. 1. Museum of Zoology, The University of Michigan, Ann Arbor, MI.

Andersson, M. (1984). The evolution of eusociality. *Annual Review of Ecology and Systematics* 15: 165–189.

Arnold, K. E., & Owens, I. P. F. (1998). Cooperative breeding in birds: A comparative test of the life history hypothesis. *Proceedings of the Royal Society B* 265: 739–745.

Binford, L. R. (2001). *Constructing Frames of Reference: An Analytical Method for Archaeological Theory Building Using Hunter Gatherer and Environmental Data Sets*. University of California Press, Berkeley, CA.

Blumstein, D. T., & Møller, A. P. (2008). Is sociality associated with high longevity in North American birds? *Biology Letters* 4: 146–148.

Brewer, M. B. (1999). The psychology of prejudice: Ingroup love or outgroup hate? *Journal of Social Issues* 55: 429–444.

Brockmann, H. J. (1984). The evolution of social behaviour in insects. In *Behavioural Ecology: An Evolutionary Approach* (eds. J. R. Krebs & N. B. Davies), pp. 340–361. Sinauer Associates, Inc., Sunderland, MA.

Brown, J. L. (1987). *Helping and Communal Breeding in Birds: Ecology and Evolution*. Princeton University Press, Princeton, NJ.

Carney, D. R., Jost, J. T., Gosling, S. D. et al. (2008). The secret lives of liberals and conservatives: Personality profiles, interaction styles, and the things they leave behind. *Political Psychology* 29: 807–840.

Cashdan, E. (2001a). Ethnic diversity and its environmental determinants: Effects on climate, pathogens, and habitat diversity. *American Anthropology* 103: 968–991.

Cashdan, E. (2001b). Ethnocentrism and xenophobia: A cross-cultural study. *Current Anthropology* 42: 760–765.

Cashdan, E., & Steele, M. (2013). Pathogen prevalence, group bias, and collectivism in the standard cross-cultural sample. *Human Nature* 24: 59–75.

Census Bureau. (2005). http://www.census.gov/hhes/migration/data/acs/state-to-state.html.

Census Bureau. (2005). http://www.census.gov/popest/data/historical/2000s/vintage_2005/index.html.

Charnov, E. L. (1993). *Life History Invariants: Some Explorations of Symmetry in Evolutionary Ecology*. Oxford University Press, Oxford, U.K.

Clay, R., Terrizzi Jr., J. A., & Shook, N. J. (2012). Individual differences in the behavioral immune system and the emergence of cultural systems. *Journal of Social Psychology* 43: 174–184.

Cohen, D. (2001). Cultural variation: considerations and implications. *Psychological Bulletin* 127: 451–471.

Cukur, C. S., De Guzman, M. R. T., & Carlo, G. (2004). Religiosity, values, and horizontal and vertical individualism–collectivism: A study of Turkey, the United States, and the Philippines. *Journal of Social Psychology* 144: 613–634.

Dunbar, R., & Shultz, S. (2007). Evolution in the social brain. *Science* 317: 1344–1347.

Dunn, R. R., Davies, T. J., Harris, N. C. et al. (2010). Global drivers of human pathogen richness and prevalence. *Proceedings of the Royal Society B* 27: 2587–2595.

Ekman, J. (2006). Family living among birds. *Journal of Avian Biology* 37: 289–298.

Emlen, S. T. (1994). Benefits, constraints and the evolution of the family. *Trends in Ecology and Evolution* 9: 282–285.

Emlen, S. T. (1995). An evolutionary theory of the family. *Proceedings of the National Academy of Sciences USA* 92: 8092–8099.

Emlen, S. T. (1997). The evolutionary study of human family systems. *Social Science Information Sur Les Sciences Sociales* 36: 563–589.

Eppig, C., Fincher, C. L., & Thornhill, R. (2010). Parasite prevalence and the worldwide distribution of cognitive ability. *Proceedings of the Royal Society of London B* 277: 3801–3808.

Eppig, C., Fincher, C. L., & Thornhill, R. (2011). Parasite prevalence and the distribution of intelligence among the states of the USA. *Intelligence* 39: 155–160.

Faulkner, J., Schaller, M., Park, J. H. et al. (2004). Evolved disease-avoidance mechanisms and contemporary xenophobic attitudes. *Group Processes and Intergroup Relations* 7: 333–353.

Fincher, C. L., & Thornhill, R. (2008a). A parasite-driven wedge: Infectious diseases may explain language and other biodiversity. *Oikos* 117: 1289–1297.

Fincher, C. L., & Thornhill, R. (2008b). Assortative sociality, limited dispersal, infectious disease and the genesis of the global pattern of religion diversity. *Proceedings of the Royal Society of London, Biological Sciences* 275: 2587–2594.

Fincher, C. L., & Thornhill, R. (2012). Parasite-stress promotes in-group assortative sociality: The cases of strong family ties and heightened religiosity. *Behavioral and Brain Sciences* 35: 61–79.

Fincher, C. L., Thornhill, R., Murray, D. R. et al. (2008). Pathogen prevalence predicts human cross-cultural variability in individualism/collectivism. *Proceedings of the Royal Society of London Biological Sciences* 275: 1279–1285.

Flinn, M. V., Geary, D. C., & Ward, C. V. (2005). Ecological dominance, social competition, and coalitionary arms races: Why humans evolved extraordinary intelligence. *Evolution and Human Behavior* 26: 10–46.

Freedom House. (2008). http://www.freedomhouse.org.

Freeland, W. J. (1976). Pathogens and the evolution of primate sociality. *Biotropica* 8: 12–24.

Freeland, W. J. (1979). Primate social groups as biological islands. *Ecology* 60: 719–728.

Fumagalli, M., Sironi, M., Pozzoli, U. et al. (2011). Signatures of environmental genetic adaptation pinpoint pathogens as the main selective pressure through human evolution. *PLoS Genetics* 7: e1002355.

Gangestad, S. W., & Buss, D. M. (1993). Pathogen prevalence and human mate preference. *Ethology and Sociobiology* 14: 89–96.

Gangestad, S. W., Haselton, M. G., & Buss, D. M. (2006). Evolutionary foundations of cultural variation: Evoked culture and mate preferences. *Psychological Inquiry* 17: 75–95.

Gelfand, M. J., Bhawuk, D. P. S., Nishii, L. H. et al. (2004). Individualism and collectivism. In *Culture, Leadership, and Organizations: The GLOBE Study of 62 Societies* (eds. R. J. House, P. J. Hanges, M. Javidan et al.), pp. 437–512. Sage Publications, Thousand Oaks, CA.

Gordon, R.A. (1968). Issues in Multiple Regression. *The American Journal of Sociology* 73: 592–616.

Greger, M. (2007). The human/animal interface: Emergence and resurgence of zoonotic infectious diseases. *Critical Review of Microbiology* 33: 243–299.

Guernier, V., Hochberg, M. E., & Guégan, J. -F. (2004). Ecology drives the worldwide distribution of human diseases. *PLoS Biology* 2: 740–746.

Gupta, V., & Hanges, P. J. (2004). Regional and climate clustering of societal cultures. In *Culture, Leadership, and Organizations: The GLOBE Study of 62 Societies* (eds. R. J. House, P. J. Hanges, M. Javidan et al.), pp. 178–218. Sage Publications, Thousand Oaks, CA.

Gurven, M., Allen-Arave, W., Hill, K. et al. (2000). "It's a Wonderful Life": Signaling generosity among the Ache of Paraguay. *Evolution and Human Behavior* 21: 263–282.

Haidt, J., & Graham, J. (2007). When morality opposes justice: Conservatives have moral intuitions that liberals may not recognize. *Social Justice Research* 20: 98–116.

Heine, S. J. (2008). *Cultural psychology.* Norton, New York, NY.

Hill, K., & Hurtado, A. M. (2009). Cooperative breeding in South American hunter-gatherers. *Proceedings of the Royal Society B* 276: 3863–3870.

Hofstede, G. (2001). *Culture's Consequences: Comparing Values, Behaviors, Institutions, and Organizations Across Nations*, 2nd ed. Sage Publications, Thousand Oaks, CA.
House, R. J., Hanges, P. J., Javidan, M. et al., Eds. (2004). *Culture, Leadership, and Organizations: The GLOBE study of 62 Societies.* Sage Publications, Thousand Oaks, CA.
Inbar, Y., Pizarro, D. A., Iyer, R. et al. (2012). Disgust sensitivity, political conservatism, and voting. *Social Psychological and Personality Science* 5: 537–544.
Jetz, W., & Rubenstein, D. R. (2011). Environmental uncertainty and the global biogeography of cooperative breeding in birds. *Current Biology* 21: 72–78.
Jones, K. E., Patel, N. G., Levy, M. A. et al. (2008). Global trends in emerging infectious diseases. *Nature* 451: 990–993.
Kaplan, H. S., & Gangestad, S. W. (2005). Life history theory and evolutionary psychology. In *The Handbook of Evolutionary Psychology* (ed. D. M. Buss), pp.68–95. John Wiley and Sons, Inc., New York, NY.
Kashima, E. S., & Kashima, Y. (1998). Culture and language: The case of cultural dimensions and personal pronoun use. *Journal of Cross-Cultural Psychology* 29: 461–486.
Kitayama, S., Ishii, K., Tmada, T. et al. (2006). Voluntary settlement and the spirit of independence: Evidence from Japan's "Northern Frontier." *Journal of Personality and Social Psychology* 91: 369–384.
Loehle, C. (1995). Social barriers to pathogen transmission in wild animal populations. *Ecology* 76: 326–335.
Low, B. S. (1988). Pathogen stress and polygyny in humans. In *Human Reproductive Behavior: A Darwinian Perspective* (eds. L. Betzig, M. Borgerhoff Mulder, & P. Turke), pp. 115–127. Cambridge University Press, Cambridge, U.K.
Low, B. S. (1990). Marriage systems and pathogen stress in human societies. *American Zoologist* 30: 325–339.
Low, B. S. (1994). Pathogen severity cross-culturally. *World Cultures* 8: 24–34.
Low, B. S. (2000). *Why Sex Matters: A Darwinian Look at Human Behavior*. Princeton University Press, Princeton, NJ.
Maudlin, I., Eisler, M., & Welburn, S. (2009). The neglected zoonoses. *Philosophical Transactions of the Royal Society B* 364: 2777–2787.
Mayr, E. (1982). *The Growth of Biological Thought: Diversity, Evolution, and Inheritance*. Harvard University Press, Cambridge, MA.
McNeill, W. H. (1998). *Plagues and Peoples*. Anchor, Harpswell, ME.
Minkov, M. (2011). *Cultural Differences in a Globalizing World.* Emerald Group Publishing Ltd., Bingley, U.K.
McNeill, W. H. (1981). Migration patterns and infection in traditional societies. In *Changing Disease Patterns and Human Behavior* (eds. N. F. Stanley & R. A. Joske), pp. 27–36. Academic Press, Salt Lake, UT.
Møller, A. P. (1998). Evidence of larger impact of parasites on hosts in the tropics: Investment in immune function within and outside the tropics. *Oikos* 82: 265–270.
Møller, A. P. (2008). Interactions between interactions: Predator–prey, parasite–host, and mutualistic interactions. *Annals of New York Academy Sciences* 1133: 180–186.
Møller, A. P., Dufva, R., & Allander, K. (1993). Parasites and the evolution of host social behaviour. *Advances in the Study of Behavior* 22: 65–102.
Murdock, G. P. (1949). *Social Structure*. MacMillan, New York, NY.
Murdock, G. P., & White, D. R. (1969). Standard cross-cultural sample. *Ethnology* 8: 329–69.
Murray, D. R., & Schaller, M. (2010). Historical prevalence of infectious diseases within 230 geopolitical regions: A tool for investigating origins of culture. *Journal of Cross-Cultural Psychology* 41: 99–108.
Murray, D. R., Trudeau, R., & Schaller, M. (2011). On the origins of cultural differences in conformity: Four tests of the pathogen prevalence hypothesis. *Personality and Social Psychology Bulletin* 37: 318–329.

Newson, L., Postmes, T., Lea, S. E. G. et al. (2005). Why are modern families small? Toward an evolutionary and cultural explanation for the demographic transition. *Journal of Personality and Social Psychology* 9: 360–375.

Oishi, S., Schimmack, U., Diener, E. et al. (1998). The measurement of values and individualism–collectivism. *Personality and Social Psychology Bulletin* 24: 1177–1189.

Pearce-Duvet, J. M. C. (2006). The origin of human pathogens: evaluating the role of agriculture and domestic animals in the evolution of human disease. *Biological Reviews* 81: 369–382.

Pollet, T. V., Tybur, J. M., Frankenhuis, W. E., & Rickard, I. J. (in press). What can cross-cultural correlations teach us about human nature? *Human Nature*

Quinlan, R. J. (2007). Human parental effort and environmental risk. *Proceedings of the Royal Society B* 274: 121–125.

Robinson, W. S. (1950). Ecological correlations and the behavior of individuals. *American Sociological Review* 15: 351–357.

Schaller, M., & Murray, D. R. (2011). Infectious disease and the creation of culture. In *Advances in Culture and Psychology* (eds. M. Gelfand, C.-y. Chiu, & Y.-y. Hong), pp. 99–151. Oxford University Press, New York, NY.

Sherman, P. W., Lacey, E. A., Reeve, H. K. et al. (1995). The eusociality continuum. *Behavioral Ecology* 6: 102–108.

Sherman, P. W., & Billing, J. (1999). Darwinian gastronomy: Why we use spices. *BioScience* 49: 453–463.

Shultz, S., & Dunbar, R. I. M. (2007). The evolution of the social brain: Anthropoid primates contrast with other vertebrates. *Proceedings of the Royal Society B: Biological Sciences* 274: 2429–2436.

Smith, K. F., Sax, D. F., Gaines, S. D. et al. (2007). Globalization of human infectious disease. *Ecology* 88: 1903–1910.

Spottiswoode, C. N. (2008). Cooperative breeding and immunity: A comparative study of PHA response in African birds. *Behavioral Ecology and Sociobiology* 62: 963–974.

Stearns, S. C. (1976). Life-history tactics: A review of the ideas. *Quarterly Review of Biology* 51: 3–47.

Stevens, G. C. (1989). The latitudinal gradient in geographical range: How so many species coexist in the tropics. *American Naturalist* 133: 240–256.

Subramanian, S. V., Jones, K., Kaddour, A. et al. (2009). Revisiting Robinson: the perils of individualistic and ecological fallacy. *International Journal of Epidemiology* 38: 342–360.

Sugiyama, L. S. (2004). Illness, injury, and disability among Shiwiar forager–horticulturalists: Implications of human life history. *American Journal of Physical Anthropology* 123: 371–389.

Sugiyama, L. S., & Sugiyama, M. S. (2003). Social roles, prestige, and health risk: Social niche specialization as a risk-buffering strategy. *Human Nature* 14: 165–190.

Suh, E., Diener, E., Oishi, S. et al. (1998). The shifting basis of life satisfaction judgments across cultures: Emotions versus norms. *Journal of Personality and Social Psychology* 74: 482–493.

Terrizzi Jr., J. A., Shook, N. J., & McDaniel, M. A. (2013). The behavioral immune system and social conservatism: A meta-analysis. *Evolution and Human Behavior* 34: 99–108.

Thornhill, R. (1984). Scientific methodology in entomology. *Florida Entomologist* 67: 74–96.

Thornhill, R., & Fincher, C. L. (2013). The comparative method in cross-cultural and cross-species research. *Evolutionary Biology* 40: 480–493.

Thornhill, R., Fincher, C. L., & Aran, D. (2009). Parasites, democratization, and the liberalization of values across contemporary countries. *Biological Reviews* 84: 113–131.

Thornhill, R., Fincher, C. L., Murray, D. R. et al. (2010). Zoonotic and non-zoonotic diseases in relation to human personality and societal values: Support for the parasite-stress model. *Evolutionary Psychology* 8: 151–169.

Thornhill, R., & Palmer, C. T. (2000). *A Natural History of Rape: Biological Bases of Sexual Coercion*. MIT Press, Cambridge, MA.

Triandis, H. C. (1995). *Individualism and Collectivism*. Westview Press, Boulder, CO.

U.S. Bureau of Economic Analysis (BEA), U.S. Department of Commerce. http://www.bea.gov.
U.S. Census Bureau, U.S. Department of Commerce, http://www.census.gov.
Vandello, J. A., & Cohen, D. (1999). Patterns of individualism and collectivism across the United States. *Journal of Personality and Social Psychology* 77: 279–292.
Van den Berghe, P. L. (1981). *The Ethnic Phenomenon*. Elsevier, New York, NY.
Vanhanen, T. (2003). *Democratization: A Comparative Analysis of 170 Countries*. Routledge, New York, NY.
Van Leeuwen, F., Park, J. H., Koenig, B. L. et al. (2012). Regional variation in pathogen prevalence predicts endorsement of group-focused moral concerns. *Evolution and Human Behavior* 33: 429–437.
Wilson, E. O. (1971). *The Insect Societies*. Belknap Press of Harvard University Press, Cambridge, MA.
Wilson, E. O. (1998). *Consilience: The Unity of Knowledge*. Knopf, New York, NY.
World Bank. (2008). http://data.worldbank.org/topic/education.
World Cultures. http://www.worldcultures.org.
World Health Organization. (2004). *Global Burden of Disease: 2004 Update,* http://www.who.int. World Health Organization, Geneva, Switzerland.

# Chapter 6
# Mating Systems, Mate Choice, Marriage, Sexual Behavior, and Inbreeding

## 6.1 Introduction

There is a considerable body of research on the relationship between parasite stress and a range of values involved in human mate choice, romantic relationships, competition for mates, sexual behavior, and marriage systems. This chapter provides an overview of this fascinating and growing research area.

## 6.2 Polygyny

### *6.2.1 Low's Research*

Bobbi Low (1988, 1990, 1994) did the earliest pioneering research connecting parasite stress with human marital relationships. She made an index of historical human parasite severity (the number of disease cases) comprised of seven major human disease categories, with parasite severity measured on a three-level scale that ranged from "disease absent" to "present and serious" to "widespread or endemic." Her three levels of parasite severity correspond to the three levels of parasite severity we used in our parasite severity measures made from the GLOBE data. (See, for example, Chap. 5 for discussion of the *Contemporary Parasite Severity* measure.) Her sample of human cultures was the 186 indigenous societies in the Standard Cross-Cultural Sample, a data set often used in cross-cultural anthropological research. The scores for the seven categories of parasites were summed for each society's geographic location leading to a parasite-severity variable that ranged from seven (corresponding to the situation where none of the parasites were present in a society's geographic range) to 21 (all seven parasite categories were endemic). She had multiple measures of polygyny across societies of the Standard Cross-Cultural Sample that had been reported already in the literature. All of the polygyny

R. Thornhill and C.L. Fincher, *The Parasite-Stress Theory of Values and Sociality: Infectious Disease, History and Human Values Worldwide*,
DOI 10.1007/978-3-319-08040-6_6, © Springer International Publishing Switzerland 2014

measures were intercorrelated and each reflects variation in the success of men in obtaining wives. Wife number for men ranged from zero to seven in the sample. (In this sample, there were no highly despotic indigenous societies, which, as shown by the historian Laura Betzig, are characterized by the huge number of wives and concubines obtained by a very small number of elite men (Betzig 1986)).

Low (1990) reported that, across the indigenous societies in the Standard Cross-Cultural Sample, parasite-severity scores are correlated positively with the degree of polygyny. Hence, as parasite severity increases across these societies, an increasingly small percentage of men monopolize more of the direct reproduction (offspring production) of women. As mentioned in Chap. 3, Low predicted this finding based on the parasite theory of sexual selection. Accordingly, among men, high parasite stress generates high variance in genetic and phenotypic quality, and thus it can be more adaptive for a woman to pair with a man who has a wife (or wives) than to engage in a monogamous marriage. Most men in these societies, however, are monogamously pair-bonded; only men of very high quality are polygynous. Such men provide high genetic quality to their offspring, plus resources that are attractive to women, even though the men are already paired with a mate(s). Typically, in these polygynous societies, female mate choice for marriage is a family affair (Chagnon 1992; Low 2000), so the decision of women to pair polygynously is not independent of the reproductive interests and decisions of the woman's family, especially the more influential and socially powerful male members of the family. Frank Marlowe (2003) tested the parasite-stress hypothesis of human polygyny by examining the subset of cultures in the Standard Cross-Cultural Sample that are foragers or hunter–gatherers, and repeated for this subsample Low's general finding across the entire sample.

### 6.2.2 *Collectivisim and Polygyny*

We hypothesize that collectivism is the value system that serves as a mediator of the relationship between the degree of polygyny and parasite stress. That is to say, collectivism is evoked by parasite adversity (as shown in Chap. 5) and hence is the value system that is a proximate cause of much of the polygyny in the ethnographic record of anthropology. We predict, therefore, that the degree of polygyny will correlate positively with the degree of collectivism across the societies in the Standard Cross-Cultural Sample. Our hypothesis is consistent with the greater role of the family in marriage arrangements under collectivism than under individualism (Buunk et al. 2010). Moreover, in-depth studies of modern polygynous households in the West (e.g., among Mormons) reveal highly collectivist values in those homes and in these polygynous communities in general (Hales 2007).

Pratto and Hegarty's (2000) research findings are consistent also with an important role of collectivism in promoting polygynous marriage systems. They studied among Western unmarried college students the relationship between social dominance orientation (SDO) and desire for multiple, simultaneous mating partners. SDO scores are correlated highly and positively with conservatism scores among individuals (see Chap. 4). People with high SDO scores have an ideology that is more traditional than do people with low scores; this includes traditional values of social inequality, hierarchy, and sex roles, with male superiority and privilege. Pratto and Hegarty found that men scoring high on SDO were much more approving of polygynous sexual relationships (multiple female mates simultaneously) than were low-SDO-scoring men. The same pattern was not apparent among women—that is, women's SDO scores did not correspond to their approval of men having multiple, simultaneous mates. However, high SDO women, compared to low SDO women, placed more value on obtaining a high-status, high-earning mate. In sum, this research supports an abundance of other research, including that on family ties (discussed in Chap. 5), showing that conservatives prefer traditional values of family life; in particular, this study indicates that conservative women value resource-rich men to a greater degree than do individualistic (and thus more independent) or liberal women, and that conservative men have a greater passion for polygyny than liberal men do. The values of traditionalism and female dependency on a male pair-bond partner and simultaneity of mates of a man underlie polygynous marriage systems.

Male slave owners in the parasite-rich, collectivist Old South USA sometimes reproduced by polygyny despite monogamous marriage being the law of the land. Among historians, there are mixed opinions about the frequency of married slave owners fathering children by slaves, but, according to some accounts, male plantation masters oftentimes sired more children by slaves than by their legal wife (Betzig and Weber 1993). Van den Berghe's (1981) review of slavery systems, both in the South and in other parts of the world, led him to conclude that the mating system of the slave plantation, whenever and wherever it occurs, is polygyny. In regard to the pre-Civil War South, he points out that the owner, his sons, and overseers were paired and reproduced polygynously because "they had access not only to the white women whom they married—but also the pick of young slaves whom they took as concubines" (p. 132). He reports, too, that it was not until after the American Civil War and with Reconstruction that white men's access to African-American women was normatively and legally restricted.

## 6.3 Mate Choice

### *6.3.1 Gangestad and Buss' Research*

Soon after Low's research linking marital polygyny and parasite adversity, cross-national research by Gangestad and Buss (1993) linking mate choice and parasite severity was published. It, like Low's, was inspired by Hamilton and Zuk's (1982) parasite theory of sexual selection. According to that theory, physical attractiveness is a certification of genetic resistance to parasites—good genes for parasite resistance. Hamilton and Zuk provided data showing that, across species of birds, male plumage brightness was correlated positively with the prevalence of blood parasites in the birds. This finding indicates that sexual selection resulting from both female mate preference for bright-plumaged males and male–male competition favoring bright-plumage males was stronger in the evolutionary histories of species with high parasite prevalence than in species with low infectious-disease stress.

Gangestad and Buss (1993) tested the parasite theory of sexual selection as applied to human mate choice. Specifically, they tested for a positive correlation across countries between human parasite severity and the importance people place on physical attractiveness in mate choice. They used data from 29 of 37 countries studied by Buss (1989) and Low's parasite-severity scores (mentioned earlier) for those 29 countries. Each participant in Buss' huge study—there were many thousands of participants—completed a questionnaire and rated the importance of 18 attributes as criteria for choosing a mate, including the variable of interest here, "good looks." Certain variables across nations (e.g., average income, world region) were statistically controlled. Gangestad and Buss (1993) reported a robust, positive correlation between the value people attribute to good looks in a mate and parasite severity, and the correlation was found in each sex. Gangestad and Buss (1993) argued that the psychological machinery responsible for their results may be a combination of (1) species-wide facultative (conditional) adaptation enabling people to track local parasite stress and, based on that assessment, to adjust the priority given to good looks, and (2) adaptations that are genetically different across regions and set a region-specific value on good looks of a mate. Both (1) and (2) are processes of the ontogeny of values that we discussed in earlier chapters. For reasons discussed in Chap. 3, (1) is likely to be the most general and widespread enculturation process, and even when (2) is involved, so, too, is conditional adoption of values.

More recently, Gangestad et al. (2006) expanded the 1993 Gangestad and Buss study by statistically controlling for some additional potential confounding variables (e.g., gender inequality); the 1993 results were upheld. Hence, there is considerable evidence for humans showing that high parasite stress evokes an elevated valuation given to physical attractiveness in a mate; correspondingly, low parasite stress evokes a lower priority placed on physical attractiveness. According to the parasite-stress theory of values, the high value given to physical attractiveness in parasite-rich areas is part of the behavioral immune system. As parasite stress increases across regions, men and women place increased importance on obtaining a mate that is physically attractive, because physical attractiveness is a marker of high phenotypic and genetic quality pertaining, in part, to the ability to resist local infectious diseases.

### 6.3.2 *Women's Mate-Choice Trade-Off*

There is also considerable evidence that women engage in a trade-off in mate choice because physically attractive men—men of high phenotypic and genetic quality—invest less in their romantic partners and in offspring than do physically unattractive men (reviewed in Thornhill and Gangestad 2008). Lee and Zietsch (2011) have studied this trade-off by priming women with either a questionnaire that contained parasite-relevant cues (the perceived-vulnerability-to-disease (PVD) scale) or a questionnaire containing cues of food and other basic resource limitations. Following the primes, women indicated the importance to them in mate choice of a mate with earning potential and willingness to invest versus a mate with masculinity, creativity, and other traits thought to correspond to male genetic quality. (See Thornhill and Gangestad 2008 for a review of the evidence for certain male traits as markers of genetic quality.) For a control prime, they used a questionnaire unrelated to the two trade-off variables. Women primed by the parasite-prevalence cues prioritized a mate of high genetic quality more than women primed by the control or the resource-scarcity condition; women primed by parasite-salient cues showed the least preference for a male with the ability and willingness to invest. The women primed by the resource-scarcity condition showed the reverse pattern, with a maximum priority for an investing mate. This research indicates that women's perception of contagion in their current environment immediately activates the aspect of their behavioral immune system that enhances their psychological preference for a mate of high genetic quality—a mate that ancestrally would sire offspring with above average resistance to infectious disease.

### *6.3.3 Other Attractiveness Research*

Lisa DeBruine et al. (2012) have provided additional evidence that high parasite stress evokes in women an enhanced preference for masculine men. They used a website that presented to women from many different countries men's faces that varied in testosteronization/masculinity and hence in facial markers of male phenotypic and genetic quality. (See Thornhill and Gangestad 2008 for a review of the evidence for male testosteronization as a signal of phenotypic and genetic quality, and Rantala et al. (2012) and Pawlowski et al. (2014) that have shown links between testosteronization and immunocompetence in men.) DeBruine et al. (2012) found that women's preference for testosteronization of the male faces correlated significantly and positively with our measure *Combined Parasite Stress* across 30 countries. They reported, too, that the same pattern occurs across the 50 states of the USA, using our measure of *Parasite Stress USA*. (These two measures of parasite stress are explained in Chap. 5.) Debruine et al. (2012) also discuss two earlier studies, one across the 30 countries and one across the 50 USA states, that they and colleagues conducted using the same facial stimuli, but somewhat different measures of parasite stress. These studies gave results similar to those with our measures of parasite stress. Hence, in countries and USA states with high parasite stress, women value men's good looks in terms of masculinity of the face more than in regions of low parasite stress. In related research, Moore et al. (2013) studied cross-cultural differences in women's preferences of composite images of male faces constructed to represent different combinations of features associated with high versus low cortisol and testosterone in men. With respect to pathogen stress variation across countries, they found that women from countries with higher levels of pathogen stress strongly preferred testosteronization in male faces, which is consistent with the findings of Debruine et al. (2012).

Complementary research by Jones et al. (2013a) investigated the role of a measure of disgust sensitivity in women's judgments of the attractiveness of masculine facial features of men. They reported that individual differences in disgust sensitivity positively predicted these attractiveness judgments. Their other studies reported in the same paper examined women's disgust sensitivity in relation to their preferences for men's masculine voices and nonfacial bodily features. The same results were found for men's voices and bodily features as found with faces. In a separate study, Jones et al. (2013b) found that men's judgments of the attractiveness of feminine facial features (facial estrogenization) in women's faces positively related to the men's disgust sensitivity.

Little et al. (2010) experimentally manipulated cues of contagion risk presented to research participants using pictures of high and no parasite salience. They then measured the participants' attractiveness ratings attributed to human faces that varied in bilateral symmetry and hormone markers. Symmetry of bodily features, like sex-specific hormone markers in the face (estrogen markers in women's faces and testosterone markers in men's), probably depicts phenotypic and genetic quality and hence resistance to infectious disease (Thornhill and Gangestad 2008). Little et al. (2010) found that people who were exposed to cues of high contagion risk, compared to those seeing no contagion risk, showed increased facial attractiveness preferences for opposite-sex individuals with greater facial sex-specific hormone markers and symmetry.

A study by Welling et al. (2007), like those of DeBruine et al. (2012), Jones et al. (2013a, b), and Little et al. (2010), provided evidence that facial attractiveness judgments are a component of the behavioral immune system. Welling et al. linked facial attractiveness judgments of research participants of both sexes to the participants' PVD scores. As we have discussed, PVD, an aspect of the behavioral immune system, is an individual-difference measure created from people's responses to questionnaire items about their concern with contagion. Men and women who perceived themselves to be more vulnerable to infectious disease attributed higher attractiveness to faces separately rated as healthy (by nonexperimental participants) than did individuals who perceived themselves to be relatively less vulnerable to infectious disease.

A recent investigation by Young et al. (2011) built on those of Little et al. (2010) and Welling et al. (2007) just described. Young and colleagues conducted two studies to test the hypothesis that variation in people's disease concerns is related positively to the variation in which they value symmetrical faces versus asymmetric faces. Both studies involved both sexes judging faces that varied in symmetry. In one study, they measured individual differences among research participants in PVD. In the second study, they primed participants with infectious-disease-relevant pictures as well as control pictures and then measured their facial symmetry preferences immediately thereafter. In both studies, increased sensitivity to infectious diseases was associated with heightened value placed on symmetric faces. Notably, their results did not generalize to symmetric versus asymmetric nonfacial images.

Hence, these studies indicate that people have a specific preference for faces that are symmetric when infectious disease becomes more salient.

de Barra et al. (2013) investigated in Bangladesh the ontogeny of facial attractiveness judgments in relation to infectious diseases experienced during childhood. They reported that childhood illness, particularly episodes of diarrhea, positively relate to individual men and women's facial attractiveness ratings of sexually dimorphic faces of the opposite sex. Thus, an ontogeny of high infectious-disease encounters evokes an elevated preference in women for masculine faces and an elevated preference in men for feminine/estrogenized faces.

An interesting line of research has developed that seeks to determine whether people who have greater behavioral immune system reactivity avoid unattractive, unfit mating partners because they represent a potentially large cost of association or whether they are drawn to physically attractive mating partners because they represent an especially beneficial association. At this point, evidence has been found for both conjectures. Park et al. (2012) reported that people high in pathogen disgust assigned lower attractiveness ratings for otherwise determined unattractive targets, while for attractive targets, pathogen disgust was uncorrelated with attractiveness ratings. Meanwhile, Cantú (2013) reported that when pathogen prevalence was made temporarily salient, women showed a strong preference for physically attractive men; for men, this result did not emerge. The contrasting findings are intriguing and certainly point to the need for more research in this area.

### *6.3.4 Overview: Mate-Choice Studies*

The research studies discussed earlier conducted by Gangestad and colleagues, Lee and Zeitsch, Debruine and colleagues, Moore and colleagues, Jones and colleagues, Little and colleagues, Welling and colleagues, Young and colleagues, and de Barra et al. provide mutually reinforcing evidence. In high-parasite regions, both sexes value a mate's looks more than in low-parasite regions, and men's masculinity becomes increasingly important in women's attractiveness judgments as parasite stress increases across regions. The research on the ontogeny of facial-attractiveness mate preferences indicates that individual people who have more infectious diseases during childhood possess a psychological mate preference for enhanced hormonal effects in faces that relates to a genetic immunity which results in offspring with elevated immunocompetence. In high parasite regions, more people experience an ontogeny of high infectious-disease encounters than in low parasite-stress regions. Furthermore, masculinity preferences in women are predicted by their disgust sensitivity. Regional variation in ontogenetic experience with infectious disease and the associated evoked degree of disgust sensitivity apparently combine to account for the regional differences in the importance of physical attractiveness of a mate across countries and states of the USA. Thus, it seems that a childhood of high parasite stress evokes higher disgust and simultaneously evokes more value placed on physical attractiveness than a childhood of low parasite stress. In addition to the evoked stable differences in valuation of looks, people respond immediately upon viewing contagion risk and specifically by increasing their attractiveness ratings of two categories of traits (symmetry and hormone markers) that relate to increased health and genetic quality of a mate. Also, upon perceiving a parasite threat, women respond immediately by prioritizing male physical attractiveness and other good-genes markers in a mate over male resourcefulness and investment. Nevertheless, it is not clear whether people with higher levels of contagion concerns are particularly drawn to attractive people or avoid unattractive people. Overall, then, as parasite stress increases, so does the value of obtaining a mate with phenotypic and genetic quality pertaining to dealing with parasite adversity. Finally, people with high PVD value health cues and symmetry in faces more than people with low PVD.

All these studies are supportive of the enculturation process we described earlier in our book. People contingently adopt attractiveness values based on their local utility in defense against parasite adversity. This begins during childhood and tracks individuals to a given level of attractiveness valuation, creating individual differences as well as regional differences. These individuals, however, retain conditionality in their sensitivity to cues of local parasite threat, allowing them to modify their attractiveness-preference values as current circumstances of parasite threat wax and wane. These studies, as well as many others we discuss throughout the book, reveal the exquisite functional organization of the behavioral immune system. We anticipate that future research on values other than attractiveness judgments that are also features of the behavioral immune system will be shown to possess the same degree of exquisite functional design. Evidence we have discussed already in this book certainly indicates this is the case across values of collectivism–individualism.

The research by Welling et al. and Young et al. discussed earlier on PVD suggests a relationship between attractiveness judgments and collectivism–individualism. Although physical attractiveness is a significant social asset everywhere (Thornhill and Gangestad 1993), it may be especially salient in collectivist cultures, which are characterized by high parasite stress.

We hypothesize that the importance of having a physically attractive mate is correlated positively with conservatism across cultures and people. This hypothesis is consistent with (a) the positive relationship between parasite stress and importance attributed to physical attractiveness and (b) the greater value placed on facial health of a mate and on facial symmetry by high scorers on PVD than by low scorers. PVD scores across individuals positively correlate with conservatism (and negatively with liberalism) (see Chap. 4). However, as of yet, the direct relationship between the salience-of-looks variables and collectivism has not been studied.

## 6.4 Marital Divorce

We mentioned in Chap. 4 the relationship between people's core values and marital durability. Gelfand et al. (2004) discussed evidence that has shown that, across countries, collectivism–individualism is related to variation in marital divorce rates.

Vandello and Cohen (1999) provided evidence of the same pattern across states of the USA: the ratio of marriages to divorces was an item they used to measure states' collectivism (Chap. 5). Collectivists' marriages are more durable than are those of individualists. On the one hand, this pattern seems contrary to expectations. Collectivist marriages are based more often on family honor, duty and preference, norm conformity, and in-group social preference, and less so on romantic love between a man and woman. As the basis of long-term pairings and marriage, romantic love is more characteristic of individualist cultures and thus one might infer that love-based marriages would be more durable. On the other hand, individualists are more self-sufficient and independent, which appears to override the effect of love on duration of individualists' marriages and generate the overall higher divorce rate among individualists than among collectivists. Individualists' greater risk-proneness (acceptance of uncertainty, Chap. 4) may affect also their higher divorce rate.

Another factor that might contribute to higher divorce rates in individualistic cultures than in collectivist cultures is the greater sexual liberation of women (from traditional sexual values of caution and continence) and their associated reduced sexual restrictiveness in individualistic societies. In women, the value of reduced sexual restrictiveness may give them an opportunity to experience sexual intimacy with multiple men, which could contribute to partner desertion and switching. Given this, increased sexual liberation of women, characteristic of individualistic cultures, is expected to increase the divorce rate. In contrast to women's sexual restrictiveness, men's sexual restrictiveness is uncorrelated, or only weakly correlated, with collectivism–individualism. Probably contributing also to the higher divorce rate in individualistic societies is the relaxation of legal grounds for divorce in such societies, which was initiated during the sexual revolution of the 1960s in the West (Chap. 10). The relationships between sexual restrictiveness and both collectivism–individualism and parasite adversity have received attention from researchers, as we describe next.

## 6.5 Female Sociosexual Orientation

The Sociosexual Orientation Inventory (SOI; Simpson and Gangestad 1991) is a self-report questionnaire commonly used to assess a behavioral disposition toward unrestricted sexuality (e.g., willingness to engage in sexual relations in the absence of a long-term commitment). Based on data collected from 14,059 adults worldwide, Schmitt (2005) reported sex-specific average SOI scores for nearly 50 countries. High SOI scores indicate a more unrestricted approach to sexual behavior.

Using these data, Schaller and Murray (2008) found that parasite prevalence predicted both male and female mean SOI scores—more parasites, greater restricted sexuality—although only the relation with female SOI scores remained statistically significant after controlling for additional variables. Thornhill et al. (2010) analyzed the cross-national SOI scores again, but with the three types of human disease richness discussed in Chap. 5: human-specific, multihost, and zoonotic. Thornhill et al.'s analyses focused exclusively on average female SOI. Across 45 countries, female SOI was correlated negatively with indices of both human-specific parasite richness ($r=-0.38$, $p=0.01$) and multihost parasite richness ($r=-0.47$, $p<0.001$). This pattern means that as parasite adversity declines, women become increasingly sexually unrestricted. The relation with zoonotic parasite richness was negligible and nonsignificant ($r=-0.12$, $p=0.44$). Hence, as in the research on collectivism–individualism (described in Chap. 5), human infectious diseases that are transmissible from human to human impact people's values, in this case women's sexual values, to a far greater extent than zoonotic diseases.

Furthermore, research done on individuals has found that people who have greater behavioral immune system reactivity, either by measuring constructs such as PVD or by manipulating parasite salience, demonstrated greater sexual restrictiveness in attitude and behavior, most especially women (Duncan et al. 2009; Murray et al. 2013).

It also has been shown that, across many countries of the world, women's sexual unrestrictiveness correlates positively with (a) individualism, (b) gender equality in opportunity and participation in societal matters, and (c) democratization (Thornhill et al. 2009). Hence, women's sexual liberation from traditional values of sexual continence is a component of overall liberalization of values. In Chap. 10, we treat further the so-called sexual revolution and women's liberation that occurred in the West in the 1960s and 1970s.

## 6.6 Inbreeding

### 6.6.1 *The Hypothesis and Hoben et al. (2010)*

It may appear counterintuitive to anchor the parasite-stress theory of sociality on the proposition that parasite stress leads individuals to remain in the local community and avoid distant members of the same species. Indeed, one hypothesized benefit of sexual reproduction and outbreeding is the genetic diversification of a brood of offspring in order to combat parasite threats (Tooby 1982; Trivers 1985; Hamilton et al. 1990; Ridley 1993). Shields' (1982) and others' ideas (Kokko and Ots 2006), however, indicate that inbreeding may be adaptive under a range of circumstances that give rise to outbreeding depression. We hypothesized that the costs and benefits of inbreeding and outbreeding will vary regionally primarily in accordance with variation in parasite severity (Fincher and Thornhill 2008a; b). In areas of high parasite severity, inbreeding is costly because it reduces variation in the molecular milieu from which an individual can mount an immune response (Penn and Potts 1999), but beneficial through the maintenance of locally adaptive, genetically based immune responses as well as the avoidance of infectious diseases that are not harbored by the local group. Thus, although close inbreeding is maladaptive under high parasite severity levels, distant outbreeding is too. This is because host–parasite coevolutionary races build complex, locally adaptive host immune adaptations including coadapted gene complexes that often work best against infectious disease if not disrupted by distant outbreeding. Both coadapted gene complexes and local adaptation render some degree of inbreeding adaptive under high parasite prevalence, as first hypothesized by William Shields (1982).

We discovered that Abraham Buunk, Ashley Hoben, and Mark Schaller were independently thinking along the same lines we were about how parasite stress may promote inbreeding. We began collaboration in 2006 with them to test the prediction that parasite stress would show a positive relationship with inbreeding across contemporary countries of the world. That work is reported in detail in Hoben et al. (2010). People have psychological adaptation that functions to avoid mating with very close relatives, such as parents, and full- and half-siblings (Lieberman et al. 2003; Lieberman et al. 2007). Our hypothesis pertains to relatives more distant than these categories and especially cousins. Here we give a brief sketch of the research in Hoben et al. (2010).

As we were conducting our research, some evidence for the parasite-stress hypothesis for inbreeding appeared in the literature. Denic and Nicholls (2007; also Denic et al. 2008a, b) reported a positive correlation between malaria endemism and the frequency of consanguineous marriages across many countries of the world. Denic and Nicholls' theoretical reasoning was similar to our own. Their evidence is specific to just one of the many human infectious diseases that, historically, has been of differential prevalence worldwide. Hence, we wanted to test the hypothesis more generally by looking at human infectious diseases widely.

Consanguineous marriage occurs between genetic relatives (Thornhill 1991). Despite the negative effects for offspring produced from this type of marriage (e.g., inbreeding depression), it continues to occur at high rates in various parts of the world and typically involves various categories of cousins, and, in some cases, uncle–niece unions (e.g., Rao and Inbaraj 1977). Consanguineous marriage is especially prevalent among Muslim and Hindu populations in Africa and Asia (Jaber et al. 1998). For example, in Jordan, 50 % of all marriages are between genetic relatives, and in Kuwait and Saudi Arabia 54 %, whereas in India consanguineous marriages vary from 5 to 61 % depending on socioeconomic status, religion, and caste (Jaber et al. 1996). These rates are in sharp contrast to the low rate in western countries such as the United States (Jaber et al. 1998).

Our colleagues and we wanted to see if regional variation in consanguineous marriages might be explained by variable parasite stress. From Bittles (1998) we obtained data from 381 surveys that include information for millions of marriages (see www.consang.net) and data on the proportion of consanguineous marriages in 72 different countries. We used the index *Historical Parasite Severity* described in Chap. 5. In addition to the two variables of primary conceptual interest, we also assessed additional variables in order to address possible alternative causal explanations for the hypothesized relationship between historical pathogen severity and consanguineous marriages.

In support of the parasite-stress hypothesis of inbreeding, we found that *Historical Parasite Severity* positively and significantly predicts regional differences in the percentage of consanguineous marriages: $r = 0.40$ ($p < 0.001$). As we have mentioned earlier in our book, parasite severity is negatively correlated with economic wealth within a region—wealthy nations have less parasite adversity than poor nations (Fincher et al. 2008, Chap. 11). Therefore, the significant positive correlation between parasite severity and consanguineous marriages might be a spurious result of a negative relation between economic wealth and consanguineous marriage; however, we found no support for this alternative explanation. When *Historical Parasite Severity* and gross domestic product per capita (GDP per capita; 2009 data obtained from the *CIA World Factbook*; www.cia.gov) were entered jointly as predictors of consanguineous marriages, the effect of GDP per capita was weak and nonsignificant (std. $\beta = -0.17$, $p = 0.13$); the effect of pathogen severity was reduced slightly, but still remained stronger than that of GDP per capita (std. $\beta = 0.37$, $p = 0.001$).

Moreover, because it has been observed previously that consanguineous marriage is more common in regions in which malaria is endemic (Denic and Nicholls 2007), we felt it is important to see if the effect we found is driven primarily or only by malaria's prevalence. This was not the case. We conducted an analysis in which *Historical Parasite Severity* was entered along with the specific prevalence of malaria as predictors of consanguineous marriage. Results indicated no unique effect of malaria prevalence ($\beta = 0.02$, $p = 0.92$); but the overall index of historical pathogen severity remained a significant predictor of consanguineous marriage (std. $\beta = 0.35$, $p = 0.04$). Hoben et al. (2010) examined analytically a number of other alternative conjectures about the causes of cousin marriage and found that, when they are considered, the positive relationship between the prevalence of consanguineous marriage and parasite severity remains statistically robust.

We also computed the historical severity of parasites and the percentage of consanguineous marriages within each of the six cultural areas of the world identified by Murdock (1949), and conducted additional analyses in which we treated these six world regions as the units of analysis. Although the sample size of six is very small, it is noteworthy that the correlation between pathogen severity and consanguineous marriage replicates the positive correlation observed in our cross-national analysis ($r = 0.56$). Thus, in cultural regions with a higher historical severity of parasites, there is a higher frequency of consanguineous marriage, and there are probably no exceptional regions in this overall pattern.

In sum, historical pathogen prevalence was a substantial and significant predictor of regional differences in consanguineous marriages, and this effect does not appear to be a spurious result of other variables associated with pathogen prevalence. These results are consistent with the hypothesis that regional differences in consanguineous marriage emerged, in part, because consanguineous marriages conferred immunological resistance to local pathogens and that these adaptive benefits accrued primarily in geographical regions with high levels of parasite severity.

In our study with colleagues on inbreeding, we felt it was of interest to consider the effect of historical pathogen severity on consanguineous marriage alongside other cross-cultural differences that are predicted by parasite adversity. In regions with high levels of pathogens, people (especially women) have more restricted approaches to mating, as discussed in Sect. 6.5 on female sociosexual orientation. People in such regions also are more likely to endorse collectivist values that emphasize within-group interdependence, extended-family nepotism, neophobia, xenophobia, and philopatry, as documented in Chap. 5. All of these characteristics are conceptually consistent with increased inbreeding. As expected from this consistency, additional analyses in Hoben et al. (2010) revealed that correlations between consanguineous marriage and cultural values result, in part, from shared variance with parasite severity. For example, the two measures of collectivism examined (*Gelfand In-group Collectivism* and *Kashima Collectivism*; see Chap. 5) were correlated positively with consanguineous marriage ($r$'s = 0.31 and 0.29), but, when controlling for historical pathogen severity, these correlations were reduced (partial $r$'s = 0.22 and 0.09). Thus, while the broader cultural value system in a region may reinforce behavioral norms promoting or inhibiting inbreeding (and, in turn, be reinforced by inbreeding norms), evidence indicates that the cultural values of collectivism–individualism are tied to patterns of inbreeding–outbreeding by way of underlying parasite stress.

### 6.6.2 Hoben's Dissertation

Ashley Hoben wrote her doctoral dissertation on human inbreeding and outbreeding while she was a student at the University of Groningen in the Netherlands (Hoben 2011). The published paper Hoben et al. (2010) (discussed earlier) was one chapter of her dissertation. Here we describe her other doctoral studies contained in her dissertation. First, using parasite-severity data from Low's research (discussed in Sect. 6.2.1), she discovered that the cross-national pattern of inbreeding marital frequency showing a positive relationship with parasite stress is repeated in the sample of indigenous human societies comprising the Standard Cross-Cultural Sample. Hoben used "first-cousin-marriage allowed" versus "not allowed" as her inbreeding variable. Her analysis controlled for variables that she felt were potential confounders. The end result was that parasite stress significantly and positively predicts the value of permitting versus the value of not permitting first-cousin marriage across the indigenous societies in the ethnographic record of anthropology. Said differently, in ethnographic societies facing low parasite stress, there is greater outbreeding than in societies facing high parasite stress.

The question of why consanguineous marriage occurs in some indigenous societies and not others has been a long-standing one in anthropology. (See a review of this research history in Thornhill and Thornhill 1987.) Hoben's research places this question in a bigger intellectual framework and provides a more encompassing and convincing answer. Evidence indicates that the answer is the same for inbreeding variation across indigenous societies as for the variation in inbreeding across contemporary countries. Even more synthetic is that both patterns of inbreeding predicted by parasite stress are now patterns that provide partial support for an overarching theory of sociality—the parasite-stress theory of values.

Hoben made some additional relevant discoveries that are reported in her dissertation. She obtained the opinions of the Dutch students at her university about various scenarios describing romantic relationships or marriage among cousins.

Based on the relatively low parasite stress in the region of her study, she predicted from the parasite-stress theory of values that there would be a general negativism toward such relationships—that is, a general preference for outbreeding. The prediction was supported significantly across a range of experimental designs. The liberal-minded students who attend university in the Netherlands are open-minded in general, as is typical of liberals, but that openness stops when it comes to inbreeding. Their preference was for outbreeding, with inbreeding viewed negatively or even as immoral sexuality.

## 6.7 Male Sexual Competition

### 6.7.1 *Out-Group Men as Sexual Competitors*

It is well established that conservatism and its associated high PVD are related to many types of prejudice, including prejudices against foreigners and immigrants (Chaps. 3–5). Klavina et al. (2011) expanded knowledge of this pattern by investigating the relationship between out-group prejudice and competition for mates in the Netherlands. In one study of their investigation, they presented Dutch research participants of Dutch ancestry with a bogus newspaper article that described Italians (out-group immigrants) as a mating threat by being attractive, in high numbers, and marrying Dutch-in-group members. Men read an article that depicted out-group men as sexual competitors; women read an article that depicted out-group women as sexual competitors. A control group of participants of each sex read a bogus article that described the absence of interest among in-group people in dating and marrying Italians. They also measured the research participants' PVD and their out-group prejudice using standard and validated questionnaires. They hypothesized that the mating-threat prime would increase the male research participants prejudice toward out-group men and especially so in high-PVD male research participants. Their data supported this hypothesis. They found, however, no statistically significant patterns for the women research participants.

Klavina et al. (2011) proposed that males only are expected to show the pattern because they have been under stronger sexual selection for high mate number than women have, and hence value mate number more than women do. The main out-group threat for women is violent out-group men, not out-group women. Out-group men pose the potential threats to women of kidnap and sexual coercion (Thornhill and Palmer 2000; Navarrete et al. 2009, 2010). Sexual coercion circumvents female mate choice and thus reduces female reproductive success. McDonald et al. (2011) reported that women's bias against out-group men, who they perceive as physically formidable, increases during estrus (the several fertile days of an ovulatory menstrual cycle, Thornhill and Gangestad 2008). Estrus is associated with maximum cost to women of sexual coercion, because it is the time of conception and hence of sire determination of an offspring. PVD was predicted by Klavina et al. (2011) to influence men's prejudice toward out-group mating threat because prior research has shown PVD's positive relationship with prejudice (Chap. 4).

Klavina et al. (2011) conducted a second study in their investigation. In this experiment, they gave Dutch men a series of infectious-disease salient pictures known to increase PVD. They showed other Dutch men control pictures that depicted either neutral situations or dangerous situations unrelated to contagion threat. Then, all of these participants read the bogus article that described Italian men as a mating threat. After that, prejudice toward out-group men was measured. This experiment repeated the results in the first study discussed earlier. Men primed with infectious-disease cues in pictures, but not men not so primed, increased prejudice toward out-group men. The results of the two studies revealed that it is the Dutch men's assessment that Italian men are competing successfully for Dutch men's potential mates, in combination with heightened perceived vulnerability to contagion, that result in increased negative attitudes toward Italian (out-group) men.

Klavina et al. (2011) mention that their results imply that the perception of mate-competition threat from out-group and associated out-group prejudice is stronger in collectivist men than in individualist men. The separate finding that conservatism and PVD are robustly and positively correlated (see Chap. 4) supports this link and thus their suggestion. They mention, too, the implication of their finding for inter-group coalitional aggression, which is almost exclusively perpetrated by men. Collectivist men's negativism about out-group men poaching in-group women may be an important factor causing the documented positive association between collectivism and frequency of civil conflicts across nations (see Chap. 12 on warfare, parasite stress, and values).

### 6.7.2 *White-on-Black Violence*

Klavina et al.'s (2011) findings allow a deeper understanding of some collectivist men's prejudicial custom of segregating and perpetrating violence on out-group men. A well-studied example is the USA, primarily in the South, from the period of Reconstruction (following the American Civil War) and into the 1960s. Historians have documented the racist practices of white supremacist mobs of men castrating, mutilating, shooting, burning, and lynching African-Americans and destroying their property and communities over this period (e.g., Brundage 1993). The documentation shows that most of the violence was directed at black men, less commonly black boys (pre-reproductive-age males), and the least at females. Violence was perpetrated on females when they tried to defend the men who were the focus of white supremacist prejudice and violence. Violence was perpetrated on black men as a result of alleged homicide, rape, attempted rape, or the disrespect of whites or other violations of Old South customs of racial segregation.

Historians have documented that this violence contributed importantly to the great mass migrations of blacks from the South that began after the USA Civil War and continued into the 1960s (Wilkerson 2010). A combination of economic dire straits for southern blacks and threatened physical pain or death at the hands of racist mobs is a strong incentive to immigrate, even for these highly philopatric people. Both of these emigration factors were generated by the highly conservative values held by southern whites that dehumanized and disenfranchised black people. Collectivism's value of inequality not only restricts resources, opportunity, and participation in society to those of the privileged group (by birth, color, or class), but also can lead the privileged group to more directly exterminate out-groups.

The have-and-have-not classist culture of the Old South also impacted negatively a large portion of the white population. The plight of the average white southerner during this period was dire, given the overall undemocratic societal system and extreme wealth inequality generated by the conservative political regimes of this era. Key (1949) has documented that two central and persistent issues anchored the politics and policies of the governing bodies of the Old South: economic and political disenfranchisement of blacks and poor whites.

Historians typically explain the white-on-black violence of the era in terms of white racism. This is certainly correct, but the Klavina et al. (2011) study discussed earlier adds new depth to this explanation. We suggest, the southern men who perpetrated the violence not only hated blacks, they also feared them, because they perceived them as both a contagion risk and a mating threat. The unbridled sexual motivation of black men to sexually access white women was a widespread ideological deduction among whites in the conservative culture of the Old South. Then and there it was normatively taboo for a black man to be alone with or touch a white woman. Also, romantic or sexual interactions between white women and black men were illegal. White men in the Old South were often quick to condemn black men for perceived sexually inappropriate behavior toward white women (Hodes 1999).

We have investigated a prediction from our thinking here: lynchings across USA states of blacks will be correlated positively with parasite adversity and collectivism. Data exist to allow an initial test. Data on lynching of blacks by whites, collected from newspaper articles, are archived at Tuskegee University (Tuskegee Archive: http://law2.umkc.edu/faculty/projects/ftrials/shipp/lychingstate.html). The data extend over the period 1882–1968, by state. As predicted, both relationships were found, and are statistically significant. Lynchings of blacks by whites are correlated with *Parasite Stress USA* very strongly: $r=0.71$, $p<0.0001$, $n=44$ states for which data are available. Collectivism (based on Vandello and Cohen 1999) shows a moderate relationship with the lynchings: $r=0.46$, $p=0.0021$, $n=44$. Note that the lynching data are not rates based on population estimates as are the rates of other types of homicides we analyze in Chap. 8. Rates would be impossible to compute because of the changing social structure over the period covered by the lynching data. Although lynchings and other white-on-black violence has been much discussed by historians of the South, to our knowledge this is the first analysis that ties the regional variation in the violence to an evolutionary theory of human sociality.

Racism-inspired lynchings are right up there with Nazi and other holocausts in anchoring the most horrific pole of the dimension of "man's inhumanity to man," as the poet Robert Burns famously labeled hurtful prejudice in general. According to the parasite-stress theory of values, much of the variation across the entire dimension of intolerance and hate is caused proximately by variable parasite stress and associated evoked values. According to the same theory, it is caused ultimately by past Darwinian selection in the context of variable parasite adversity that crafted condition-dependent values-adopter psychological adaptation.

## 6.8 Disgust and Sexual Arousal

In earlier chapters, we discussed the emotion of disgust as an adaptation that defends against contagion. A range of stimuli cues disgust: diseased people, vermin that imply contagion risk, and even certain norm or moral violations. Evolution-minded researchers interested in the functional design of disgust have proposed that disgust should be organized functionally to show little or no activation under conditions that have infectious-disease risk, but if these conditions are not tended to or pursued (rather than avoided) would reduce inclusive reproductive success (Oaten et al. 2009; Curtis et al. 2011). Imminent sexual opportunity is a context in which disgust sensitivity to sexual risks of contagion should show design for reduction in order to promote sexual intercourse and hence male mating success.

### *6.8.1 Women's Sexual Arousal*

Borg and de Jong (2012) investigated women's disgust sensitivity in relation to their sexual arousal. Female participants were placed into one of three groups: sexual arousal, nonsex positive arousal, or a control group. After watching video clips meant to stimulate the proper mood, participants were faced with participating in sex-related disgust activities or nonsex disgust activities in order to measure their avoidance behavior. Women who were sexually aroused participated in more disgusting activities (both sex-related and nonsex), suggesting a reduction in disgust sensitivity due to sexual arousal. Fleischman (2014) reported on an ongoing study of women's sexual arousal and disgust that also used films to elicit different mood states. Rather than measure sexual arousal subjectively as in Borg and de Jong (2012), they used a vaginal photoplethysmograph to detect sexual arousal. Additionally, to control for menstrual cycle variation, they tested all participants at about the same point in their respective menstrual cycles. Fleischman and colleagues found that women who were more disgust sensitive prior to the manipulation actually became more disgusted after sexual arousal, and women who were less disgust sensitive became less disgusted after becoming sexually aroused. Fleischman suggests this may mean that women who have well-functioning immune systems (those with low baseline disgust sensitivity) and can afford exposure to disease cues reduce the disgust sensitivity during sexual arousal, while those women with less effective immune systems—those that cannot afford exposure to disease cues—will actually increase disgust sensitivity during sexual arousal.

### 6.8.2 *Men's Sexual Arousal*

Stevenson et al. (2011) examined the conditional expression of disgust in men under sexual arousal. They assigned men to one of four groups that viewed pictures: one group viewed erotic women's pictures, a second group viewed nonerotic images of women, a third group viewed pleasant, but nonerotic-arousing images, and the fourth group viewed unpleasant, but arousing images (e.g., a gun aimed at them). Then, all the men viewed paired images of actual disgust elicitors; each pair contained a sexual disgust image and a nonsexual disgust image. The paired stimuli (sexual and nonsexual) were repeated with validated tactile and auditory disgust elicitors. Finally, the men's disgust level was measured with the Disgust Sensitivity Scale, a widely used, validated metric measuring disgust intensity. Men in the erotic-image group reported more sexual arousal than men in the other three groups; they also showed less disgust from sex-related disgust elicitors across the three types of sexual disgust elicitors than men in the other groups, and the lowering of disgust sensitivity in the erotic-image group was specific to sex-related disgust elicitors, not disgust in general. Hence, sexually aroused men adjust by lowering their disgust sensitivity, whether cued visually, tactilely, or auditorily, specifically to the sex-related component of disgust. Also, types of arousal other than sexual used in the experiment did not affect men's sexual disgust sensitivities. General arousal or even fear did not change men's sensitivity to sexual disgust elicitors.

Stevenson et al.'s (2011) findings provide evidence that men's sexual motivation is designed to interface with a part of the behavioral immune system—disgust sensitivity—in the way predicted by the parasite-stress theory of sociality. These findings imply something quite remarkable about disgust's functional design when it is remembered that people are functionally designed to accept many false positives in terms of stimuli suggestive of contagion risk (Chap. 3). Although this acceptance is well established, disgust's design shows nuances that prevent its typical activation from interfering with men's mating success, even though mating carries contagion risk. Though undeveloped, Fleischman (2014) offered the intriguing hypothesis that sexually transmitted diseases could benefit by somehow reducing disgust in order to increase their transmission rate. She was agnostic on this point, but such diseases would likely get the best payoff by focusing on male sexuality.

## 6.9 Sex Differences in Sickness Behavior and Suppression of Sexual Activity

Sickness behavior refers to the suite of behaviors that ensue once an individual becomes sick. This can include weakness, malaise, depression, lethargy, and the like (reviewed in Avitsur and Yirmiya 1999). One intriguing research finding (Avitsur and Yirmiya 1999) is that, under experimental exposure to infectious disease cues such as injections of lipopolysaccharide (an important constituent of the outer membrane of Gram-negative bacteria), female rats show sickness behavior and a suppression of sexual activity. Male rats, on the other hand, do not suppress sexual activity, though they show other symptoms of sickness such as fever and reduced food consumption. Avitsur and Yirmiya (1999) suggested that the female suppression of sexual activity when sick may be a protective mechanism for avoiding conception while infected. They said that males "… seem to conceal their sickness when presented with an estrous female" (p. 793), and argued this was a strategy that functions in increasing male mating opportunities. Currently, it is unknown how carrying a current infection affects the sexual psychology of humans, but it is possible to investigate. A researcher could "infect" participants with endotoxins (e.g., lipopolysaccharide, Olsson et al. 2014; Schedlowski et al. 2014) and measure subsequent changes in their sexual attitudes or mate preferences.

## 6.10 Summary

The parasite theory of sexual selection, first discussed in detail and tested by Hamilton and Zuk in 1982, inspired research on the variation in marital systems across human indigenous societies. The variation in marital systems studied was the presence of monogamy versus polygyny and the degree of polygyny. Low (1988, 1990, 1994) and Marlowe (2003) reported that, across the traditional societies archived by anthropologists in the Standard Cross-Cultural Sample, parasite-stress scores are correlated positively with the degree of polygyny observed and recorded by anthropologists. Hence, as parasite stress increases across these societies, an increasingly small percentage of men monopolize more of the direct reproductive success of women. According to the parasite theory of sexual selection, this pattern is expected because high parasite stress generates high variation in the phenotypic and genetic quality of men, which then makes polygynous marriage adaptive for women, even though they must share a husband's resources with harem wives.

We hypothesize that collectivism is the value system that mediates the relationship between polygyny and parasite stress in the ethnographic record of anthropology. This is consistent with ethnographic accounts of Western polygyny and with certain research findings in social psychology.

Gangestad and Buss (1993) and Gangestad et al. (2006) conducted cross-national research inspired by the parasite theory of sexual selection that empirically linked human mate choice and parasite stress. They reported a positive correlation across countries between human parasite stress and the importance people place on physical attractiveness in mate choice. This finding was made from their reasoning that physical attractiveness is a certification of genetic resistance to parasites—good genes for parasite resistance—and hence is expected to be valued more in high than low parasite-stress regions.

The parasite theory of sexual selection is a subcategory of the more general and encompassing parasite-stress theory of values. The research studies reviewed in the chapter show that, as parasite stress increases, so does the value of obtaining a mate with phenotypic and genetic quality pertaining to dealing with parasite adversity.

Collectivists' marriages are more durable than are marriages of individualists. We argue that this difference is caused by the distinct values of conservatives versus liberals related to family ties, women's sexual restrictiveness, and adherence to traditional sex roles.

Parasite stress predicts positively women's sexual restrictiveness (continence) across many countries of the world. Specifically, as parasite stress declines, women show increased importance placed on having sexual relationships without commitment (less sexual restrictiveness). Nonzoonotic human parasites affect this pattern much more strongly than do zoonotic human parasites. Cross-national evidence indicates that women's sexual liberation from traditional values of sexual continence is a component of overall liberalization of values.

The frequency of consanguineous marriage varies greatly across contemporary countries and indigenous societies in the ethnographic record. We have hypothesized that a benefit of consanguineous marriage is keeping coadapted alleles together that defend against local parasites. Hence, as parasite stress increases across cultures, the parasite-stress theory of values predicts that people will engage in increased marital inbreeding and reduced marital outbreeding. Research findings support this hypothesis both cross-nationally and across indigenous cultures.

Klavina et al. (2011) showed that aspects of men's prejudice against out-group, immigrant men derive from the perception that the out-group men will mate with in-group women. They found, too, that this prejudice is related positively to men's PVD, which implies conservative men show more of the prejudice than liberal men. The violence and terrorism perpetrated by white men on African-American men in the Old South are informed importantly by Klavina et al.'s study. Stevenson et al. (2011) showed that men's sensitivity to sexual disgust is condition dependent. Sexually aroused men show reduced sexual disgust, but not reduced disgust in other disgust domains. Also, that effect was specific to sexual arousal, not general arousal. Their research implies sophisticated functional design of men's sexual disgust. Like other kinds of disgust, it functions to protect against contagion risk, but is moderated when men's mating opportunities arise.

## References


Avitsur, R., & Yirmiya, R. (1999). The immunobiology of sexual behavior: Gender differences in the suppression of sexual activity during illness. *Pharmacology Biochemistry and Behavior* 64: 787–796.

Betzig, L. (1986). *Despotism and Differential Reproduction: A Darwinian View of History*. Aldine, New York, NY.

Betzig, L., & Weber, S. (1993). Polygyny in American politics. *Politics and the Life Sciences* 12: 45–52.

Bittles, A. H. (1998). *Empirical Estimates of the Global Prevalence of Consanguineous Marriage in Contemporary Societies*. Morrison Institute for Population and Resource Studies, Working Paper 0074. Stanford University, Stanford, CA.

Borg, C., & de Jong, P. J. (2012). Feelings of disgust and disgust-induced avoidance weaken following induced sexual arousal in women. *PLoS One* 7: e44111.

Brundage, W. F. (1993). *Lynching in the New South: Georgia and Virginia, 1880–1930*. University of Illinois Press, Champaign, IL.
Buss, D. M. (1989). Sex-differences in human mate preferences–evolutionary hypothesis tested in 37 cultures. *Behavioral and Brain Sciences* 12: 1–14.
Buunk, A. P., Park, J. H., & Duncan, L. A. (2010). Cultural variation in parental influence on mate choice. *Cross-Cultural Research* 44: 23–40.
Cantú, S. M. (2013). When are women especially attracted to attractive men? Human mate preferences in a pathogen prevalent ecology. Ph.D. Dissertation submitted to University of Minnesota.
Chagnon, N. A. (1992). *Yanamamo: The Last Days of Eden*. Harcourt, Brace and Co., San Diego, CA.
Curtis, V., de Barra, M., & Aunger, H. (2011). Disgust as an adaptive system for disease avoidance behaviour. *Philosophical Transactions of the Royal Society B* 366: 389–401.
de Barra, M., DeBruine, L. M., Jones, B. C. et al. (2013). Illness in childhood predicts face preferences in adulthood. *Evolution and Human Behavior* 34: 384–389.
DeBruine, L. M., Little, A. C., & Jones, B. C. (2012). Extending parasite-stress theory to variation in human mate preferences. *Behavioral and Brain Sciences* 35: 86–87.
Denic, S., & Nicholls, M. G. (2007). Genetic benefits of consanguinity through selection of genotypes protective against malaria. *Human Biology* 79: 145–158.
Denic, S., Nagelkerke, N., & Agarwal, M. M. (2008a). Consanguineous marriages and endemic malaria: Can inbreeding increase population fitness? *Malaria Journal* 7: 150.
Denic, S., Nagelkerke, N., & Agarwal, M. M. (2008b). Consanguineous marriages: Do genetic benefits outweigh its costs in populations with alpha(+)-thalassemia, hemoglobin s, and malaria? *Evolution and Human Behavior* 29: 364–369.
Duncan, L. A., Schaller, M., & Park, J. H. (2009). Perceived vulnerability to disease: Development and validation of a 15-item self-report instrument. *Personality and Individual Differences* 47: 541–546.
Fincher, C. L., & Thornhill, R. (2008a). A parasite-driven wedge: Infectious diseases may explain language and other biodiversity. *Oikos* 117: 1289–1297.
Fincher, C. L., & Thornhill, R. (2008b). Assortative sociality, limited dispersal, infectious disease and the genesis of the global pattern of religion diversity. *Proceedings of the Royal Society of London Biological Sciences* 275: 2587–2594.
Fincher, C. L., Thornhill, R., Murray, D. R. et al. (2008). Pathogen prevalence predicts human cross-cultural variability in individualism/collectivism. *Proceedings of the Royal Society of London Biological Sciences* 275: 1279–1285.
Fleischman, D. S. (2014). Women's disgust adaptations. In *Evolutionary Perspectives on Human Sexual Psychology and Behavior* (eds. V.A.Weekes-Shackelford & T.K. Shackelford), pp. 277–296. Springer, New York, NY.
Gangestad, S. W., & Buss, D. M. (1993). Pathogen prevalence and human mate preference. *Ethology and Sociobiology* 14: 89–96.
Gangestad, S. W., Haselton, M. G., & Buss, D. M. (2006). Evolutionary foundations of cultural variation: Evoked culture and mate preferences. *Psychological Inquiry* 17: 75–95.
Gelfand, M. J., Bhawuk, D. P. S., Nishii, L. H. et al. (2004). Individualism and collectivism. In *Culture, Leadership, and Organizations: The GLOBE Study of 62 Societies* (eds. R. J. House, P. J. Hanges, M. Javidan et al.), pp. 437–512. Sage Publications, Thousand Oaks, CA.
Hales, B. C. (2007). *Modern Polygamy and Mormon Fundamentalism: The Generations after the Manifesto*. Greg Kofford Books, Draper, UT.
Hamilton, W. D., & Zuk, M. (1982). Heritable true fitness and bright birds: A role for parasites? *Science* 218: 284–387.
Hamilton, W. D., Axelrod, R., & Tanese, R. (1990). Sexual reproduction as an adaptation to resist parasites (a review). *Proceedings of the National Academy of Science USA* 87: 3566–3573.
Hoben, A. D. (2011). An evolutionary investigation of consanguineous marriages. Doctoral dissertation, University of Groningen.
Hoben, A. D., Buunk, A. P., Fincher, C. L. et al. (2010). On the adaptive origins and maladaptive consequences of human inbreeding: Parasite prevalence, immune functioning, and consanguineous marriage. *Evolutionary Psychology* 8: 658–676.

Hodes, M. (1999). *White Women, Black Men: Illicit Sex in the Nineteenth-Century South.* Yale University Press, New Haven, CT.

Jaber, L., Shohat, M., & Halpern, G. J. (1996). Demographic characteristics of the Israeli Arab community in connection with consanguinity. *Israel Journal of Medical Sciences* 32: 1286–1289.

Jaber, L., Halpern, G. J., & Shohat, M. (1998). The impact of consanguinity worldwide. *Clinical Genetics* 1: 12–17.

Jones, B. C., Feinberg, D. R., Watkins, C. D. et al. (2013a). Pathogen disgust predicts women's preferences for masculinity in men's voices, faces, and bodies. *Behavioral Ecology* 24: 373–379.

Jones, B. C., Fincher, C. L., Welling, L. L. M. et al. (2013b). Salivary cortisol and pathogen disgust predict men's preferences for feminine shape cues in women's faces. *Biological Psychology* 92: 233–240.

Key, V. O. (1949). *Southern Politics: In State and Nation.* A. F. Knopf, New York, NY.

Klavina, L., Buunk, A. P., & Pollet, T. V. (2011). Out-group mating threat and disease threat increase implicit negative attitudes toward the out-group among men. *Frontiers in Psychology* 2: 1–8.

Kokko, H., & Ots, I. (2006). When not to avoid inbreeding. *Evolution* 60: 467–475.

Lee, A. J., & Zietsch, B. P. (2011). Experimental evidence that women's mate preferences are directly influenced by cues of pathogen prevalence and resource scarcity. *Biology Letters* 7: 892–895.

Lieberman, D., Tooby, J., & Cosmides, L. (2003). Does morality have a biological basis? An empirical test of the factors governing moral sentiments relating to incest. *Proceedings of the Royal Society of London B* 270: 819–826.

Lieberman, D., Tooby, J., & Cosmides, L. (2007). The architecture of human kin detection. *Nature* 445: 727–731.

Little, A. C., DeBruine, L. M., & Jones, B. C. (2010). Exposure to visual cues of pathogen contagion changes preferences for masculinity and symmetry in opposite-sex faces. *Proceedings of the Royal Society of London B* 278: 2032–2039.

Low, B. S. (1988). Pathogen stress and polygyny in humans. In *Human Reproductive Behavior: A Darwinian Perspective* (eds. L. Betzig, M. Borgerhoff Mulder, & P. Turke), pp. 115–127. Cambridge University Press, Cambridge, U.K.

Low, B. S. (1990). Marriage systems and pathogen stress in human societies. *American Zoologist* 30: 325–339.

Low, B. S. (1994). Pathogen severity cross-culturally. *World Cultures* 8: 24–34.

Low, B. S. (2000). *Why Sex Matters: A Darwinian Look at Human Behavior*. Princeton University Press, Princeton, NJ.

Marlowe, F. W. (2003). The mating systems of foragers in the standard cross-cultural sample. *Cross-Cultural Research* 37: 282–306.

McDonald, M. M., Asher, B. D., Kerr, N. L. et al. (2011). Fertility and intergroup bias in racial and minimal group contexts: Evidence for shared architecture. *Psychological Science* 22: 860–865.

Moore, F. R., Coetzee, V., Contreras-Garduño, J. et al. (2013). Cross-cultural variation in women's preferences for cues to sex- and stress-hormones in the male face. *Biology Letters* 9: 20130050.

Murdock, G. P. (1949). *Social Structure*. MacMillan, New York, NY.

Murray, D. R., Jones, D. N., & Schaller, M. (2013). Perceived threat of infectious disease and its implications for sexual attitudes. *Personality and Individual Differences* 54: 103–108.

Navarrete, C. D., McDonald, M. M., Molina, L. E. et al. (2010). Prejudice at the nexus of race and gender: An outgroup male target hypothesis. *Journal of Personality and Social Psychology* 98: 933–945.

Navarrete, C. D., Olsson, A., Ho, A. et al. (2009). Fear extinction to an outgroup face: The role of target gender. *Psychological Science* 20: 155–158.

Oaten, M., Stevenson, R. J., & Case, T. I. (2009). Disgust as a disease-avoidance mechanism. *Psychological Bulletin* 135: 303–321.

Olsson, M. J., Lundstrom, J. N. Kimball, B. A. et al. (2014). The scent of disease: Human body odor contains an early chemosensory cue of sickness. *Psychological Science* 25: 817–823.

Park, J. H., van Leeuwen, F., & Stephen, I. D. (2012). Homeliness is in the disgust sensitivity of the beholder: relatively unattractive faces appear especially unattractive to individuals higher in pathogen disgust. *Evolution and Human Behavior* 33: 569–577.

Pawlowski, B., Nowak, J., Borkowska, B. et al. (2014). Human body morphology, prevalence of nasopharyngeal potential bacterial pathogens, and immunocompetence handicap principal. *American Journal of Human Biology* (published online, DOI: 10.1002/ajhb.22510).

Penn, D. J., & Potts, W. K. (1999). The evolution of mating preferences and major histocompatibility complex genes. *American Naturalist* 153: 145–164.

Pratto, F., & Hegarty, P. (2000). The political psychology of reproductive strategies. *Psychological Science* 11: 57–62.

Rantala, M. J., Moore, F. R., Skrinda, I., et al. (2012). Evidence for the stress-linked immunocompetence handicap hypothesis in humans. *Nature Communications* 3: 694.

Rao, P. S. S., & Inbaraj, S. G. (1977). Inbreeding effects on human reproduction in Tamil-Nadu of South-India. *Annals of Human Genetics* 41: 87–98.

Ridley, M. (1993). *The Red Queen: Sex and the Evolution of Human Nature*. Macmillan Publishing Company, New York, NY.

Schaller, M., & Murray, D. (2008). Pathogens, personality, and culture: Disease prevalence predicts worldwide variability in sociosexuality, extraversion, and openness to experience. *Journal of Personality and Social Psychology* 95: 212–221.

Schedlowski, M., Engler, H., & Grigoleit, J.-S. (2014). Endotoxin-induced experimental systemic inflammation in humans: A model to disentangle immune-to-brain communication. *Brain, Behavior, and Immunity* 35: 1–8

Schmitt, D. P. (2005). Sociosexuality from Argentina to Zimbabwe: A 48–nation study of sex, culture, and strategies of human mating. *Behavioral and Brain Sciences* 28: 247–311.

Shields, W. M. (1982). *Philopatry, inbreeding, and the evolution of sex*. State University of New York Press, Albany, NY.

Simpson, J. A., & Gangestad, S. W. (1991). Individual differences in sociosexuality: Evidence for convergent and discriminant validity. *Journal of Personality and Social Psychology* 60: 870–883.

Stevenson, R. J., Case, T. I., & Oaten, M. J. (2011). Effect of self-reported sexual arousal on responses to sex-related and non-sex-related disgust cues. *Archives of Sexual Behavior* 40: 79–85.

Thornhill, N. W. (1991). An evolutionary analysis of rules regulating human inbreeding and marriage. *Behavioral and Brain Sciences* 14: 247–260.

Thornhill, N. W. & Thornhill, R. (1987). Evolutionary theory and rules of mating and marriage pertaining to relatives. In *Psychology and Sociobiology: Ideas, Issues, and Applications* (eds. C. Crawford, M. Smith, & D. Krebs), pp. 373–400. Lawrence Erlbaum Associates, London, U.K.

Thornhill, R., & Gangestad, S. W. (1993). Human facial beauty: Averageness, symmetry, and parasite resistance. *Human Nature: An Interdisciplinary Biosocial Perspective* 4: 237–269.

Thornhill, R. & Gangestad, S. W. (2008). *The Evolutionary Biology of Human Female Sexuality*. Oxford University Press, New York, NY.

Thornhill, R. & Palmer, C. (2000). *A Natural History of Rape: Biological Bases of Sexual Coercion*. MIT Press, Cambridge, MA.

Thornhill, R., Fincher, C. L., & Aran, D. (2009). Parasites, democratization, and the liberalization of values across contemporary countries. *Biological Reviews* 84: 113–131.

Thornhill, R., Fincher, C. L., Murray, D. R. et al. (2010). Zoonotic and non-zoonotic diseases in relation to human personality and societal values: Support for the parasite-stress model. *Evolutionary Psychology* 8: 151–169.

Tooby, J. (1982). Pathogens, polymorphism, and the evolution of sex. *Journal of Theoretical Biology* 97: 557–576.

Trivers, R. L. (1985). *Social Evolution*. Benjamin/Cummings, Menlo Park, CA.

Tuskegee Archive. http://law2.umkc.edu/faculty/projects/ftrials/shipp/lychingstate.html.
Vandello, J. A., & Cohen, D. (1999). Patterns of individualism and collectivism across the United States. *Journal of Personality and Social Psychology* 77: 279–292.
Van den Berghe, P. L. (1981). *The Ethnic Phenomenon*. Elsevier, New York, NY.
Welling, L. L. M., Conway, C. A., DeBruine, L. M. et al. (2007). Perceived vulnerability to disease is positively related to the strength of preferences for apparent health in faces. *Journal of Evolutionary Psychology* 5: 131–139.
Wilkerson, I. (2010). *The Warmth of Others Suns: The Epic Story of America's Great Migration*. Random House, New York, NY.
Young, S. G., Savvo, D. F., & Hugenberg, K. (2011). Vulnerability to disease is associated with a domain-specific preference for symmetrical faces relative to symmetrical non-face stimuli. *European Journal of Social Psychology* 41: 558–563.

# Chapter 7
# Personality

## 7.1 Introduction

The scientific study of human personality and the resulting published literature comprise a large and empirically rich research area. Personality researchers use various classifications and definitions of personality; in general, however, these researchers view a personality trait as one that has a considerable degree of permanence or repetition in a person's cognitions, motivations, and behavior across a range of contexts of experience. For example, a person with a high score on the personality trait "openness to experience" is open-minded to new ideas as well as to interaction with new and different people, with this openness frequently shown across times and places. The sociosexual orientation inventory (SOI) variable, discussed in Chap. 6, is sometimes considered a personality variable (e.g., Schaller and Murray 2008; Thornhill et al. 2010), because it shows consistency in individual differences. Many of the topics we have discussed in previous chapters could be considered personality dimensions because of the chronicity of the traits within individuals (e.g., perceived vulnerability to disease (PVD) and associated cognition and behavior, disgust sensitivity, collectivism–individualism or conservatism–liberalism, philopatry, xenophobia, importance placed on physical attractiveness, adherence to traditional values, and so on). Traditionally in psychological research, values were often considered personality traits (see Carney et al. 2008 for historical review).

As we have discussed earlier in this book, however, many of these same personality dimensions are known to be subject to significant change, even immediately, and in ways predicted by the parasite-stress theory of values. Indeed, as shown in this chapter, certain classical personality dimensions also are conditionally expressed within the individual as expected from the parasite-stress theory of values. This theory, with its new perspective on personality, promises to provide a general theory of personality, because it allows an understanding of the proximate and ultimate causes of traits that are classical personality factors and those less unanimously considered personality traits by psychologists. According to the parasite-stress theory of values, the components of the behavioral immune system, including those

R. Thornhill and C.L. Fincher, *The Parasite-Stress Theory of Values and Sociality: Infectious Disease, History and Human Values Worldwide*,
DOI 10.1007/978-3-319-08040-6_7, © Springer International Publishing Switzerland 2014

related to factors typically labeled personality, will be designed for stability as well as conditionality to meet the local degree of parasite adversity. Also, the theory may provide a framework for understanding the degree of conditionality of personality traits as well as when during ontogeny they become relatively consistent individual differences. The parasite-stress theory predicts that as parasite stress increases and simultaneously becomes temporally stable across a region, individuals will adopt behavioral immunity defenses earlier in ontogeny and will be more resistant to subsequent liberalization of pathogen-defense values. Correspondingly, it is predicted that, as parasite stress decreases and this low parasite adversity simultaneously becomes temporally stable across a region, individuals will adopt individualistic values earlier in ontogeny and will be more resistant to subsequent illiberal changes in these values. The parasite-stress theory also predicts that, in regions with temporal fluctuations in parasite adversity (e.g., seasonal monsoons), individuals may more easily shift between conservative and liberal values compared to areas with stable parasite adversity. Hence, the parasite-stress theory of values may resolve the debate in psychological research about whether conservative versus liberal values have sufficient permanence to warrant the label of individual differences in ideology (see Carney et al. 2008 for discussion of this debate).

In this chapter we focus on the standard personality factors in relation to parasite stress. Most contemporary personality researchers restrict their analyses to five personality factors, the so-called big five—agreeableness, conscientiousness, extraversion, neuroticism, and openness to experience. The initial research that began to interrelate and synthesize the fields of evolutionary biology, parasitology, and big-five personality was conducted by Schaller and Murray (2008) and Thornhill et al. (2010), who addressed cross-national variation in personality, and by Mortensen et al. (2010), who examined individual differences in personality and within-individual changes in personality.

## 7.2 Schaller and Murray (2008)

Schaller and Murray (2008) first tested the application of the parasite-stress theory to extraversion (and, conversely, introversion) and openness to experience (conversely, closed-mindedness to the new and different). Many behaviors can increase individuals' exposure to infectious diseases. The costs to individuals of contacting disease, however, are weighed against the potential benefits to individuals associated with the behaviors. A behavioral preference for extraversion and associated social interactions with a diversity of people brings specific kinds of benefits—e.g., more diverse social networks, including mating pools, and exposure to new ideas—but, as first suggested by Daniel Nettle (2005), also implies greater exposure to new infectious diseases. There is much evidence that social network diversity and composition affects people's health relative to infections. For instance, diversity in one's social network is costly in terms of exposure to infectious disease, such as upper respiratory viruses, and this effect is especially strong in individuals experiencing stressful life events (Hamrick et al. 2002 and studies referenced therein). According to the parasite-stress theory of values, it follows that human populations will be characterized by extraversion primarily under ecological conditions of low infectious-disease adversity, whereas a more introverted personality will emerge when the prevalence of infectious disease is high.

Similarly, advantages may accrue to individuals who are curious, adventurous, and generally "open" to unfamiliar experiences and new ideas. Openness, however, also may be associated with increased risk of contracting infectious disease. Individuals who are curious and adventurous may be more likely to violate rituals and norms (such as those pertaining to hygiene, manners, and food preparation; e.g., Sherman and Billing 1999; Curtis et al. 2011) that serve as defenses against contact with local parasites. In addition, openness, like extraversion, is associated with increased contact with out-group members and other unfamiliar people who may be hosts to novel parasites to which the immune defenses of one's self and one's fellow group members are not adapted.

As we have emphasized, host–parasite coevolutionary races are localized on a microgeographical scale (Chap. 3). Thus, defenses of both the classical and behavioral immune systems are most suited to local infectious diseases, but not those outside one's in-group or typical social milieu. In short, neophobia reduces the risk of parasite-based morbidity and mortality, whereas neophilia increases the risk of contracting infectious diseases. The parasite-stress theory of values implies that human cultures are expected to show higher levels of openness (neophilia) within ecologies characterized by low parasite adversity and will show lower levels of openness (neophobia) within ecologies characterized by a higher prevalence of parasites.

Schaller and Murray (2008) used *Historical Parasite Severity*, a variable described in Chap. 5, as their measure of parasite adversity across nations. Their data on personality traits for countries included five measures of personality traits across cultures from the published literature and their own combination of published measures. The sample size for the relationships of parasite stress and personality varied, depending upon the personality measure, and ranged from 23 to 56 countries.

They reported that extraversion and openness to new experience are negatively and significantly correlated with parasite stress. Hence, as parasite stress increases, so does introversion and closed-mindedness to novel experiences. They found, however, no consistent relationships between parasite stress and the other three personality traits of the big five (conscientiousness, agreeableness, and neuroticism). In follow-up analyses, they chose to control multiple potentially confounding variables, and found that variation in life expectancy, economic factors (gross domestic product), and climate (absolute latitude, mean temperature) did not eliminate the predicted significant relationships between extraversion or openness to new experience and parasite stress. They also provided evidence that the global pattern is replicated across Murdock's six world regions.

They proposed that their finding of international variation in personality is the result of a conditional adoption by people of personality values based on a local assessment of parasite stress, or genetically different localized adaptation for such adoption, or both. As explained in Chap. 3, the existence of regional variation in genetically differentiated psychological adaptations that function in adopting and using values locally does not imply the absence of conditionality in the adaptations.

## 7.3 Experimental Studies

Mortensen et al.'s (2010) research clearly identifies the conditionality of personality adoption within individuals, and that the personality adopted is related to the individual's chronic, hence, stable, degree of concern about contagion risk (one's PVD). These researchers studied how people's self-perception, as related to certain personality variables, changed upon viewing pictures with parasite-relevant cues. The research found that when the participants were primed with stimulus pictures simulating a contagion threat, compared to viewing control stimuli (architectural objects), they shifted their self-perception toward an avoidance of interaction with people: they were less agreeable, less open to experience, and less extraverted. The shifts were especially strong in individuals who were more concerned about infectious diseases in their environment (high PVD participants).

Mortensen et al. (2010), in the same research report, followed up with a second experiment to determine if the psychological changes in self-perception were manifested in behavioral avoidance of other people; they were. Compared to reactions to control stimuli, parasite-primed participants reacted more quickly with avoidant arm movements when observing facial photographs of strangers. Also, the avoidant reactions were most rapid among high PVD subjects. Combined, the two experiments by Mortenson et al. provide evidence for previously unknown features of the behavioral immune system: perceived contagion risk causes people to change their attitudes about self such that they reinterpret their self as avoidant of social interactions and then behave in avoidant ways toward others.

As we discussed in Chap. 3, Schaller et al. (2010) and Stevenson et al. (2011) reported functional interactions between the behavioral immune system and the classical immune system. Schaller et al. reported that, compared to stress-inducing control stimuli (i.e., viewing photographs of people pointing guns at them), infectious-disease-relevant stimuli caused subjects to mount an immune response related to white blood cell activity that is defensive against parasites. Stevenson et al. (2011) showed that disgust induction in research participants causes an oral immune response. Prior to Schaller and Murray's (2010) and Stevenson et al.'s (2011) recent work, these psychological and behavioral reactions of people to parasite-relevant stimuli were not known to be connected to an immediate response of the classical immune system. The research described earlier on personality as a component of the behavioral immune system includes the immediate activation of the personality-defense aspect of behavioral immunity to visual stimuli simulating contagion risk in the immediate environment. These same stimuli also are known to evoke disgust, various aversions, and social prejudices in research participants, which reduce interpersonal contact (Park et al. 2003; Curtis et al. 2004; Park et al. 2007; Oaten et al. 2009).

## 7.4 Thornhill et al. (2010)

We now return to the topic of regional differences in personality. Thornhill et al. (2010) conducted a follow-up, cross-national study of the Schaller and Murray's (2008) cross-national research discussed earlier. The cross-national differences in personality traits are predicted by a parasite-stress theory of human sociality that emphasizes the potential infection risks associated with interactions with conspecifics. The risks associated with extraversion, for instance, refer specifically to the risk of human-to-human transmission. The infection risks associated with openness to new experience are not quite so specific, but many of the behaviors associated with openness (e.g., increased contact with out-group members and new ideas) do imply a higher risk of human-to-human transmission. According to the parasite-stress theory of sociality applied to humans, worldwide differences in the domains of human personality are unlikely to correlate with the presence of zoonotic parasites (which cannot be transmitted from human to human), but should correlate strongly with the presence of nonzoonotic parasites (which involve human-to-human transmission). Empirical evidence consistent with this prediction would provide unique support for the parasite-stress theory of human sociality. In a paper with Murray and Schaller, we examined this prediction by using the data discussed in Chap. 5 on the richness (number) of the types of human infectious diseases in terms of their mode of transmission (Thornhill et al. 2010).

The NEO Personality Inventory-Revised (NEO-PI-R) questionnaire is a widely employed and well-validated instrument available for assessing the five fundamental trait dimensions that account for most of the variability in the five classic types of human personality (see McCrae 2002); extraversion is one of these dimensions. Two different investigations have employed the NEO-PI-R questionnaire to assess and describe worldwide differences in extraversion. McCrae (2002) summarized results from several dozen independent investigations that used the NEO-PI-R questionnaire to assess the self-reported personality traits in about 30 different countries. Separately, McCrae et al. (2005) used the NEO-PI-R questionnaire and an observer-report methodology to assess the personality traits of 11,985 individuals living in about 50 different countries. Both investigations produced average extraversion scores for each country included in their analyses. Schaller and Murray (2008) found that parasite stress significantly and negatively predicted extraversion for both of these measures of extraversion. Openness to experience is also one of the five fundamental trait dimensions assessed by the NEO-PI-R questionnaire. McCrae (2002) and McCrae et al. (2005) reported average openness scores for each country included in their analyses. Schaller and Murray (2008) found that parasite stress significantly and negatively predicted measures of openness from both studies.

For the personality variables of extraversion and openness, Thornhill et al. (2010) computed correlations with each of the three parasite-richness indices (zoonotics and each of the two nonzoonotic indices). Recall that *Disease Richness* is a measure of the number of kinds of infectious diseases, not their severity. We treat parasite severity in relation to personality dimensions separately in the next section of this

**Table 7.1** Pearson zero-order correlations and $p$-values between each measure of parasite richness and each measure of extraversion and openness to experience; $n$ = the number of countries in each analysis (results originally reported in Thornhill et al. 2010)

| | Parasite richness measure | | | | | | |
|---|---|---|---|---|---|---|---|
| Values measure | Human-specific | $p$ | Multihost | $p$ | Zoonotic | $p$ | $n$ |
| Extraversion (McCrae 2002) | −0.58 | 0.001 | −0.49 | 0.006 | −0.28 | >0.10 | 30 |
| Extraversion (McCrae et al. 2005) | −0.54 | <0.001 | −0.34 | 0.02 | −0.31 | 0.03 | 48 |
| Openness (McCrae 2002) | −0.43 | 0.02 | −0.35 | 0.06 | −0.29 | >0.10 | 30 |
| Openness (McCrae et al. 2005) | −0.31 | 0.03 | −0.28 | 0.06 | −0.11 | >0.10 | 48 |

chapter. In addition, because the three indices of parasite richness were positively intercorrelated ($r$'s = 0.56–0.66), additional regression analyses were conducted to assess more rigorously the unique predictive effects associated with each index.

Table 7.1 reports the correlations between each parasite richness index and the two measures of extraversion. A clear pattern is evident: extraversion was predicted most strongly by human-specific parasite richness, somewhat less strongly by multihost parasite richness, and least strongly by zoonotic parasite richness. In follow-up regression analyses that included all three parasite-richness indices as simultaneous predictors of each extraversion measure, only the human-specific indices remained a statistically significant predictor (for the McCrae 2002 and McCrae et al. 2005 measures of extraversion, $p$'s = 0.02 and 0.001, respectively). There were negligible unique effects associated with the other two indices of parasite richness ($p$'s > 0.18).

A similar pattern emerged in the correlations between the parasite-richness indices and the two measures of openness, although the correlations involving the human-specific and multihost indices were not substantially different in magnitude (see Table 7.1). An additional set of regression analyses was created in which the human-specific and multihost parasite-richness indices were summed to create a broader index of nonzoonotic parasite richness; this nonzoonotic index was entered along with the zoonotic index as predictors of openness. The results revealed that the nonzoonotic index was a significant predictor of the McCrae et al. (2005) openness measure ($p$ = 0.03) and a near-significant predictor of the McCrae (2002) openness measure ($p$ = 0.10). In contrast, zoonotic parasite richness exerted no predictive effect whatsoever (std. $\beta$ = 0.00 and 0.04, $p$'s > 0.85).

In sum, the results were illuminating. Both human-specific and multihost parasite richness predicted uniquely cross-national differences in personality traits. Zoonotic parasite richness contributed little, if at all. Thus, the worldwide variation in certain personality dimensions is predicted almost entirely by the prevalence of nonzoonotic diseases. Several cross-national differences were predicted especially strongly by human-specific parasite richness. For example, only human-specific parasite richness (but not multihost parasite richness) uniquely predicted differences in extraversion. This highly specific effect fits the functional logic that informs the parasite-stress theory of sociality, given that extraversion is defined by behavioral interactions with other humans, but has little bearing on the broader range of behaviors that may expose individuals to between-species pathogen transmission.

## 7.5 Personality and Parasite Severity

Thornhill et al. (2010) were based on parasite richness measures. Subsequently, we computed for each country the parasite severity (number of cases) of each of the three types of human infectious diseases (see Chap. 5 for details). Basically, the above results for parasite richness were repeated with the new parasite severity scores. The analyses revealed that nonzoonotic severity correlated more robustly with each of the two personality measures provided by McCrae et al. than did zoonotic severity. The two extraversion data sets reported by McCrae et al. (2002 and 2005, respectively) show the following: nonzoonotics: $r=-0.50$, $p=0.005$, $r=-0.46$, $p=0.001$; zoonotics: $r=-0.28$, $p=0.13$, $r=-0.32$, $p=0.03$; sample sizes as in Table 7.1. The two openness data sets of McCrae et al. (2002 and 2005, respectively) show: nonzoonotics: $r=-0.44$, $p=0.02$, $r=-0.32$, $p=0.03$; zoonotics: $r=-0.29$, $p=0.12$, $r=-0.08$, $p=0.56$; sample sizes as in Table 7.1. All four of the correlations between nonzoonotics and personality were negative and statistically significant with moderate to strong effect sizes. Only one of the four correlations with zoonotics was statistically significant. Hence, as parasite severity increases across nations, people generally become more introverted and less open to experiences, and nonzoonotic infectious diseases show these patterns to a greater extent than do zoonotic diseases.

## 7.6 Condition-Dependent Personality

Our proximate theory for how humans obtain their value system and personality emphasizes condition-dependent psychology, which is supported by a diversity of evidence (Chap. 3; also Sect. 7.3). This contingency implies the presence of variation in infectious-disease stress on both a local scale and a short time scale (within the lifetime of individuals) in the deep-time generations of human evolutionary history. Thus, we are not saying that humans are primarily introverted and closed-minded, nor are they basically extraverted and open to experiences. Fundamentally, in terms of species-typical psychological design, humans are all of these, and how someone's personality develops depends on experiences pertaining to parasite adversity in the local ecological setting. The developmental ancestral cues affecting personality continue to act after adulthood and can shift a person's personality immediately.

As discussed in Chap. 3, the conditionality of values is not negated by evidence of genetic differences in value systems among regions should such evidence be found in future research. It is established that personality and political-values dimensions all show heritability, i.e., individual differences due to, in part, genetic differences among individuals (Hatemi et al. 2012; Ludeke et al. 2013). We emphasize that such heritability may arise from localized adaptation in genetic immunity, rather than in localized, genetically differentiated adaptation for adopting and using values. In this view, individual differences in genetic immunity are ontogenetic cues affecting the optimal values one chooses during ontogeny. Recent research has provided evidence for this view. There is within-population covariation at the individual level between certain features of behavioral immunity and alleles involved in classical immunity. Individuals with alleles for infectious-disease susceptibility have higher introversion and less exploratory behavior than individuals with resistance alleles (MacMurray et al. 2014). In this case, the susceptibility alleles may affect the adoption of behavioral immunity in either of two ways: they may do so because (a) bearers of the alleles contract the infectious disease and this causes bearers to increase behavioral immunity, or (b) allele bearers don't get the disease first, but their genetic susceptibility per se causes them to choose behaviorally defensive values.

As we mentioned at the beginning of this chapter, the parasite-stress theory of values promises a theoretical framework for predicting the degree of conditionality of personality traits and when in ontogeny they become relatively stable as individual differences. Temporally stable, high parasite adversity should evoke ontogenetically an early onset of behavioral-immunity defenses resistant to change toward increased liberalization. And temporally stable, low parasite adversity is expected to evoke ontogenetically an early onset of liberal values resistant to subsequent changes toward increased conservatism. And, temporally fluctuating parasite adversity is expected to evoke a transitional value system. The parasite-based chronicity of values, coupled with the ability of individuals to shift values immediately in relation to stimuli of contagion threat, we suggest gave rise to the debate in psychology about whether people actually have stable individual differences in values (see Carney et al. 2008 on this debate).

## 7.7 Implications for Synthesis and Future Research

Most generally, evolutionary thinking and research is the greatest generator and synthesizer of knowledge about all aspects of life on Earth. The evolutionary biological research discussed in this chapter provides indications that major parts of the research areas of parasitology, personality, and immunology are actually one field, and thus the research is unifying fields of scholarship that previously were thought to be unconnected. We anticipate that these findings and their synthetic implications will lead to an explosion in research that will produce a bounty of new discoveries that would not have been thought about or examined scientifically from the traditional approach to scholarship of distinct and independent disciplines. The hope of Francis Bacon and the visionaries of the Enlightenment and Scientific Revolution were a unified knowledge about nature—what Ed Wilson called consilience (Wilson 1998). In this big-picture perspective, to paraphrase Bacon, the globe of explanation is small—meaning that explanation is accomplished and unified by a small set of theoretical concepts. The unification of disciplines by the parasite-stress theory of sociality is apparent in each of the empirical chapters of this book.

In future research on human personality, it will be important to consider fully the parasite-stress theory's main features and their relationships to the psychometric scales that have been used to measure personality. For example, agreeableness, as used and measured currently in personality research, is not a factor or a trait that fully fits with the parasite-stress theory, because this theory is based on preferences or values related to in-group versus out-group sociality. In high parasite-stress areas with accompanying collectivism, in-group members will be agreeable with one another through conformity with traditional values, but not extend agreeableness to out-group members. In low parasite-stress regions, individualists are expected to show agreeableness widely and with a diversity of kinds of people. A psychometric tool (a questionnaire) is needed that allows people to respond to in-group and out-group attitudes regarding cooperation and agreeableness, while also measuring their basic values of conservatism–liberalism. The same considerations apply fully to the personality factor of conscientiousness and less so, but still significantly, to openness to experience. Extraversion, as measured in the currently used questionnaire, would need the least amount of revision to maintain it as a part of a parasite-stress-theory-driven psychometric tool. It is hard to know what the personality trait neuroticism means in light of the parasite-stress theory; it may have to be reevaluated entirely. Depressive mood, a component of neuroticism, may be caused, in part, by certain infectious diseases because depressive mood may function to analyze and solve socially complex problems and diseases may exacerbate these problems (Andrews and Thomson 2009; Watson and Andrews 2002). Aspects of depression, however, may be alleviated by the dutiful and reliable in-group social support seen primarily in collectivist cultures (see Chiao and Blizinsky 2010), which co-occur with high infectious-disease stress. Also, aspects of depression may be costly by-products of alleles that provide immunity to parasites in some generations or regions (Raison and Miller 2012).

The five-factor perspective on personality is not derived from evolutionary theory. Instead, the perspective is an analytical framework for describing measured variation in personality in a sample (see review in Gurven et al. 2013). Research supporting a big-five factor structure in personality has been conducted largely on educated, affluent samples, especially the personality research conducted in the West. There is no reason that the big-five view should apply to cultures in general. Gurven et al. (2013) conducted the first major study of personality in a small-scale indigenous society, the Tsimane, a forager–horticulturalist people of Bolivia. Their results do not support the five-factor model of personality. They report a two-factor structure of prosociality and industriousness. Gurven et al. (2013) describe the Tsimane as conservative, closed to new experiences and ideas, and under high infectious-disease adversity. The parasite-stress theory of values may inform future studies of personality in indigenous peoples by defining situations in which certain factors of five-factor personality will be absent or restricted or require a new interpretation.

Another domain of research that we suggest would be enriched by the application of the parasite-stress theory of values to personality is that of personality in nonhuman animals. Behavioral ecologists have described individual and average group differences in personality (also called behavioral syndromes) in a range of animal taxa (Barber and Dingemanse 2010; Sih et al. 2012; Wolf and Weissing 2012). Behavioral-immunity traits may be commonly what behavioral ecologists refer to as personality traits of nonhuman animals, especially the personality traits referred to as adventurous, dispersal-prone, shy, sedentary, and aggressive. We anticipate that, depending upon the similar personality traits compared between humans and nonhuman animals, in some cases the personality traits will be the result of convergence (independent evolution) and in other cases the result of common ancestry (homology). If these ideas are correct, the values dimension of conservatism–liberalism, which is thought to be human-specific, can be applied to nonhuman animals in light of a common theoretical basis in variable infectious-disease adversity (see too Chaps. 5 and 13).

## 7.8 Summary

This chapter focuses on three investigations of the relationship between human personality and parasite stress: Schaller and Murray (2008), Thornhill et al. (2010), and Mortensen et al. (2010). The first two treat cross-national variation in personality; the third is a study of individual differences in personality and within-individual changes in personality. These three studies indicate that the parasite-stress theory of sociality may provide a new and general theory of personality.

Schaller and Murray (2008) reported that as parasite stress increases, so does introversion and closed-mindedness to new experiences across moderate samples of countries. The various potential confounding variables they included in follow-up analyses did not eliminate the significant relationships between extraversion or openness to new experience and parasite stress. They also reported that the global pattern is replicated across the six world regions. Finally, they did not find significant, consistent relationships between parasite stress and the other three personality traits of the big five (conscientiousness, agreeableness, and neuroticism).

Thornhill et al. (2010) extended the study by Schaller and Murray (2008) by studying cross-national variation in extraversion and openness to experiences in relation to the richness of the three types of human infectious diseases: human-specific, multihost, and zoonotic. The two nonzoonotic categories predicted significantly cross-national differences in the personality traits; zoonotic parasite richness did so only marginally at best. Thus, the cross-national variation in the personality dimensions appears to be attributable almost entirely to the prevalence of nonzoonotic diseases. A follow-up study involving nonzoonotic versus zoonotic disease severity (rather than richness) provided results similar to those based on pathogen richness.

Mortensen et al. (2010) studied how people's self-perception of their own personality changed immediately after viewing pictures with parasite-relevant cues. They reported that when people were primed with stimuli simulating a contagion threat, they shifted their self-perception toward an avoidance of interaction with people: they were less agreeable, less open to experience, and less extraverted. Also, parasite-primed participants reacted with avoidant arm movements when observing strangers. Separate studies have shown that people's classical immune system is activated by the same types of parasite-salient cues that cause shifts in personality.

These three studies are consistent with the hypothesis that important aspects of personality are components of the behavioral immune system. Future research in human personality will benefit from consideration of the parasite-stress theory of values and its relationship to each of the big five personality factors. Future research in personalities of nonhuman animals may benefit from studying personality traits as traits of behavioral immunity.

## References


Andrews, P. W., & Thomson, J. A. (2009). The bright side of being blue: Depression as an adaptation for analyzing complex problems. *Psychological Review* 116: 620–654.

Barber, I., & Dingemanse, N. J. (2010). Parasitism and the evolutionary ecology of animal personality. *Philosophical Transactions of the Royal Society B–Biological Sciences* 365: 4077–4088.

Carney, D. R., Jost, J. T., Gosling, S. D. et al. (2008). The secret lives of liberals and conservatives: Personality profiles, interaction styles, and the things they leave behind. *Political Psychology* 29: 807–840.

Chiao, J. Y., & Blizinsky, K. D. (2010). Culture–gene coevolution of individualism–collectivism and the serotonin transporter gene. *Proceedings of the Royal Society B* 277: 529–537.

Curtis, V., Aunger, R., & Rabie, T. (2004). Evidence that disgust evolved to protect from risk of disease. *Proceedings of the Royal Society B* (Supplement) 271: 17–31.

Curtis, V., de Barra, M., & Aunger, H. (2011). Disgust as an adaptive system for disease avoidance behaviour. *Philosophical Transactions of the Royal Society B* 366: 389–401.

Gurven, M., von Rueden, C., Massenkoff, M. et al. (2013). How universal is the Big Five? Testing the Five-Factor Model of Personality Variation among forager–farmers in the Bolivian Amazon. *Journal of Personality and Social Psychology* 104: 354–370.

Hamrick, N., Cohen, S., & Rodriguez, M. S. (2002). Being popular can be healthy or unhealthy: Stress, social network diversity, and incidence of upper respiratory infection. *Health Psychology* 21: 294–298.

Hatemi, P. K., Byrne, E., & McDermott, R. (2012). Introduction: What is a 'gene' and why does it matter for political science? *Journal of Theoretical Politics* 24: 305–327.

Ludeke, S., Johnson, W., & Bouchard, Jr., T.J. (2013). "Obedience to traditional authority:" A heritable factor underlying authoritarianism, conservatism, and religiousness. *Personality and Individual Differences* 55: 375–380.

MacMurray, J., Comings, D.E., & Napolioni, V. (2014). The gene-immune-behavioral pathway: Gamma-interferon(INF-) simultaneously coordinates susceptibility to infectious disease and harm avoidant behaviors. *Brain, Behavior, and Immunity* 35: 169–175.

McCrae, R. R. (2002). NEO-PI-R data from 36 cultures: Further intercultural comparisons. In *The Five-Factor Model Across Cultures* (eds. R. R. McCrae and J. Allik), pp. 105–126. Kluwer Academic/Plenum Publishers, New York, NY.

McCrae, R. R., Terracciano, A., & 78 Members of the Personality Profiles of Cultures Project. (2005). Universal features of personality traits from the observer's perspective: Data from 50 cultures. *Journal of Personality and Social Psychology* 88: 547–561.

Mortensen, C. R., Becker, D. V., Ackerman, J. M. et al. (2010). Infection breeds reticence: The effects of disease salience on self-perceptions of personality and behavioral avoidance tendencies. *Psychological Science* 21: 440–447.

Nettle, D. (2005). An evolutionary approach to extraversion continuum. *Evolution and Human Behavior* 26: 363–373.

Oaten, M., Stevenson, R. J., & Case, T. I. (2009). Disgust as a disease-avoidance mechanism. *Psychological Bulletin* 135: 303–321.

Park, J. H., Faulkner, J., & Schaller, M. (2003). Evolved disease avoidance processes and contemporary anti-social behavior: Prejudicial attitudes and avoidance of people with disabilities. *Journal of Nonverbal Behavior* 27: 65–87.

Park, J. H., Schaller, M., & Crandall, C. S. (2007). Pathogen-avoidance mechanisms and the stigmatization of obese people. *Evolution and Human Behavior* 28: 410–414.

Raison, C. L., & Miller, A. H. (2012). The evolutionary significance of depression in pathogen host defense (PATHOS-D). *Molecular Psychiatry* 18: 15–37.

Schaller, M., & Murray, D. (2008). Pathogens, personality, and culture: Disease prevalence predicts worldwide variability in sociosexuality, extraversion, and openness to experience. *Journal of Personality and Social Psychology* 95: 212–221.

Schaller, M., & Murray, D. R. (2010). Infectious diseases and the evolution of cross-cultural differences. In *Evolution, Culture, and the Human Mind* (eds. M. Schaller, A. Norenzayan, S. J. Heine et al.), pp. 243–256. Psychology Press, New York, NY.

Schaller, M., Miller, G. E., Gervais, W. M. et al. (2010). Mere visual perception of other people's disease symptoms facilitates a more aggressive immune response. *Psychological Science* 21: 649–652.

Sherman, P. W., & Billing, J. (1999). Darwinian gastronomy: Why we use spices. *BioScience* 49: 453–463.

Sih, A., Cote, J., Evans, M. et al. (2012). Ecological implications of behavioural syndromes. *Ecological Letters* 15: 278–289.

Stevenson, R. J., Hodgson, D., Oaten, M. J. et al. (2011). The effect of disgust on oral immune function. *Psychophysiology* 48: 900–907.

Thornhill, R., Fincher, C. L., Murray, D. R. et al. (2010). Zoonotic and non-zoonotic diseases in relation to human personality and societal values: Support for the parasite-stress model. *Evolutionary Psychology* 8: 151–169.

Watson, P. J., & Andrews, P. W. (2002). Toward a revised evolutionary adaptationist analysis of depression: The social navigation hypothesis. *Journal of Affective Disorders* 72: 1–14.

Wilson, E. O. (1998). *Consilience: The Unity of Knowledge*. Knopf, New York, NY.

Wolf, M., & Weissing, F. J. (2012). Animal personalities: Consequences for ecology and evolution. *Trends in Ecology and Evolution* 27: 452–461.

# Chapter 8
# Interpersonal Violence

## 8.1 Introduction

In Chap. 6, we analyzed homicides involving lynchings of black Americans by white American men. As predicted by the parasite-stress theory of values, the numbers of such murders across US states are related positively to parasite stress and collectivism. This chapter treats some other types of interpersonal violence and homicide. We examine here interpersonal violence and homicide between romantic partners, between men, and perpetrated on children by caretakers. Later in the book (Chap. 12), we treat warfare, violence that is characteristically intergroup. The types of interpersonal violence analyzed in this chapter have been the focus of a great deal of research. The parasite-stress theory of values offers new perspectives and research directions for the study of interpersonal violence, and provides a theoretical and empirically synthetic foundation that promises to be more encompassing than those used by previous interpersonal-violence researchers.

## 8.2 Nonlethal Physical Aggression Between Romantic Partners, and Honor Ideology

In this section, we examine nonlethal physical aggression between romantic partners. This aggression includes events such as physical abuse, battering, slapping, kicking, and so on. Both men and women perpetrate such acts of physical aggression against their romantic partner. Men do this much more frequently than women, and women are injured much more frequently than men in aggressive interactions, but at least in Western countries such as the USA, the UK and New Zealand, a detectable proportion of romantic partner violence involves women as perpetrators. Hence, the degree of sex difference in the perpetrator and victim in nonlethal, romantic-partner violence varies across cultures of the world.

R. Thornhill and C.L. Fincher, *The Parasite-Stress Theory of Values and Sociality: Infectious Disease, History and Human Values Worldwide*,
DOI 10.1007/978-3-319-08040-6_8, © Springer International Publishing Switzerland 2014

John Archer (2006) examined this sex-difference variation in some detail. He reported from data across 16 countries that as the ideology of gender equality (the opposite of gender inequality) increased, the sex ratio of perpetrator-to-victim in romantic partner aggression changed such that more and more men were victims. He also reported that the same pattern was seen with individualism scores across these nations. Said differently, Archer (2006) found that as gender inequality increased (more sexism with assumed male superiority) or as collectivism increased, the sex ratio of male-perpetrator-to-female-victim increased.

Women's freedom across regions covaries with collectivism–individualism. Archer (2006) reviewed multiple, separate studies that found that gender inequality and collectivism are robustly co-occurring values across countries; colleagues and we also have shown this in recent publications (Thornhill et al. 2009; Thornhill et al. 2010). Moreover, Vandello and Cohen (1999) provide evidence of the positive relationship between collectivism and gender inequity across states of the USA.

A detailed discussion of gender equality–inequality is delayed until our chapter on democracy (Chap. 10), because this unidimensional value system is an aspect of democratization—of women with civil liberties and rights becoming political participants in society. For the purposes of our discussion in this chapter, it is only necessary to understand that gender inequality–equality entails the way males and females interrelate socially and hierarchically. Gender inequality reflects the value that women are inferior to men, whereas gender equality is the ideology of sexual egalitarianism. In countries with high gender inequality, the beating of wives by husbands is normative and endorsed by both sexes. For instance, a study in the collectivist country of Egypt reported that most women (86 % of those sampled) agreed that a man was morally justified in beating his wife if she opposed his will (see Archer 2006). (Discussions of violence in evolutionary terms can be misinterpreted by those who are unaware of the difference between a scientific analysis and a moral judgment. The "endorsement" we just referred to connotes that the victims and perpetrators are making strategic decisions, weighing costs and benefits of endorsement versus opposition. Under some circumstances, high parasite stress for example, endorsement of violence may be the most successful social strategy in spite of its costs. This is true for both the perpetrator and victim. While the victim may suffer from the immediate damage and the perpetrator experiences his or her own costs, doing something different, such as desertion of the mateship, may be even more costly. These benefits and costs, though measured in the pain and agony of the here-and-now, are the result of a legacy of decisions measured solely in the metric of the reproductive success of individuals in human evolutionary history.)

In the same paper, Archer (2006) analyzed data from 52 countries; he showed that three different, but related measures of rates of women's victimization by their romantic partner were significantly and negatively correlated with gender equality and individualism. He also showed that cross-national scores measuring sexist attitudes and approval of wife beating are related positively to women's rates of aggressive victimization by their romantic partner. Moreover, Archer (2006) reviewed numerous other studies, contributed to the scientific literature prior to his 2006 study, that support the patterns he found. These studies included cross-national analyses of various samples of countries as well as analyses across the states of the USA. Combined, then, the research by Archer (2006) and the authors of other studies summarized in his paper strongly support that collectivism and its component value, gender inequality, are associated with higher female romantic partner aggressive victimization and male partner aggressive perpetration than is seen under individualism and its associated gender equality, and that this finding is robust across both nations and the US states.

Archer (2006) interpreted his findings using so-called social-role theory. He emphasized that traditional sex roles and associated sex-specific behavioral expectations are normative in collectivist regions, whereas more liberalized egalitarian sexual roles are normative under individualism. Hence, in collectivist regions, women are believed to be inferior to men and are expected to be subservient to men. In such regions, women also are expected to be sexually reserved and faithful and respectful of their male romantic partner. As described in Chap. 4, in conservative cultures, these are major aspects of female honor. As the traditional sexism relaxes across cultures, or, said differently, as one moves across nations or US states from high collectivism to high individualism, gender equality increasingly becomes normative. Concomitant with this trend, women are more respected by men, which may, in part, reduce male-perpetrated aggression against female partners. With increasing individualism, women increasingly become autonomous agents behaving according to their personal interests rather than following the expectations of traditional sexist values, which, in part, may result in their greater willingness to resist male aggression with their own aggression and to act assertively in general. (Also see Thornhill et al. 2009, 2010 for additional empirical support of Archer's interpretations.)

Archer (2006) also mentions the strategic role of male aggression in mate guarding (as emphasized earlier by Daly and Wilson 1988). The research he reports (summarized above) is consistent with the view that the more collectivist the general cultural setting, the more men view their romantic partner as property to whom only the pair-bond male has legitimate sexual access. Male aggression toward the partner is for sexual control of the female, which functions to increase the male's probability of paternity (Platek and Shackelford 2006). Furthermore, men's aggression, both nonlethal and lethal, toward sexual rivals also functions, in part, as paternity protection (Daly and Wilson 1988). The combination of male aggression toward a romantic partner and male aggression toward sexual rivals is a salient part of what has been labeled "male honor" by researchers (Daly and Wilson 1988; Nisbett and Cohen 1996).

The research reported by Archer (2006) reveals that male honor is an ideology more characteristic of collectivist cultures than of individualist ones. This pattern, when coupled with the documented robust positive relationship between collectivism and parasite stress (Chap. 5), implies that high rates of physical aggression by men toward romantic partners and toward sexual competitors are more typical of regions of high than of low parasite stress. This prediction is confirmed later in this chapter.

We do not take issue with Archer's (2006) empirical findings or those of the various scholars he summarizes. Archer's explanations based on social-role theory, however, only go so far. Explanations that stop with cultural features of variation in ideological sex roles, patriarchy, or cultural history do not address basic or comprehensive causal understanding, because these cultural features are caused by the values of people and, in turn, by the causes of peoples' values. For full explanations of romantic-partner and male-honor violence, one must know what causes people's values, including the variation in those values across cultures. It is this kind of encompassing proximate causal understanding that the parasite-stress theory of values potentially provides. The parasite-stress theory of values also is anchored in the ultimate causation of evolutionary selection as the designer of the psychological adaptations that function in acquiring values.

As mentioned, women romantically paired to collectivist men are expected to honor and obey their husband—"to stand by their man"—and to not engage in any activity that might challenge his paternity or authority. When a collectivist, romantically paired man suspects or confirms that his romantic partner has violated the feminine normative expectations for her, he may perpetrate violence against her. Supporting this are findings by Vandello et al. (2009) reported in the literature after Archer's review in 2006. Vandello et al. (2009) investigated the effect of male- and female-honor ideology on people's opinions about a woman leaving an abusive romantic relationship and the moral correctness of a husband's physical abuse of his wife. Participants from strong honor-based cultures (Latinos, southern USA Anglos, and Chileans) rated as more favorable that a woman should remain in an abusive romantic relationship than did participants from cultures with low valuation of honor (the northern region of the USA and Anglo-Canadians). These researchers also reported that high-honor-culture participants were more positive than low-honor-culture participants about a husband perpetrating violence against his wife when he perceives her as flirting with another man. Although Vandello et al. (2009) distinguished the two types of cultures in terms of honor-based ideology, and not collectivism–individualism, the cultures they studied are distinct in this way, too (see Fincher et al. 2008; Fincher and Thornhill 2012).

Thus, strong honor cultures, i.e., highly collectivist cultures, have values that, in combination, can cause male-perpetrator–female-victim violence within romantic relationships. The male- and female-honor aspects of the ideology include: (a) a devaluation of females (relative to males), (b) the female partner is sexually controlled property of her male partner and hence is off-limits to sexual rivals and to be defended against them, (c) the female partner should stay in the relationship regardless of her treatment by the male, and (d) abuse of the female partner by the male for actual or suspected sexual activity outside of the romantic relationship is regarded as morally correct behavior (see also Vandello et al. 2009).

There is another part of this perspective on male violence toward a romantic partner that we emphasize. According to the parasite-stress theory of values, the collectivist values of gender inequality and female sexual control by male partners are causes of the violence. We stress that the same theory purports that men's high concern about paternity in collectivist cultures is a value directed at a very real adaptive problem for men in such cultures. In Chap. 6, we discussed the diverse evidence across countries and US states showing that regions of high parasite adversity, compared to low parasite-stress regions, coincide with increased importance of good looks in a mate. Good looks reflect phenotypic and genetic quality, as explained in Chap. 6. Thus, in high parasite-stress settings, women are expected to place more value on a good-genes sire for their offspring than in low parasite-stress settings. This, in turn, may lead pair-bonded women under high parasite adversity, compared to the same category of women in low parasite-adversity regions, to perceive more frequently that the benefits of extra-pair copulation to obtain good genes for their children exceed its costs. The enhanced effort and concern that men in collectivist cultures, in comparison to men in individualist cultures, place on female fidelity is consistent with this hypothesis. So, too, is the higher male-perpetrated violence against romantic partners in collectivist than in individualistic cultures—this violence is an extreme manifestation of collectivist men's elevated mate guarding. This effort and concern of men ranges from norms and laws of gender inequality that disenfranchise and limit women's activities and interaction with men to sexual control of the female partner by male-partner violence.

Estrus is the time in a woman's menstrual cycle when a sire's genetic quality becomes paramount, because estrus is the time when conception occurs. A considerable literature indicates that women's sexual preferences for phenotypic traits of men that likely connote genetic quality increase during estrus. Outside of estrus, however, women's mate preferences prioritize male resources and investment (Thornhill and Gangestad 2008). To date, multiple studies indicate that men's mate guarding increases when a partner is in estrus, compared to when she is not (Thornhill and Gangestad 2008). As of yet, it is unknown whether during estrus collectivist, romantically paired men mate-guard more than individualistic men. We predict that they do because they face greater paternity threats given that in collectivist cultures women highly value male genetic quality for offspring and may be motivated to obtain good genes through extra-pair mating.

## 8.3 Romantic-Partner Homicide

Now we turn to empirical investigation of homicide—lethal aggression—against a romantic partner, after which we treat some other types of homicide. When we initiated our research on homicide in relation to the parasite-stress theory of values, we found in the scientific literature only one prior report of homicide in relation to values: a brief, unpublished conference abstract by Lester (2002) that stated a strong, positive relationship ($r = 0.70$, $p < 0.0001$) between the overall homicide rate and collectivism across the 48 continental states of the USA. Lester's homicide data were for 1992 (National Center for Health Statistics) and the collectivism–individualism scores were the Vandello and Cohen's (1999) measures discussed in Chap. 5. Our analyses reported in Thornhill and Fincher (2011) (discussed below) replicated Lester's finding and extended it by showing that it occurs in each of multiple categories of homicide involving adult perpetrators and victims in the USA. Our analyses in Thornhill and Fincher (2011) also related these categories of homicide to parasite stress and thus to the parasite-stress theory of values.

We were unable to find suitable cross-national homicide-rate data that distinguish types of adult-on-adult homicides in terms of perpetrator, victim, and context. There are, however, data for these homicide types across the US states, and we used these data. Vandello (2007) provided these homicide-rate data (homicides per 100,000 people), which are divided into two categories of romantic-partner homicide (male- and female-perpetrated) as well as other homicide categories (which are analyzed later in this chapter). These data derive from reports of the US Federal Bureau of Investigation (FBI) in 1980, 1990 and 2000; two US states (Florida and Mississippi) did not provide data. Collectivism–individualism data for the USA are from Vandello and Cohen (1999). Rates of infectious disease across each of the US states were compiled from US Centers of Disease Control data for the years 1993–2006 as described in Chap. 5 and referred to there as *Parasite Stress USA*. Subsequent to the publication of Thornhill and Fincher (2011), we tabulated the severity (number of cases) of nonzoonotic and zoonotic human diseases in each of the states of the USA. (See Chap. 5 for discussion of methods and data.) We include below analyses that examine the covariation of adult-on-adult homicide types with *Parasite Stress USA* and each of these two other measures of human diseases.

### 8.3.1 Men Killing Female Romantic Partner

These homicides involved men killing wives, ex-wives, common-law wives or girlfriends, an exhaustive list of types of male–female romantic relationships in the FBI homicide data compiled by Vandello (2007). We tested for the following two predicted relationships: parasite stress and collectivism will be correlated positively with these homicides. The predicted relationships were found and were strong patterns: men killing a romantic partner and *Parasite Stress USA*, $r=0.70$, $p<0.0001$, $n=48$ states; men killing a romantic partner and collectivism, $r=0.53$, $p<0.002$, $n=48$. The parasite-stress theory proposes that collectivism, in part, mediates the relationship between these homicides and parasite stress. Consistent with this, an analysis of these homicides in relation to parasite stress with collectivism controlled resulted in a reduced effect; the zero-order $r$ of 0.70 declines to a partial $r$ of 0.57, $p<0.0001$. Also, as theoretically expected, the effect size of the relationship between collectivism and these homicides declined when parasite stress was controlled: partial $r=0.20$, $p=0.18$, not significant. Hence, collectivism appears to be a partial mediator of the relationship between the homicides and parasite stress and the relationship between the homicides and collectivism seems to be mediated largely by parasite stress. These results support the parasite-stress theory's application to the category of homicide in which a man kills his romantic partner.

In Chap. 5 we showed that collectivism is much more strongly predicted (positively) from nonzoonotic human diseases than from zoonotic human diseases in a region, which is consistent with the parasite-stress theory of values. Hence, we tested for the comparative effect of the two disease categories on male-perpetrated homicide of a romantic partner. As anticipated, the relationship of these homicides is much stronger with nonzoonotic disease severity ($r=0.65$, $p<0.0001$, $n=48$ states) than with zoonotic disease severity ($r=-0.09$, $p=0.54$, $n=48$). The severity of nonzoonotic diseases best accounts for rates of homicides in which men kill a romantic partner.

### *8.3.2 Women Killing Male Romantic Partner*

This category of homicide is the least common in Vandello's (2007) homicide data. Daly and Wilson (1988, 2010) evaluated the circumstances leading up to the homicidal behavior of women against their male romantic partner. Typically, these cases involve a long history of male nonlethal violence against the female, likely promoted in part by the female honor-based value mentioned above: a woman should remain in a relationship regardless of her partner's treatment of her. Although the female in these homicides tolerate the violence for long periods of time, 1 day they take their revenge. That the long-term, nonlethal aggression of the mate leading up to the female-perpetrated homicide is more characteristic of collectivist cultures than of individualist ones is supported by the research reported by Vandello et al. (2009) and Archer (2006) (discussed above).

We predicted and found that female-perpetrated homicide against the romantic partner is related positively to parasite stress and to collectivism across the US states: for *Parasite Stress USA*, $r=0.73$, $p<0.0001$, $n=48$ states; for collectivism, $r=0.46$, $p<0.008$, $n=48$. Both of these relationships, like those for male-perpetrated homicide of a romantic partner, are substantial in size. Controlling collectivism in analysis of the relationship between these homicides and parasite stress reduces the effect; the partial $r$ is 0.64, $p<0.0001$. The relationship between collectivism and these homicides with parasite stress controlled is near zero (partial $r=0.07$, $p=0.64$). The rates of these homicides are substantially, positively related to nonzoonotic disease severity ($r=0.61$, $p<0.0001$, $n=48$), but insignificantly and negatively related to zoonotic disease severity ($r=-0.15$, $p=0.31$, $n=48$).

## 8.4 Potential Confounders

The adult-on-adult homicide literature emphasizes the role of wealth inequality, as measured by the so-called Gini index, as a cause of homicidal behavior. Gini is a measure of the extent of harsh and dire circumstances experienced by individuals in a region as reflected in income variation, and the higher the Gini, the greater the variation in income. A robust positive relationship between homicide and the Gini index, as measured by income inequality at the family or the household level, has been documented cross-nationally, as well as across US cities and states, Canadian provinces, and Chicago neighborhoods. Income inequality is a stronger predictor than average or median income and various other variables that have been analyzed in traditional homicide research (see the multiple studies reviewed in Daly et al. 2001). The theoretical focus on income disparity is certainly warranted, as it is based on sexual selection theory. Accordingly, wealth disparity, not absolute or average resource level, should motivate men to compete violently, even lethally, for resources and mates (Daly and Wilson 1988; Daly et al. 2001).

Income inequality, however, is not a variable that is independent of parasite stress and associated collectivism–individualism. First, across the 50 US states and including the District of Columbia, *Parasite Stress USA* and Gini are strongly positively correlated: $r=0.76$, $p<0.0001$, $n=51$. The positive relationship across the 50 US states between collectivism and Gini is also substantial: $r=0.45$, $p=0.0001$, $n=50$. Second, we have argued for, and empirically supported (Thornhill et al. 2009), the following view. As collectivism increases, or said differently, individualism decreases, income inequality also increases. Conservatives value human inequality, viewing some people as more human and deserving than others, which, when combined with the conservative value of authoritarianism, promote and morally validate economic and social inequality across the populace. In contrast, liberals value human equality and support a more equitable resource and opportunity distribution across the populace (see Chap. 4). We have much more to say about values in relation to resource distribution across the people in different cultures in our chapter on democracy (Chap. 10) and our chapter on economics (Chap. 11).

Although the parasite-stress theory of values implies that resource inequality is caused, in part, by parasite stress, given the central role that prior homicide researchers have placed on the Gini index, we controlled it in additional analyses of the two types of homicide across US states treated above. (The Gini scores for US states and the District of Columbia are available from the US Census Bureau for two measures, household Gini and family Gini. The two measures are almost perfectly correlated: $r=0.97$, $p<0.0001$, $n=51$, for the census data of 1999.)

A multiple regression with the rate of male-perpetrated homicide of a romantic partner predicted by *Parasite Stress USA* and Gini (household) showed only parasite stress had a significant effect: $R^2=0.49$, $F=21.75$, $p<0.0001$, $n=48$; parasite-stress std. $\beta=0.72$, $p<0.0001$; Gini std. $\beta=-0.04$, $p=0.73$, not significant. A multiple regression with male-perpetrated homicide of a partner predicted by collectivism and Gini showed only collectivism had a significant effect: $R^2=0.29$, $F=9.03$, $p=0.0005$, $n=48$; collectivism std. $\beta=0.49$, $p=0.0008$; Gini std. $\beta=0.11$, $p=0.41$, not significant.

Similar results were found with female-perpetrated homicide of a romantic partner, with parasite stress and collectivism each predicting the rate of homicide, but Gini did not: for parasite stress and Gini in relation to the female-perpetrated homicide, $R^2=0.53$, $F=25.39$, $p<0.0001$, $n=48$; parasite-stress std. $\beta=0.72$, $p<0.0001$; Gini std. $\beta=0.01$, $p=0.94$, not significant. For collectivism and Gini in relation to the female-perpetrated homicide: $R^2=0.25$, $F=7.46$, $p=0.0016$; collectivism std. $\beta=0.39$, $p=0.007$; Gini std. $\beta=0.20$, $p=0.16$, not significant.

Hence, parasite stress and collectivism significantly explain variation in each of the two types of romantic-partner homicide independent of Gini. Gini itself seems not to influence the two homicide rates independent of parasite stress and collectivism.

Vandello (2007) hypothesized that the ratio of adult males to females across US states may cause variation in homicide rates by affecting the degree that men compete for women—the higher the ratio, the more homicides. His analyses of sex ratio in relation to homicide, however, yielded no good evidence for sex ratio as a significant predictor of variation in homicide rates at the conventional level of recognizing statistical significance ($p \leq 0.05$). Hence, we did not consider sex ratio to be a potential confounding variable. Vandello (2007) concluded that some of the relationships he did find, although not statistically significant, might achieve significance in large samples; his sample size was restricted by the fact that there were only 48 states for which he had both homicide and sex ratio data. Even with the restricted sample size, however, both collectivism and parasite stress predict romantic partner homicide at highly significant levels.

Scholars have proposed that warm or humid weather conditions cause human aggression (e.g., Simister and Cooper 2005; Van de Vliert 2009). We have criticized this claim on evolutionary theoretical grounds: the high cost of aggression leads to the expectation that aggressive acts will be engaged in primarily when individuals perceive benefits can be gained and not merely when they are uncomfortable from ambient conditions (see Letendre et al. 2010). The positive association across regions between aggression and temperature is undisputed. We propose, however, that this relationship occurs because each of the variables covaries positively with parasite stress. Hence, in this chapter, we do not examine climatic factors as potential confounds. For further treatment of climatic factors in relation to values, see Chap. 14.

To summarize, parasite stress and collectivism–individualism may be the strongest correlates of rates of romantic-partner homicide known to date. Also, nonzoonotic human diseases—those infectious diseases that can be transmitted human-to-human—are far better predictors of romantic partner homicide than are zoonotic diseases. The overall empirical picture implies that variation in parasite stress causes a region's values of collectivism–individualism, as well as its wealth inequality/equality, which then cause the region's rates of the two types of romantic-partner homicide. This picture is supported, too, by the earlier research reported in this chapter dealing with gender inequality and sexist ideology in general and their interrelationships with romantic-partner aggression and collectivism–individualism.

## 8.5 Male-Honor Homicides

### *8.5.1 Context*

The parasite-stress theory's application to homicide is not restricted to romantic-partner murders. We now examine the application of the theory to argument-related homicide—also called honor-related homicide—across US states. In this type of homicide, both perpetrator and victim are male. Typically, protagonists are acquainted, unrelated young adults (Daly and Wilson 1988). Honor-related homicide is the most common type of homicide. The context here is a brawl due to the influence of alcohol or narcotics, a lovers' triangle, an argument about money or property, or other arguments (Vandello 2007). It is called honor-related homicide because the issues boil down to argument and then violent escalation between men when their status is disrespected or their sexual exclusivity to a woman is threatened. The contest is about male face, status and reputation within the local community—in a nutshell, the contest centers on local male honor.

The label "honor homicide" also has been applied to homicides perpetrated by a family member on another family member who has violated family values. We discuss such homicides later in this chapter. We use the term honor homicide in this section in reference to homicides in the context of male–male competition for status, reputation, and mates in the local community.

## *8.5.2 Hypothesis*

The hypothesis we tested, which was derived from the parasite-stress theory of values, is as follows: honor homicide is caused by collectivist values and the high parasite stress that evokes such values. First, collectivists' values place a premium on family and other in-group's reputation and honor. Second, collectivists are embedded in the in-group and do not understand self as separate from in-group harmony and well-being; that is, collectivists have an interdependent self-concept, whereas individualists have an independent or autonomous self-concept. Collectivists, relative to individualists, have what researchers have called "low emotional and intellectual autonomy." (These features of collectivism are discussed in more detail in Chap. 4.) It follows, then, that when a collectivist is disrespected, simultaneously and necessarily that collectivist's family and other in-group members are disrespected. It is easy to understand then why collectivists become angry when "dissed" (vernacular for disrespected), and then may become lethally violent. We suggest, too, that the conservative values of parochialism and philopatry generate the typical context for these homicides of local competition between acquainted men.

## *8.5.3 The South's Culture of Male Honor*

The strong connection between male-honor ideology and status-related homicide among men has been documented by many scholars (see reviews in Nisbett and Cohen 1996 and Vandello et al. 2008), but most extensively by Nisbett and Cohen in their 1996 book, *Culture of Honor: The Psychology of Violence in the South*, which has been widely reviewed and cited. Scientific critiques of that book also have been published (Chu et al. 2000; Shackelford 2005; Barber 2009; Daly and Wilson 2010). Shackelford's (2005) and Daly and Wilson's (2010) critiques argued that Nisbett and Cohen's proposal in their book that culture is transmitted across generations by non-material means is scientifically wrong. Of course, Daly and Wilson and Shackelford are very likely correct (see Chap. 2). What Nisbett and Cohen were surely correct about, however, is their claim that there is a culture of male honor in the US South, and that it causes male–male homicide. In their book, Nisbett and Cohen review other scholars' similar claims and associated evidence, and provide important and copious new evidence of the South's culture of male honor and its relationship to male-honor homicide.

Nisbett and Cohen argue in their book that the culture of honor in the South exists because the South was settled by Britons who had the cultural heritage of herding grazing animals, and that in past generations the culture of honor was functionally important to the defense of the herds by their owners. They go on to say that as generations passed, herding disappeared from the South except in a few areas not suitable for the more productive farming and agricultural endeavors of the people. Moreover, they claim that cultural inertia retained the culture of honor in the South long after its material or ecological basis in herding was no longer present. Nisbett and Cohen's thinking is an example of using a region's cultural history as a cause of current culture in the region, a view we criticized in Chap. 2. In reality, culture is not transmitted by its own forces (inertia)—instead, its maintenance in a region is caused by people's value-adopter psychological adaptation, as we have emphasized (see Chap. 2). A region's cultural history is not an explanation; instead, it is what needs to be explained. Daly and Wilson (2010) advanced a similar criticism, too, and then proposed that wealth inequality (Gini) is the material basis (ecological/proximate cause) of the maintenance of the culture of male honor. As discussed above, they are likely correct, but apparently the more encompassing proximate causes of the culture of honor are collectivism and parasite stress. High parasite stress evokes collectivism, and collectivism includes the values of authoritarianism and social and economic inequality, which then account for wealth disparity across the population of a collectivist culture.

In their book, Nisbett and Cohen emphasized the culture of honor of southern white men due to the British herding ancestry of such men. The culture of the South, however, was caused by the values of southern blacks as well as by those of southern Anglos. The southern institution of slavery transported many Africans, with their own traditions of collectivism, from tropical and semi-tropical Africa to the South. With them, they brought numerous new infectious diseases, such as malaria and certain hookworm parasites, which became established in the South (Faust 1955; McGuire and Coelho 2011). As a result, the historically high parasite stress of the South, resulting from its climatic conditions that favor parasitic diseases, increased even more with slavery. This increase in even greater parasite adversity, in turn, maintained the collectivist ideology in the displaced Africans and their American descendants. Moreover, the historically high parasite stress of the South, combined with the virulent parasites brought in by slaves, evoked collectivism in the white immigrants and their descendants. The white immigrants came from higher latitudes, and thus relatively low parasite-stress regions, so, in their new home in the South, parasite stress was undoubtedly greater than they had encountered traditionally. Of course, even today, parasite stress in the South remains higher than in other regions of the USA, which (as documented in Chap. 5) corresponds to the high collectivism seen currently in the South in its people in general, not just Anglos.

To be explicit and clear here: we are hypothesizing that the culture of honor is the culture of collectivism. Individualistic men—liberal men—settle their differences of opinion with cool arguments and diplomacy. Collectivist men do not, as documented by the high honor-related homicide rates of the US states comprising the South, as compared to other states (Nisbett and Cohen 1996). When a collectivist man is offended, so are his father, mother, siblings, cousins, and other in-group allies—and all hell may break loose, including lethal violence.

### *8.5.4 Analyses*

If the hypothesis, stated immediately above, is reasonable, there must exist the following two correspondences: a positive relationship between male-honor homicide and parasite stress, and a positive relationship between male-honor homicide and collectivism. The hypothesis is supported by a robust, positive relationship in each case. There is a strong, positive relationship between honor-related homicide rates and *Parasite Stress USA*: $r=0.74$, $p<0.0001$, $n=48$ states. Also, there is a strong, positive correlation between male-honor homicide and collectivism: $r=0.58$, $p<0.0001$, $n=48$ states. The relationship between these homicides and parasite stress declined somewhat when collectivism was controlled: partial $r=0.64$, $p<0.0001$. The relationship between the homicides and collectivism declined substantially and became only marginally significant when parasite stress was controlled: partial $r=0.28$, $p<0.06$.

A multiple regression with male-honor homicide predicted by *Parasite Stress USA* and Gini showed parasite stress had a stronger effect, but both predictor variables were statistically significant: $R^2=0.62$, $F=36.71$, $p<0.0001$, $n=48$; parasite-stress std. $\beta=0.59$, $p<0.0001$, Gini std. $\beta=0.32$, $p=0.004$. A multiple regression with male-honor homicide by collectivism and Gini revealed that the two predictors had equal effects: $R^2=0.50$, $F=22.77$, $p<0.0001$, $n=48$; collectivism std. $\beta=0.43$, $p=0.0004$, Gini std. $\beta=0.43$, $p=0.0004$. Hence, wealth inequality does affect the rate of honor homicide across the states of the USA independently of the variable parasite stress, whereas this was not the case with the two types of romantic-partner homicides above—there, Gini had no independent effect on the two rates.

In the case of honor homicides, as in the two types of romantic-partner homicides, nonzoonotic disease severity strongly predicts homicide rates, whereas zoonotic disease severity does not. The positive correlation with nonzoonotics is substantial ($r=0.68$, $p<0.0001$, $n=48$ states). The correlation with zoonotics is negative in sign and near zero ($r=-0.07$, $p=0.61$, $n=48$).

## 8.6 Felony-Related Homicide

The US FBI distinguishes a category of homicide called felony-related homicide, which predominantly involves male–male homicides that happen during robbery, burglary, vehicle theft, or which are associated with narcotic drug law violations or other felonies or suspected felonies (Vandello 2007). In general, these homicides are not distinct in provocation and motivation from male-honor killings. Felony murders frequently involve a similar scenario to that of male-honor killings: insult or disrespect leads to escalation of conflict and the motivation of the parties to defend their honor (see Daly and Wilson 2010). Thus, rates of felony-related homicides should correlate with collectivism and parasite stress about as strongly as honor-related homicides; this is the case: with *Parasite Stress USA*, $r=0.65$, $p<0.0001$, $n=48$ states; with collectivism, $r=0.53$, $p<0.0001$, $n=48$ states. The relationship between these homicides and parasite stress declined when collectivism was controlled: partial $r=0.49$, $p=0.0005$. That between the homicides and collectivism declined considerably and became insignificant when parasite stress was controlled: partial $r=0.24$, $p=0.10$.

In multiple-regression analyses, parasite stress and Gini each significantly explained felony-related homicide, but parasite stress had a stronger effect: $R^2=0.49$, $F=21.90$, $p<0.0001$; parasite-stress std. $\beta=0.51$, $p<0.0001$, Gini std. $\beta=0.30$, $p=0.01$. Collectivism and Gini are about equal predictors of felony-related homicide: $R^2=0.42$, $F=16.03$, $p<0.0001$; collectivism std. $\beta=0.38$, $p=0.003$, Gini std. $\beta=0.40$, $p=0.002$.

The rates of felony-related homicide were correlated strongly with nonzoonotic disease severity ($r=0.56$, $p<0.0001$, $n=48$ states). These rates, however, were uncorrelated with zoonotic disease severity ($r=-0.02$, $p=0.89$, $n=48$).

## 8.7 Male Lifespan

In addition to the Gini index, in the homicide scientific literature, there is an emphasis on reduced male life expectancy as a cause of male–male homicide perpetration (Wilson and Daly 1997). This stems from life-history theory (Kaplan and Gangestad 2005; Figueredo et al. 2006). Accordingly, when men face short lives, they engage in more risk-taking, including violence, than when their lifespan is extended. It is important to understand, however, that male life expectancy is, in large part, a product of infectious disease morbidity and mortality; thus, life expectancy is not a variable independent of parasite stress. First, medical and sanitation improvements that reduce parasite stress were apparently the major cause of dramatically reduced mortality and increased longevity across many regions of the world over the last 100 years (Cutler et al. 2006). Second, the relationship between parasite stress and male life expectancy across the USA is −0.77 ($n=51$, including the District of Columbia), and the same robust pattern is seen cross-nationally: *Combined Parasite*

*Stress* (see Chap. 5) by male life expectancy (*Factbook* 2008), $r = -0.78$, $p < 0.0001$, $n = 190$ countries (for additional analyses, see Thornhill et al. 2009). Hence, parasite stress may be an ultimate and a proximate cause of men's life-history decisions. For completeness, we examined the relationship across US states between male-honor homicide and *Parasite Stress USA* with male life expectancy controlled. The zero-order relationship reported above in Sect. 8.5.4 ($r = 0.74$) was considerably reduced, but remained strong and statistically significant: partial $r = 0.50$, $p = 0.0003$. In Chap. 14, we discuss in more detail the relationship between life-history theory and the parasite-stress theory of values.

## 8.8 Sex Ratio

Barber (2009) has published cross-national evidence that total violent crime (murders, rapes, and assaults) is related negatively to sex ratio—that is, countries with *fewer* males have more violent crime. We hypothesize that this effect is caused by cross-national variation in parasite stress affecting birth sex ratio and men's life-history decisions pertaining to engaging in costly and risky criminal behavior. With Anders Møller, we have shown that sex ratio at birth and in adulthood correspond to variation in parasite stress across countries of the world. As parasite stress increases, sex ratios become more female-biased across 121 countries (Møller et al. 2008). These patterns were discovered as a result of our thinking about the high cost of male production by parents, coupled with the relatively poor condition of mothers, under high parasite stress, compared to low parasite stress. Infectious diseases negatively affect male survival more than female longevity. Hence, males are relatively more expensive to produce as parasite stress increases, and thus mothers are expected to produce fewer of them, especially given that the condition of mothers declines as parasite stress increases (see also Barber 2008). This setting, we argue, is why the empirical pattern Møller et al. found was that of female bias in birth sex ratio corresponding positively to parasite stress across regions. Hence, we are saying that Barber's (2009) finding is explained as follows. Although males are fewer in number in high-parasite-stress areas as a result of fewer being produced and surviving, those surviving are more risk-prone, because they face early mortality from parasite stress and hence engage in more violent and illegal activity.

We note that Dama (2012) recently conducted a study of cross-national variation in birth sex ratio in relation to parasite stress involving data from 226 countries. This study was inspired by the same theoretical reasoning as used in our study mentioned above: parasites cause morbidity and hence reduce the ability of mothers to invest in relatively expensive sons. Dama (2012) found the same cross-national result we found—parasite stress negatively predicts the birth sex ratio. Said differently, as parasite stress increases, the birth sex ratio becomes increasingly female-biased. Dama also showed that parasite stress was a much stronger predictor than a range of other variables. Thus, considering the Dama study and our study, there is evidence from two independent studies that variation in parasite stress causes variation in birth sex ratio.

## 8.9 Medical Care Affects Homicide Rates

Duntley and Buss (2008) have discussed findings indicating that emergency medical care, antibiotics, and other medical practices reduce greatly the immediate mortality following escalated, violent interpersonal aggression. They emphasize that this means that homicide rates in the Western world would be much higher without its modern medicine. Evidence-based medical treatment for wounds reduces the likelihood that a person inflicted with wounds will die and hence become a homicide victim. In Chap. 11, we discuss the evidence provided by medical care researchers showing that the use of a wide range of medical practices administered to Medicare beneficiaries varies greatly across the states of the USA. We argue that collectivism's reduced openness to new ideas and ways limits the quality of delivered healthcare by medical professionals. We show there that as both collectivism and parasite stress increase across the states, there is a corresponding decline in the use of evidence-based medical innovations. It follows that a man shot by a sexual competitor in Alabama is more likely to die from the shooting than a man similarly shot in New Hampshire. Hence, more collectivist states are expected to have higher homicide rates than individualist states because of reduced use of modern medicine in collectivist states. The variation in use of modern medicine, then, becomes a potential confound in the relationships presented above between homicide and both parasite stress and collectivism. We realized this potential confound after the publication of Thornhill and Fincher (2011).

Here, we addressed this issue by statistically controlling the effect of variable use of medical innovations across US states in the relationships of *Parasite Stress USA* and collectivism with each of the four types of homicides analyzed above. The medical variable involves the frequency of use in 2000–2001 of 22 medical interventions that have been empirically documented as lifesavers. The higher a state's rank, the less frequent the sum of these interventions is used in medical practice (see Chap. 11). The medical incompetency variable correlates highly with each of the four types of homicide across states: male-honor homicide, $r=0.77$, $p<0.0001$; felony-based homicide, $r=0.56$, $p<0.0001$; male-perpetrated partner-homicide, $r=0.52$, $p<0.0001$; female-perpetrated partner-homicide, $r=0.69$, $p<0.0001$; $n=48$ states for each analysis. Thus, the lower the quality of medical treatment in a state, the higher the rates of all four types of homicides.

The quality-of-medical-treatment variable, however, does not confound our conclusions that parasite stress and collectivism show significant relationships with the homicides. Statistically controlling the medical-use variable shows the following results for *Parasite Stress USA* and each of the homicides across 48 states: male-honor homicide, partial $r=0.51$, $p=0.0003$; felony-related homicide, partial $r=0.46$, $p=0.001$; male-perpetrated partner-homicide, partial $r=0.57$, $p<0.0001$; female-perpetrated partner-homicide, partial $r=0.53$, $p=0.0001$. Although the magnitude of the relationship between parasite-stress and each of the homicide types declined when modern medical use is controlled (see above sections for the zero-order statistics), the relationships remain moderately ($r=0.46$) to strongly ($r$'s about 0.5–0.6) correlated. Similar patterns are seen with collectivism (based on Vandello and Cohen's (1999) measure, see Chap. 5). When the medical-use variable is controlled: male-honor homicide, partial $r=0.46$, $p=0.001$; felony-related homicide, partial $r=0.39$, $p=0.007$; male-perpetrated partner-homicide, partial $r=0.40$, $p<0.005$; female-perpetrated partner-homicide, partial $r=0.28$, $p=0.057$. The partial-correlation effect sizes for collectivism, despite being lower than those for parasite stress, are statistically significant. The female-perpetrated partner-homicide partial $r$ is marginally significant with the two-tailed probability we use, but with a one-tail probability is $p=0.03$, and the one-tail is appropriate for inference because the positive direction of the relationship is predicted from the parasite-stress theory.

## 8.10 Cross-National Homicide

### *8.10.1 Data Issues*

We have focused above on homicide data across the US states because the rates of homicide types of interest are separated. To our knowledge, the separation of these types of murders is not available for international rates of homicide. The US data are also more likely to be reliable and accurate estimates than international homicide data (also see Vandello 2007). We hypothesize that under-reporting of homicides is correlated positively with the degree of collectivism across countries. First, compared to individualist countries, collectivist countries have more governmental corruption (see Chap. 11), which may allow elites to hide their political acts of lethal violence from the public eye or record. Highly conservative governments are highly autocratic, sometimes achieving a despotic rule by governmental elites (Chap. 10), which, by definition of despotism, gives those in authority license to kill political opponents with impunity. One can safely assume that autocratic governments do not keep accurate records of this carnage. Second, our analyses in this chapter suggest that officials in collectivist countries are more likely to interpret certain homicides (e.g., honor-based ones) as morally justified, which would reduce these being recorded as illegal violence. Third, the people in power in collectivist countries show low interest and monetary investment in public goods and services (Chap. 10), which would reduce record keeping and police activity that benefits the wellbeing of the public at large.

For these reasons, collectivist countries, relative to individualistic countries, are expected to show biased records of homicides, with the specific direction of the bias being under-reporting. If this is the case, then report-bias is systematic across countries in a way that counters our prediction that there will be a positive relationship between homicide rate and parasite stress or collectivism (and a negative relationship with individualism). We cannot discount that there is not the same systematic bias, but to a lesser degree, across the states of the USA. It seems that US states can opt out of reporting homicides, as two collectivist states, Florida and Mississippi, did in the FBI homicide data Vandello (2007) reported. However, even with error in the data, systematic or random, there are strong relationships between the homicide types and parasite stress and collectivism–individualism across the US states.

Error in data, systematic or random, reduces the likelihood of detecting a predicted relationship and the magnitude of the detected relationship if one is detected. There is, of course, error in all the variables we used in analyzing homicides, as well as in all other variables in this book. No credible scientist would claim to have an error-free data set. The reported relationships in the book are those detectable *despite* the error, and it is highly unlikely that any relationship we report is caused by error in data. Moreover, we emphasize that the homicide data were not tabulated by people aware of the parasite-stress theory of values, and thus there can be no unconscious bias in data collection toward supporting the theory.

With the caveats mentioned above in mind about the international homicide data, when conducting our research on interpersonal violence that eventually was published in Thornhill and Fincher (2011), we searched for some cross-national data that might be reasonably accurate. We obtained publicly available, national homicide rates (per 100,000 people) from Public Health Sources of the United Nations Office on Drugs and Crime (www.unodc.org/unodc/en/data-and-analysis/homicide.html). These data are rates of "the intentional killing of a person by another" for 2008. Unknown is how countries may differ in reporting of different types of homicide. Presumably, the United Nations data include to varying degrees all the homicide types we have analyzed above, as well as other types of homicides we treat later in this chapter. Also unknown is how variation in availability and application of modern medical knowledge affects cross-national homicide rates. It can be assumed safely that modern medical care is least used in highly collectivist countries and most used in highly individualistic countries. This would increase the relationships of parasite stress and collectivism with homicide. Further research is necessary to untangle the effects of cross-national variation in report-bias, differences in classifications of homicide types, and medical care on the relationships of interest here.

### 8.10.2 Analyses

The United Nations homicide variable was analyzed in relation to the variable *Combined Parasite Stress* described in Chap. 5. The relationship is strong: $r=0.54$, $p<0.0001$, $n=179$ countries. A multiple regression with this homicide variable by *Combined Parasite Stress* and Gini (*World Factbook* 2008) reveals a sizeable effect of parasite stress, but a smaller effect of Gini: $R^2=0.36$, $F=34.47$, $p<0.0001$, $n=126$; parasite-stress std. $\beta=0.43$, $p<0.0001$; Gini std. $\beta=0.24$, $p=0.01$. The relationship between homicide and individualism across countries is moderate in magnitude: $r=-0.40$, $p=0.001$, $n=64$. This analysis used the measure of individualism provided by Hofstede (see Chap. 5) as this measure has a relatively large sample size. A multiple regression with individualism and Gini as predictors of the international homicide rate indicated a robust effect of Gini, but an insignificant effect of individualism: $R^2=0.35$, $F=15.67$, $p<0.0001$, $n=61$ countries; individualism std. $\beta=-0.13$, $p=0.34$, not significant; Gini std. $\beta=0.51$, $p=0.0003$.

In sum, the findings of parasite stress and of collectivism in relation to homicide across US states are repeated at the cross-national scale. Homicide rates increase with both parasite stress and collectivism. Unlike the US states analyses, the national analysis indicates that collectivism is not a significant predictor of homicide independent of Gini.

Additional evidence that the parasite-stress theory is relevant to cross-national variation in homicide comes from the classic study of homicide by Archer and Gartner (1984), which provided the first data for homicide rates across many countries for the period of 1900 to early 1970s. They emphasized that their data suffered from variation in the definition, reporting, and labeling of homicides. We used in Thornhill and Fincher (2011) their category for "homicide" for the last year reported after World War II under the assumption that the latest reports would be most accurate. The reports used ranged from the 1950s to early 1970s, depending on the country. This variable for homicide we created and *Combined Parasite Stress* showed a strong, positive correlation: $r=0.53$, $p=0.004$, $n=27$. The covariation of *Hofstede Individualism* and the homicide variable we created was negative and moderate in effect size, but not statistically significant given the small sample of countries: $r=-0.37$, $p=0.11$, $n=20$.

We examined the United Nations homicide variable as well as the Archer and Gartner homicide variable in relation to cross-national estimates of zoonotic human disease severity and nonzoonotic human disease severity. Our findings, in general, are the same as with the analyses across states discussed above: nonzoonotics show the stronger relationships. For the United Nations data, across 184 countries, nonzoonotics are correlated strongly and positively with homicide ($r=0.52$, $p<0.0001$), whereas nonzoonotics have a marginally significant relationship with homicide ($r=0.14$, $p=0.06$). The same contrast is seen with the Archer and Gartner data, but both relationships are positive and statistically significant ($n=31$ countries). Nonzoonotics are correlated strongly and positively with homicide ($r=0.63$, $p<0.0001$), and zoonotics showed a moderate effect size ($r=0.36$, $p=0.05$).

## 8.11 Family-Honor-Based Homicide

The research presented so far in this chapter documents that parasite stress and collectivism–individualism are critical variables to consider in future research on aggression in romantic mateships, as well as the types of homicide we focused on above. The killing of family members by their relatives as a result of the victim's violation of family honor—often called honor killings—is another type of homicide that we propose is caused by collectivism and parasite stress. This type of homicide is linked closely to family honor and embeddedness, as is collectivism. Sexual and marital norm violations are frequent types of family-honor violations. If a girl or a young woman violates these values, it can be a serious threat to family honor. The perpetrator of the honor violation may be ostracized or disinherited from the family, requested by the family to commit suicide, or killed directly by the family, or, in some cases, killed by the local in-group outside of the family (as in community stoning of the norm violator). We are unaware of data that specifically address the frequency of this type of homicide cross-nationally, but abundant anecdotal reports in the literature indicate that it is most common in collectivist family settings (for a review, see Chap. 9 of Parrot and Cummings 2009).

## 8.12 Child Maltreatment and Parasite Stress

### *8.12.1 Hypothesis*

Researchers have investigated child maltreatment—neglect, abuse, and murder—by caretakers, typically parents, extensively. Three proximate causes of the maltreatment have received strong research support: (a) low levels of available resources to invest in dependent offspring, (b) a step-parent in the household in which a child is reared, and (c) when offspring are relatively low in phenotypic quality as a result of compromised health or disability (Daly and Wilson 1988). In regard to (c), it is established that unhealthy children and children with disabilities are at much greater risk of maltreatment than are healthy children (e.g., US Department of Health and Human Services, see website below; also Daly and Wilson 1988). Like other parental animals, human parents exhibit discriminative parental solicitude, as Daly and Wilson (1988) call this selective investment. Parents are designed by evolutionary selection to invest primarily when the investment would have resulted ancestrally in reproductive success. When conditions are dire, parents divest and thereby save their investment for later times when ecological conditions for offspring are improved. Step-parents lack the genetic overlap with their step-children that genetic parents have, hence step-parents are less investing in step-children than genetic parents are in their genetic children. Offspring of low phenotypic quality would have had low reproductive value in evolutionary ancestral settings and hence receive parental divestment.

Of the three above causes of child maltreatment, in Thornhill and Fincher (2011), we proposed that the effect of offspring phenotypic quality is related most directly to parasite stress. First, in research with Anders Møller we showed that the higher the parasite stress across nations, the higher the child mortality (Møller et al. 2009). It would follow from this that the higher the parasite stress, the greater the proportion of offspring born with low phenotypic quality. Second, pregnant females infected with parasites will be in poorer condition than unparasitized mothers, which will limit the ability of the unhealthy mothers to produce high-quality offspring. The maternal-condition effect is seen in the strong, positive relationship between *Parasite Stress USA* and the proportion of low birth-weight births to total births across the US states and the District of Columbia (2007, http://www.statehealthfacts.org): $r=0.69$, $p<0.0001$. Moreover, in a multiple regression, parasite stress has a larger influence on the percentage of low birth-weight infants than does Gini: $R^2=0.54$, $F=28.40$, $p<0.0001$; parasite-stress std. $\beta=0.45$, $p=0.0016$, Gini std. $\beta=0.35$, $p=0.013$. It follows from the two factors mentioned that the higher the parasite stress, the more offspring will be born with low phenotypic quality and associated reduced health and reproductive value. Hence, parasite stress is expected to be a proximate cause of child maltreatment by parents. We tested this hypothesis in Thornhill and Fincher (2011).

### *8.12.2 Evidence*

One well-established pattern consistent with the parasite-stress theory is that both nonlethal and lethal maltreatment by caretakers are focused on very young children, with rates declining strongly across increasing age categories. (See various data sources in Daly and Wilson 1988; also the US National Child Abuse and Neglect Data System website URL below.) The discriminative-parental-solicitude psychological adaptation of parents will be designed by evolutionary selection to detect low phenotypic quality early in the lives of offspring and then divest to reduce the costs of continuing parental care. Also, as offspring increase in age up to puberty and adolescence, their reproductive value increases, which causes parents to perceive them as more worthy recipients of parental investment (Daly and Wilson 1988).

Furthermore, if the parasite-stress theory applies to child maltreatment by parents, parasite stress will predict positively both homicide and nonlethal abuse and neglect of children by parents. Data available for the US states allow a test of this. The US government data on child maltreatment are replete with errors and inconsistencies (e.g., Crume et al. 2002; also see URLs cited below). States differ in how cases of the maltreatment are counted and in other procedures for its assessment and recording. Also, states differ in definitions of the maltreatment (see http://www.childwelfare.gov/pubs/factsheets/fatality.cfm). There is some indication, however, that the data on child maltreatment by caretakers have improved through the years. As of 2008, many states employed an improved standardized system of compiling information about child maltreatment cases (see URL just cited). Hence, in Thornhill and Fincher (2011) we used data for 2008 (the most recent year of reporting) from the National Child Abuse and Neglect Data System (NCANDS), US Department of Health and Human Services (http://www.childwelfare.gov/systemwide/statistics/can.cfm). According to NCANDS, most of these data derive from US child welfare agencies. They are submitted voluntarily by the states and the District of Columbia to NCANDS. In 2008, in the USA, 772,000 children were reported victims of abuse and neglect. About 80 % of child maltreatment perpetrators were parents, and another 7 % were other relatives of the victim. Of perpetrators who were parents, 90 % were the assumed genetic parent(s) of the child. In 2008, 1,740 children died from the maltreatment, a rate of 2.33/100,000 children. About 80 % of child-fatality victims were younger than 4 years old; 45 % were younger than 1 year.

First, we treat the patterns of rates of child death per 100,000 children due to abuse and neglect across the US states that reported data, including the District of Columbia. The relationship between *Parasite Stress USA* and child murders was strongly positive: $r=0.67$, $p<0.0001$, $n=48$. Gini is a measure of widespread economic stress in a state, and hence relates to the causal variable in child maltreatment of a limitation of resources available to caretakers to invest in their offspring—the higher the Gini, the higher the proportion of parents with resource limitation. As we have mentioned, economic indicators such as Gini seem to be driven, in part, by parasite stress and its effects on the values of people in a region. Relatively collectivist US states show higher wealth inequality (Gini) than do individualist states (see Sect. 8.4). Gini in relation to the child fatalities was $r=0.58$, $p<0.0001$, $n=48$. A multiple regression with the child fatalities predicted by *Parasite Stress USA* and Gini yielded a strong effect of parasite stress, but an insignificant effect of Gini: $R^2=0.46$, $F=19.46$, $p<0.0001$, $n=48$; parasite-stress std. $\beta=0.53$, $p=0.002$; Gini std. $\beta=0.20$, $p=0.21$. Gini alone has little predictive power to explain interstate variation in the child murder rate beyond Gini's covariation with parasite stress. Parasite stress is a stronger predictor than Gini of rates of child fatalities resulting from caretaker abuse and neglect.

Rates of nonlethal caretaker's abuse and neglect of their children also is related positively and significantly to parasite stress, but less strongly than the pattern for child murders by caretakers. Data for the nonlethal category derive from the 2008 reports to NCANDS from 49 states plus the District of Columbia. The relationship is $r=0.28$, $p=0.05$, $n=50$. In this case, however, there is no statistically significant effect of parasite stress independent of Gini: nonlethal abuse and neglect by parasite stress with Gini controlled, partial $r=0.10$, $p=0.50$.

As mentioned early in this chapter, subsequent to Thornhill and Fincher (2011) we derived parasite severity scores for the two transmission categories of human parasitic diseases, nonzoonotic and zoonotic, for each state. Nonzoonotic human disease severity is much more strongly related to child maltreatment than zoonotic disease severity. For child murders and nonzoonotics, the correlation is 0.71 ($p<0.0001$, $n=48$ states), but with zoonotics it is negative and statistically insignificant ($r=-0.23$, $p=0.11$, $n=48$). For nonlethal abuse, the relationship with nonzoonotics is $r=0.24$ ($p=0.09$, $n=50$), but with zoonotics is near zero ($r=0.0003$, $p=0.97$, $n=50$); with a one-tailed $p$, the correlation with nonzoonotics is statistically significant ($p=0.05$).

In sum, parasite stress is a strong, positive predictor of child-murder-by-caretaker rates across the states of the USA. This effect is robust to the effects of certain economic conditions. Parasite stress is also a significant positive predictor of nonlethal abuse and neglect by caretakers, but this effect is smaller and seems to be mediated, in part, through the interrelationship of parasite stress and economic factors. Nonzoonotic human disease adversity is a stronger predictor of both child murders and nonlethal abuse of children than is zoonotic disease adversity.

## 8.13 Child Maltreatment and Collectivism

The relationship between the human value dimension of collectivism–individualism and child maltreatment is conceptually complex. From one perspective, it might be argued that collectivism will co-vary positively with child maltreatment (and with individualism, negatively). The reasoning is as follows. Cultures of honor—collectivist ones—condone, endorse, and value violence as a mechanism to control the behavior of social intimates and allies as well as enemies. In this chapter, we have documented this most completely with data reviewed by Archer (2006) on nonlethal aggression in romantic relationships across countries and the states of the USA, and data we analyzed on adult-on-adult homicides across US states and nations. The anecdotal evidence on within-family honor killings by relatives also is consistent with the use of violence as a morally correct mechanism of within-family social control and manipulation. Under collectivism, the strategic use of violence is potentially socially encompassing. Hence, cultures of honor also may use violence to socialize, control, and manipulate children. This is suggested from the South's relatively high value placed on the use of corporal punishment, both by parents at home and by schools, to control and punish children (Nisbett and Cohen 1996). Compared to other regions of the USA, the South places more value on the expectation that children will follow the traditional, normative rules of behavior. Also, in the USA, there are more positive attitudes about and endorsement of corporal punishment of children in Republican-party states than in Democratic-party states (see Chap. 4). From another perspective, however, one might predict that child maltreatment will be related negatively to collectivism if collectivist norms and associated behaviors are effective in controlling children such that excessively abusive and neglectful tactics are relatively unnecessary. A third perspective is that the value dimension of collectivism–individualism is not related at all to child maltreatment; in this case, child maltreatment is caused by parasite stress, economic factors, step-parent divestment in step-children, and possibly other proximate causes.

The third perspective is the only one of the three supported by the evidence (Thornhill and Fincher 2011). Across the 47 states for which data exist for both collectivism (scores from Vandello and Cohen 1999) and child death resulting from abuse and neglect, the relationship between these two variables is $r=0.14$, $p=0.35$, not significant. For the 49 states with data for both collectivism and nonlethal child abuse and neglect, the relationship is $r=0.05$, $p=0.72$, not significant.

Given these findings that collectivism does not covary significantly with child maltreatment, it is unlikely that quality of medical care, which declines as collectivism increases, is a major factor in generating the significant correlations we present above between parasite stress and child murder and nonlethal maltreatment of children. The role of variable medical care across states, however, may be a variable for study in future research on child maltreatment.

Various other factors contribute to the conceptual complexity of a possible relationship between collectivism–individualism and child maltreatment. The effect of collectivism–individualism on child maltreatment may be dependent upon the children's ages. Older children, perhaps especially those of pubertal ages, are more likely to oppose parental rules of conduct and thereby engage in activity that would provoke maltreatment by caretakers. Thus, it is in this older age group that maltreatment may be associated positively with collectivism (and negatively with individualism). Although comparative data are anecdotal at this time, the honor killings of family members by other family members (see above) appear to support this conjecture. This type of homicide seems to be concentrated on pubertal or older children, especially females, who violate norms of family honor. Finally, collectivists have more durable marriages (a lower divorce rate) than individualists (Gelfand et al. 2004, Chap. 4), which would reduce the frequency of collectivist families containing a step-parent and thus reduce the rate of child maltreatment caused by step-parent divestment.

The data do not allow a study of parasite stress and child maltreatment while controlling for the presence of a step-parent in the home or a study of whether step-parent predicts the maltreatment when other variables are controlled. We hypothesize, however, that the relationship between parasite stress and step-parent-perpetrated child fatalities is positive and significant, given that high parasite stress will have a negative effect on the phenotypic quality of children, both in the genetic-child-parent context and the step-child-parent context. Relatively more often, low-phenotypic-quality step-children may be the target of maltreatment by step-parents, because of these children's greater maintenance costs and limited ability to provide benefits to step-parents.

## 8.14 Overall Summary of Findings

The parasite-stress theory of values appears to account for considerable amounts of the variation in the rates of the types of human adult-on-adult violence across the states of the USA examined in this chapter. As predicted from the theory, across the US states, parasite stress and collectivism correlate positively and consistently with male- and female-perpetrated murder of a romantic partner, male-honor homicide, and felony-related homicide. Gini (wealth inequality) seems to have a significant role, independent of its interaction with parasite stress and collectivism, in increasing rates of male-honor and felony-related homicides, but not in increasing rates of romantic-partner homicide. Parasite stress and collectivism also positively predict international rates of overall homicide. The severity of nonzoonotic human diseases is a stronger predictor (positive) of adult-on-adult homicide across states than is zoonotic disease severity; this same pattern is seen with cross-national measures of homicide.

Each of two types of child maltreatment by caretakers (primarily parents), child murder and nonlethal abuse and neglect, across the US states is predicted positively by parasite stress. Gini had little, if any, predictive power for rates of child murder by caretakers independent of the covariation of Gini with parasite stress. Nonlethal abuse and neglect of children by caretakers is not predicted by parasite stress independent of Gini. Neither type of child maltreatment is predicted by collectivism. As with adult-on-adult homicides, the rates of each of the two types of child maltreatment are more strongly explained by nonzoonotic diseases than by zoonotic diseases.

Parasite stress appears to be the most empirically verified variable accounting for many types of interpersonal violence and homicide. The research presented in this chapter contributes to a unification of understanding of the causes of major types of interpersonal violence and homicide and anchors them in infectious diseases and the values these diseases cause.

After Thornhill and Fincher (2011) was published, other researchers published their own studies of the relationship between parasite adversity and violence. In general, this subsequent work supports the parasite-stress theory of values as applied to violence, but also offers certain criticisms of our published research on this topic. We treat this work, including the criticisms, in Chap. 14.

## 8.15 Other Tests

The parasite-stress theory's application to interpersonal violence could be tested in ways other than the cross-national and cross-states comparative methodology used in this chapter. As discussed in Chap. 2, using multiple methods is an important aspect of routine scientific investigation, because each method has strengths and weaknesses for determining causation. An experimental approach involving variation in parasite stress, say in the USA, immediately comes to mind as an additional test ground for the hypothesis that parasite stress causes interpersonal violence.

Such a test could be modeled after the eradication effort against hookworm parasites in the South (see Bleakley 2007). Health improvements against hookworms were applied in some but not other regions, providing a sample of manipulated versus comparison (control) regions that were compared in terms of educational and economic outcomes (see Miguel and Kremer 2004). Another source of data was from comparing regions of high initial hookworm prevalence with adjacent regions of low initial hookworm prevalence in terms of outcomes subsequent to health intervention across all the regions (Bleakley 2007).

In the case of hookworm in the South, eradication efforts started about 1910 when more than 40 % of the school-age children in the South were infected with this virulent parasitic worm. John D. Rockefeller, a wealthy man, initially supported the eradication effort financially. Various eradication efforts were made eventually, ranging from providing widespread publicity and information to people, free shoes, financing the building of outhouses at homes and public places, and administering antihelminthic drug treatment orally. Follow-up efforts from humanitarian groups outside the South, as well as from Southern state and local agencies, were seen across some regions of the South (Bleakley 2007). Amazingly, the South's government itself provided some resources for the eradication effort, because the hookworm issue became part of the public consciousness and thus a Southern political issue of importance. Yet, most of the major financial backing came from outside the South. The basic scientific knowledge of the hookworm life cycle, methods for measuring hookworm incidence, and documentation of hookworm-caused human morbidity were accomplished externally to the South. This scientific knowledge was essential for the visionaries involved in the eradication to understand how to intervene so as to reduce the hookworm health problem.

This was a very successful eradication program: major progress was made within a decade and continued across successive decades, as measured by the reduced incidence of the worm and by a corresponding large increase in educational motivation, school attendance, and school performance of children in areas receiving eradication intervention compared to control areas, as well as in areas with relatively high initial incidences of the parasite. The long-term benefits of the eradication program were seen in analyses of data from the 1950 census: economic productivity of adults increased greatly, as measured by occupational income and socioeconomic indices (Bleakley 2007).

This same approach could be used to test for the effect of the reduction of the impact of infectious diseases on homicide. Adding comprehensive, free, easily accessible and sustained healthcare and monitoring for all adults and children in states with a high homicide rate (such as Alabama or South Carolina; Vandello 2007) would provide data of scientific value for the test. Bleakley (2007) provides procedures for analyzing data to determine immediate and long-term effects of the health intervention. The parasite-stress theory of values leads to the prediction that homicide rates would begin to decline after one generation (about 20 years) and continue to decline in successive generations. Evidence indicates that healthcare improvements that target infectious diseases (e.g., child vaccination programs, vector control, chlorinated and fluoridated water, antibiotic availability, and sanitation legislation) begin to change the values of people in a region in one to two generations: people show increased liberalization in their values (Thornhill et al. 2009, Chap. 10). The reduction in collectivism (with a corresponding increase in individualism) following the health improvements is expected to cause reductions in rates of all types of homicides related to collectivist ideology. Child maltreatment rates also are predicted to decline with infectious disease reduction.

Another test of the parasite-stress theory of values as applied to violence would be historical research to investigate our hypothesis that reduced parasite adversity was a cause of the significant decline in violence in the West over the last several centuries. This decline has been studied by Pinker (2011), and we discuss it in relation to the parasite-stress theory in the section in Chap. 14 on future research topics.

## 8.16 Prospect for Societal Control of Homicide

If we are scientifically correct about parasite stress as a proximate cause of homicide, then a simple and obvious solution to the social problem of homicide presents itself. We assume the preference for lower homicide rates is widespread among the peoples of the world. This assumption is supported by the illegality of homicide across much of the world. And Western politicians and humanitarian groups voice the morality of low murder rates. The scientific results presented in this chapter do not allow the conclusion that a particular rate of homicide is morally superior to another rate. Instead, these results suggest a basis for an evidence-based approach to homicide if people are serious about reducing its rate. According to the parasite-stress theory of values, the reduction of homicide can be achieved by reducing the proportion of people who are collectivist in core values, and more basically by reducing infectious-disease prevalence. Moreover, the reduction in collectivism and parasite stress would reduce the wealth inequality that also contributes to certain types of homicide. The focus on parasite-stress reduction would require full, unimpeded and permanent healthcare for all people, coupled with suitable plumbing, parasite-free water, disease-vector control, and sanitary living and working environments for all. Such interventions, we argue, would liberalize the value system of people widely in a few or several generations (see Chap. 10), and thereby reduce homicide rates. Potentially, these interventions would be total solutions and hence there would not be any need for simultaneous increased investment in the legal mechanisms of law enforcement, criminal courts, jails and prisons to control homicide.

There is convincing evidence that public-health expenditures that target infectious diseases actually reduce the severity of human parasitic diseases in general (i.e., reduce the number of cases of the diseases), but do not reduce the number of kinds of infectious diseases. Robert Dunn and colleagues (2010) recently reported these important findings derived from their analyses across most of the countries of the world. In the same paper, they also reported that local ecological conditions, such as climatic factors, account for virtually all the variation in parasite richness across regions (number of kinds of infectious diseases). In only very rare cases will public health measures drive a parasite to extinction in a region; for example, this was seen with small pox. Hence, the successful employment of health interventions to combat infectious diseases is not limited to the classic success involving hookworm parasites. The findings by Dunn and colleagues, when coupled with the results presented in this chapter, offer the promise of success to humanitarians who want to reduce interpersonal violence in the world.

## 8.17 Summary

In this chapter, the parasite-stress theory of values is applied to some major categories of interpersonal violence, and the empirical findings are supportive of the theory. Parasite stress may be the strongest predictor of interpersonal violence to date.

We review research presented by Archer (2006) indicating that collectivism includes values that can promote romantic-partner violence. We argue that the collectivist values evoked by high parasite stress cause adult-on-adult interpersonal violence. Across the US states, parasite stress and collectivism each positively predict rates of men's and women's slaying of a romantic partner, as well as the rate of male-honor homicide and of felony-related homicide. Of these four types of homicide, Gini—a measure of resource inequality—has an independent effect only on rates of male-honor and felony-related homicides. Parasite stress and collectivism also positively predict cross-national homicide rates.

Child maltreatment by caretakers is caused, in part, by divestment in offspring of low phenotypic quality, and high parasite stress produces more such offspring than low parasite stress. Rates of each of two categories of the child maltreatment—lethal and nonlethal—across the US states are predicted positively by parasite stress. Gini is not a predictor of child murder rates independent of Gini's correlation with parasite stress. However, parasite stress is not an independent predictor of nonlethal child maltreatment when Gini is controlled statistically.

Nonzoonotic human diseases explain a large amount of the positive relationship between each of the types of human violence and parasite adversity we addressed in this chapter. Zoonotic human diseases play a relatively minor role, and in some types of the violence appear to play no role.

We suggest additional tests of the parasite-stress theory's application to interpersonal human violence. Health improvements in areas could be tracked to determine if homicide rates subsequently decline. If the parasite-stress theory of values applied to human violence is accurate, it could be useful in reducing rates of interpersonal violence across the world.

By no means is the application of the parasite-stress theory limited only to interpersonal violence. It also appears to help elucidate the other major category of violence, intergroup violence or warfare (Letendre et al. 2010; Letendre et al. 2012). The topic of warfare is treated in Chap. 12.

## References


Archer, J. (2006). Cross-cultural differences in physical aggression between partners: A social-role analysis. *Personality and Social Psychology* 10: 133–153.

Archer, D., & Gartner, R. (1984). *Violence and Crime in Cross-national Perspective*. Yale University Press, New Haven, CT.

Barber, N. (2008). Explaining cross-national differences in polygyny intensity: Resource-defense, sex-ratio, and infectious diseases. *Cross-Cultural Research* 42: 103–117.

Barber, N. (2009). From steroids to nation states: An integrated evolutionary approach to violent crime. *Aggression and Violent Behavior* 14: 415–422.

Bleakley, H. (2007). Disease and Development: Evidence from hookworm eradication in the American South. *The Quarterly Journal of Economics* 122: 73–117.

Chu, R., Craig, R., & Loftin, C. (2000). Herding and homicide: An examination of the Nisbett-Reaves Hypothesis. *Social Forces* 78: 971–987.

Crume, T. L., DiGuiseppi, C., Byers, T. et al. (2002). Underascertainment of child maltreatment fatalities by death certificates, 1909–1989. *Pediatrics* 110: 18.

Cutler, D., Deaton, A., & Lleras-Muney, A. (2006). The determinants of mortality. *Journal of Economic Perspectives* 20: 97–120.

Daly, M., & Wilson, M. (1988). *Homicide*. Aldine de Gruyter, New York, NY.

Daly, M., & Wilson, M. (2010). Cultural inertia, economic incentives and the persistence of "Southern violence." In *Evolution, Culture, and the Human Mind* (eds. M. Schaller, A. Norenzayan, S. Heine et al.), pp. 229–241. Psychology Press, New York, NY.

Daly, M., Wilson, M., & Vasdev, S. (2001). Income inequality and homicide rates in Canada and the United States. *Canadian Journal of Criminology* 43: 219–236.

Dama, M. S. (2012). Parasite stress predicts offspring sex ratio. *PLoS ONE* 7: e46169. doi:10.1371/journal.pone.0046169

Dunn, R. R., Davies, T. J., Harris, N. C. et al. (2010). Global drivers of human pathogen richness and prevalence. *Proceedings of the Royal Society B* 27: 2587–2595.

Duntley, J. H., & Buss, D. M. (2008). The origins of homicide. In *Evolutionary Forensic Psychology: Darwinian Foundations of Crime and Law* (eds. J. Duntley & T. Shackelford), pp. 41–64. Oxford University Press, New York, NY.

Faust, E. C. (1955). History of human parasitic infections: Parasitism in Southeastern United States. *Public Health Reports* 70: 958–965.

Figueredo, A. J., Vasquez, G., Brumbach, B. H. et al. (2006). Consilience and life history: From genes to brain to reproductive strategy. *Developmental Review* 26: 243–275.

Fincher, C. L., & Thornhill, R. (2012). Parasite-stress promotes in-group assortative sociality: The cases of strong family ties and heightened religiosity. *Behavioral and Brain Sciences* 35: 61–79.

Fincher, C. L., Thornhill, R., Murray, D. R. et al. (2008). Pathogen prevalence predicts human cross-cultural variability in individualism/collectivism. *Proceedings of the Royal Society of London Biological Sciences* 275: 1279–1285.

Gelfand, M. J., Bhawuk, D. P. S., Nishii, L. H., et al. (2004). Individualism and collectivism. In *Culture, Leadership, and Organizations: The GLOBE Study of 62 Societies* (eds. R. J. House, P. J. Hanges, M. Javidan et al.), pp. 437–512. Sage Publications, Thousand Oaks, CA.

Kaplan, H. S., & Gangestad, S. W. (2005). Life history theory and evolutionary psychology. In *The Handbook of Evolutionary Psychology* (ed. D. M. Buss), pp. 68–95. John Wiley and Sons, Inc., New York, NY.

Lester, D. (2002). Collectivism and rates of personal violence (suicide and homicide). *Psychological Reports* 90: 300.

Letendre, K., Fincher, C. L., & Thornhill, R. (2010). Does infectious disease cause global variation in the frequency of intrastate armed conflict and civil war? *Biological Reviews* 85: 669–683.

Letendre, K., Fincher, C. L., & Thornhill, R. (2012). Infectious disease, collectivism, and warfare. In *The Oxford Handbook on Evolutionary Perspectives on Violence, Homicide, and Warfare* (eds. T. Shackelford & V. Weekes-Shackelford), pp. 351–371. Oxford University Press, New York, NY.

McGuire, R. A., & Coelho, P. R. P. (2011). *Parasites, Pathogens, and Progress: Diseases and Economic Development*. MIT Press, Cambridge, MA.

Miguel, E. & Kremer, M. (2004). Worms: Identifying impacts on education and health in the presence of treatment externalities. *Econometrica* 72: 159–217.

Møller, A. P., Fincher, C.L., & Thornhill, R. (2008). Sex ratio, economic prosperity and pathogens. Chapter 8. Corey L. Fincher's Ph.D. Dissertation, The University of New Mexico, Albuquerque NM.

Møller, A. P., Fincher, C. L., & Thornhill, R. (2009). Why men have shorter lives than women: Effects of resource availability, infectious disease, and senescence. *American Journal of Human Biology* 21: 357–364.

Nisbett, R. E., & Cohen, D. (1996). *Culture of Honor: The Psychology of Violence in the South*. Westview, Boulder, CO.

Parrot, A., & Cummings, N. (2009). *Forsaken Females: The Global Brutalization of Women*. Rowman and Littlefield Publishers, New York, NY.

Pinker, S. (2011). *The Better Angels of our Nature: The Decline of Violence in History and its Causes*. Allen Lane, London, UK.

Platek, S. M., & Shackelford, T. K. (2006). *Female Infidelity and Paternal Uncertainty: Evolutionary Perspectives on Male Anti-cuckoldry Tactics*. Cambridge University Press, Cambridge, U.K.

Shackelford, T. K. (2005). An evolutionary psychological perspective on cultures of honor. *Evolutionary Psychology* 3: 381–391.

Simister, J., & Cooper, C. (2005). Thermal stress in the U.S.A.: Effects on violence and on employee behaviour. *Stress and Health* 21: 3–15.

Thornhill, R., & Fincher, C. L. (2011). Parasite stress promotes homicide and child maltreatment. *Philosophical Transactions of the Royal Society: Biological Sciences* 366: 3466–3477.

Thornhill, R. & Gangestad, S. W. (2008). *The Evolutionary Biology of Human Female Sexuality*. Oxford University Press, New York, NY.

Thornhill, R., Fincher, C. L., & Aran, D. (2009). Parasites, democratization, and the liberalization of values across contemporary countries. *Biological Reviews* 84: 113–131.

Thornhill, R., Fincher, C. L., Murray, D. R. et al. (2010). Zoonotic and non-zoonotic diseases in relation to human personality and societal values: Support for the parasite-stress model. *Evolutionary Psychology* 8: 151–169.

United Nations Office on Drugs and Crime, www.unodc.org/unodc/en/data-and-analysis/homicide.html

U.S. National Child Abuse and Neglect Data System website U.S. Department of Health and Human Services, http://www.childwelfare.gov/systemwide/statistics/can.cfm.

Vandello, J. A. (2007). Sex ratios and homicide across the U.S. *International Journal of Psychology Research* 1: 59–80.

Vandello, J. A., & Cohen, D. (1999). Patterns of individualism and collectivism across the United States. *Journal of Personality and Social Psychology* 77: 279–292.

Vandello, J. A., Cohen, D., & Ransom, S. (2008). U.S. southern and northern differences in perceptions of norms about aggression—Mechanisms for the perpetuation of a culture of honor. *Journal of Cross-Cultural Psychology* 39: 162–177.

Vandello, J. A., Cohen, D., Grandon, R. et al. (2009). Stand by your man: Indirect prescriptions for honorable violence and feminine loyalty in Canada, Chile, and the United States. *Journal of Cross-Cultural Psychology* 40: 81–104.

Van de Vliert, E. (2009). *Climate, Affluence, and Culture*. Cambridge University Press, New York, NY.

Wilson, M., & Daly, M. (1997). Life expectancy, economic inequality, homicide, and reproductive timing in Chicago neighborhoods. *British Medical Journal* 314: 1271–1274.

World Factbook, http://www.cia.gov.

# Chapter 9
# Religiosity

## 9.1 Introduction

The peoples of the world vary greatly in degree of religiosity, that is, the degree to which they manifest and value religious commitment and participation (Norris and Inglehart 2004; McCleary and Barro 2006). In this chapter, we argue that a major reason for this variation stems from a central phenomenon of social life we have called "in-group assortative sociality" that varies in degree in relation to parasite stress. As explained in Chap. 5, such sociality refers to the preferential association among similar individuals who comprise an in-group versus an out-group or dissimilar others. Phenotypic features such as dress and formal costumes, tattooing and scarification, culinary preference, language and dialect, religion and other belief systems, normative behavior, social displays, rituals, and body scent mark in-group similarity. Assortative sociality's three general social components are (a) philopatry or limited dispersal for reproduction from the natal locale, (b) ethnocentrism or in-group favoritism and association, and (c) xenophobia or out-group dislike and avoidance; neophobia, the avoidance and dislike of out-group ideas and ways, is a component of xenophobia. In prior chapters, we provided numerous lines of convergent evidence that the parasite-stress theory of sociality fundamentally ties the intensity and extent of these three phenomena to varying levels of parasite stress experienced by people, both within a region and across geographic space. These phenomena are features of the behavioral immune system and function in avoidance and management of infectious disease. In this chapter, we show that religious affiliation and commitment are linked functionally to these phenomena as an aspect of in-group assortative sociality. Accordingly, religiosity functionally is a component of the behavioral immune system. We conducted our empirical analyses both cross-nationally and across states of a single polity, the USA. We reported the findings in Fincher and Thornhill (2012).

R. Thornhill and C.L. Fincher, *The Parasite-Stress Theory of Values and Sociality: Infectious Disease, History and Human Values Worldwide*,
DOI 10.1007/978-3-319-08040-6_9, © Springer International Publishing Switzerland 2014

## 9.2 Religiosity: A Costly Signal of In-Group Allegiance

Participation in a religion has certain costs for the participant, which include the time and effort involved in learning a religion and practicing it, the loss of opportunity to engage in other beneficial activities (opportunity costs), and risks such as the avoidance of modern medical care or extended fasting (Sosis et al. 2007). To learn the emotionality and associated language of a religion requires a long developmental (ontogenetic) exposure to the belief system. Opportunity costs include the inability to associate with other groups because one's specific beliefs may be considered irrational or contra-evidentiary to out-group members. (On irrationality as a functional component of religiosity, see Irons 2008.)

This setting—that religious participation has costs to participants—is the basis for the study of religiosity from both economic and evolutionary science points-of-view. From the economics viewpoint, Iannaccone (1994) used the rational choice theory of economics to examine the relationship between a church's religious strictness and its strength or permanence, and concluded that, "Strictness reduces free riding. It screens out members who lack commitment and stimulates participation among those who remain" (p. 1204). In other words, paying-in with costs indicates commitment, but it also precludes desertion to other churches, because it is too costly to desert and develop the same level of embeddedness in a new church. Therefore, individuals in strict churches exhibit higher rates of participation, because they are assured, in comparison to individuals in less strict churches, a higher level of return on their investment through the reduction of free-riders (those who gain benefits without paying the costs) and a higher level of investment by other individuals in the church. Iannaccone (1994) established that there is variation among churches in strictness and hence in the average religiosity among its members. That same research established that the strictest churches—those that require the highest costs for continued membership—have the tightest and most permanent collectives.

Many researchers have applied evolutionary costly signaling theory to the understanding of religion and religious behavior (Cronk 1994; Irons 1996, 2001; Sosis 2000, 2003, 2005; Wilson 2002; Sosis and Alcorta 2003; Sosis and Bressler 2003; Sosis and Ruffle 2003; Bulbulia 2004a, b; Johnson 2008; Steadman and Palmer 2008; Henrich 2009). This approach builds on the same foundation as the economic study of religious behavior—that religious participation has costs. The evolution-minded researchers using costly-signaling theory propose that individuals' membership in a religious group is necessary to accrue social benefits not accessible by independent living, and engaging in religious behavior is a signal of in-group allegiance to other individuals (both in-group and out-group individuals). The greater the costs of religious participation, the more honestly the participation signals allegiance to the religious in-group. In this view, religious groups adopt their own distinct costly versions of supernatural beliefs in order to heighten costs of participation and distance themselves from out-groups.

Sosis (2000) and Sosis and Bressler (2003) provided supportive evidence for the costly signaling theory of religion by studying the longevity of nineteenth century United States communes. Sosis (2000) found that religious communes, in comparison to secular communes, had longer life spans. More specifically, Sosis and Bressler (2003) found that longevity for religious communes was related positively to the magnitude of the costly acts required for membership within a commune (e.g., costly acts included restriction from alcohol and sex). A logical prediction from the costly signaling perspective, put in evolutionary theoretical terms, is that the adaptive value of religious costly signaling to signalers, and hence the magnitude and associated costs of the signal, will vary from place to place based on the underlying ecological necessity of in-group assortative sociality for inclusive fitness maximization (also see Sosis et al. 2007). The parasite-stress theory of sociality emphasizes infectious-disease avoidance and management as the ecological adversity for which in-group assortative sociality and associated religiosity provide adaptive (ancestrally) solutions.

## 9.3 The Parasite-Stress Hypothesis of Religiosity

According to the parasite-stress theory of sociality, the formation and maintenance of in-group assortative sociality by practiced and signaled religious allegiance provides two benefits: (a) the protective barrier provided by isolation from out-group individuals who may harbor novel infectious diseases as well as perform non-normative behavior with associated contagion risks, and (b) in-group embeddedness and its associated reliable social network that reduce morbidity and mortality caused when infectious disease invades the in-group. Hence, measures of the strength of the importance of religion for the people in a region (religiosity) should be predictable based on the region's position along the parasite-stress gradient reflecting the average infectious disease stress experienced by people in the region. Therefore, we hypothesized that religious participation and commitment, reflecting the importance of in-group assortative sociality, would be related positively to parasite stress across regions (Fincher and Thornhill 2012).

One assumption of our hypothesis is that there is a positive relationship between religiosity and out-group dislike and in-group preference. Prior studies provide evidence supporting this. For example, Jackson and Hunsberger (1999) conducted a study of the relationships between individuals' religiosity and their prejudicial attitudes toward religious and non-religious others, and found that the religious participants showed significant positive attitudes toward in-group religious others, but negative attitudes toward non-religious others. As well, the magnitude of the prejudice corresponded to the individual's own level of religiosity. That is, a participant who scored highly on religious fundamentalism also scored highly on out-group prejudice. In a separate study, Bulbulia and Mahoney (2008) demonstrated that New Zealand Christians were more altruistic toward Canadian Christians than were

New Zealand citizens to other New Zealand citizens. Similarly, Widman et al. (2009) showed that individuals with strong Christian beliefs were more likely to rate others displaying a symbol of Christianity (a cross) as more kind and moral than others not displaying such a symbol. Besides supporting our assumption, these studies also suggest, on the one hand, the importance of religiosity as a marker of in-group membership, and, on the other hand, an underlying mental mechanism within individuals to detect and measure religious or some other form of in-group similarity. Such a mechanism was indicated also by Park and Schaller's (2005) finding that when people experienced attitudinal similarity, rather than attitudinal dissimilarity, with another person, they considered the person more like family. Furthermore, there is convincing separate evidence that religious prosociality is primarily in-group altruism (Norenzayan and Shariff 2008).

Also supportive of a positive relationship between religiosity and prejudice against out-groups are the recent findings of Terrizzi et al. (2012). In their study of individual differences, they found that the degree of religiosity positively predicted prejudice against certain sexual minorities.

In another supportive study of our perspective, Saroglou et al. (2004) conducted a meta-analysis of the relationships between Schwartz's model of 10 cross-culturally stable, core values (Schwartz 1992) and religiosity. Saroglou et al.'s (2004) meta-analysis focused on 21 samples from 15 countries (total $n=8{,}551$ people). They discovered that religious people favored values that promoted social order (mainly the values *Tradition* and *Conformity*), but disliked values that promoted openness to change and autonomy (*Stimulation* and *Self-Direction*). This was true across a variety of religions (i.e., Christians, Jews and Muslims) and countries from Europe, North America and the Middle East. Interestingly, the positive correlation they reported between religiosity and "conservation" (=conservative) values (*Conformity*, *Tradition* and *Security*) and the negative relationship between religiosity and openness to change and autonomy (*Stimulation* and *Self-Direction*) showed greater effect sizes in a sample of Mediterranean countries than in a sample of Western European countries. Mediterranean countries have greater levels of parasite stress than Western European countries (Guernier et al. 2004; Fincher and Thornhill 2008b).

## 9.4 Religiosity and Signal Redundancy

Signaling systems across species typically show redundancy among components or modalities (Searcy and Nowicki 2005; Thornhill and Gangestad 2008). This is thought to enhance communication, given that each signal is imperfect in information content, but, when combined, give greater accuracy. Redundancy is seen in human signaling of in-group affiliation and boundary. A combination of signals involving religiosity, language or dialect, word use, dress, music, smell, and so on comprise a redundant suite of honest signals about one's group membership and embeddedness.

## 9.5 Other Hypotheses to Explain Cross-Cultural Differences in Religiosity

McCleary and Barro (2006) explored the validity of the hypothesis that economic development causes lower levels of religiosity among individuals—known as the secularization hypothesis. Separately, Inglehart and Baker (2000) and Norris and Inglehart (2004) suggested that individuals reduce religiosity when conditions of living are benign, but become religiously embedded under dire conditions of hardship and high mortality salience—known as the existential security hypothesis. In both these hypotheses, people are less religious in regions where they have less "need" of a religion and the benefits it offers. We refer collectively to the secularization hypothesis and the existential security hypothesis as "the conditions-of-living model." The conditions-of-living model has been tested and supported, in part, by examining the relationships between religiosity and economic conditions across countries. McCleary and Barro (2006) focused on Gross Domestic Product per capita as a marker of economic development, and documented a significant negative relationship between economic development and religiosity. Norris and Inglehart (2004) showed large differences in religiosity between wealthy and poor nations, providing support for their hypothesis that people living in poor conditions also show greater religiosity. More recent tests found support for the conditions-of-living model: Rees (2009) discovered a positive relationship between income inequality (used as a proxy for personal insecurity) and religiosity across many nations, and Delamontagne (2010) found that social inequality (measured by inequalities in education, income, and race) was highly, positively predictive of religiosity across regions of the United States.

Clearly, there is an overlap between the conditions-of-living model and the parasite-stress theory of values applied to religiosity, because high levels of infectious diseases are a part of "dire conditions" and low economic development. Indeed, both Inglehart and Baker (2000) and McCleary and Barro (2006) mention disease differences across countries and explicitly try to treat disease in their analyses by using a country's latitude as part of their analyses (latitude is negatively correlated with infectious disease stress, e.g., Guernier et al. 2004). Our approach is different in that it incorporates the evolutionary history of *Homo sapiens* into the research framework for generating hypotheses and predictions. For example, we make arguments based on the fact that out-group conspecifics who carry novel infectious diseases can be potentially dangerous to an individual's reproductive success, which leads to predictions about the design of human psychology that is expected to have evolved and its manifestations in ideology that are not generated from the models presented by Inglehart and Baker (2000), Norris and Inglehart (2004) or McCleary and Barro (2006). Furthermore, the conditions-of-living model assumes that, under conditions of stress, individuals will turn to their in-group rather than to an out-group. This assumption, however, does not consider the costs and benefits of seeking support from an out-group under ecological stress. As discussed earlier in this book, contact with an out-group can provide many benefits unattainable from an in-group. The parasite-stress theory of sociality erects an encompassing framework by providing a fundamental explanation for the relative costs of interacting with in-groups versus out-groups under different degrees of parasite stress. Finally, as we have mentioned earlier in our book, the parasite-stress theory includes a hypothesis of economic development. High parasite stress is a proximate cause of low economic progress across regions and hence the conditions-of-living model is not an exclusive alternative to the parasite-stress hypothesis for religiosity. The parasite-stress theory of values in relation to economics is treated in detail in Chap. 11.

An additional model of religiosity that has received attention in the literature, called "the supply-side model," proposes that religious vitality (typically measured by some aspect of religiosity) is positively associated with religious pluralism, because the increased choice or commodity possibilities under high religious pluralism allow an individual to better find the religion that suits him or her best (for a review, see Chaves and Gorski 2001). Because people can find such great fits, they will tend to engage in greater religious behavior, leading to the prediction of a positive association between religiosity and religious pluralism. This model was supported with some empirical patterns, but was largely dismissed by Chaves and Gorski (2001) on the grounds that the empirical evidence was overwhelmingly unsupportive of the basic general contention that religious pluralism was positively associated with religious vitality. For completeness, we correlated each of our three cross-national measures of religiosity (described below) with the religious pluralism index produced by McCleary and Barro (2006). Our findings are: *Proportion of Religionists*: $r=-0.14$, $p=0.26$, $n=67$; *Proportion of Believers*: $r=-0.02$, $p=0.90$, $n=63$; *Religious Participation and Value*: $r=0.05$, $p=0.69$, $n=63$. These insignificant effects do not support the supply-side model. Specifically, two of the relationships are in the direction opposite that predicted by the supply-side model, and the other is positive, but essentially zero.

## 9.6 Methods for Establishing an Empirical Link Across Regions Between Religiosity and Parasite Stress

We predicted a positive association between religiosity and parasite stress cross-nationally and across the states of the USA. To test this, we indexed religiosity with two measures: (a) religious affiliation and (b) religious participation and value. In the next sections, we describe briefly how these variables were constructed for the cross-national and interstate analyses. The parasite-stress variables used in the analyses presented in the Results section below are described in Chap. 5. For details about the methods we used, see Fincher and Thornhill (2012), which also links to supplementary files containing the data for many of the variables used in this chapter.

## *9.6.1 Religious Affiliation*

According to the parasite-stress theory of values, people in areas with more parasite stress will adhere to local religious systems to a greater extent than individuals in areas with low parasite stress. This is because the liberal values of people in areas with low parasite stress provide them with greater flexibility in whether they adhere to a religion or not, or they may make up their own system of secular beliefs. Also, the benefits of heightened in-group assortative sociality are predicted to be greater in high parasite-stress areas than in low parasite-stress areas. Hence, we predicted that the proportion of religionists in an area would be correlated positively with parasite adversity, because higher levels of parasite stress can potentially increase the costs of infectious-disease contact associated with non-conformity to in-group values and norms. We describe next our measures of religious affiliation for both the cross-national and the US analyses.

### 9.6.1.1 Cross-National: Proportion of Religionists

To construct this variable, we extracted the proportion of non-religionists for the year 2000 from the *World Christian Encyclopedia* (Barrett et al. 2001), an oft-used and highly regarded resource in religion scholarship (Grim and Finke 2006). Non-religionists include the two forms of non-believers: agnostics and atheists. The proportion of non-religionists within nations ranged from zero (e.g., Afghanistan) to 55.6 % (Democratic People's Republic of Korea, a.k.a. North Korea, a communist country) ($n=230$ countries). The proportion of non-religionists was subtracted from one to yield our analytical variable, *Proportion of Religionists*, which was arcsine-square-root transformed.

### 9.6.1.2 Cross-National: Proportion of Believers

For the *Proportion of Believers*, we used the inverse of "the proportion of non-believers in God," as presented in Lynn et al. (2009; this is a tabulation of data described in Zuckerman 2007). This measure relies, in part, on values from the *World Christian Encyclopedia* (Barrett et al. 2001), but incorporates many other survey sources and therefore likely provides more reliable estimates. The proportion of non-believers ranged from 0.5 % (e.g., Cameroon) to 81 % (Vietnam, a communist country) ($n=137$ countries). The values were subtracted from one to represent the *Proportion of Believers*, which was arcsine-square-root transformed. The *Proportion of Religionists* and the *Proportion of Believers* were strongly, positively correlated ($r=0.67$, $p<0.0001$, $n=137$ countries).

#### 9.6.1.3 United States: Proportion of Religious Adherents

The Association of Statisticians of American Religious Bodies conducted a study of 149 religious bodies in the United States over the years 1999–2001 to assess the number of congregations in each state within the USA The study produced a measure of the total adherents of each congregation, providing a comprehensive measure of the total religious adherents in each state. These data comprised our US interstate variable, *Proportion of Religious Adherents*, which was arcsine-square-root transformed.

#### 9.6.1.4 United States: Proportion of Religionists USA

The 2001 American Religious Identification Survey (ARIS; Kosmin et al. 2001) was a telephone survey of 50,281 households. The survey asked, "What is your religion, if any?" From this, we obtained the proportion of respondents that indicated "no religion" for each state (Hawaii and Alaska were not included in the ARIS 2001). The "no religion" proportion/state score was subtracted from one to represent the *Proportion of Religionists USA* and then arcsine-square-root transformed. *Proportion of Religionists USA* was correlated positively and highly with *Proportion of Religious Adherents* ($r=0.66$, $p<0.0001$, $n=48$ states).

### 9.6.2 Religious Participation and Value

We predicted that the magnitude of time and effort dedicated to religious practice and the value placed on religious practice and ideals would correlate positively with parasite adversity. Next, we describe our measures of religious participation and value cross-nationally and within the United States.

#### 9.6.2.1 Cross-National: Religious Participation and Value

We created an index of *Religious Participation and Value* based on items contained in the World Values Survey collected 1981–2007 from about 344,000 individuals in 95 countries. Also, we created a variable, *Proportion Who Prayed Everyday*, from the same survey. The *Proportion Who Prayed Everyday* was correlated positively and highly with *Religious Participation and Value*, $r=0.93$ ($p<0.0001$, $n=59$). And *Religious Participation and Value* correlated positively and strongly with the *Proportion of Religionists* ($r=0.74$, $p<0.0001$, $n=90$) and the *Proportion of Believers* ($r=0.83$, $p<0.0001$, $n=82$). Also, the *Proportion Who Prayed Everyday* correlated positively and highly with the *Proportion of Religionists* ($r=0.64$, $p<0.0001$, $n=57$) and the *Proportion of Believers* ($r=0.85$, $p<0.0001$, $n=51$).

#### 9.6.2.2 United States: Religious Participation and Value USA

The Pew Forum on Religion and Public Life produced the report *US Religious Landscape Survey, Religious Affiliation: Diverse and Dynamic* (February 2008). From this survey, we collected data for eight items and generated the variable *Religious Participation and Value USA*. *Religious Participation and Value USA* was correlated positively with *Proportion of Religious Adherents and Proportion of Religionists USA* (adherents: $r=0.42$, $p=0.0041$, $n=45$; religionists: $r=0.56$, $p<0.0001$, $n=44$).

### *9.6.3 In-Group Assortative Sociality Variables for Countries and US States*

In Chap. 5, we showed that family ties, a measure of collectivism, when indexed as a cross-national variable and separately as a cross-states variable, is robustly and positively related to parasite stress across nations and US states. Here, we revisit the family ties measures in order to construct a composite variable of in-group assortative sociality. There is considerable conceptual overlap between religiosity and family ties/collectivism that, we argue, reflects the importance of in-group assortative sociality within societies (e.g., the cross-national *Religious Participation and Value* was correlated positively and strongly with *Strength of Family Ties*, $r=0.79$, $p<0.0001$, $n=72$ countries). Because of this conceptual overlap, we made synthetic in-group assortative sociality variables, one cross-national that was called *In-Group Assortativeness* and one for the US states called *In-Group Assortativeness USA*, to capture the common variation among our multiple dependent variables tapping in-group assortative sociality. The cross-national data and the US interstate data for in-group assortativeness and the evidence of the validity of the two composite indices can be found in Fincher and Thornhill (2012).

## 9.7 Results

### *9.7.1 Cross-National Analyses*

According to the parasite-stress theory of values, nonzoonotic human infectious diseases should have a stronger relationship with values and associated behavioral immunity (such as religiosity) than zoonotic human infectious diseases. This is expected because nonzoonotic human diseases are transmitted human-to-human, whereas zoonotic human diseases are not (see Chap. 5). Earlier in our book, we have shown this prediction to be supported across nations for family ties and other measures of collectivism (Chap. 5), certain personality traits (Chap. 7), and values related to homicides (Chap. 8). Here, we examine the findings reported in Fincher and Thornhill (2012) about the cross-national relationship of the severity (the number of cases) of each of these two categories of human diseases with religiosity and related variables. The methods we used for generating the severity scores of the two disease categories are described in detail in Fincher and Thornhill (2012; also see Chap. 5).

Across nations, nonzoonotic infectious diseases are much more important in explaining assortative sociality than are zoonotics (Table 9.1). Each of the four dependent variables measuring religiosity was correlated positively and significantly with nonzoonotic parasite severity (correlation coefficients ranged from 0.40 to 0.64), while zoonotic parasite severity was insignificantly correlated with all but one of the dependent variables (correlation coefficients ranged from –0.17 to 0.15). The only significant relationship between a dependent variable (*Proportion of Religionists*) and zoonotic parasite severity showed a negative sign (–0.17) and therefore was in the opposite direction expected if zoonotics cause increased religiosity or assortative sociality. The synthetic measure, *In-Group Assortativeness*, was correlated positively with both nonzoonotic parasite severity and zoonotic parasite severity. However, the relationship with nonzoonotic severity was strong ($r=0.65$, $p<0.0001$), whereas that with zoonotic severity was insignificant. These findings reveal that zoonotic diseases are not predictive of heightened religiosity or *In-Group Assortativeness* cross-nationally. Therefore, zoonotic disease adversity, as a variable, is not considered in our international analyses below.

Religious affiliation, as well as religious participation and value and *In-Group Assortativeness*, exhibited robust positive correlations with the other measures of parasite stress we examined—*Infectious Disease DALY* and *Combined Parasite Stress* (Table 9.1). Correlation coefficients ranged from 0.48 to 0.73. Figure 9.1 shows the data for the relationship between *Combined Parasite Stress* and *In-Group Assortativeness*.

These findings were repeated in analysis by world regions. When considering the correlation between the dependent variables and *Combined Parasite Stress* at the world regional level, all correlations were positive and thus in the direction predicted by the parasite-stress theory: *Proportion of Religionists*: $r=0.70$; *Proportion of Believers*: $r=0.82$; *Religious Participation and Value*: $r=0.76$; *Proportion Who Prayed Everyday*: $r=0.46$; *In-Group Assortativeness*: $r=0.89$; $n=6$ world regions for each relationship. When nested within world regions, *Combined Parasite Stress* predicted significantly *Proportion of Religionists* ($r^2=0.25$, $n=191$); *Proportion of Believers* ($r^2=0.44$, $n=136$); *Religious Participation and Value* ($r^2=0.55$, $n=89$); *Proportion Who Prayed Everyday* ($r^2=0.47$, $n=57$); and *In-Group Assortativeness* ($r^2=0.57$, $n=65$). All regressions were significant ($p<0.0001$).

The basic findings reported above are not confounded by variation in other variables such as human freedom or civil liberties, GDP per capita (an average of 1960–2008), resource distribution (Vanhanen's wealth disparity metric, see Chap. 10), or unidentified variables. Our procedure for treatment of potential confounds here is the same as that for *Strength of Family Ties* in Chap. 5. Among the three focal, potentially confounding variables, only the Freedom House *Civil Liberty* scores

**Table 9.1** Pearson zero-order correlations across countries between parasite stress, religiosity variables, *In-Group Assortativeness*, and potentially confounding variables

| | Proportion of religionists | Proportion of believers | Religious participation and value | Proportion who pray every day | In-group assortativeness |
|---|---|---|---|---|---|
| Nonzoonotic parasite severity | 0.40, <0.0001, 216 | 0.58, <0.0001, 137 | 0.64, <0.0001, 93 | 0.50, <0.0001, 59 | 0.65, <0.0001, 66 |
| Zoonotic parasite severity | –0.17, 0.0139, 216 | 0.03, 0.7712, 137 | 0.14, 0.1819, 93 | 0.11, 0.4124, 59 | 0.17, 0.1646, 66 |
| *Infectious Disease DALY* | 0.51, <0.0001, 191 | 0.64, <0.0001, 136 | 0.73, <0.0001, 89 | 0.57, <0.0001, 57 | 0.72, <0.0001, 65 |
| *Combined Parasite Stress* | 0.48, <0.0001, 191 | 0.63, <0.0001, 136 | 0.70, <0.0001, 89 | 0.55, <0.0001, 57 | 0.71, <0.0001, 65 |
| Civil liberty[a] | 0.14, 0.0570, 196 | 0.52, <0.0001, 137 | 0.48, <0.0001, 93 | 0.35, 0.0065, 59 | 0.60, <0.0001, 66 |
| Vanhanen Resource Distribution (see Chap. 10) | –0.37, <0.0001, 170 | –0.67, <0.0001, 137 | –0.63, <0.0001, 88 | –0.46, 0.0004, 56 | –0.66, <0.0001, 66 |
| GDP per capita | –0.37, <0.0001, 194 | –0.60, <0.0001, 135 | –0.61, <0.0001, 92 | –0.44, 0.0006, 59 | 0.65, <0.0001, 65 |

Data are the correlation coefficient, associated *p*-value, and sample size, respectively (results originally reported in Fincher and Thornhill 2012; reprinted with permission)

[a]Higher scores correspond to lower freedom

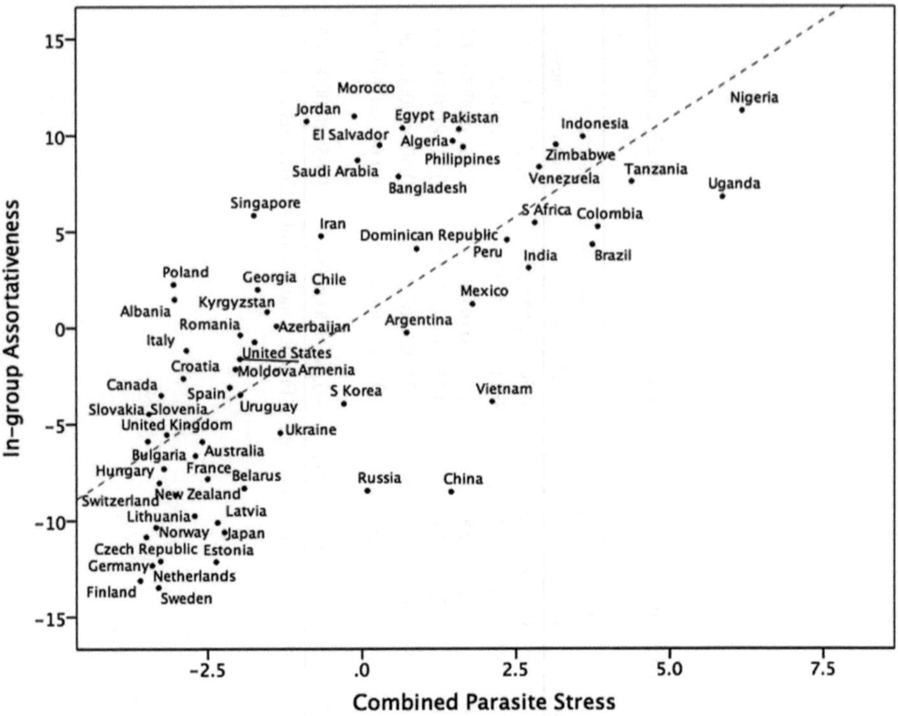


**Fig. 9.1** The correlation between *Combined Parasite Stress* and *In-Group Assortativeness* for the 65 nations with correspondent data for all 11 items that comprise the two variables ($r = 0.71$, $p < 0.0001$). The line is the regression line (results originally reported in Fincher and Thornhill 2012; reprinted with permission)

(see Chap. 10) were correlated non-significantly with the *Proportion of Religionists*; the other two potentially confounding variables had significant correlations with the two religious affiliation variables, and the two religious participation and value variables (Table 9.1). Therefore, each relevant, potentially confounding variable was checked to see if it accounted for the correlation between parasite stress and religiosity variables; none did (Table 9.2). In a series of multiple regressions, *Combined Parasite Stress* remained a significant, positive predictor of each of the four religion variables after controlling the effects of the potential confounders (standardized β coefficients for parasite stress ranged from 0.28 to 0.59). Thus, the positive association between parasite stress and religiosity was robust to the effects of freedom, resource distribution, or economic development as captured by *Civil Liberty*, Vanhanen's Resource Distribution, and GDP per capita.

All but one of the correlations between the residuals of the regression of lifespan on *Combined Parasite Stress* and religiosity variables were statistically insignificant: *Proportion of Religionists*: $r = -0.13$, $p = 0.082$, $n = 186$; *Proportion of Believers*: $r = -0.18$, $p = 0.031$, $n = 136$; *Religious Participation and Value*: $r = -0.13$, $p = 0.215$, $n = 91$; *Proportion Who Prayed Everyday*: $r = -0.05$, $p = 0.727$, $n = 59$; and

**Table 9.2** Results of multiple regression analyses with two model specifications considered as independent predictors: (1) *Combined Parasite-Stress*, *Civil Liberty*, and Vanhanen Resource Distribution; and (2) *Combined Parasite Stress* and GDP per capita (results originally reported in Fincher and Thornhill 2012; reprinted with permission)

| | Proportion of religionists | | Proportion of believers | | Religious participation and value | | Proportion who prayed every day | | In-group assortativeness | |
|---|---|---|---|---|---|---|---|---|---|---|
| Model | 1 | 2 | 1 | 2 | 1 | 2 | 1 | 2 | 1 | 2 |
| *Combined Parasite Stress* | 0.59*** | 0.54*** | 0.28** | 0.39*** | 0.53*** | 0.56*** | 0.45** | 0.45** | 0.49*** | 0.49*** |
| *Civil Liberties* | ni | – | 0.02 | – | –0.01 | – | 0.04 | – | 0.23 | – |
| Vanhanen Resource Distribution | 0.08 | – | –0.44** | – | –0.22 | – | –0.11 | – | –0.14 | – |
| GDP per capita | – | 0.03 | – | –0.29** | – | –0.18 | – | –0.14 | – | –0.30* |
| $R^2$ | 0.28 | 0.26 | 0.47 | 0.42 | 0.50 | 0.50 | 0.31 | 0.32 | 0.57 | 0.55 |
| $F$ | 32.5 | 31.8 | 39.6 | 47.5 | 28.0 | 43.8 | 7.9 | 12.5 | 27.1 | 37.2 |
| $N$ | 168 | 183 | 136 | 135 | 87 | 89 | 56 | 57 | 65 | 65 |

The rows for the predictor variables contain standard βs. All regressions ($R^2$) are statistically significant at $p<0.001$

*$p<0.05$; otherwise, the coefficient was not statistically significant

**$p<0.01$; otherwise, the coefficient was not statistically significant

***$p<0.001$; otherwise, the coefficient was not statistically significant

ni=(variable) not included, because zero-order correlation with the dependent variable was not significant at $p<0.05$

*In-Group Assortativeness*: $r=0.02$, $p=0.874$, $n=65$. The *Proportion of Believers* showed a significant correlation, but the effect size was much reduced and in the opposite direction, compared to the correlation between *Combined Parasite Stress* and the same variable (–0.18 versus 0.63). In general, especially considering the synthetic variable *In-Group Assortativeness*, the variation in life expectancy independent of that explained by parasite-stress was minimal.

### 9.7.2 United States Analyses

As first reported in Fincher and Thornhill (2012), across the US states, religiosity and in-group assortative sociality are positively and significantly related to parasite stress (Table 9.3). *Parasite Stress USA* was correlated positively and significantly with each of the two religious affiliation variables, *Proportion of Religionists USA* and *Proportion of Religious Adherents*. Also, *Parasite Stress USA* was correlated positively and significantly with *Religious Participation and Value USA*. Finally, *Parasite Stress USA* was correlated positively and significantly with the synthetic measure of *In-Group Assortativeness USA* (Fig. 9.2).

These findings were repeated in analyses based on the nine US census regions (Fincher and Thornhill 2012). When considering the correlation between the dependent variables and *Parasite Stress USA* at the regional level, all correlations were in the direction predicted by the parasite-stress theory: *Proportion of Religionists USA*: $r=0.60$; *Proportion of Religious Adherents*: $r=0.40$; *Religious Participation and Value USA*: $r=0.85$; *In-Group Assortativeness USA*: $r=0.89$; $n=9$ for all. Moreover, when nested within US regions, *Parasite-Stress USA* predicted significantly *Proportion of Religionists USA* ($r^2=0.61$, $p<0.0001$, $n=48$); *Proportion of Religious Adherents* ($r^2=0.39$, $p=0.0106$, $n=50$); *Religious Participation and Value USA* ($r^2=0.54$, $p=0.0004$, $n=46$); and *In-Group Assortativeness USA* ($r^2=0.66$, $p<0.0001$, $n=43$).

The basic findings are not confounded by wealth or wealth disparity (Fincher and Thornhill 2012). Of the potentially confounding variables, Gini was correlated significantly with *In-Group Assortativeness USA*, but not with *Proportion of Religionists USA*, *Proportion of Believers*, or *Religious Participation and Value USA*; *GDP per capita* was correlated significantly with *Religious Participation and Value USA* and *In-Group Assortativeness USA* (Table 9.3). GDP per capita was entered in a multiple regression with *Parasite Stress USA* as predictors of *Religious Participation and Value USA*: $R^2=0.60$, *Parasite Stress USA* std. $\beta=0.55$, $p<0.001$, GDP per capita std. $\beta=-0.57$, $p<0.001$, $n=46$. Both Gini and GDP per capita were included with *Parasite Stress USA* as predictors of *In-Group Assortativeness USA*: $R^2=0.63$, *Parasite Stress USA* std. $\beta=0.63$, $p<0.001$, GDP per capita std. $\beta=-0.48$, $p<0.001$, Gini std. $\beta=0.00$, not significant, $n=43$. In all cases, *Parasite Stress USA* remained a significant, positive predictor of the dependent variables. Thus, the correlation between parasite stress and religiosity and the correlation between parasite stress and in-group assortativeness were not confounded with the effects of economic inequality and development as captured by the *Gini* index and *GDP per capita*.

**Table 9.3** Pearson zero-order correlations between measures of parasite stress, religiosity, in-group assortativeness, and potentially confounding variables within the USA

| | Proportion of religionist USA | Proportion of religious adherents | Religious participation and value USA | In-group assortativeness USA |
|---|---|---|---|---|
| *Parasite Stress USA* | 0.45, 0.0012, 48 | 0.36, 0.0103, 50 | 0.53, 0.0002, 46 | 0.66, <0.0001, 43 |
| Zoonotic parasite stress | 0.28, 0.0577, 48 | 0.39, 0.0046, 50 | 0.03, 0.8573, 46 | 0.12, 0.4572, 43 |
| Nonzoonotic parasite stress | 0.36, 0.0113, 48 | 0.33, 0.0187, 50 | 0.51, 0.0003, 46 | 0.63, <0.0001, 43 |
| Human-specific parasite stress | 0.46, 0.0009, 48 | 0.34, 0.0155, 50 | 0.55, <0.0001, 46 | 0.67, <0.0001, 43 |
| Multi-host parasite stress | 0.06, 0.6943, 48 | 0.18, 0.2216, 50 | 0.22, 0.1414, 46 | 0.28, 0.0675, 43 |
| Gini | 0.17, 0.2385, 49 | 0.27, 0.0594, 50 | 0.28, 0.0591, 46 | 0.37, 0.0143, 43 |
| GDP per capita | −0.18, 0.2290, 49 | −0.12, 0.3984, 50 | −0.54, <0.0001, 46 | −0.52, 0.0004, 43 |

Data are the correlation coefficient, associated *p*-value, and sample size, respectively (only results in the first, last and next-to-last rows were reported in Fincher and Thornhill (2012) (reprinted with permission) other rows were not (see text))

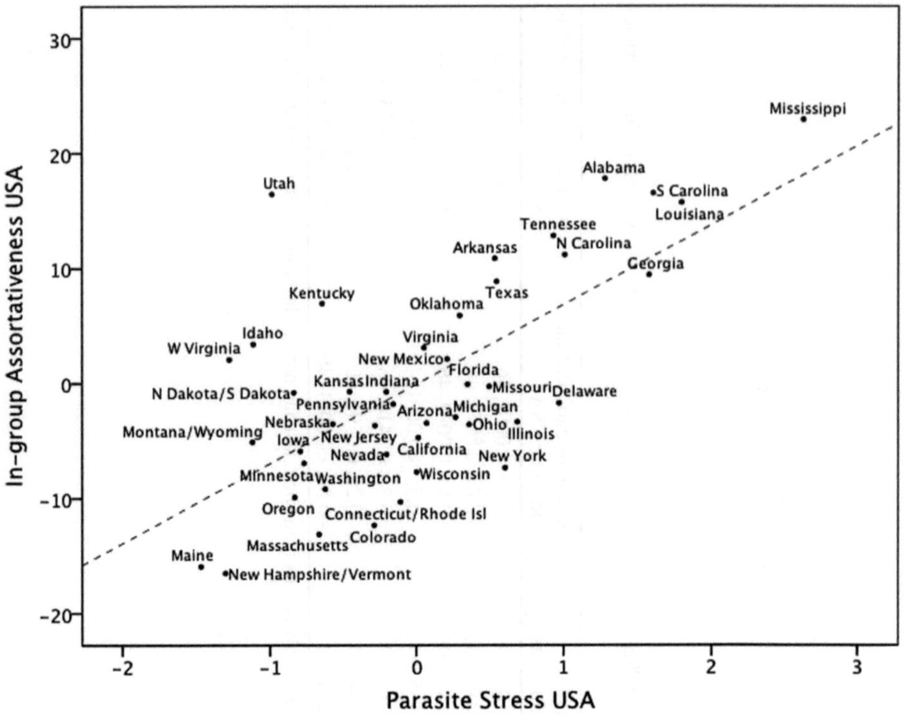


**Fig. 9.2** The correlation between *Parasite Stress USA* and *In-Group Assortativeness USA* for the 43 states/state combinations with correspondent data for all 14 items that comprise the two variables ($r = 0.66$, $p < 0.0001$) (results originally reported in Fincher and Thornhill 2012; reprinted with permission)

As shown in Fincher and Thornhill (2012) the residuals from regressing state-level life expectancy on *Parasite Stress USA* were not correlated significantly with *Proportion of Religionists USA* ($r = -0.10$, $p = 0.484$, $n = 48$) or *Proportion of Religious Adherents* ($r = 0.09$, $p = 0.518$, $n = 50$). The residuals were correlated significantly with *Religious Participation and Value USA* ($r = -0.38$, $p = 0.008$, $n = 46$) and *In-Group Assortativeness USA* ($r = -0.35$, $p = 0.021$, $n = 43$). Therefore, parasite stress accounts for much of the state-level variation in the dependent variables that reflect religious affiliation as they relate to life expectancy. Other variables besides parasite stress, however, can account for a significant proportion of the variation in life expectancy among the states, which may be the result of the greater mortality from non-infectious diseases such as forms of diabetes, heart disease, and cancer in the United States as compared to other countries that have lower income levels (see Lopez et al. 2006).

Subsequent to the publication of Fincher and Thornhill (2012) we indexed the severity (number of cases) of nonzoonotic and zoonotic human infectious diseases in each of the states of the USA. (See Chap. 5 for discussion of methods and data.) We consider here for the first time the ability of zoonotic versus nonzoonotic human diseases to explain religiosity and in-group assortativeness among the states of the USA.

The findings are reported in Table 9.3. (Also listed in Table 9.3 are separate scores for the multihost and human-specific categories of nonzoonotic diseases.) For the *Proportion of Religionists USA*, zoonotic infectious diseases were correlated at about the same level as nonzoonotic infectious diseases, but slightly less so (0.28 versus 0.36). For the *Proportion of Religious Adherents*, zoonotic infectious diseases were correlated at a higher level than nonzoonotic infectious diseases (0.39 versus 0.33). For *Religious Participation and Value USA*, zoonotic infectious diseases were correlated at a much lower level than nonzoonotic infectious diseases (0.03 versus 0.51). And, for *In-Group Assortativeness USA*, zoonotic infectious diseases were correlated at a much lower level than nonzoonotic infectious diseases (0.12 versus 0.63). Thus, for the two measures focused on the proportion of religionists within US states, zoonotic infectious diseases were similarly or more greatly correlated than the nonzoonotic infectious diseases, and for each of the two measures focused on the magnitude of religiosity and in-group assortativeness, nonzoonotic infectious diseases showed a greater effect. We conducted regression analyses with nonzoonotic and zoonotic infectious diseases as simultaneous predictor variables. For the *Proportion of Religionists USA*, only nonzoonotic diseases were associated significantly (nonzoonotic std. $\beta=0.33, p=0.02$, zoonotic std. $\beta=0.22, p=0.11$). For the *Proportion of Religious Adherents*, both nonzoonotic and zoonotic infectious diseases were associated significantly, but the effect size was slightly larger for zoonotic diseases (nonzoonotic std. $\beta=0.28$, $p=0.04$, zoonotic std. $\beta=0.35$, $p=0.009$). For *Religious Participation and Value USA*, only nonzoonotic diseases were associated significantly (nonzoonotic std. $\beta=0.53$, $p=0.0003$, zoonotic std. $\beta=-0.09, p=0.50$). And, for *In-Group Assortativeness USA*, only nonzoonotic diseases were associated significantly (nonzoonotic std. $\beta=0.64$, $p<0.0001$, zoonotic std. $\beta=-0.03$, $p=0.79$). That the zoonotic infectious diseases were important for explaining the proportion of religionists and adherents was an unusual finding considering the overall patterns we have presented in the book, which show nonzoonotics being more important in predicting people's values.

We also conducted regression analyses with the three different types of infectious diseases (zoonotic, multihost, and human-specific) to assess their unique predictive effects. For the *Proportion of Religionists USA*, only human-specific infectious diseases were significantly associated (zoonotic std. $\beta=0.16$, $p=0.25$; multihost std. $\beta=-0.09, p=0.51$; human-specific std. $\beta=0.45, p=0.003$). For the *Proportion of Religious Adherents*, only zoonotic infectious diseases were significantly associated (zoonotic std. $\beta=0.35, p=0.01$; multihost std. $\beta=0.11, p=0.45$; human-specific std. $\beta=0.23, p=0.12$). For *Religious Participation and Value USA*, only human-specific infectious diseases were significantly associated (zoonotic std. $\beta=-0.14$, $p=0.29$; multihost std. $\beta=0.05$, $p=0.72$; human-specific std. $\beta=0.58$, $p=0.0002$). And, for *In-Group Assortativeness USA*, only human-specific infectious diseases were significantly associated (zoonotic std. $\beta=-0.10, p=0.42$; multihost std. $\beta=0.07, p=0.58$; human-specific std. $\beta=0.69, p<0.0001$). Across all four of the dependent variables, human-specific diseases have a greater effect than the multi-host diseases, but in no case were multihost diseases a significant predictor. For only one dependent variable, *Proportion of Religious Adherents,* was zoonotic infectious diseases a significant predictor, and for this variable neither multihost nor human-specific diseases were significant predictors.

## 9.8 Overview of Findings

Our cross-national analyses showed that religiosity, as measured by religious affiliation and religious participation and value, was correlated positively with all the measures of parasite stress that we used. Also as predicted by the parasite-stress theory of values, cross-national religiosity was correlated more strongly with the prevalence of nonzoonotic infectious diseases than with zoonotic infectious diseases. That is, religiosity was correlated more strongly with diseases that can be transmitted human-to-human than those infectious diseases that are not transmitted between humans. Within the United States, religiosity also was correlated positively with parasite stress, and generally nonzoonotic diseases, especially human-specific nonzoonotics, were more predictive. Our results support the hypothesis that religiosity functions as an honest signal of in-group commitment and boundary, and that religiosity is a fundamental component of in-group assortative sociality and hence of behavioral immunity. Moreover, the results reveal that the parasite-stress theory of sociality potentially can explain all societal degrees of religiosity from the irreligious to the ultra-religious as arising from cultures' relative position along a parasite-stress gradient and corresponding collectivism–individualism gradient.

Our treatment of religiosity in this chapter ignored some important aspects of religion, such as beliefs in the afterlife and attribution to supernatural causation. We focused on features (religious affiliation and commitment) that had been measured comparably across regions in all types of people including the irreligious. The predictive power of the parasite-stress theory may not end with these features. We anticipate that certain unique elements of religion may be disentangled with an eye toward the human history of contending with parasites. For example, ancestor worship is a widespread component of many religions, but variation in its extent and nature does exist (Rossano 2007). We hypothesize that the strength of family ties/collectivism of the living, which is evoked by parasite stress, may provide a foundation for the strength of worship of ancestors.

## 9.9 The Findings' Relationships to Some Other Research Topics

### 9.9.1 *The USA Is Not an Exception*

Wall and Shackelford (2012) have criticized our cross-national findings on religiosity discussed above. They suggested that the USA is more religious than can be accounted for by measures of infectious diseases. This, they propose, is the result of high immigration rates that create a hyper-activation of assortative sociality, including religiosity, to the point that the extraordinarily high assortative sociality mismatches the actual parasite stress. They mention the common media treatment of immigration in general and the widely broadcasted conservative media that includes prejudicial language against immigrants from some regions. In the last chapter of the book (Chap. 14), we address empirically the idea that the USA is an outlier country with regard to the general cross-national relationship between parasite stress and religiosity. Evidence does not convincingly support this idea. Basically, the religiosity of the USA is consistent with its level of parasite adversity, which is quite high relative to other Western nations. Chapter 14 addresses and responds to various other scholarly criticisms of the parasite-stress theory of values applied to religiosity.

### 9.9.2 Economics

While we found that multiple religiosity variables were correlated with societal wealth variables, as predicted by the models in Inglehart and Baker (2000), Norris and Inglehart (2004) and McCleary and Barro (2006), the effect of parasite stress in explaining variation in religiosity was still significant when the effects of societal wealth and resource inequality were removed. Furthermore, parasite stress was significant after removing the effects of differences in freedoms as measured by the Freedom House *Civil Liberty* scores; this was as predicted. However, societal wealth, resource inequality, and freedom are hardly separable from parasite stress, because the values that lead to democracies versus autocracies or wealth versus dearth appear to arise causally from different infectious disease ecologies (Thornhill et al. 2009, Chaps. 10 and 11). The long-standing tradition in economics, historical scholarship, political science, and sociology is to view economic measures, such as gross domestic product (GDP) per capita and wealth inequality, as encompassing causal factors. According to the parasite-stress theory of sociality, however, variation in parasite stress is a proximate cause of variation in GDP and resource inequality. Parasite stress not only strongly and negatively influences human labor capacity (e.g., Price-Smith 2002) and cognitive ability (Eppig et al. 2010, 2011; Hassall and Sherratt 2011), the undemocratic values generated by high parasite stress cause widespread economic dearth and inequality. High parasite stress yields philopatry and localized/ethnocentric economic priorities and investment, devaluation and divestment outside the dominant in-group, and limitations on innovation, private property rights, willingness to adopt new ideas and technologies; low parasite stress has the opposite effects. Hence, we propose that these effects of variable parasite-stress level manifest cross-nationally as economic variables (Thornhill et al. 2009, Chap. 11).

### 9.9.3 Secularization

Norris and Inglehart (2004) have described patterns of secularization across the planet (the declining religiosity in many countries of the world). They noted that religiosity has declined most in wealthy nations, but very little, if at all, in poor nations.

They attributed this to the importance of the lack of wealth (poor living conditions) for promoting heightened religiosity. We emphasize that the poorest nations are also those that have the highest parasite stress. This is evident in the very strong negative correlation across nations between GDP per capita and *Combined Parasite Stress* ($r = -0.77$, $p < 0.0001$, $n = 184$) presented in Fincher and Thornhill (2012).

We have hypothesized that the reason that the poorest nations have maintained similar levels of high religiosity over time is because of the salience in such nations placed on tradition, conformity, parochialism and other conservative values, which, in turn, are caused by high parasite stress (Fincher and Thornhill 2012). We tested this hypothesis by a cross-national comparison of the religiosity of people born before 1945 versus during or after 1945. This temporal break is very meaningful in terms of the marked reduction in parasite adversity in regions with health interventions such as widespread sanitation, antibiotic availability, child vaccination programs, and disease-vector control (Thornhill et al. 2009, Chap. 10). We employed one question from the World Values Survey that is a component of our cross-national *Religious Participation and Value* measure (rating the importance of God from 1 to 10). We subtracted the proportion of those born in 1945 and later who rated their response a 10 from the proportion of those born before 1945 who rated their response a 10 within each country, and then correlated this difference with *Combined Parasite Stress*. The correlation was significantly negative ($r = -0.32$, $p = 0.002$, $n = 91$; Fincher and Thornhill 2012). Assuming a general persistence of country-level differences in parasite stress from the earlier part of the 1900s to the early part of the 2000s, this indicates that, in the countries with high pathogen stress, there was little difference between those born before or after 1945 in how they felt about God's importance; however, in countries with low parasite stress, there was a significantly larger difference between people born before 1945 and those born more recently in how they perceived God's importance. Thus, consistent with our hypothesis, lower levels of parasite stress positively predicts secularization across the world.

### 9.9.4 Ontogeny of Religiosity

It is clear that religiousness, religious identities, and related beliefs start forming at early ages (Finkel et al. 2009). It is also clear that people exhibit an awareness of contagion risks and contamination in early childhood (Siegal 1988; Stevenson et al. 2010; Siegal et al. 2011). The correspondence between the ontogenetic onset of religiosity and infectious-disease awareness suggests an important aspect of the developmental ecology of values. At this point, the ancestral cues that are relevant are incompletely understood, but appear to include infection frequency experienced by individuals (Stevenson et al. 2009; de Barra et al. 2013). In Chaps. 2, 3 and 6, we discussed various potential ontogenetic causes of values.

### 9.9.5 Health

The literature indicates that typically in the West the relationships are positive between religiosity and mental health and freedom from coronary disease and certain cancers (Koenig 1997; George et al. 2002). Future research could focus on the covariation of religiosity and health affected by infectious diseases per se. According to the parasite-stress model, religiosity will reduce followers' recent infectious disease problems via its associated ethnocentrism, xenophobia, and philopatry. As George et al. (2002) emphasized, despite a great deal of research, little is known about the mediators of the positive relationship between religiosity and health. Our approach suggests that future research would find that the relationship between contracted infectious diseases and religion is mediated by collectivism/conservatism and disgust and contamination sensitivity. According to the parasite-stress theory of values, although high disease severity in childhood is expected to produce high in-group assortative sociality, once those values are acquired ontogenetically, they will reduce the incidence in people of recent infectious diseases. Such research would add a new empirical approach to the study of the relationship between health and religiosity.

### 9.9.6 Geographical Expansions of Religions

Colonialism, imperialism, large-scale intergroup conquest, and related forms of societal expansion have large benefits (primarily reaped by elites) in the acquisition of land and other resources and the enslavement of conquered people. During such events, expansionists often coercively force their value systems on the original inhabitants of the acquired region. Typically, this involves committed and encompassing efforts by the conquerors, with religious beliefs being central to ideological reformation. This colonialist effort, we hypothesize, spreads and enforces the conquerors' behavioral norms and thereby reduces contact with out-group beliefs that the conquerors perceive to be disgusting (dangerous in terms of contagion risk). If the conquered have the same value system as the conquerors, then the cost of the conquest, in terms of contagion risk, is perceived by the conquerors to be reduced. Such perception will be maladaptive when the native's ideology is a better behavioral immunity than the ideology of the conquerors. Additional research could examine whether this speculation applies to conquest events in the historical record.

A related issue is the geographical pattern of large-scale historical conquests by Eurasian imperialism, described by Diamond (1997, *Guns, Germs and Steel*), whose thesis focused on unique aspects of geography, such as the east–west orientation of the Eurasian continent and the distribution of domesticable animals and plants. We (with Kenneth Letendre) have suggested a complementary, and in part alternative, model for this history (Letendre et al. 2010). First, conservative or collectivist values are correlated positively with severity of infectious disease. Second, such values include philopatry, parochialism and dislike and avoidance of innovations. Third, collectivism is concentrated at low latitudes. Fourth, collectivism is related negatively with societal wealth and associated technology. We have argued that, as humans migrated from Africa to higher latitudes in Eurasia, they moved into climates less hospitable to human infectious diseases, which, in turn, generated relatively individualistic cultures. Such cultures are characterized by an increased openness to and value of innovation and place a positive value on long-range dispersal. Thus, the accumulation of wealth and technology, the domestication of plants and animals, and the large population sizes that enabled the imperial domination of impoverished and less innovative cultures resulted not from aspects of Eurasian geography, but from the relative emancipation from parasites, which allowed and promoted the rise of cultures that were more individualistic than their forbearers. The technological dominance and individualism of these cultures motivated and fostered their expansion to obtain the benefits of conquest of other peoples.

### *9.9.7 In-Group Assortative Sociality*

Taken together, the findings presented in this chapter showing that strong family ties and collectivism are robustly associated with heightened religiosity in the face of parasite threat provide further support for the parasite-stress theory of sociality. Collectivism and religiosity are components of in-group assortative sociality, which is an adaptive response to heightened parasite stress. This adaptive response of in-group assortative sociality to parasite adversity likely is an ancient feature in *Homo sapiens*. Furthermore, the evidence that similar types of adaptive responses to parasite adversity apparently are found in other animals (discussed in Chaps. 5 and 13) suggests parasites had important impacts to social life early in evolutionary history.

The social isolation of groups is a concomitant of the elevated in-group assortative sociality evoked by high parasite-stress. The social isolation of groups is not without costs to individuals in the groups. For example, under conditions of social isolation, significant inbreeding can take place, possibly generating inbreeding depression. However, adaptive inbreeding is possible (Shields 1982; Kokko and Ots 2006). This seems especially likely under the ecology of high infectious-disease stress, as discussed and empirically supported in Chap. 6. Social isolation also can limit access to trade with out-groups, innovations generated by out-groups, and out-group social alliances. As we have explained, out-group interaction and affiliation is a benefit of individualism/liberalism, but one that is only widely optimal under relatively low parasite stress.

Our findings that infectious disease stress promotes in-group assortative sociality potentially can inform the study of epidemiology or spread of infectious diseases. For example, it is the case that the prevalence of many types of parasites is greater in large or denser populations (Altizer et al. 2003; Guégan and Constantin de

Magny 2007). At the same time, under conditions of high parasite stress, groups are comparatively more isolated (via in-group assortative sociality) than groups in conditions of low parasite stress (Chap. 13). This suggests that an important consideration for understanding parasite transmission is to recognize the difference between out-group versus in-group contact. Contact rates between individuals may be high within a group that is socially isolated from other groups. This is, indeed, an implication from the research presented in our book. Thus, high rates of contact in low pathogen areas are different from high contact rates in high pathogen areas. Based on our research, in low pathogen areas (individualistic locales), a high contact rate implies high rates of contact between genetically different, and differently infected, individuals whereas high contact rates in high pathogen areas (collectivistic locales) occur between individuals that are genetically close and likely carry similar infectious diseases and have similar classical immunity.

## 9.10 Life History and Assortative Sociality

Gladden et al. (2009) explored the interactions of religiosity, moral intuitions and life-history patterns, and showed that both the strength of moral intuitions (automatic emotional reactions brought on by norm and other rule violations (Gladden et al. 2009)) and religiosity coincide with a slow life-history strategy. That is, both are signs of a life-history strategy of people that is focused on somatic investment or an investment in survival in contrast to an investment in immediate reproduction. They suggested their findings are consistent with the fact that pathogen stress and collectivism are related positively, presumably because much of moral intuitions tap into cognition about pathogen avoidance (Oaten et al. 2009). In other work, Figueredo and Wolf (2009) showed that slow life-history people assortatively pair, sexually and socially, more strongly than fast life-history strategists. Both sets of findings are consistent with what we have presented here: in-group assortative sociality is strongly and positively associated with pathogen stress.

In Chaps. 5 and 8, we mentioned other applications of the parasite-stress theory of values to human life history. In Chap. 14, we return to the topic of human life history.

## 9.11 Other Tests of the Parasite-Stress Hypothesis of Religiosity

One limitation of our country-level research reported in this chapter is that the empirical tests of the parasite-stress theory's application to religiosity relied on contemporary measures of pathogen prevalence. Thus, one potentially useful line of research would be to explore how historical pathogen stress (e.g., Murray and Schaller 2010) relates to modern levels of religiosity.

Another limitation of our research reported in this chapter is that the empirical tests of the parasite-stress theory's application to religiosity were at the macroscale across countries of the world, or on a finer, but still large scale, within a single polity, the USA In order to understand better how parasite stress relates to differences in religiosity, it would be useful to conduct additional tests of this application within more localized regions. One such method of testing the hypotheses is to record people's changes in religiosity and other features of assortative sociality after infectious disease levels are reduced locally (e.g., by greater access to modern medicine and safe water) or increased locally (e.g., by the emergence of a new infectious disease). Evidence we have discussed indicates that changes in people's values can occur immediately (Chaps. 3 and 7) and may change and stabilize across one or a few generations (see Thornhill et al. 2009, Chap. 10). Easily administered, brief, valid questionnaires that could measure the relevant value changes are available, including those discussed above that measure religiosity (also see questionnaires in Faulkner et al. 2004; Gelfand et al. 2004; Thornhill and Fincher 2007).

At another microscale, we propose that a questionnaire-based ethnographic study would find that degree of religious commitment within churches and between churches in a restricted region, such as a US county or large city, will correlate positively with childhood infectious disease rates, perceived vulnerability to disease, philopatry, involvement with extended family and collectivism in general, and will correlate negatively with the two factors of personality, openness and extraversion.

One prediction of the parasite-stress theory of values as applied to religiosity is that concern about infectious diseases and religious commitment will be related positively within individuals in a region. Terrizzi et al. (2012) have studied this. In multiple samples, they found that people of high religiosity are more disgust sensitive and cognizant of infectious disease threat than are people of low religiosity. We predict, too, that these sensitivities will predict negatively recent history of infectious disease.

## 9.12 Summary

Researchers have studied regional variation in religiosity extensively. Such research, whether based on economic theory or evolutionary theory, emphasizes the high costs to individuals of participation in religious activities. We have offered a new hypothesis of religiosity based on the parasite-stress theory of values. It relies on the theory of honest signaling in biology. We propose that religiosity is one important way that people engage in and display their in-group allegiance and boundary in order to avoid and manage infectious disease threats. In support of this, we provide evidence that religiosity is an aspect of in-group assortative sociality—and therefore an aspect of the behavioral immune system—and that religiosity correlates positively with parasite adversity, both cross-nationally and across states of the USA We suggest additional tests of the parasite-stress theory's application to religiosity.

Other hypotheses of religiosity in the literature are discussed. The parasite-stress theory of values appears to best account for religiosity and its diversity across regions.

Our findings on religiosity have implications for a multitude of other areas of research such as secularizaton, health, ontogeny of religious values, life history, and geographical expansions of religion.

## References

Altizer, S., Nunn, C. L., Thrall, P. H. et al. (2003). Social organization and parasite risk in mammals: Integrating theory and empirical studies. *Annual Review of Ecology Evolution and Systematics* 34: 517–547.

Barrett, D. B., Kurian, G. T., & Johnson, T. M., Eds. (2001). *World Christian Encyclopedia: A Comparative Survey of Churches and Religions in the Modern World, Volume 1: The World by Countries: Religionists, Churches, Ministries*, 2nd ed. Oxford University Press, Oxford, U.K.

Bulbulia, J. (2004a). Religious costs as adaptations that signal altruistic intention. *Evolution and Cognition* 10: 19–42.

Bulbulia, J. (2004b). The cognitive and evolutionary psychology of religion. *Biology and Philosophy* 19: 655–686.

Bulbulia, J., & Mahoney, A. (2008). Religious solidarity: The hand grenade experiment. *Journal of Cognition and Culture* 8: 295–320.

Chaves, M., & Gorski, P. S. (2001). Religious pluralism and religious participation. *Annual Review of Sociology* 27: 261–281.

Cronk, L. (1994). Evolutionary theories of morality and the manipulative use of signals. *Zygon* 29: 81–101.

de Barra, M., DeBruine, L. M., Jones, B. C. et al. (2013). Illness in childhood predicts face preferences in adulthood. *Evolution and Human Behavior* 34: 384–389.

Delamontagne, R. G. (2010). High religiosity and societal dysfunction in the United States during the first decade of the twenty-first century. *Evolutionary Psychology* 8: 617–657.

Diamond, J. (1997). *Guns, Germs and Steel: The Fates of Human Societies*. W. W. Norton and Co., New York.

Eppig, C., Fincher, C. L., & Thornhill, R. (2010). Parasite prevalence and the worldwide distribution of cognitive ability. *Proceedings of the Royal Society of London B* 277: 3801–3808.

Eppig, C., Fincher, C. L., & Thornhill, R. (2011). Parasite prevalence and the distribution of intelligence among the states of the USA. *Intelligence* 39: 155–160.

Faulkner, J., Schaller, M., Park, J. H. et al. (2004). Evolved disease-avoidance mechanisms and contemporary xenophobic attitudes. *Group Processes and Intergroup Relations* 7: 333–353.

Figueredo, A. J., & Wolf, P. S. A. (2009). Assortative pairing and life history strategy: A cross-cultural study. *Human Nature* 20: 317–330.

Fincher, C. L., & Thornhill, R. (2012). Parasite-stress promotes in-group assortative sociality: The cases of strong family ties and heightened religiosity. *Behavioral and Brain Sciences* 35: 61–79.

Fincher, C. L., & Thornhill, R. (2008b). Assortative sociality, limited dispersal, infectious disease and the genesis of the global pattern of religion diversity. *Proceedings of the Royal Society of London, Biological Sciences* 275: 2587–2594.

Finkel, D. N., Swartwout, P., & Sosis, R. (2009). The socio-religious brain: A developmental model. *Proceedings of the British Academy* 158: 287–312.

Gelfand, M. J., Bhawuk, D. P. S., Nishii, L. H., et al. (2004). Individualism and collectivism. In *Culture, Leadership, and Organizations: The GLOBE Study of 62 Societies* (eds. R. J. House, P. J. Hanges, M. Javidan et al.), pp. 437–512. Sage Publications, Thousand Oaks, CA.

George, L. K., Ellison, C. G., & Larson, D. B. (2002). Explaining the relationships between religious involvement and health. *Psychological Inquiry* 13: 190–200.
Gladden, P. R., Welch, J., Figueredo, A. J. et al. (2009). Moral intuitions and religiosity as spuriously correlated life history traits. *Journal of Evolutionary Psychology* 7: 167–184.
Grim, B. J., & Finke, R. (2006). International religion indexes: Government regulation, government favoritism, and social regulation of religion. *Interdisciplinary Journal of Research on Religion* 2: 1–40.
Guégan, J. -F., & Constantin De Magny, G. (2007). Epidemiology in a changing world: The need for a bigger picture! *Encyclopedia of Infectious Diseases: Modern Methodologies* (ed. M. Tibayrenc), pp. 569–589. John Wiley & Sons, Inc., San Francisco, CA.
Guernier, V., Hochberg, M. E., & Guégan, J. -F. (2004). Ecology drives the worldwide distribution of human diseases. *PLoS Biology* 2: 740–746.
Hassall, C., & Sherratt, T. N. (2011). Statistical inference and spatial patterns in correlates of IQ. *Intelligence* 39: 303–310.
Henrich, J. (2009). The evolution of costly displays, cooperation and religion: Credibility enhancing displays and their implications for cultural evolution. *Evolution and Human Behavior* 30: 244–260.
Iannaccone, L. R. (1994). Why strict churches are strong. *American Journal of Sociology* 99: 1180–1211.
Inglehart, R., & Baker, W. E. (2000). Modernization, cultural change, and the persistence of traditional values. *American Sociological Review* 65: 19–51.
Irons, W. (1996). In our own self image: The evolution of morality, deception, and religion. *Skeptic* 4: 50–61.
Irons, W. (2001). Religion as a hard-to-fake sign of commitment. *Evolution and the Capacity for Commitment* (ed. R. M. Nesse), pp.292–309. Russell Sage Foundation, New York.
Irons, W. (2008). Why people believe (what other people see as) crazy ideas. *The Evolution of Religion: Studies, Theories, and Critiques* (eds. J. Bulbulia, R. Sosis, C. Genet et al.), pp. 51–57. Collins Foundation Press, Santa Margarita, CA.
Jackson, L. M., & Hunsberger, B. (1999). An intergroup perspective on religion and prejudice. *Journal for the Scientific Study of Religion* 38: 509–523.
Johnson, D. (2008). Gods of war: The adaptive logic of religious conflict. In *The Evolution of Religion: Studies, Theories, and Critiques* (eds. J. Bulbulia, R. Sosis, C. Genet et al.), pp.111–117. Collins Foundation Press, Santa Margarita, CA.
Koenig, H. G. (1997). *Is Religion Good for Your Health? Effects of Religion on Mental and Physical Health.* Haworth Press, New York.
Kokko, H., & Ots, I. (2006). When not to avoid inbreeding. *Evolution* 60: 467–475.
Kosmin, B. A., Mayer, E., & Keysar, A. (2001). *American Religious Identification Survey 2001*, pp. 1–46. The Graduate Center of the City University of New York, New York.
Letendre, K., Fincher, C. L., & Thornhill, R. (2010). Does infectious disease cause global variation in the frequency of intrastate armed conflict and civil war? *Biological Reviews* 85: 669–683.
Lopez, A. D., Mathers, C. D., Ezzati, M. et al. (2006). Global and regional burden of disease and risk factors, 2001: Systematic analysis of population health data. *Lancet* 367: 1747–1757.
Lynn, R., Harvey, J., & Nyborg, H. (2009). Average intelligence predicts atheism rates across 137 nations. *Intelligence* 37: 11–15.
McCleary, R. M., & Barro, R. J. (2006). Religion and political economy in an international panel. *Journal for the Scientific Study of Religion* 45: 149–175.
Murray, D. R., & Schaller, M. (2010). Historical prevalence of infectious diseases within 230 geopolitical regions: A tool for investigating origins of culture. *Journal of Cross-Cultural Psychology* 41: 99–108.
Norenzayan, A., & Shariff, A. F. (2008). The origin and evolution of religious prosociality. *Science* 322: 58–62.
Norris, P., & Inglehart, R. (2004). *Sacred and Secular: Religion and Politics Worldwide*. Cambridge University Press, New York.

Oaten, M., Stevenson, R. J. & Case, T. I. (2009). Disgust as a disease-avoidance mechanism. *Psychological Bulletin* 135: 303–321.

Park, J. H. & Schaller, M. (2005). Does attitude similarity serve as a heuristic cue for kinship? Evidence of an implicit cognitive association. *Evolution and Human Behavior* 26: 158–170.

Price-Smith, A. T. (2002). *The Health of Nations: Infectious Disease, Environmental Change, and Their Effects on National Security and Development*. MIT Press, Cambridge, MA.

Rees, T. J. (2009). Is personal insecurity a cause of cross-national differences in the intensity of religious belief. *Journal of Religion and Society* 11: 1–24.

Rossano, M. J. (2007). Supernaturalizing social life: Religion and the evolution of human cooperation. *Human Nature* 18: 272–294.

Saroglou, V., Delpierre, V., & Dernelle, R. (2004). Values and religiosity: A meta-analysis of studies using Schwartz's model. *Personality and Individual Differences* 37: 721–734.

Schwartz, S. H. (1992). Universals in the content and structures of values: Theoretical advances and empirical tests in 20 countries. *Advances in Experimental Social Psychology* 25: 1–65.

Searcy, W. A., & Nowicki, S. (2005). *The Evolution of Animal Communication: Reliability and Deception in Signaling Systems*. Princeton University Press, Princeton, NJ.

Siegal, M. (1988). Children's knowledge of contagion and contamination as causes of illness. *Child Development* 59: 1353–1359.

Siegal, M., Fadda, R., & Overton, P. G. (2011). Contamination sensitivity and the development of disease-avoidant behavior. *Philosophical Transactions of the Royal Society B: Biological Sciences* 366: 3427–3432.

Sosis, R. (2000). Religion and intragroup cooperation: Preliminary results of a comparative analysis of utopian communities. *Cross-Cultural Research* 34: 70–87.

Sosis, R. (2003). Why aren't we all Hutterities? Costly signaling theory and religious behavior. *Human Nature-An Interdisciplinary Biosocial Perspective* 14: 91–127.

Sosis, R. (2005). Does religion promote trust? The role of signaling, reputation, and punishment. *Interdisciplinary Journal of Research on Religion* 1: 1–30.

Sosis, R., & Alcorta, C. (2003). Signaling, solidarity, and the sacred: The evolution of religious behavior. *Evolutionary Anthropology* 12: 264–274.

Sosis, R., & Bressler, E. R. (2003). Cooperation and commune longevity: A test of the costly signaling theory of religion. *Cross-Cultural Research* 37: 211–239.

Sosis, R., Kress, H. C., & Boster, J. S. (2007). Scars for war: Evaluating alternative signaling explanations for cross-cultural variance in ritual costs. *Evolution and Human Behavior* 28: 234–247.

Sosis, R., & Ruffle, B. J. (2003). Religious ritual and cooperation: Testing for a relationship on Israeli religious and secular kibbutzim. *Current Anthropology* 44: 713–722.

Steadman, L. B., & Palmer, C. T. (2008). The Supernatural and Natural Selection: The Evolution of Religion. Paradigm Publishers, Boulder, CO.

Stevenson, R. J., Case, T. I. & Oaten, M. J. (2009). Frequency and recency of infection and their relationship with disgust and contamination sensitivity. *Evolution and Human Behavior* 30: 363–368.

Stevenson, R. J., Oaten, M. J., Case, T. I. et al. (2010). Children's response to adult disgust elicitors: Development and acquisition. *Developmental Psychology* 46: 165–177.

Terrizzi Jr., J. A., Shook, N. J., & Ventis, W. L. (2012). Religious conservatism: an evolutionarily evoked disease-avoidance strategy. *Religion, Brain and Behavior* 2: 105–120.

Thornhill, R. & Gangestad, S. W. (2008). *The Evolutionary Biology of Human Female Sexuality*. Oxford University Press, New York, NY.

Thornhill, R., Fincher, C. L., & Aran, D. (2009). Parasites, democratization, and the liberalization of values across contemporary countries. *Biological Reviews* 84: 113–131.

Thornhill, R., & Fincher, C. L. (2007). What is the relevance of attachment and life history to political values? *Evolution and Human Behavior* 28: 215–222.

Wall, J. N., & Shackelford, T. K. (2012). Immigration, parasitic infection, and United States religiosity. *Behavioral and Brain Sciences* 35: 97–98.

Widman, D. R., Corcoran, K. E., & Nagy, R. E. (2009). Belonging to the same religion enhances the opinion of others' kindness and morality. *Journal of Social, Evolutionary, and Cultural Psychology* 3: 281–289.

Wilson, D. S. (2002). *Darwin's Cathedral: Evolution, Religion, and the Nature of Society*. University of Chicago Press, Chicago, IL.

Zuckerman, P. (2007). Atheism: Contemporary Numbers and Patterns. In *The Cambridge Companion to Atheism* (ed. M. Martin), pp. 47–65. Cambridge University Press, NY.

# Chapter 10
# Democracy and Other Governmental Systems

## 10.1 Introduction

Human politics is basically social competition among individuals and groups using held and behaviorally manifested values as the competitive tools. It is a huge and boundless arena of human affairs, ranging from politicians competing with their expressed values for supporters, to male versus female antagonisms arising from conflicting sex-specific goals, to coercive and legal restrictions of property ownership, to governmental corruption versus transparency, and to intergroup relations. Western scholarship in political science extends back to discussions by Plato in *The Republic* and Aristotle in *Politic*. In this chapter, our goal is to provide an evidentiary and comprehensive explanation, based on the parasite-stress theory of values, of some major topics traditionally studied by political scientists and other scholars interested in political behavior. A diversity of evidence supports our application of the parasite-stress theory of values to forms of governance and associated topics. We cover here the autocracy–democracy dimension of governance across countries and the related topics of gender relations, women's sexual liberation, property rights, and diffusion of innovations. We also document that collectivism–individualism is related to these topics. Moreover, we discuss the early democratic transitions in Western governance, censorship and freedom in the media across nations, variation in autocracy across indigenous societies, and prejudices in voting caused by values. We first presented the core ideas and most analyses found in this chapter in Thornhill et al. (2009). We will investigate interregional economic relations and governmental corruption in Chap. 11, and civil conflicts and warfare in Chap. 12.

R. Thornhill and C.L. Fincher, *The Parasite-Stress Theory of Values and Sociality: Infectious Disease, History and Human Values Worldwide*,
DOI 10.1007/978-3-319-08040-6_10, © Springer International Publishing Switzerland 2014

## 10.2 Autocracy and Democracy in Political Science

Across the countries of the world, what are the causes of democracy and its antipole, autocracy? For many decades, this question has been a major focus of political science research. Researchers agree in their basic conceptualization of governmental democracy, but differ in their emphasis on the various components of democracy. Some researchers, for instance, emphasize how resources are distributed among people of a region, whereas others emphasize political liberties and participation. Most generally speaking, political democracy and autocracy refer to the extensiveness of empowerment, entitlement, and freedom of the people of a geopolitical region. Across the populace in such a region, the degree of democracy is precisely how widespread empowerment, entitlement, and freedom are. Hence, high democratization entails emancipation of the majority and expansive social and economic opportunity across socially different groups, care-taking of the health and other welfare of the many, widespread freedom of expression and of political organization, fair elections involving suffrage extended across the population, and widespread civil liberties and political rights (Lipset 1959; Bollen 1980; Betzig 1986; Somit and Peterson 1997; Zweifel and Navia 2000; Lake and Baum 2001; Perry and Robertson 2002; Inglehart 2003; Vanhanen 2003; Franco et al. 2004). The opposite of democracy is autocracy, which is the restriction of social and economic power and entitlement to a small portion of the people; the minority in power has omnipotent authority, with opportunity, privilege and rights limited to them and their kith and kindred. At the highest end of this pole are the political systems of authoritarianism, despotism, dictatorship, fascism, monarchy, oligarchy, and totalitarianism. Some scholars (e.g., Diamond 2002) attach descriptive labels to several points along the continuum of governmental types from highly autocratic to highly democratic; other researchers have sought to derive a continuous metric of democratization–autocratization that places all governments along the continuum.

Vanhanen (2003) provided such a metric in his Index of Democracy, which is an empirically verifiable, continuous variable of autocracy–democracy across most contemporary countries. Other metrics of democratization in the literature are less objective by involving opinions of panels of experts who assign countries' rankings; examples are Freedom House's Comparative Survey of Freedom (www.freedomhouse.org; e.g., Karatnycky 1998) and the Human Freedom Index of the *World Christian Encyclopedia* (Barrett et al. 2001). Scholars have used all these metrics repeatedly in efforts to investigate the causes of variation in democratization across the countries of the world. In these efforts, early hypotheses for autocracy–democracy have been expanded, refined, and tested (reviews in Inglehart 2003; Vanhanen 2003).

## 10.3 Traditional Theories of Democracy and Associated Empirical Patterns

Lipset (1960) advocated that democracy most likely arises and persists in countries of highest economic development, whereas autocracy arises and persists chiefly in the poorest countries. Later political scholars adopted and expanded Lipset's hypothesis that economic development causes democracy (e.g., Dahl 1989; Gurr et al. 1990; Muller 1997; Barro 1999; Bunce 2000; Perry and Robertson 2002; Wejnert 2005). Certainly, there is convincing evidence of a positive relationship between democratization and a variety of economic-development indicators, including the health of the citizenry or, said differently, a negative relationship between the degree of autocracy and economic development (e.g., Lipset 1983; Dahl 1989; Marks and Diamond 1992; Perry and Robertson 2002; Franco et al. 2004; Wejnert 2005). Some investigators have cast the causal relationship between economic development and democracy in terms of the broader framework of modernization (i.e., industrialization, urbanization, education, technology, and secularity) (e.g., Lernery 1958; Inglehart 2003; Norris and Inglehart 2004; Teorell and Hadenius 2006). As Welzel (2007) showed, the robust positive covariation between the two variables, economic development and modernization, arises because these variables are similar and each describes major features of democratization.

Vanhanen (1968, 1971) hypothesized that neither a country's economic development nor its modernization are the primary causes of democratization, but rather the salient cause is the distribution of economic resources, social opportunities, and political power among the people in a region; a component of this is individuals' rights of ownership of property, which allow widespread wealth accumulation and capital investment across the population of a geopolitical region (e.g., Vanhanen 2003; Miller and Diamond 2006). Related to the same thinking, Rueschemeyer et al. (1992) focused on economic developments' influence on interclass relations, especially the increased power and access to resources of the working and middle classes under industrialization that coincided with these classes' increased usefulness to the upper class. Similarly, Betzig (1986) proposed that the elites' sharing of social power with lower classes is the means by which all classes achieve more power, with elites always having the most power. To these scholars, then, and consistent with the general consensus of democracy's meaning in political science, democracy is the sharing of resources (economic and otherwise) and power among the people in a society. As the analyses by Vanhanen (2003) demonstrated, the degree of this distribution (and hence of resource equality) corresponds positively to the degree of empirically verifiable democratization across the world.

Vanhanen (2003) tied his resource-distribution hypothesis to evolutionary biology by emphasizing that human political psychology and behavior are caused ultimately by past natural selection favoring individuals who effectively competed for and secured resources, often by alliances with others to form socially competitive groups; for similar treatments, see Tiger and Fox (1971) and Betzig (1986). Somit and Peterson (1997) discuss a similar evolutionary framework, but focused on the rarity of a highly democratic governmental system among the more common system of authoritarianism.

Authoritarian or autocratic political systems are characterized by socially powerful resource-holders' unwillingness to share with others, whereas democracy rests on governmental officials' willingness to share and thereby create public goods and services for citizens such as educational opportunities, healthcare, and infrastructure promoting sanitation and recreation (Zweifel and Navia 2000; Lake and Baum 2001; Franco et al. 2004). Authoritarianism is based on a traditional and immutable hierarchy, with high-ranking individuals viewing themselves as superior humans and others as a subordinate out-group(s) undeserving of social and economic benefits. Authoritarianism includes an acceptance of low status by those with few resources, because of their respect for and obedience of those in power (the elites) and enforcement of the hierarchy by the elites through legal and coercive means. Authoritarian values of non-elites include the want and perceived need of their domination and guidance by elites (Altemeyer 1996).

Wejnert (2005) separated causes of democratization into two general categories: socioeconomic processes (such as resource distribution and equity) and diffusion processes, and reported that each of the two processes covaries positively with democratization across countries. Diffusion is the initial adoption and spread of an idea or technology within a social network, country or other region. As documented by Rogers (1995), diffusion stems from communication among a variety of people or groups, including through media channels, and by governmental agencies and other facilitators of information transfer. Such communication is promoted and widespread in democracies and restricted legally and coercively in autocracies.

## 10.4 The Parasite-Stress Hypothesis of Democratization

We acknowledge that economic development, modernization, widespread resource and social power equity, and the diffusion of information, innovations and technology are salient in the rise, maintenance and increase of political democratization. Indeed, the parasite-stress theory of values identifies these factors as salient components of democratization. This theory proposes that these factors, along with democratization in general, are effects or consequences of a proximate causal framework previously unrecognized by political scientists, and that the factors become causal within this framework. We seek an explanation of democratization at the most general and inclusive level that applies to all political systems across contemporary and historical geopolitical areas. Our approach centers on evolved human psychology modulating in-group and out-group emotions, behavior, and interactions that function to cope with the level of infectious disease present. We argue that the risk of human infectious disease, which historically and currently exhibits large variation across countries of the world, is the most encompassing, proximate cause of global variation in three core aspects of democratization: (1) the willingness of people in power to extend economic and social resources and opportunities outside their own in-group, and support the rights, liberties and political involvement of the populace; (2) the validity of rank/authority, as perceived by the general population, and thus the authoritarian–anti-authoritarian dimension of values; and (3) attitudes about non-traditional ideas and ways of life that determine whether innovations occur and are rewarded, as well as whether innovations diffuse within and across a region. According to the parasite-stress theory, these three aspects of democratization are closely tied to the value dimension collectivism–individualism; these ties were documented in Chaps. 4, 5, and 7. Compared to collectivists, individualists place more value on out-groups, less worth on the views of authority figures, and are more open to and rewarding of new ideas and ways of doing.

Hence, the parasite-stress hypothesis for autocracy–democracy has two conceptually related parts. First, high parasite stress evokes a value system of collectivism in which individuals are authoritarian, xenophobic and ethnocentric, with a disregard for—and in extreme, a moral disgust about—the rights, liberties and well-being of out-group members, including those lower in the established social hierarchy. Such out-group individuals are viewed as invalid members of the in-group(s) in power, and may be dehumanized and disenfranchised. This conservative ideology includes negativism toward and oppression of ideas and other innovations perceived to threaten traditional norms and values, and hence is a barrier to the creation and diffusion of novel ideas and technologies. As we have emphasized, collectivist ideology is a defense against novel contagion harbored in out-groups and a means to manage infectious agents that arise within the group. Second, low parasite stress evokes a value system of individualism that includes anti-authoritarianism, and tolerance, validity and trust of out-groups; a willingness to interact with, support, and empathize with different others; and a high regard for the rights and freedoms of the majority, including those of lower social class or with different values, religious convictions, languages, and ethnic identities. In extreme form, this is an ideology without prejudice toward any people, because all are considered to be sentient beings with morality, autonomy, and inalienable worth. Individualist ideology promotes and rewards innovations and includes a willingness to consider and adopt them, even those from out-groups. Liberal values serve to provide benefits to individuals via promoting out-group interaction and associated exchange of goods and services, the adoption of functional innovations of out-groups, and the establishment of social alliances with out-groups. The empirical prediction that follows from these considerations is that, across the globe, the degree of democratization will correlate negatively with parasite stress and correlate positively with individualism (and negatively with collectivism). Said differently, as parasite stress and collectivism increase across nations, so, too, will the degree of autocracy.

As Betzig (1986) and Rueschemeyer et al. (1992) have emphasized (see above), a disenfranchised out-group becomes increasingly important to the ruling group as the benefit the out-group provides to the ruling group increases. Effective use of a disenfranchised out-group by the socially powerful, however, can occur with the simultaneous total segregation between in-group and out-group and casting the out-group as a pariah (e.g., the autocratic USA. Old South during Reconstruction following the American Civil War). Clearly, then, the utility of subordinate out-groups to the elites does not in itself solely give rise to the adoption of democratic values toward out-groups by elites. The shift toward significant democracy requires more than the utility of the out-group to the ruling group. According to the parasite-stress theory of values, the required ideological changes for democracy to take root are: (1) the socially powerful adopt a view of the out-group as being comprised of people much like those in power, and equally human, and (2) the disenfranchised out-group devalues authority and considers themselves to be as valuable as the elites—they stop seeing themselves as needing the domination and guidance of the elites. Both (1) and (2) are evoked only in the condition of a low risk of contagion. Correspondingly, as infectious-disease risk increases, socially powerful people increasingly devalue and disenfranchise the out-group members, and these out-group members adopt more authoritarian values and the belief of their inferiority and inability to govern their own activities.

In the parasite-stress hypothesis of democratization, as parasite adversity declines within a region, there is a concomitant evocation, spread, and legalization of liberalized or individualistic attitudes and values that embrace traditionally disenfranchised groups. Consequently, wealth equality, social welfare, economic and educational opportunities, healthcare, safe public water, sanitation, and rights of holding private property also become widespread; infant mortality declines and adult longevity increases. Furthermore, simultaneously, the personality factor of openness toward new ideas and experiences is evoked and becomes widespread, increasing the creation and adoption of useful innovations arising from within and outside the group. Thus, global variation in economic development, social equality, emancipation from oppression by elites, and invention and diffusion of innovations are not independent of parasite adversity, but instead are variables caused by regional variation in infectious disease problems. Indeed, regional variation in both parasite stress and collectivism–individualism are proximate causes of the political democratization continuum, as well as its associated economic, educational, health, innovation, and diffusion components.

Furthermore, we propose that there is a bidirectional, proximate causal feedback between parasite stress, economic and social factors, and related liberalization of values. As parasite stress declines and peoples' values shift to affect openness as well as egalitarian and widespread economic and other well-being in a region, the changes will reduce parasite stress further through increases in widespread nutrition, sanitation and access to medical care and educational information, all of which cycle back to reduce mortality and morbidity from parasites. Thus, as parasite stress declines, democratization factors increase correspondingly, which, in turn, further reduce infectious disease. The opposite also holds: as parasite stress increases or maintains high levels, the conservative values of allegiance to traditional ideas and methods, prejudice, inequality and authoritarianism that arise further magnify the morbidity and mortality from infectious disease.

As an example of this bidirectionality, consider the implementation by a government of widespread sanitation service such as potable water for the people of a region. For those in power (and hence in control of resources) to view such implementation as morally correct requires their respect for people in general, which, in the parasite-stress theory, is caused by the perception by those in social power of lowered risk of contagion in the local environment and resultant liberalization of values. Widespread healthier water subsequently reduces parasite adversity across the populace, which, in turn, expands democratic values and the humanitarian cultural infrastructure these values generate, which, in turn, further reduce parasite stress.

In terms of ultimate causation, the parasite-stress theory of values as applied to democratization argues that, in human evolutionary history, natural selection crafted species-wide conditional psychological adaptation instantiated in the human nervous system that functions in adopting values with local utility in social navigation. Under conditions of high parasite stress, this conditional psychological adaptation yielded widespread high devaluation of out-groups, high parochialism, high authoritarianism, and closed-mindedness to innovations. In contrast, under conditions of low parasite stress, the same psychological adaptation yielded widespread positive valuation of out-groups, reduced ethnocentrism, high anti-authoritarianism, and openness to innovations. In some cases, this conditionality may co-occur with region-specific genetic adaptation functionally designed to adopt and use values of high local utility (Chap. 3).

As explained in Chap. 2, the parasite-stress theory of values does not minimize the importance of social learning in the ontogeny of personal values; indeed, social learning is a proximate cause of the values. The theory does suggest, however, that this cultural learning will be different in high-parasite-stress areas compared to low-parasite-stress areas. For example, imitation of authority figures and anti-imitation of low status people is expected to increase in importance with increased parasite stress.

Somit and Peterson (1997) cast democratic values as late-comers in human history; the parasite-stress theory of values argues against this view. The condition-dependent design of the psychological adaptation in the theory implies that a large range of values from high collectivism to high individualism were ancestrally adaptive outputs of the species-typical psychology, including within-population, individual variation along the values' continuum. Hence, the basic values of individuals in general that correspond to what is now called democracy or autocracy at the societal level were present during human evolutionary history at least from the debut of *Homo sapiens*, and perhaps earlier in the hominin lineage. Yet, certain Western extremes in the core values of individualism/liberalism (e.g., extremes in xenophilia and nuclear-family restriction of nepotism) may be evolutionarily novel, because of the evolutionary novelty of recent public health advances (see below). The parasite-stress hypothesis of democratization, however, does not imply, rest on, or deny that values and associated behaviors are adaptive currently (currently result in high inclusive fitness) (Chap. 2).

As explained above, according to our thinking, democratization importantly involves the liberalization of many types of values under the ecological condition of low infectious-disease stress. The widespread distribution of suffrage and political participation of women and their rights and freedoms in general are considered appropriately by many scholars to be a component of democratization and a liberalization of the traditional conservative value of men's assumed superiority over women (e.g., Inglehart 2003; Wejnert 2005). Hence, women and men's political interrelations are expected to covary with democratization—specifically, the higher the democratization, the more equitable or democratic these relations are, and the lower the democratization, the more patriarchal or conservative are male–female relations. Consistent with this, Inglehart et al. (2003) showed that a few items from the World Values Survey that measure gender equality correlated positively with Freedom House scores of democratization (see below) across countries, and Gelfand et al. (2004) reported that cross-national collectivism correlated negatively with gender equality values taken from the same survey. In earlier chapters, numerous research findings were discussed that link gender equality with individualism and gender inequality with collectivism. As multiple researchers have correctly pointed out, patriarchy, or the exclusion of women from political participation, as well as from many other social freedoms (e.g., restrictions on movement, social interactions with males, dressing behavior, and attainment of education and other status bars), is directed fundamentally at sexual control of females so as to promote investing men's paternity reliability (Chap. 8). As documented in Chap. 8, patriarchy is an aspect of the culture of honor that can promote and ideologically validate men's violent control of women's sexuality.

Gangestad et al. (2006) were the first researchers to document a cross-national, positive relationship between gender inequality and parasite stress. They used a measure of parasite stress related to one we use in this chapter and the Gender Empowerment Measure (see below) for a sample of about 30 countries. Our analyses below expand considerably this earlier finding.

Beyond the various aspects of general gender relations and equity, democratization is expected to include more acceptance of and pursuit of sexual activity without long-term romantic involvement by women—in effect, a sexual liberation of women from traditional sexual continence—because, under the conditions of low parasite stress and thus, increased democratization, this is in women's reproductive interests. In contrast, the conservative and patriarchal authoritarian ideology of gender differentiation and associated assumed male superiority has a sexual double-standard, with strict female sexual continence. This traditional sex-role ideology imposes costs on women for engaging in premarital sex and short-term sexual activity such as brief affairs. Women, however, often have much to gain by these activities—both genetic benefits for offspring and non-genetic material benefits received from men in exchange for sexual access (Thornhill and Gangestad 2008). Supporting this perspective, various studies are reviewed in Chap. 6 showing that, across countries, parasite stress is correlated positively with cultural norms of sexual restrictiveness, particularly as applied to women, a pattern seen especially for nonzoonotic infectious diseases, but not for zoonotics. In previous chapters, some research findings are mentioned briefly that link cultural norms of women's sexual restrictiveness to collectivism; the analyses below document this relationship.

In this chapter, we test the parasite-stress hypothesis of democratization by examining, across many contemporary countries of the world, the relationships between measures of parasite stress and measures of democratization, and between democratization and collectivism–individualism. Evidence presented in Chaps. 4 and 5 suggested that collectivism is low democratization and individualism is high democratization. In this chapter, the relationship between democratization and collectivism–individualism is directly examined, as well as the interrelationships of parasite stress, gender equality, democratization, and private-property rights across the globe. Related to gender relations, we treat the covariation between sociosexual orientation (i.e., attitudes about having sex without commitment), democratization and collectivism–individualism across countries. Moreover, we tie the parasite-stress hypothesis of democratization to the increase in liberalism, including women's sexual liberation, in the West following the major public health advances in the twentieth century that dramatically reduced the negative impact of infectious diseases in the region. Related to this, we propose that the parasite-stress theory explains the early democratic transitions in Britain, France and the USA as a result of reduced parasite prevalence at high latitudes. Finally, we discuss a recent study by Murray and colleagues (2013) that reported a positive association between parasite stress and autocracy across a large number of the aboriginal societies from the Standard Cross-cultural Sample.

Discussed first is our first study of cross-national democratization (Thornhill et al. 2009), and then we discuss our second study of this topic (Thornhill et al. 2010). Details of the methods and empirical results can be found in those two publications.

## 10.5 Thornhill et al. (2009)

### *10.5.1 Measures of Democratization*

In this study, we used validated political democratization measures that are in the public domain and accessible. One was Vanhanen's (2003) Index of Democratization for 170 countries during 1999–2001. This index is comprised of two basic components of democracy: *Competition* and *Participation*. *Competition* is computed by subtracting the percentage of votes won by the largest political party from 100. When these data on votes were unavailable, *Competition* was calculated from the number of seats in parliament. *Participation* reflects data on voter involvement in elections and is computed from the total population. Referendums (a measure of direct democracy) were included in *Participation*. Vanhanen's Index of Democratization (ID)—a continuous variable from zero to its highest score—combines *Competition* and *Participation* with equal weight by multiplying them and dividing the product by 100, and depicts the range of political systems from highly autocratic to highly democratic, i.e., lowest democratization to highest democratization (e.g., Pakistan and Somalia, 0; Ethiopia, 3.5; Madagascar, 12.3; Poland, 20.8; Norway, 39.2).

Vanhanen's ID does not include a direct measure of civil and political rights and liberties, a major component of democratization. He emphasized that the legal competition between groups for power through elections (measured by his *Competition* variable) and the legal participation of people in elections (his *Participation*) correspond closely to civil and political rights and liberties. In fact, Vanhanen's ID is highly correlated with the Freedom House ratings (see below) of political rights and civil liberties (high scores equal low freedom) ($r=-0.80$, $n=162$ countries) (Vanhanen 2003).

Vanhanen (2003) reported that, as he hypothesized, ID correlated positively and highly with his five measures of resource distribution (the more democracy, the more widely and equitably resources are distributed across people in the country). His resource-distribution measures include Gross Domestic Product (GDP) per capita, percentages of university students and literates (related to educational opportunism and egalitarianism), the degree to which land ownership is widespread and evenly distributed, and the degree of decentralization of non-agricultural economic resources. As an additional measure of democratization, we used his resource distribution variable—the average per country of his five resource-distribution scores—which we refer to as *Vanhanen's Resource Distribution*, a variable introduced earlier in this book. Each of Vanhanen's five scores taps democratization and thus *Vanhanen's Resource Distribution* including all five gives a robust measure of democratization and associated values related to our theoretical framework.

We used two additional democracy measures that are less objective than Vanhanen's two measures described above. Vanhanen's are based on verifiable data, whereas the other two are based on subjective judgments of experts (see Vanhanen 2003 for discussion). The Freedom House Survey (FH) measures for 193 countries for 2007 (taken from www.freedomhouse.org) provide separate ratings on political rights and civil liberties on a seven-category scale, with 1 representing the most freedom and 7 the least (see Karatnycky 1998 for a discussion of rating methods). We summed the two categories of ratings into one democratization score per country. The two categories are highly intercorrelated ($r=0.94$, $p<0.0001$, $n=193$ countries). The final measure we used was the Human Freedom Index (HFI), obtained from the *World Christian Encyclopedia* (Barrett et al. 2001), which is a composite score of several variables related to individual freedom in 233 countries; the highest scores correspond to the most individual freedom.

In certain analyses, we included two variables of distribution of economic resources that are commonly included in research in political science and related disciplines. We collected each country's wealth as GDP per capita from the World Factbook 2007 (www.cia.gov). For many countries, Gini scores for families also were taken from the *World Factbook* for the same year. Whereas GDP per capita measures the average resource holdings per capita in a country, Gini measures the degree of inequality in the distribution of family income in a country. A higher Gini score indicates a greater inequality in a country's income distribution. Presumably, the Gini score is a reasonable measure of the degree of competition for economic resources in each country (Chap. 8).

A between-regions analysis was used to examine separately the predicted relationships in temperate and tropical regions of the world. We conducted this analysis with each of the four democratization measures by parasite stress in the two regions. We defined tropical countries as those whose latitudinal center was located between 30° north and 30° south; temperate countries are outside of this latitudinal range.

### 10.5.2 Parasite Stress

We used the cross-national *Contemporary Parasite Severity* measure described in Chap. 5. It is based on a scale of the number of cases of 22 important human infectious diseases in each country.

### 10.5.3 Measures of Collectivism–Individualism

Gelfand et al. (2004) reported the relationships across countries between one measure of collectivism–individualism and several political values from the World Values Survey. This included *Dislike for Democracy*, which is relevant to the democratization ideology among the people in a country. They found a significant, positive, moderate correlation ($n=27$ countries) between *Dislike for Democracy* and collectivism, which is consistent with the parasite-stress theory.

Here, we greatly expanded this early research by including the multiple measures of democratization (discussed above), multiple measures of collectivism–individualism, and a large sample of countries. We used three conceptually related collectivism–individualism measures: *Gelfand In-group Collectivism, Hofstede Individualism*, and *Suh Individualism* (see Chap. 5 for descriptions of these variables).

### 10.5.4 Property Rights

We used mean scores from 2004 to 2008 of the private-property-rights item for 162 countries from the 10-item Index of Economic Freedom (www.heritage.org/Index/). These scores reflect the ability of people to own and accumulate private property, which is secured by laws enforced by the government of the country. High scores reflect high property rights, whereas low scores reflect governmental restriction or the illegality of private property. We tested the relationships between private-property rights, democratization, collectivism–individualism, and parasite stress. This property-rights variable excludes, of course, men's right to own and control their romantic partners and daughters, a right of men associated with collectivism and the culture of honor (Chap. 8).

### 10.5.5 Gender Equality

We also tested the relationship of gender equality with democratization, parasite stress, and collectivism–individualism. We used two measures of gender relations reported for numerous countries. The Gender Empowerment Measure (GEM) (reported in the U.N. Human Development Report (http://hdrstats.undp.org/indicators/280) for 93 countries in 2007) is an index comprised of gender inequalities in political and economic participation and decision-making and power over economic resources. High scores indicate increased gender equality and low scores indicate increased gender inequality.

An index of gender-equality values, based on 2007 data from the World Values Survey, was our second measure of gender relations across the world; it was computed by summing five items, and is scaled in the same direction as the GEM; data were available for 50 countries (for a discussion of items, methods, and validity as a gender-equality measure, see Javidan and Hauser 2004). This second gender equality index was correlated highly at $r=0.84$ with GEM ($p<0.0001$, $n=34$).

Gelfand et al. (2004) reported a significant negative correlation between our second measure of gender-equality values and *Gelfand In-group Collectivism*. Inglehart et al. (2003) showed that certain items of this measure of gender equality positively correlate with one measure of democratization that we use (Freedom House). Our study greatly expands these earlier findings.

### 10.5.6 Sexual Liberalization

The Sociosexual Orientation Inventory (SOI, Simpson and Gangestad 1991) measures individual attitudes and behaviors related to permissiveness–restrictiveness about casual sex, i.e., sexual relations without long-term romantic love and commitments. High SOI scores indicate a more permissive or unrestricted approach to sexual behavior. As mentioned in Chap. 6, Schmitt (2005) reported mean sex-specific SOI scores for 48 countries. We use these means for our measures across countries of unrestricted and restricted sexual behavior (high SOI = unrestricted). Chapter 6 discussed the evidence provided by multiple studies for the negative relationship between parasite stress and SOI, which is especially robust for women. As also detailed in Chap. 6, this pattern is seen for nonzoonotic infectious diseases but not for zoonotics. We report below the relationships between SOI, democratization, gender equality, and collectivism–individualism.

### 10.5.7 Results

The results reported in this section originally were reported in Thornhill et al. (2009).

#### 10.5.7.1 The Four Democratization Measures Tap the Same Thing

The measures of democratization we employed were highly intercorrelated. For the sample of 169 countries shared by Vanhanen's Index of Democratization, Freedom House scores, and Human Freedom Index, the relationships were: Index of

Democratization (ID) and Freedom House (FH), $r = -0.80$ (FH, lowest scores = most freedom) and Human Freedom Index (HFI), $r = 0.64$; HFI and FH, $r = -0.63$ (all $p$-values < 0.0001). *Vanhanen's Resource Distribution* correlated strongly in the predicted direction with each of the other three measures of democratization: ID, $r = 0.81$; FH, $r = -0.65$; HFI, $r = 0.69$; all $p$-values < 0.0001, all $n$'s = 168.

#### 10.5.7.2 Parasite Stress Positively Predicts Autocracy

Parasite stress showed the predicted relationship with each of the democratization measures: ID and parasite stress, $r = -0.52$, $n = 169$ countries; FH and parasite stress, $r = 0.45$, $n = 192$ countries; HFI and parasite stress, $r = -0.52$, $n = 212$ countries; all $p$-values < 0.0001. Vanhanen's Resource Distribution correlated highly with parasite stress ($r = -0.67$, $p < 0.0001$, $n = 169$ countries). As parasite stress increases across countries, democracy, including the equitability of resource distribution among people, declines—i.e., resources and educational opportunities become more concentrated in smaller numbers of people. In sum, parasite stress was moderately (0.45) to highly (0.67) correlated with the four democratization measures across a large sample of countries of the world, and in each of the four cases, more parasite adversity corresponds to lower democratization.

The same patterns were found when each of the democratization variables was analyzed by parasite stress, but separately for temperate and tropical countries: ID temperate, $r = -0.57$, $p < 0.0001$, $n = 66$; ID tropical, $r = -0.17$, $p < 0.10$, $n = 103$; FH temperate, $r = 0.51$, $p < 0.0001$, $n = 71$; FH tropical, $r = 0.36$, $p < 0.0001$, $n = 121$; HFI temperate, $r = -0.53$, $p < 0.0001$, $n = 72$; HFI tropical, $r = -0.55$, $p < 0.0001$, $n = 140$; *Vanhanen's Resource Distribution* temperate, $r = -0.47$, $p < 0.0001$, $n = 66$; *Vanhanen's Resource Distribution* tropical, $r = -0.55$, $p < 0.0001$, $n = 103$. For both temperate and tropical regions, all the patterns were highly statistically significant with the one exception of ID tropical, which showed a probability of 0.10. However, with one-tail probability, this is significant ($p = 0.05$); one-tailed probability is appropriate, given the directional relationship as predicted by the parasite-stress theory. Hence, in two ecologically distinct regions of the world, parasite stress is associated negatively with democratization. The analyses of empirical patterns presented below involve the world-wide sample of countries unseparated into temperate and tropical categories.

For completeness, we include here the relationships between parasite stress and each of the measures of democratization with the Gini (among-family income inequality) and GDP per capita statistically controlled. As described above, a country's GDP per capita and wealth inequality (Gini) are apparently largely effects of the region's level of parasite adversity and the values evoked by the given parasite adversity in the region. All partial correlation coefficients remained highly significant (≤0.0004). The partial correlation coefficients between democratization and parasite stress with Gini partialled out were similar in magnitude to the zero-order correlations: ID, partial $r = -0.52$, $p < 0.0001$, $n = 120$ countries; FH, partial $r = 0.42$, $p < 0.0001$, $n = 120$ countries; HFI, partial $r = -0.39$, $p < 0.0001$, $n = 121$ countries;

**Table 10.1** Pearson zero-order correlations between three measures of collectivism–individualism and four measures of democratization across countries

| Democratization | Hofstede individualism | Suh individualism | Gelfand in-group collectivism |
|---|---|---|---|
| Index of democratization (Vanhanen 2003) | 0.65 (66) | 0.76 (54) | –0.66 (56) |
| Freedom House | –0.55 (67) | –0.57 (54) | 0.58 (56) |
| Human Freedom Index | 0.54 (66) | 0.70 (56) | –0.79 (56) |
| *Vanhanen's Resource Distribution* | 0.67 (66) | 0.74 (54) | –0.70 (56) |

All $p$-values are <0.0001. Sample sizes of number of countries are in parentheses following correlation coefficients (results originally reported in Thornhill et al. 2009)

*Vanhanen's Resource Distribution*, partial $r=-0.59$, $p<0.0001$, $n=120$ countries. With GDP per capita controlled, the partial correlation coefficients between democratization and parasite stress were lower than the zero-order coefficients, but highly statistically significant in all cases: ID, partial $r=-0.27$, $p=0.0004$, $n=169$ countries; FH, partial $r=0.30$, $p<0.0001$, $n=192$ countries; HFI, partial $r=-0.34$, $p<0.0001$, $n=207$ countries; *Vanhanen's Resource Distribution*, partial $r=-0.40$, $p<0.0001$, $n=169$ countries. (We remind the reader that Vanhanen's Resource Distribution measure actually includes variation due to differences in GDP per capita, so this partialling is not strictly a removal of GDP per capita.) Moreover, as expected, in a zero-order correlation, *Vanhanen's Resource Distribution* (higher=more equitability of resources in general) is related negatively with Gini across countries: $r=-0.45$, $p<0.0001$, $n=120$ countries.

#### 10.5.7.3 Autocracy Is Positively Predicted by Collectivism

Each of the measures of collectivism–individualism was related strongly to democratization in the manner predicted by the parasite-stress theory (Table 10.1). Correlations ranged from 0.54 (HFI and *Hofstede Individualism*) to –0.79 (HFI and *Gelfand In-group Collectivism*). Hence, as democratization increased, individualism increased, and, correspondingly, collectivism decreased across all measures of democracy and of collectivism–individualism. This includes Vanhanen's Resource Distribution. So, as individualism increases, resources become more equitably distributed among people, or said differently, as collectivism increases, resources are less equitably distributed.

#### 10.5.7.4 Property Rights

Property rights from the Index of Economic Freedom had the predicted relationships, all strong, with parasite stress, democratization and collectivism–individualism. Across the world, property rights showed a strong, negative relationship with parasite stress ($r=-0.53$, $p<0.0001$, $n=153$), and, hence, are most restricted in the highest parasite-stress countries. Property rights and democratization were strongly and positively related: ID, $r = 0.65$, $n = 152$; FH, $r = -0.68$, $n = 152$; HFI, $r = 0.69$, $n = 151$; Vanhanen's Resource Distribution, $r = 0.78$, $n = 152$; all $p$-values $< 0.0001$. Collectivism–individualism, in relation to property rights, showed the following: *Hofstede Individualism*, $r = 0.66$, $n = 67$; *Gelfand In-group Collectivism*, $r = -0.75$, $n = 57$; *Suh Individualism*, $r = 0.69$, $n = 55$; all $p$-values $< 0.0001$. Thus, across the world, the property rights of people are maximized under the lowest parasite stress and the highest democratization and individualism, and minimized under the highest parasite stress, autocracy and collectivism.

#### 10.5.7.5 Gender Equity

The Gender Equality Measure (GEM) showed the predicted and strong relationships (all $n$'s $= 91$) with parasite stress ($r = -0.50$, $p < 0.0001$) and with democratization (ID, $r = 0.66$; FH, $r = -0.65$; HFI, $r = 0.74$; Vanhanen's Resource Distribution, $r = 0.74$; all $p$-values $< 0.0001$). Hence, the subordination of women's civil rights and freedoms, relative to men's, is greatest under high parasite stress, autocracy and inequitable resource distribution, whereas women's rights and freedoms are most recognized and validated under low-parasite prevalence, high democracy, and equitable resource distribution.

Corroborating this finding, the gender equality index for 50 countries (from the World Values Survey) showed a similar relationship with parasite stress ($r = -0.39$, $p = 0.005$). This gender-equality measure also showed strong correlations with each of the democratization measures (ID, $r = 0.50$, $n = 48$; FH, $r = -0.51$, $n = 50$; HFI, $r = 0.62$, $n = 49$; Vanhanen's Resource Distribution, $r = 0.58$, $n = 48$; all $p$-values $< 0.0003$).

Furthermore, as predicted by the parasite-stress hypothesis of democratization, GEM and the gender-equality index measure each showed a strong, negative correlation with collectivism and a strong, positive relationship with individualism: GEM and *Suh Individualism*, $r = 0.72$, $p < 0.0001$, $n = 44$; and *Hofstede Individualism*, $r = 0.62$, $p < 0.0001$, $n = 57$; and *Gelfand In-group Collectivism*, $r = -0.75$, $p < 0.0001$, $n = 47$; gender equality index and *Suh Individualism*, $r = 0.62$, $p < 0.0009$, $n = 25$); and *Hofstede Individualism*, $r = 0.57$, $p < 0.0008$, $n = 31$; and *Gelfand In-group Collectivism*, $r = -0.74$, $p < 0.0001$, $n = 27$).

#### 10.5.7.6 Sexual Liberty of Women

GEM and the gender equality index each showed a positive relationship with the Sociosexual Orientation Inventory (SOI) (sexes combined) across societies (GEM and SOI, $r = 0.43$, $p = 0.006$, $n = 39$; gender equality index and SOI, $r = 0.38$, $p = 0.120$, not significant, $n = 18$). Hence, high gender egalitarianism corresponded to less restricted casual sexual behavior, whereas low gender egalitarianism was associated with more restricted casual sexual activity. As anticipated, these relationships were stronger and significant only for women (GEM and women's SOI, $r=0.62$, $p<0.0001$, $n=38$; and men's SOI, $r=0.16$, $p=0.34$, $n=38$; gender equality index and women's SOI, $r=0.63$, $p=0.005$, $n=18$; and men's SOI, $r=0.09$, $p=0.73$, $n=18$). Hence, values pertaining to equitability of gender relationships have a more liberalizing effect on women's sexual attitudes and behaviors than on men's.

Women's SOI was related positively and significantly to all the measures of democratization: ID, $r=0.39$, $p=0.008$, $n=44$; FH, $r=-0.51$, $p=0.0002$, $n=45$; HFI, $r=0.42$, $p=0.005$, $n=44$; Vanhanen Resource Distribution, $r=0.40$, $p=0.006$, $n=44$. Similarly, women's SOI was related positively and significantly to all measures of individualism: *Hofstede Individualism*, $r=0.52$, $p=0.001$, $n=36$; *Gelfand In-group Collectivism*, $r=-0.64$, $p=0.0001$, $n=28$; *Suh Individualism*, $r=0.41$, $p=0.023$, $n=30$. Said differently, women's SOI was related negatively to collectivism.

## 10.6 Thornhill et al. (2010)

### *10.6.1 Democracy*

We collaborated with Damian Murray and Mark Schaller in this follow-up of Thornhill et al. (2009). In the follow-up, we directly tested the differential predictive effects of zoonotic and nonzoonotic (both human-specific and multihost) parasite adversity on democratization. In Thornhill et al. (2010), the parasite adversity arising from each of the three types of human diseases—zoonotic, human-specific, and multihost—was based on the number of kinds of each type, and thus, on parasite richness. As explained in Chap. 5, the parasite-stress theory of values predicts that nonzoonotic human infectious diseases will have a stronger relationship with values and their effects such as democracy than will zoonotic human infectious diseases. We have shown this to be the case cross-nationally for collectivism–individualism (Chap. 5), certain personality factors (Chap. 7), homicides (Chap. 8), and religiosity (Chap. 9); and for collectivism–individualism (Chap. 5), interpersonal violence (Chap. 8) and religiosity (Chap. 9) across the states of the USA.

As described above, Thornhill et al. (2009) reported the significant correlations between pathogen stress and four measures employed by scholars to describe the variation in democratization of political systems across the globe. Thornhill et al. (2010) used the same four and added a fifth measure of democratization from the Economist Intelligence Unit (EIU; www.eiu.com). This cross-national democracy index was based on 60 indicators assessing five defining components of democratization (electoral process and political pluralism, civil liberties, the functioning of the government, political participation, and political culture). We employed the EIU Democracy Index for 2008; higher scores reflect higher levels of democratization.

**Table 10.2** Pearson zero-order correlations and *p*-values between each measure of parasite richness and each measure of democratization; *n* = the number of countries in each analysis (results originally reported in Thornhill et al. 2009)

| | Parasite richness measure | | | | | | |
|---|---|---|---|---|---|---|---|
| | Human-specific | *p* | Multihost | *p* | Zoonotic | *p* | n |
| Democracy index (EIU) | −0.48 | <0.001 | −0.26 | 0.001 | −0.02 | >0.10 | 163 |
| Index of Democracy (Vanhanen 2003) | −0.55 | <0.001 | −0.30 | <0.001 | −0.02 | >0.10 | 168 |
| *Vanhanen's Resource Distribution* | −0.70 | <0.001 | −0.43 | <0.001 | −0.08 | >0.10 | 168 |
| Human Freedom Index (Barrett et al. 2001) | −0.51 | <0.001 | −0.41 | <0.001 | −0.39 | <0.001 | 214 |
| Restrictions on rights[a] (Freedom House 2008) | 0.43 | <0.001 | 0.32 | <0.001 | 0.22 | 0.002 | 190 |

[a]Freedom House measure is scored with higher scores meaning greater restriction on civil liberties and political rights

In addition, gender equality, a measure of the gender-relations component of democracy, was examined in relation to each of the three disease types in Thornhill et al. (2010). We used the Gender Empowerment Measure (GEM) scores described above.

## 10.6.2 Results

The results reported in this section originally were reported in Thornhill et al. (2010). Across all five democratization measures, a clear pattern emerged in the relative magnitude of correlations (see Table 10.2). Human-specific parasite richness had the strongest correlations with democratization measures, followed by multihost parasite richness, and all correlations were statistically significant and showed that as these infectious diseases increase, democracy simultaneously declines. In contrast, zoonotic parasite richness had relatively weaker relations with democratization measures. In follow-up regression analyses, with all three parasite richness indices entered simultaneously as predictors, human-specific parasite richness remained a robust and statistically significant predictor of all five outcome measures (all *p*'s $< 0.001$). These regression analyses revealed a unique effect of multihost parasite richness on one of the five democratization measures (*Vanhanen's Resource Distribution*; $p = 0.05$). These same analyses revealed that any apparent effect of zoonotic parasite richness disappeared entirely or, if anything, reversed in sign. Zoonotic parasite richness was significantly positively, rather than negatively, correlated with the EIU democracy index, and with both of Vanhanen's indices, when controlling for shared variance with the other parasite richness indices; *p*'s $< 0.005$.

Across 93 countries, gender equality was correlated negatively with indices of both human-specific parasite richness ($r = -0.52$, $p < 0.001$) and multihost parasite richness ($r = -0.35$, $p < 0.001$). The relation with zoonotic parasite richness was negligible and non-significant ($r = -0.09$, $p = 0.37$). In a follow-up regression analysis with the three parasite-richness indices simultaneously entered as predictors of gender equality, only the human-specific index remained a statistically significant predictor ($p < 0.001$).

In sum, supporting the parasite-stress theory of values, cross-national differences in democratization and gender equality are predicted specifically by nonzoonotic parasite adversity. Zoonotic adversity has little, if any, effect on these two related societal value systems across the globe. In Chap. 6, we provided the cross-national evidence that women's sexual restrictiveness (SOI) is more strongly associated with nonzoonotic than with zoonotic human diseases. Hence, reduction in nonzoonotic diseases in particular liberates women from their traditional and collectivist sex role of sexual continence.

## 10.7 Nonzoonotic and Zoonotic Disease Severity Compared

Subsequent to the publication of Thornhill et al. (2010), we computed the nonzoonotic and zoonotic severity scores for essentially all the countries of the world (these scores and the methods are in Fincher and Thornhill 2012). Severity refers to a scale of the number of cases of the diseases. The analyses above comparing these disease types used richness, not severity. Severity and richness measures are correlated almost perfectly: zoonotic severity and richness, $r = 0.98$, $p < 0.0001$, $n = 222$; nonzoonotic severity and richness, $r = 0.96$, $p < 0.0001$, $n = 222$.

Not surprisingly, given this strong covariation between richness and severity of each of the two categories of disease, the above analyses comparing the association of each of the two categories with democratization and related variables are repeated with the severity measures—that is, nonzoonotics only or more robustly predict these variables. We give here a few examples. *Vanhanen's Resource Distribution* in relation to severity reveals for nonzoonotics an $r = -0.66$, $p < 0.0001$, $n = 169$; for zoonotics, $r = -0.09$, $p = 0.26$, $n = 169$. The same pattern is seen for Vanhanen's Index of Democracy: nonzoonotics, $r = -0.50$, $p < 0.0001$, $n = 169$; zoonotics, $r = -0.00$, $p = 0.94$, $n = 169$. GEM shows the pattern as well: nonzoonotics, $r = -0.43$, $p < 0.0001$, $n = 93$; zoonotics, $r = -0.13$, $p = 0.21$, $n = 93$.

## 10.8 F-Scale and Autocracy

As we explained earlier in this chapter, authoritarianism is a fundamental ingredient of autocratic governance. A recent study by Murray et al. (2013) investigated the relationship between authoritarianism and a measure of historical parasite severity in a sample of 31 countries for which authoritarianism personality scores of individuals were available. (The measurement of historical parasite severity is discussed in Chap. 5.) The authoritarianism scores of individuals were based on responses on the F-scale (F stands for fascist) of Adorno et al. (1950). High scores equal high fascist/authoritarian values and low scores correspond to high democratic values. The F-scale is a valid instrument and measures personality traits such as authoritarian submission, conventionalism, political power exclusively held by elites, and ethnocentrism; thus, it overlaps and complements the measures of autocratic values we used in the analyses above. Murray et al. (2013) showed that parasite stress strongly and positively predicted authoritarian values of people across the 31 countries.

Their additional analyses in the same paper showed that individuals' authoritarianism is a mediator of the correlation across these countries between parasite stress and multiple measures of authoritarianism at the level of the federal government of the countries. Their measures of authoritarian versus democratic governance were ones we have discussed above: property rights, Vanhanen's Index of Democracy and Freedom House scores. Hence, they document that autocratic/authoritarian governments arise in high-parasite-stress countries in which the populations of people, on average, are autocratic, whereas democratic governments arise in low-parasite-stress countries in which people are democratic in general. Murray et al. (2013) concluded that their research supports the parasite-stress theory of values applied to democratization as proposed by Thornhill et al. (2009) and explained above.

## 10.9 Authoritarianism in the Standard Cross-Cultural Sample

In the same paper, Murray et al. (2013) used a measure of authoritarianism available for 90 societies in the Standard Cross-Cultural Sample in order to test the prediction that parasite stress will covary positively with authoritarianism in the small-scale indigenous societies comprising the ethnographic record of anthropology. Their authoritarian measure was a published index derived from 12 variables related to degree of exclusive power concentration in societal leaders, an important characteristic of authoritarian governance. They used Cashdan and Steele's (2013) and Low's (1988) measures of parasite stress across the societies in their sample. The prediction was supported.

## 10.10 Cross-National Freedom of the Media

A topic fundamentally related to both democracy and diffusion is the freedom to access accurate and uncensored information from the media. This freedom varies greatly across countries, and is measured by the Press Freedom Index (http://en.rsf.org/press-freedom-index-2013,1054.html). Press-freedom scores are compiled annually from reports by journalists, researchers, and news organizations. The reports reflect data and opinions on governmental control of media and penalties and punitive actions (including violence and imprisonment) against those who collect, produce, and circulate accurate news and other information. Considering the 2013 scores, the media in Finland, Netherlands and Norway are most free, whereas the media in Turkmenistan and North Korea are among the least free.

As expected from the parasite-stress theory of values, freedom in the media is strongly, positively related to individualism (and negatively with collectivism) across countries. For example, *Suh Individualism* shows a correlation of 0.68 ($n = 55$ countries); other measures of collectivism–individualism show similar effect sizes. (See Chap. 5 for discussion of collectivism–individualism variables.) Relatedly, press freedom scores are strongly, positively related to democratization across nations ($r = 0.72$, $n = 167$, based on the EIU Democracy Index (described in Sect. 10.6.1)). The relationship between freedom in the media and parasite stress (*Contemporary Parasite Severity*) across countries is significantly negative ($r = -0.40$, $p < 0.0001$, $n = 173$ countries), but the relationship is moderate and hence less than that between collectivism-individualism and media freedom.

## 10.11 Overview of Findings

Traditionally, evolutionary thinking has been rare in research on governmental systems and related topics. As discussed earlier in this chapter, the evolutionary thought that was used merely recognized that people are evolved animals and therefore will show competition for resources and social power. Of course, this is a useful beginning for understanding politics, and those few who broke from the Darwinian agnosticism of early scholarship made important contributions to understanding human politics. Our efforts were to construct and test a more detailed application of Darwinism to political life—that is, the parasite-stress hypothesis of democratization. The hypothesis received strong support. In this chapter, we have shown that the parasite-stress hypothesis of politics offers a synthetic and empirically robust explanation of many features of autocracy and democracy that had been understood previously as unconnected or, in some cases, connected, but with inadequate explanation of the connections. The hypothesis unifies and anchors a range of political variables in proximate and ultimate causation by parasite adversity and the values evoked by variable levels of infectious disease.

### *10.11.1 Democracy*

In this chapter, we discuss and test the hypothesis for cross-cultural variation in democratization first proposed by Thornhill et al. (2009). It is compatible and consistent with earlier hypotheses for democratization in the literature that proposed causation as average wealth, resource distribution or other economic factors, health factors, modernization, or diffusion. Our hypothesis recognizes these as proximate causes and as effects and components of democratization. Our hypothesis, based on evolved psychology for dealing with infectious diseases, contains a proximate, encompassing causal framework, as well as an ultimate causal framework. We propose that past natural selection for defense against morbidity and mortality from infectious disease (ultimate causation) designed the psychology that, depending on local disease severity (proximate causation), manifests today as regional political behavior on the continuum of high autocracy to high democracy. The consequences or effects of this evolved psychology that were identified by the earlier hypotheses become proximately causal and feed back to reduce pathogen stress (effects of democratization, e.g., more widespread distribution of education or healthcare) or to increase pathogen stress (effects of autocracy, e.g., denial of education or healthcare to out-groups).

More specifically, we propose that the liberalization of the values of people, or the opposite shift in their values to increased conservatism, is explicable by psychological changes in attitudes and associated behavior toward in-groups and out-groups that are caused proximately by individuals' perception of their vulnerability to infectious disease. Across societies, as this perceived vulnerability increases, so do autocracy, authoritarianism, collectivism/conservatism and closed-mindedness about innovations; conversely, as the perceived vulnerability declines, democracy, individualism/liberalism, anti-authoritarianism, and openness to innovations increase. Undemocratic versus democratic attitudes and behaviors, we argue, are manifestations of psychological adaptation which functions to adjust group-related activity of individuals in ways that were adaptive at the individual level in human evolutionary history (maximized inclusive fitness). Furthermore, we propose that there is a bidirectional, proximate causal feedback between parasite stress and the economic, social and ideological effects it causes.

Our theoretical framework does not predict that the political behavior of individuals is adaptive currently (meaning that it currently maximizes inclusive fitness of individuals; Chap. 2). Hence, its support does not require data confirming the widespread, current adaptiveness of political activity. Modern humans are faced with widespread, evolutionarily novel settings that may render their behavior, political or otherwise, neutral or maladaptive with regard to fitness promotion because of a mismatch between the current environment and the historical environment that shaped the behavior (e.g., Thornhill 1990, 1997). According to the parasite-stress theory, regardless of current adaptiveness and *because* of evolutionary historical adaptedness, modern humans exhibit xenophobia, ethnocentrism and collectivism in general in high-parasite-stress situations, whereas, in low-parasite-stress situations, they show xenophilia, in-group activity involving primarily nuclear family, and individualism in general.

We emphasize that high parasite stress does not eliminate adaptive, intergroup interactions. The parasite-stress theory, like other evolutionary models, relies on the ecological assessment of benefits and costs. If the benefits of intergroup alliances are high and exceed costs stemming from contagion, then individuals are expected to engage in such alliances despite the costs. Alliances in warfare are one scenario in which xenophobia and disgust may be reduced adaptively, even in the face of high infectious disease locally.

The parasite-stress hypothesis of democratization is tested in the same way as any other hypothesis: by examining empirical implications of the hypothesis. This hypothesis is supported when its predicted patterns are found in the covariation between infectious-disease risk in the environment and people's values and related behavior. Supportive of this hypothesis, parasite-stress measures showed moderate to high negative covariation with each of five democratization measures across countries. The same negative relationships were seen when each of four of the measures were separated into temperate and tropical world regions. Moreover, that same pattern remained statistically significant in the worldwide sample of countries when GDP per capita and Gini were controlled statistically. However, to control these two economic variables in the analysis is not consistent with the parasite-stress theory, because, in the theory, variation in GDP per capita and Gini are caused by variation in parasite stress and hence are components of democratization. Our analysis included them simply for completeness, given the long-standing tradition in economics, historical scholarship, political science, and sociology to view economic measures as the encompassing causal factors. (See also the discussion of the partialling fallacy in Chap. 5.) Moreover, parasite stress from nonzoonotic human diseases shows a much stronger relationship (negative) with democratization across nations than does zoonotic parasite stress.

The predicted relationship between parasite stress and democratization also was seen in a focal sample of 31 countries in which an independent measure of autocracy (based on F-scale scores of authoritarian personality) was available. That research reported, too, that authoritarian values of individuals in the countries mediate this relationship. Finally, the positive relationship between authoritarianism and parasite stress was found across a large sample of the societies in the Standard Cross-cultural Sample.

### *10.11.2 Collectivism–Individualism*

Also supportive of the parasite-stress hypothesis of democratization is the relationship we reported in this chapter between collectivism–individualism and democratization when coupled with a considerable body of separate research presented in Chap. 5 that identified a strong relationship between parasite stress and collectivism–individualism. High parasite stress is associated with high collectivism (low individualism) and low parasite stress with low collectivism (high individualism). Also, nonzoonotic human disease is much more predictive (positively) of collectivism than is zoonotic disease (Chap. 5).

Collectivism–individualism is a fundamental value dimension affecting a society's interest in the welfare of the populace and hence societal democratization. Relative emancipation from parasites generates widespread individualism and associated positivism toward out-groups. An aspect of individualism is an interest in the wellbeing and support of many social groups. One route to promote widespread wellbeing is through wealth redistribution (e.g., paying taxes to support social welfare). In contrast, collectivists are less interested in distributing wealth beyond their in-group members. Alesina and Giuliano (2010) reported that, across many countries, "family ties," a variable that measures collectivism (Chap. 5), correlated negatively with the value of interest in social welfare programs. Hence, collectivism is associated with low interest in social welfare for the populace (and individualism with high interest). Furthermore, with individualism's more extended social welfare, there is concomitant infectious-disease control and investment in education and healthcare for other-group members, not just in-group ones. These phenomena (disease control, increased interest in alliances with other-group members and their success, and wealth redistribution) compound each other. Put simply, individualism promotes the redistribution of resources, which reduces disease burden (e.g., through better sanitation and health) resulting in an increased spread of individualism.

### *10.11.3 Innovation and Diffusion*

Here, we emphasize that individualism's self-expression and independent self and associated positive attitude toward creation and adoption of innovation provide the critical connection between parasite-avoidance and parasite-management psychology of people—the behavioral immune system—and the diffusion of innovations, which is a necessary component of democratization (also see Thornhill et al. 2009). Rogers' (1995) research indicates that, for diffusion of new products to start and succeed, there must be people present who are venturesome and open. The proportion of suitable people who innovate and adopt innovations is predicted by our hypothesis to increase as infectious-disease adversity declines. Hence, low parasite adversity should increase the willingness of people to think in new ways, leading to innovation, and to question and defy tradition and thereby use new ideas and technologies. Hence, via the evolved psychology we have postulated, parasite adversity may be related causally to diffusion between and within societies. Consistent with this, Alesina and Giuliano (2010), in the study mentioned just above, reported that weak family ties (hence, individualism) is associated positively with endorsement of the World Values Survey' item, "New ideas are generally better than old ones," whereas strong family ties (hence, collectivism) is related positively with endorsement of "Ideas that have stood the test of time are generally the best." Also consistent are the findings of Taylor and Wilson (2012) showing the strong, positive relationship across countries between individualism and scientific and technological innovativeness.

We also document the strong cross-national relationships between governmental censorship of media and collectivism–individualism and autocracy–democracy. Autocratic and collectivist countries control media and thereby limit available information. Freedom of the press is characteristic of liberal and democratic regions.

In Chap. 11, which deals with economics, we treat additional topics pertaining to diffusion of innovations. The diffusion of innovations and technology is a salient topic in economics, because increased diffusion widens the distribution of resources among people within a region and across regions and thereby positively influences economic indicators. Although a role for collectivism–individualism in diffusion has been recognized by some economists (e.g., Greif 1994), the fundamental and encompassing significance of this values' dimension in diffusion is seen only when the parasite-stress theory of values is considered. Our findings (reported in Chap. 11) support a central role for low parasite adversity and the liberal values it evokes in the technological and cultural innovation and diffusion that results in economic progress.

### 10.11.4 Property Rights

The rights of people other than the governing elites to own property and hence to control and accumulate wealth is highly variable across countries of the world. As expected under the parasite-stress theory, this variation is predicted by parasite adversity, collectivism–individualism and democratization. The degree of legal restriction of property rights as the privilege of only the elite in-group correlates positively with parasite stress, collectivism and autocracy. The evidence in this chapter indicates that this restriction fundamentally is based on the prejudice against out-groups and associated in-group favoritism exhibited by elites.

### 10.11.5 Gender Relations and Women's Sexual Liberties

Our findings reveal that, across countries, the degree of gender equitability or disparity is predictable from measures of parasite stress, democracy, and collectivism and individualism. As predicted by the parasite-stress hypothesis of democratization, gender equality corresponds to low parasite stress and high democratization and individualism. Relatedly, across nations, we found that the restrictiveness of women's sexual interactions to committed relationships is correlated positively to parasite stress and negatively to democratization, women's economic and political power relative to men's, and individualism. Hence, this study provides evidence consistent with variation in parasite stress among countries as being the common, underlying cause of the degrees of democratization of women's political, social and economic opportunities and of the liberalization of women's sexual behavior (also see Gangestad et al. 2006; Schaller and Murray 2008). The findings also indicate that nonzoonotic diseases are more important than are zoonotics in patterns of gender relations and women's sexual restrictiveness across the globe.

Women's disenfranchisement from civil rights and liberties has varied by region across the United States, and, as predicted by the parasite-stress theory, was greatest in the South. In Chap. 8, some aspects of gender inequality in the South pertaining to the culture of honor were discussed. The South's conservative ideology that females are inferior to males is seen, too, in the history of the 19th Amendment of the US Constitution, which gave the right of suffrage to women. Although the Amendment became US law in 1920 with its ratification by 36 states, the states of the Old South generally rejected it at that time and greatly delayed its eventual ratification in, for example, Alabama 1953, Louisiana and Georgia 1970, North Carolina 1971, and Mississippi 1984 (see http://en.wikipedia.org/wiki/Nineteenth_Amendment_to_the_United_States_Constitution). (Also see Vandello and Cohen (1999) on the negative relationship between collectivism and states' support of the 19th Amendment.)

## 10.12 Political Elections and Parasites

One approach to investigate the parasite-stress hypothesis of democratization is to examine the influence of parasite adversity on voting decisions. Inbar et al. (2012) studied the role of disgust sensitivity in relation to political conservatism and voting behavior. They found in both an American sample and a more global sample of people from 121 countries that higher levels of disgust sensitivity were associated positively with conservatism. They also showed, by measuring voting intentions prior to the 2008 USA Presidential election, that those participants with higher levels of disgust sensitivity were less likely to vote for Barack Obama, the liberal candidate. Thus, these research findings are consistent with an important role of parasites in shaping the psychology of voting decisions.

At the beginning of this chapter, we stated that "[h]uman politics is basically social competition among individuals and groups using held and behaviorally manifested values as the competitive tools." Although values-as-socially-competitive-mechanisms are the most fundamental feature of political contests, there are other factors that play into winning and losing political contests. Physical attractiveness is one such factor. Physical attractiveness positively affects outcomes of social competition in a wide range of life endeavors. Attractive infants, older children, and adults typically benefit socially from being physically attractive. Thornhill and Gangestad (1993) hypothesized that the robust pattern of the social benefit of looks is the result of human psychological adaptation that assesses health related to resistance to infectious disease and favors healthy others in mate choice as well as in nepotistic and other social interactions. In Chap. 6, we discussed some of the diverse and copious cross-regional as well as experimental evidence supporting that hypothesis as applied to mate choice. (See Thornhill and Gangestad 1993 for a discussion of early studies supporting favoritism toward attractive others in non-mating social domains.) Several recent studies involving lab experiments and real-world voting behavior have reported that people favor attractive political candidates over unattractive ones (reviewed in White et al. 2013). White et al. (2013) hypothesized that political candidates' looks will be more important in winning political races in high infectious-disease regions than in low infectious-disease regions. Certainly, there is strong evidence that looks matter more to people in regions of high parasite adversity than in regions of low parasite adversity (reviewed in Chap. 6).

White et al. (2013) tested their hypothesis in part using attractiveness ratings of the pictures of 2010 US congressional candidates and a surrogate measure of parasite stress in each of the 436 congressional districts (the combination of infant mortality per 1,000 births and average life expectancy at birth per district for 2010–2011; these data are available at Measures of America (http://www.measureofamerica.org/). They reported that the standardized scores of the two components of their surrogate parasite-adversity measure are highly and negatively correlated.

They found that the higher the parasite stress in a region, the greater the importance of rated attractiveness of a candidate in both percentage of votes obtained and winning the election. Also, these effects remained statistically robust after controlling for income, education, and sex of candidate. In the same publication, White et al. (2013) reported a second study of their hypothesis using experimental manipulations of threats, including an infectious-disease threat, and a control condition. Research participants rated the importance of a number of characteristics in political leaders. Their evidence indicated that disease threat is the primary condition that affected people's opinion that leaders should be physically attractive. In a third study, White et al. (2013) found that, among pictures of British politicians, the disease-threat condition activated in participants a favorable attitude toward casting a vote for the more physically attractive politicians. In a final study, White et al. (2013) found that the importance of physical attractiveness in a group *leader*, as opposed to just physical attractiveness in a group *member*, is most strongly activated by an infectious-disease-threat-manipulation.

Given that actual parasite adversity scores are not available for the congressional districts of the USA, we examined the surrogate measure of infectious-disease adversity that White et al. (2013) used by computing its covariation with *Parasite Stress USA*, a cross-state parasite-stress variable described in Chap. 5. The means of the surrogate measure for the congressional districts of each state were combined to give a score for each state on the surrogate measure. The state-level scores and *Parasite Stress USA* correlate highly at $r = 0.64$, $n = 50$ states. Hence, the surrogate measure used by White et al. (2013) is a valid measure of infectious disease at the congressional-district geographical scale. This is encouraging because it indicates that other variables related to people's values available at the congressional-district scale could be explored in relation to the parasite-stress theory and with the large sample size provided by the number of congressional districts (436). As explained in Chap. 2, regions are statistically independent, even if adjacent, because people are discriminative in their choice of cultural values, selecting those cultural items during ontogeny that are solutions to local problems.

The research of White et al. discussed in this section indicates that people in high parasite regions are biased toward the most attractive candidate when casting their vote. Their results further indicate that, in the lowest parasite-stress congressional districts, looks may not matter or may matter very little. Thus, in the low-parasite districts characterized by relatively liberal people, what are the biases that determine political winners and losers of political elections? We hypothesize that liberal voting will favor candidates with educational achievement and diplomatic skills, and that these traits will figure importantly in liberal voters' assessment of the competence of leaders.

## 10.13 The Western Revolution in Values

In Chap. 4, we discussed the linguistic changes in the USA that began in 1960 and extend to the present time resulting in an increased use of individualistic words, phrases, and pronouns. These changes were only a small part of the broader revolution in values toward more individualism across the West in the same time period.

Over the last 50–60 years, many people in Western cultures have increasingly accepted non-conformist and non-traditional sexual behavior. More positivism, or at least tolerance, has been documented in Western people's attitudes about unmarried cohabitation of a man and a woman, contraception usage, premarital sex, homosexuality, and sexually explicit material (pornography). In addition, this period had the decriminalization of homosexual behavior, the elimination of infidelity as grounds for marital dissolution, and the liberalization of divorce law in general (see reviews in Johnson et al. 1994; Laumann et al. 1994; Treas 2002). These attitudinal shifts in permissiveness are discussed widely as originating in the 1960s in what is called the "sexual revolution." Sociologists and historians typically attribute the sexual revolution to increased individualism, higher living standards, secularity, modernization, and education (e.g., Sigusch 1998; Treas 2002; Norris and Inglehart 2004; Petigny 2004).

The sexual revolution is a much too restrictive label to cover the more general ideological revolution that accelerated in the Western world in the 1960s and 1970s. It was a broader social revolution that encompassed not just the liberalization of sexual attitudes and values, but also decreased respect for traditional authority (anti-authoritarianism) and increased recreational drug use, as well as major changes in recognition and legalization of social and political power for women—reduction in the sexual double standard—and increased civil rights and liberties for minorities. It became widely illegal to discriminate on the basis of religious beliefs, age, sex, ethnicity, or disability. Moreover, Norris and Inglehart (2004) documented that, in the West, secularization has been proceeding more rapidly since the mid-twentieth century. Clearly, during this period in the West, there was a great increase in overall liberalization of values. We agree with the historians and sociologists that these changes were caused by increased individualism, or in other terms, increased democratization, liberalization, modernization, and secularity. We have hypothesized, however, that, at a fundamental and encompassing proximate level, reduction in the impact of infectious disease was causal (Thornhill et al. 2009).

Ward and Warren (2007; also Bud 2007; Levitt et al. 2007) review the literature on the remarkable achievements in public health in the twentieth century in the West. Sulfa drugs became available in the 1930s, but were used on a more limited basis than the newer antibiotics (e.g., penicillin) that were available a decade later. Penicillin's history has been documented thoroughly by Bud (2007). These later antibiotics had fewer negative side-effects and killed more kinds of parasitic organisms. At about the same time, antibiotics that were useful against fungal diseases, viral diseases, and protozoan and nematode parasites became available (Levitt et al. 2007). At the end of World War II, penicillin, quickly followed by other antibiotics that provided an even broader spectrum of defense against bacterial diseases, were administered widely in the West—so widely that warranted concern about the evolution of antibiotic resistance quickly arose. In Britain in 1948, as copious evidence accumulated for evolved antibiotic resistance, legislation was enacted to limit through prescription the public's access to antibiotics (Bud 2007). Additionally, in the West in the late 1940s, child vaccination programs were implemented; for example, a combined diphtheria–pertussis–tetanus vaccine was given widely to children (Levitt et al. 2007). Moreover, by the late 1940s in the USA, malaria was reduced to negligible levels through mosquito-control programs of insecticide use combined with eliminating mosquito-breeding areas (Levitt et al. 2007). These public-health landmarks greatly reduced morbidity and infant and adult mortality from infectious diseases, and concomitantly extended longevity (Armstrong et al. 1999). There were additional, important health advances in the second decade of the twentieth century: chlorinated drinking water became widely available; sanitation departments were established for public garbage removal; and indoor plumbing, public sewer systems and solid waste disposal and treatment became available. These technologies greatly reduced water-borne diseases such as typhoid and cholera. Public health food-safety practices also were implemented widely during this period. Fluoridation of drinking water began in 1945 and rapidly reached large segments of Western societies. Fluoridation has reduced the detrimental health impact of dental caries and other infectious diseases of the teeth and mouth (Burt and Eklund 2005). Thus, in the West by the 1940s, there was widespread use of technologies—antibiotics and related drugs, vector control practices, sanitation, water quality, and food safety—that dramatically reduced people's exposure to infectious diseases and thereby reduced mortality of infants and adults and extended longevity (Armstrong et al. 1999).

Bud (2007) emphasized that children, teenagers, and adults in the West with effective antibiotics available after 1945 experienced a very different world in regard to infectious diseases compared to prior generations. The same can be said with regard to the many other public health advances in the West over the same period we have mentioned. A human generation is about 20 years. Thus, these public health improvements that began to affect widely the populace in the West in the 1940s, and to a significant degree beginning a generation earlier, may account for the major increase in liberalization of values that began in the 1950s–1960s.

We have proposed that a cue that may affect a human's perception of contagion risk in the environment is the frequency and duration of his or her immune system activation. Other potential cues are observations of illness and the effects of illness in local people and other social information about the frequency of illness locally (Chaps. 2 and 3). These may serve as ancestral cues read by functionally specialized psychological adaptation that interacts with other cognitive adaptation to track values toward or away from high xenophobia, ethnocentrism, authoritarianism, and related values. This scenario is consistent not only with the dramatic increase by the 1960s of Western liberalized sexual and other social values and policy, but also their continuation and expansion up to the present as healthcare has improved. (See Levitt et al. 2007 for discussion of healthcare advances in the West in the last half of the twentieth century.)

The scenario also is consistent with the pattern now found across the globe. Secularization and associated openness to new ways and ideas and concomitant rejection of traditional values has increased only in countries that are relatively high in GDP per capita. In contrast, poorer countries, those with historically and currently high morbidity and mortality, a major portion of which is caused by infectious disease (e.g., World Health Organization 2004), remain as traditional and hence collectivist as centuries earlier (Norris and Inglehart 2004). In fact, across countries, our contemporary parasite-severity index (described in Chap. 5) covaries strongly and positively with the *2007 World FactBook* variable "infant mortality per 1,000 births" ($r=0.68$, $p<0.0001$, $n=209$) and negatively with its variable "life expectancy" in the same year ($r=-0.67$, $p<0.0001$, $n=209$).

The correspondence between increased public health and the shift toward liberalization of values in the West that accelerated in the 1960s and 1970s and continues today is as predicted by the parasite-stress hypothesis of democratization. These changes did not occur or were minor in regions outside the West where infectious disease took great tolls in terms of morbidity and mortality. These events comprise a natural experiment. Another method for testing the parasite-stress hypothesis of liberalization would be to record people's changes in values after infectious disease levels are reduced locally (e.g., by greater access to modern medicine and healthy water) or increased locally (e.g., by the emergence of a new infectious disease). If the adaptation for assessing local contagion risk is based on the cues mentioned above, people's value changes are expected to occur across one or a few generations, as is implied by the results of the natural experiment. Easily administered, brief, valid questionnaires that could measure the relevant value changes are available (Altemeyer 1996; Faulkner et al. 2004; Gelfand et al. 2004; Thornhill and Fincher 2007).

## 10.14 Reconstruction, Parasite Stress, and Civil Rights

Key (1949), after documenting in his book the corruption, exploitation and inhumanity in terms of racist and classist social policies that characterized the Old South, proposed that, despite this history to the contrary, the South has the potential capability to govern itself and become the kind of moral democracy advocated by the Constitution of the USA. In Chap. 11, we treat economics and show the robust positive relationships between governmental corruption and inefficiency and both parasite stress and collectivism. Historians of the South typically point to the Civil

Rights Movement and the associated US federal intervention in the 1960s, which led to southern racial apartheid being outlawed, as the turning point toward increased political stability, competence, and democracy in the South. This is only part of the causal story.

According to the parasite-stress theory of values, more widespread liberal values began to appear as improved healthcare and sanitation reached the South, and it was these improvements that provided the tipping point, allowing the federal "meddling" with Old South culture to have its democratizing effects. Federal efforts to democratize the Old South earlier during Reconstruction (1863–1877) were unsuccessful (Foner 2002). Some historians refer to Reconstruction as the "second civil war," because southerners fought, often violently and lethally, to keep the status quo of white-power supremacy and African American disenfranchisement. The promise of civil rights and liberties and opportunity for freed slaves in Abraham Lincoln's Emancipation Proclamation (1863) and the 13th Amendment to the US Constitution did not materialize against hard-line southern conservatism. Indeed, many aspects of race relations worsened across all the states in the region, with widespread murder of freed slaves and white liberals who supported racial equity. During Reconstruction, the Ku Klux Klan (KKK) and various other white supremacy organizations arose, and the KKK became politically powerful across the Old South. We hypothesize that the Civil Rights Movement of the 1960s gained its masses of southern African-American supporters because they were the enculturation products of two generations of health interventions that significantly reduced parasite stress. As health interventions such as chlorinated and fluoridated drinking water, vector control, hookworm eradication and other health improvements reached more and more black communities, authoritarianism and the associated acceptance of white superiority declined, and individualism increased, among blacks. There were similar liberalizing effects on the values of many white southerners from the same health improvements. The Civil Rights Movement, both regionally across the southern USA and across the nation, we argue, was caused proximately by the historically unique degree of emancipation from parasite adversity that began in the 1920s and within two generations spanned segments of the populace. The parasite-stress theory of values explains the gradual recession of apartheid after the 1960s, as well as the failure of the earlier Reconstruction to democratize the southern region of the USA.

## 10.15 Early Western Transitions Toward Democracy

Significant societal democracy preceded by a few hundred years the Western revolution in values that we discussed above and attributed to public services reducing infectious diseases. The geography of the original democratic transitions, however, supports the parasite-stress hypothesis of democratization. There is considerable latitudinal variation in human pathogens; they are especially abundant and severe at low latitudes and decrease in abundance and severity as distance increases from the equator (Guernier et al. 2004; Low 1990; Dunn et al. 2010). Moreover, the positive relationship between individualism and latitude across modern countries is well established (e.g., Hofstede 2001). The parasite-stress hypothesis of democratization predicts that widespread democratic and related individualistic values that preceded those seen under modern healthcare are more likely to arise at higher latitudes than at lower ones. This prediction is supported.

The Age of Enlightenment started in Britain, Germany and France and spread throughout much of Europe (Gay 1996). In sharp contrast to loyalty to the views of authoritarian nobility and the church, the Enlightenment involved a shift in values to using one's own intelligence as a source of knowledge. The Enlightenment was a widespread critical questioning of traditional values, and the consequent political changes included a decline in the power of royalty and religion and increased rights for ordinary people (e.g., see Kant's (1784) famous essay, "What Is Enlightenment?"; http://www.marxists.org/reference/subject/ethics/kant/enlightenment.htm). From our perspective, the Enlightenment could be called "The Age of Incipient Individualism." The Enlightenment began with, or somewhat before, Britain's Glorious Revolution of 1688. The English Bill of Rights of 1689 extended civil and political rights of citizens, limited the Crown's control of government, and established considerable political power in the Parliament, which represented the people (http://en.wikipedia.org/wiki/Bill_of_Rights_1689). We are not saying that infectious disease was not still a major issue in Europe during the Enlightenment, but we are saying that the impact of parasitic diseases was reduced, at least in many regions, compared to earlier times.

Other early ideological revolutions toward democracy occurred at higher latitude locales as well. Historians have thoroughly documented the rise of individualism and democratization as seen in the new religious philosophies that emerged during the founding of the United States of America and in other historical records. The Puritans, beginning in the mid-1600s, and other Protestant groups through the 1700s, advocated that the individual is competent to understand religious scriptures, should be free to choose religious doctrines, and is in control of his or her salvation. These religious values were the antithesis of those of the authoritarian religious institutions that their European ancestors had followed, which mandated total elite control of all matters of religion (Hatch 1991; Morgan 2006). The widespread individualism of this period in nascent America also is seen in its classical documents. The founding document of the USA—The Declaration of Independence (1776)—had a strong, liberal sentiment, including "that all men are . . . equal." It was followed by the US Constitution in 1787 and The Bill of Rights in 1791.

At about the same time, the French Revolution was underway. Liberal political values, based on Enlightenment ideology and inalienable rights of all citizens, quickly replaced the longstanding autocratic and theocratic French government. Near the start of the French Revolution (1789), the Declaration of the Rights of Man and of the Citizen was adopted in France; it asserted fundamental rights for "all men, without exception," and, like the US documents just mentioned, limited the power of traditional authority (http://en.wikipedia.org/wiki/Declaration_of_the_Rights_of_Man_and_of_the_Citizen).

These major milestones in the historical record reflect salient and widespread changes in values among the people of the regions involved. These changes were shifts from an in-group focus and in-group egalitarianism to a broader, pan-group orientation and pan-group egalitarianism, from tradition as the source of information to an openness to new and different means and ways, and from authoritarianism to anti-authoritarianism. In short, the changes reflect a general increase in democratic, individualistic, liberal values. These changes, we suggest, were made possible by the locations of the societies in which they occurred—areas of the globe relatively low in parasite stress. Among contemporary countries, democratization is strongly and positively related to latitude (absolute latitude measured at the mid-points of each country (*World Factbook* 2007, www.cia.gov) and Vanhanen's Index of Democracy, $r=0.57$, $p<0.0001$, $n=169$ countries). The correlation between the early steps toward democracy in Europe and the USA (mentioned above) and high latitude, as well as the correspondence between democratization and latitude across contemporary countries, is spurious, because the underlying cause of both relationships is variation in parasite stress.

As higher and higher latitudes were reached by ancient humans, the potential for democratic societies comprised of many individualists arose. The northward migration of *Homo sapiens* into Eurasia from the species' low-latitude origin in Africa (Templeton 2002; Finlayson 2005) led to an escape from high parasite stresses. In democratic settings, this escape continued, and has reached a zenith in high-latitude, contemporary countries such as Norway with generous healthcare and other social welfare for all.

## 10.16 Politics and Human Nature

At the University of Alabama in 1963, Thornhill's political science professor was fond of expounding during lectures various paraphrases of a famous comment made in 1787 by James Madison, who is often referred to as the father of the US Constitution: "But what is government itself but the greatest of all reflections on human nature?" (cited by Betzig 2009, p. 104). The research reported in this chapter demonstrates the evidentiary truth of this. Of course, all human affairs, not just political matters, are caused by what some refer to as human nature: human psychological adaptations ultimately caused by evolution by Darwinian selection that acted in the past. Political matters, in large part, we have argued, are the product specifically of psychological adaptations that yield in-group and out-group values of people in relation to infectious-disease adversity, and, as such, are components of the behavioral immune system.

Human nature is sometimes misunderstood to mean that genes have an autocratic or omnipotent reign on human mental activity and behavior, which eliminates any role for environmental causes. Actually, human behavioral and psychological adaptations, like all other bodily traits of all organisms, arise during their ontogenies by fully democratic developmental machinery and hence genes and environment are equally causal (Chap. 2). So, in fact, the nurture acting causally during human development is nature—nurture is a part of the natural realm. The parasite-stress theory maintains that individual humans are born with the potential to incorporate the ideology that meets the challenges of the local cultural ecology, and, in particular, personal ontogenetic challenges pertaining to contagion stress. People, then, have the potential to end up at a large range of points on the continuum of human prejudicial variation, or, said differently, at many points on the collectivism–individualism dimension. Just where the endpoint is for the individual depends on ontogenetic conditions experienced. Of course, a person's genetic immunity comprises a portion of his or her ontogeny.

This is the modern scientific version of the blank slate of human potential. People are not more fundamentally conservative than liberal or vice versa. The implication for achieving more democracy in the world is obvious. If the parasite-stress hypothesis of democratization continues to be supported by research, humanitarian efforts to reduce human rights violations and to increase human liberties and democracy in general will be most effective if focused on the most fundamental and encompassing causal level of human infectious-disease reduction. Moreover, as we document in our book, there is considerable empirical support at this time indicating that humanitarian efforts focused in particular on reducing the impact of nonzoonotic human infectious diseases would have the greatest effect on liberalizing values.

## 10.17 Summary

The countries of the world vary in their position along the autocracy–democracy continuum of governance. Traditionally, researchers explain this variation as based on resource distribution and disparity among nations. We provide a different framework for understanding the autocracy–democracy dimension and related value dimensions, one that is complementary (not alternative) to the research tradition, but more encompassing and synthetic, involving both evolutionary (ultimate) causation and proximate causation of the values. We hypothesize that the variation in values pertaining to autocracy–democracy arises fundamentally out of human species-typical psychological adaptation that manifests contingently, producing values and associated behaviors that functioned adaptively in human evolutionary history to cope with local levels of infectious diseases. We test this parasite-stress hypothesis of democratization using data measuring democratization, collectivism–individualism, gender egalitarianism, property rights, sexual restrictiveness, and parasite adversity across many countries of the world. We show that, as the hypothesis predicts, collectivism (hence, conservatism), autocracy, women's subordination relative to men's status, and women's sexual restrictiveness are features that positively covary, and that correspond with high prevalence of infectious disease. The psychology of xenophobia, ethnocentrism, traditionalism, and authoritarianism links these features to avoidance and management of parasites. Also as predicted, we show that the antipoles of each of the above features—individualism (hence, liberalism), democracy, anti-authoritarianism, and women's rights, freedom and increased participation in casual sex—are a positively covarying set of features in countries with relatively low parasite stress. We discuss evidence that the generation and diffusion of innovations (in thought, action, and technology), which is an important component of democratization, is causally related to low parasite stress and its evoked values, particularly liberalism and associated openness. The freedom of the media to provide accurate information is shown to relate strongly to democracy and individualism.

Beyond the cross-national support for the parasite-stress hypothesis of democratization, the hypothesis is consistent with the geographical location at high latitudes (and hence reduced parasite stress) of the early democratic transitions in Britain, France and the USA. It, too, is consistent with the marked increase in the liberalization of social values in the West in the 1960s (in part, the sexual revolution), regions that, a generation or two earlier, experienced dramatically reduced infectious diseases as a result of antibiotics, vaccinations, food- and water-safety practices, parasite-vector control, and increased sanitation. Further support of the parasite-stress hypothesis of democratization is seen in cross-national evidence that authoritarianism is related positively to parasite stress and that individuals' authoritarian personality (as measured by the F-scale) mediates this relationship. Finally, the positive correlation between parasite stress and autocratic governance is seen not only across nations, it is seen as well in the Standard Cross-cultural Sample.

Recent research indicates that people in high parasite regions of the USA are biased toward the most physically attractive candidate when casting their vote, and that, in the lowest parasite-stress regions, looks may not matter or may matter very little. We hypothesize that in low parasite regions, characterized by relatively liberal people, voting will favor candidates with educational achievement and diplomatic skills, and that these traits will figure importantly in liberal voters' assessment of the competence of leaders.

Overall, and most fundamentally, the parasite-stress theory of values conceptually and empirically unifies the scholarly study of political systems with the field of the ecology and evolution of infectious diseases. Hence, political systems are proximate manifestations of the human behavioral immune system and its range of values evoked by variable parasite stress. In terms of evolutionary or ultimate causation, political systems are the product of natural selection that favored the conditional expression of the features of the behavioral immune system that allow ancestrally adaptive social navigation under variable amounts of parasite stress in the local environment.

## References

Adorno, T. W., Frenkel-Brunswik, E., Levinson, D. J. et al. (1950). *The Authoritarian Personality*. Harper and Brothers, New York, NY.

Alesina, A., & Giuliano, P. (2010). The power of the family. *Journal of Economic Growth*, 15: 93–125.

Altemeyer, B. (1996). *The Authoritarian Spector*. Harvard University Press, Cambridge, MA.

Armstrong, G. L., Conn, L. A., & Pinner, R. W. (1999). Trends in infectious disease mortality in the United States during the 20th Century. *Journal of the American Medical Association* 281: 61–65.

Barrett, D. B., Kurian, G. T., & Johnson, T. M., Eds. (2001). *World Christian Encyclopedia: A Comparative Survey of Churches and Religions in the Modern World, Volume 1: The World by Countries: Religionists, Churches, Ministries*, 2nd ed., Oxford University Press, Oxford, U.K.

Barro, R. J. (1999). *Determinants of Economic Growth: A Cross-country Empirical Study*. MIT Press, Cambridge, MA.

Betzig, L. (1986). *Despotism and Differential Reproduction: A Darwinian View of History*. Aldine, New York, NY.

Betzig, L. (2009). But what is government itself but the greatest of all reflections on human nature? *Politics and the Life Sciences* 28: 102–105.

Bollen, K. A. (1980). Issues in the comparative measurement of political democracy. *American Sociological Review* 45: 370–390.

Bud, R. (2007). *Penicillin: Triumph and Tragedy*. Oxford University Press, New York, NY.

Bunce, V. (2000). Comparative democratization: Big and bounded generalizations. *Comparative Political Studies* 33: 703–734.

Burt, B. A., & Eklund, S. A. (2005). *Dentistry, Dental Practice, and the Community*, 6th ed. W.B. Saunders Co, Philadelphia, PA.

Cashdan, E., & Steele, M. (2013). Pathogen prevalence, group bias, and collectivism in the standard cross-cultural sample. *Human Nature* 24: 59–75.

Dahl, R. A. (1989). *Democracy and Its Critics*. Yale University Press, New Haven, CT.

Diamond, L. (2002). Thinking about hybrid regimes. *Journal of Democracy* 13: 21–35.

Dunn, R. R., Davies, T. J., Harris, N. C. et al. (2010). Global drivers of human pathogen richness and prevalence. *Proceedings of the Royal Society B* 27: 2587–2595.

Economist Intelligence Unit, http://www.eiu.com.

Faulkner, J., Schaller, M., Park, J. H. et al. (2004). Evolved disease-avoidance mechanisms and contemporary xenophobic attitudes. *Group Processes and Intergroup Relations* 7: 333–353.

Fincher, C. L., & Thornhill, R. (2012). Parasite-stress promotes in-group assortative sociality: The cases of strong family ties and heightened religiosity. *Behavioral and Brain Sciences* 35: 61–79.

Finlayson, C. (2005). Biogeography and evolution of the genus *Homo*. *Trends in Ecology and Evolution* 20: 457–463.

Foner, E. (2002). *Reconstruction: America's Unfinished Revolution, 1863–1877*. Harper Perennial Modern Classics, New York, NY.

Franco, A., Alvarez-Dardet, C., & Ruiz, M. (2004). Effect of democracy on health: Ecological study. *British Medical Journal* 329: 1421–1423.

Freedom House. (2008). http://www.freedomhouse.org.

Gangestad, S. W., Haselton, M. G., & Buss, D. M. (2006). Evolutionary foundations of cultural variation: Evoked culture and mate preferences. *Psychological Inquiry* 17: 75–95.

Gay, P. (1996). *The Enlightenment: The Science of Freedom*. W. W. Norton and Co., New York, NY.

Gelfand, M. J., Bhawuk, D. P. S., Nishii, L. H., et al. (2004). Individualism and collectivism. In *Culture, Leadership, and Organizations: The GLOBE Study of 62 Societies* (eds. R. J. House, P. J. Hanges, M. Javidan et al.), pp. 437–512. Sage Publications, Thousand Oaks, CA.

Greif, A. (1994). Cultural beliefs and the organization of society: A historical and theoretical reflection on collectivist and individualist societies. *The Journal of Political Economy* 102: 912–950.

Guernier, V., Hochberg, M. E., & Guégan, J. -F. (2004). Ecology drives the worldwide distribution of human diseases. *PLoS Biology* 2: 740–746.

Gurr, T. R., Jaggers, K., & Moore, W. (1990). The transformation of the Western state: The growth of democracy, autocracy, and state power since 1800. *Studies in Comparative International Development* 25: 73–108.

Hatch, N. O. (1991). *The Democratization of American Christianity*. Yale University Press, New Haven, CT.

Hofstede, G. (2001). *Culture's Consequences: Comparing Values, Behaviors, Institutions, and Organizations Across Nations*, 2nd ed. Sage Publications, Thousand Oaks, CA.

Index of Economic Freedom, http://www.heritage.org/Index/.

Inbar, Y., Pizarro, D. A., Iyer, R. et al. (2012). Disgust sensitivity, political conservatism, and voting. *Social Psychological and Personality Science* 5: 537–544.

Inglehart, R., Ed. (2003). *Human Values and Social Change: Findings from the Values Surveys*. Brill, Boston, MA.

Inglehart, R., Norris, P., & Welzel, C. (2003). Gender equality and democracy. *Human Values and Social Change: Findings from the Values Surveys* (ed. R. Inglehart), pp.121–145. Brill, Boston, MA.

Javidan, M., & Hauser, M. (2004). The linkage between Globe findings and other cross-cultural information. *Culture, Leadership and Organizations: The Globe Study of 62 Societies* (eds. R. J. House, P. J. Hanges, M. Javidan et al.), pp. 102–121. Sage Publications, Thousand Oaks, CA.

Johnson, A. M., Wadsworth, J. W., Wellings, K. et al. (1994). *Sexual Attitudes and Lifestyles*. Blackwell, Oxford, U. K.

Kant, I. (1784). *An Answer to the Question: What Is Enlightenment?* http://www.marxist.org/reference/subject/ethics/enlightenment.htm.

Karatnycky, A., Ed. (1998). *Freedom in the World: The Annual Survey of Political Rights and Civil Liberties, 1997–1998*. Transaction Publishers, New Brunswick, NJ.

Key, V. O. (1949). *Southern Politics: In State and Nation*. A. F. Knopf, New York, NY.

Lake, D. A., & Baum, M. A. (2001). The invisible hand of democracy: Political control and the provision of public services. *Comparative Political Studies* 34: 587–621.

Laumann, E. O., Gagnon, J. H., Michael, R. T. et al. (1994). *The Social Organization of Sexuality*. University of Chicago Press, Chicago.

Lernery, D. (1958, republished in 1968). *The Passing of Traditional Society: Modernizing the Middle East*. Free Press, New York, NY.

Levitt, A. M., Drotman, D. P., & Ostroff, S. (2007). Control of infectious diseases: A 20th century public health achievement. In *Silent Victories: The History and Practice of Public Health in 20th Century America* (eds. J. M. Ward & C. Warren), pp. 3–17. Oxford University Press, New York, NY.

Lipset, S. M. (1959). Some social requisites of democracy: Economic development and political legitimacy. *American Political Science Review* 53: 69–105.

Lipset, S. (1960). *Political Man*. Doubleday, Garden City, NY.

Lipset, S. M. (1983). *Political Man: The Social Basis of Politics, Expanded and Updated Edition*. Heineman, London, U.K.

Low, B. S. (1988). Pathogen stress and polygyny in humans. In *Human Reproductive Behavior: A Darwinian Perspective* (eds. L. Betzig, M. Borgerhoff Mulder, & P. Turke), pp. 115–127. Cambridge University Press, Cambridge, U.K.

Low, B. S. (1990). Marriage systems and pathogen stress in human societies. *American Zoologist* 30: 325–339.

Marks, G., & Diamond, L. (1992). Seymour Martin Lipset and the study of democracy. In *Re-examining Democracy: Essays in Honor of Seymour Martin Lipset* (eds. G. Marks and L. Diamond). Sage Publications, Newbury Park, CA.

Measures of America, http://www.measureofamerica.org/.

Miller, S., & Diamond, J. (2006). A new world of differences. *Nature* 441: 411–412.

Morgan, E. S. (2006). *The Puritan Dilemma: The Story of John Winthrop*, 3rd ed. Pearson/Longman, White Plains, NY.

Muller, E. N. (1997). Economic determinants of democracy. In *Inequality, Democracy, and Economic Development* (ed. M. I. Midlarsky), pp. 133–155. Cambridge University Press, Cambridge, U.K.

Murray, D. R., Schaller, M., & Suedfeld, P. (2013). Pathogens and politics: Further evidence that parasite prevalence predicts authoritarianism. *Plos ONE* 8: e62275. doi:10.1371/journal.pone.0062275

Norris, P., & Inglehart, R. (2004). *Sacred and Secular: Religion and Politics Worldwide*. Cambridge University Press, New York, NY.

Perry, R. L., & Robertson, J. D. (2002). *Comparative Analysis of Nations: Quantitative Approaches*. Westview Press, Boulder, CO.

Petigny, A. (2004). Illegitimacy, postwar psychology, and the reperiodization of the sexual revolution. *Journal of Social History* 38: 63–80.

The Press Freedom Index, http://en.rsf.org/press-freedom-index-2013,1054.html.

Rogers, E. M. (1995). *Diffusion of Innovations*, 4th Ed. The Free Press, New York, NY.

Rueschemeyer, D., Stephens, E. H., & Stephens, J. D. (1992). *Capitalist Development and Democracy*. Polity Press, Cambridge, U.K.

Schaller, M., & Murray, D. (2008). Pathogens, personality, and culture: Disease prevalence predicts worldwide variability in sociosexuality, extraversion, and openness to experience. *Journal of Personality and Social Psychology* 95: 212–221.

Schmitt, D. P. (2005). Sociosexuality from Argentina to Zimbabwe: A 48-nation study of sex, culture, and strategies of human mating. *Behavioral and Brain Sciences* 28: 247–311.

Sigusch, V. (1998). The neosexual revolution. *Archives of Sexual Behavior* 27: 331–359.

Simpson, J. A., & Gangestad, S. W. (1991). Individual differences in sociosexuality: Evidence for convergent and discriminant validity. *Journal of Personality and Social Psychology* 60: 870–883.

Somit, A., & Peterson, S. A. (1997). *Darwinism, Dominance, and Democracy: The Biological Bases of Authoritarianism*. Praeger Publishers, Westport, CT.

Taylor, M. Z., & Wilson, S. (2012). Does culture still matter? The effects of individualism on national innovation rates. *Journal of Business Venturing* 27:234–247.

Templeton, A. (2002). Out of Africa again and again. *Nature* 416: 45–51.

Teorell, J., & Hadenius, A. (2006). Democracy without values. *Studies in Comparative International Development* 41: 95–105.

Thornhill, R. (1990). The study of adaptation. In *Interpretation and Explanation in the Study of Behavior*, Vol. II (eds. M. Bekoff & D. Jamieson), pp. 31–62. Westview Press, Boulder, CO.

Thornhill, R. (1997). The concept of an evolved adaptation. In *Characterizing Human Psychological Adaptations* (eds. G. Bock & G. Cardew), pp. 4-13. CIBA Foundation, London, U.K.

Thornhill, R., & Fincher, C. L. (2007). What is the relevance of attachment and life history to political values? *Evolution and Human Behavior* 28: 215–222.

Thornhill, R., & Gangestad, S. W. (1993). Human facial beauty: Averageness, symmetry, and parasite resistance. *Human Nature: An Interdisciplinary Biosocial Perspective* 4: 237–269.

Thornhill, R. & Gangestad, S. W. (2008). *The Evolutionary Biology of Human Female Sexuality*. Oxford University Press, New York, NY.

Thornhill, R., Fincher, C. L., & Aran, D. (2009). Parasites, democratization, and the liberalization of values across contemporary countries. *Biological Reviews* 84: 113–131.

Thornhill, R., Fincher, C. L., Murray, D. R. et al. (2010). Zoonotic and non-zoonotic diseases in relation to human personality and societal values: Support for the parasite-stress model. *Evolutionary Psychology* 8: 151–169.

Tiger, L. & Fox, R. (1971). *The Imperial Animal*. Holt, Rinehart and Winston, New York, NY.

Treas, J. (2002). How cohorts, education, and ideology shaped a new sexual revolution on American attitudes toward nonmarital sex, 1972–1998. *Sociological Perspectives* 45: 267–283.

U.N. Human Development Report, http://hdrstats.undp.org/indicators/280.

Vandello, J. A., & Cohen, D. (1999). Patterns of individualism and collectivism across the United States. *Journal of Personality and Social Psychology* 77: 279–292.

Vanhanen, T. (1968). *Parties and Pluralism*. Werner Soderstrom Osakeyhtio, Parvoo, Finland.

Vanhanen, T. (1971). *Dependence of Power on Resources: A Comparative Study of 114 Countries in the 1950s*. Pub. 11, Institute of Social Sciences, University of Jyraskyla, Finland.

Vanhanen, T. (2003). *Democratization: A Comparative Analysis of 170 Countries*. Routledge, New York, NY.

Ward, J. W., & Warren, C. Eds. (2007). *Silent Victories: The History and Practice of Public Health in 20th Century America*. Oxford University Press, New York, NY.

Wejnert, B. (2005). Diffusion, development, and democracy, 1800–1999. *American Sociological Review* 70: 53–81.

Welzel, C. (2007). Are levels of democracy affected by mass attitudes? Testing attainment and sustainment effects on democracy. *International Political Science Review* 28: 397–424.

White, A.E., Kenrick, D. T., & Neuberg, S. L. (2013). Beauty at the ballot box: disease threats predict preferences for physically attractive leaders. *Psychological Science*. doi:10.1177/0956797613493642.

World Health Organization. (2004). *Global Burden of Disease: 2004 Update,* http://www.who.int. World Health Organization, Geneva, Switzerland.

World Factbook. (2007). http://www.cia.gov.

Zweifel, T. D., & Navia, P. (2000). Democracy, dictatorship, and infant mortality. *Journal of Democracy* 11: 99–114.

# Chapter 11
# Economics, Values, and Cognitive Ability

## 11.1 Introduction

In this chapter, we propose and empirically support the hypothesis that some major topics of the scholarly study of economics comprise a subdiscipline of the biological discipline of host–parasite ecology and evolution. Adam Smith's book, *An Inquiry into the Nature and Causes of the Wealth of Nations*, first published in 1776, is widely discussed as the foundational work that gave rise to the scholarly discipline of economics. Smith emphasized that a fundamental issue to understand is why some nations prosper while others lag behind in poverty. Smith was interested, too, in moral systems and how they relate to variation in national wealth. His book, *The Theory of Moral Sentiments*, written in 1759, provided the philosophical and moral underpinnings to his later book on the wealth of nations. Smith's book on morals advocated that self-promoting or individualistic pursuit of monetary gain—capitalism—causes economic prosperity and well-being. Both of these foundational topics of economics—variation in national wealth and its connection to values—are informed fundamentally and importantly by the parasite-stress theory of values. Accordingly, a region's degree of parasite adversity proximately causes the region's values, and a region's values proximately cause the region's economic productivity. Furthermore, a region's values feedback and causally influence the region's level of parasite adversity. In ultimate or evolutionary causal terms, natural selection's historical force resulting from variable infectious-disease adversity crafted human psychological adaptation that functions in conditionally adopting values suitable to the local parasite threats, with collectivism optimal under high parasite adversity and individualism under low parasite adversity.

On a basic empirical level, the consistency of economics research with the parasite-stress theory is evidenced in the robust correlation cross-nationally between Gross Domestic Product (GDP) per capita (a major marker of economic progress and vitality) and each of the two variables, parasite stress and collectivism–individualism. Parasite stress and collectivism are negative predictors of GDP per capita (individualism, a positive predictor), as we showed first in Chap. 5.

R. Thornhill and C.L. Fincher, *The Parasite-Stress Theory of Values and Sociality: Infectious Disease, History and Human Values Worldwide*,
DOI 10.1007/978-3-319-08040-6_11, © Springer International Publishing Switzerland 2014

We document these relationships here again and discuss further their connection to the parasite-stress theory of values.

For GDP per capita (World Factbook 2008) and parasite stress, using two measures of parasite stress—*Infectious Disease DALY* (see below) and *Combined Parasite Stress* (see Chap. 5)—in a large sample of the world's countries: GDP per capita by *Infectious Disease DALY*, $r = -0.83$, $p < 0.0001$, $n = 192$; GDP per capita by *Combined Parasite Stress*, $r = -0.74$, $p < 0.0001$, $n = 192$. Thus, across the countries of the world, parasite stress is probably the leading candidate for the status of the best predictor of GDP per capita. This applies also to wealth variation within countries, typically measured by the Gini score (high Gini equals more wealth disparity). Gini, like GDP per capita, is an important economic indicator. For instance, *Infectious Disease DALY* and Gini are correlated at $r = 0.63$, $p < 0.0001$, $n = 131$ (see also Chap. 10). It is intriguing, too, that, in an even larger sample of countries, nonzoonotic human infectious disease severity is a much stronger predictor of GDP per capita than is zoonotic human disease severity: nonzoonotic, $r = -0.63$, $p < 0.0001$, $n = 216$; zoonotic, $r = -0.22$, $p = 0.0009$, $n = 216$. The difference is seen with Gini also: nonzoonotics, $r = 0.58$, $p < 0.0001$, $n = 133$; zoonotics, $r = 0.20$, $p = 0.02$, $n = 133$. These patterns of GDP per capita and Gini, each showing a stronger relationship with nonzoonotics than with zoonotics, are predicted by the parasite stress theory, because nonzoonotics are transmitted person-to-person, whereas zoonotics are not. Earlier in this book, we have shown that nonzoonotic human diseases have a much stronger relationship with values of collectivism–individualism, religiosity, and democratization than zoonotic human diseases (Chaps. 5, 9, and 10). High collectivism (low individualism) and religiosity and high autocracy (low democracy) characterize countries with high nonzoonotic disease adversity.

GDP per capita also exhibits the predicted and strong relationship with the collectivism–individualism value dimension across countries. For example, using two measures of this value dimension, *Strength of Family Ties* (a measure of collectivism) and *Hofstede Individualism* (see Chap. 5 for details about these two measures), the patterns are: GDP per capita by *Strength of Family Ties*, $r = -0.61$, $p < 0.0001$, $n = 72$; by *Hofstede Individualism*, $r = 0.61$, $p < 0.0001$, $n = 68$. Hence, cross-nationally, GDP per capita is strongly and negatively related to collectivism, and positively related to individualism. Similarly, Gini shows the expected pattern across nations. For example, *Hofstede Individualism* in relation to Gini is $r = -0.53$, $p < 0.0001$, $n = 65$.

These analyses of the relationship of economic indicators with parasite stress and collectivism–individualism by themselves confirm the economist Robert Frank's opinion (mentioned in Chap. 2) that scholarship in economics would benefit from adopting evolutionary theory as its metatheory. More forcefully, these analyses indicate the necessity of the parasite-stress theory of values in economics research.

That necessity is illustrated as well by the links between democracy, values, and parasite stress, as documented in the previous chapter. As mentioned there, democratization traditionally has been a central topic in economics scholarship because of the robust, positive association between wealth and democracy across countries. We provided evidence there that the major features of international variation in democracy, and thus in its antipole autocracy, are caused by variable parasite stress—the higher the parasite stress, the lower the democratization. There, too, we explained why the causal interaction between democracy and parasite stress is expected to be bidirectional: reduction of parasite stress evokes liberal values, liberalized values further reduce parasite stress, which liberalizes more, and so on. Also in the previous chapter, we showed that the parasite-stress theory includes an important role for the value dimension of collectivism–individualism in international variation in democratization, because a polity's value system affects its overall wealth and wealth distribution.

We propose that the parasite-stress theory can elucidate other central topics in economics. Economics researchers have investigated the influence on economic activity and productivity of global variation in cognitive ability, within-region innovation, between-region diffusion of ideas and technology, and governmental corruption. In this chapter, we provide evidence that these topics also are causally anchored to infectious-disease stress and the values it evokes. First, we will discuss the structure of what we refer to as "the parasite-stress hypothesis of economics," which is a subtheory of the parasite-stress theory of values. Then, we reiterate some findings already mentioned in this book and link them to the parasite-stress hypothesis of economics, and after that we move to further empirical evaluation of the parasite-stress theory's application to economic issues.

## 11.2 The Parasite-Stress Hypothesis of Economics

As first mentioned in Chap. 4, freedom from infectious disease positively affects the ability of people to work and produce. Morbidity due to parasites, then, as a number of scholars have shown, affects a variety of economic factors at the regional and country level and accounts for a considerable amount of international economic variation (Gallup and Sachs 2001; Sachs and Malaney 2002; Bonds et al. 2010; reviews in Landes 1998; Price-Smith 2002, 2009; Bonds et al. 2012). An example is the reduction of infectious diseases associated with the hookworm control program in the South in the first half of the twentieth century, which is thought to account for the subsequent, immediate, and large increase in economic productivity in the South (Bleakley 2007). Certainly, it is not surprising that regions with low morbidity from infectious disease produce more economic product than regions with reduced health, and that the effect is large. This is shown on the broadest scale across nations by the strong, negative relationship between parasite stress and GDP per capita presented immediately above.

The parasite-stress theory of values greatly enriches and details the overall relationship between parasite stress and economic production. A region's economy is not determined only by how widespread health is in a region, but also by the values caused by the level of parasite stress in the region. The relationship between values and economic outcomes in regions and times is a long-standing research topic in economic scholarship (for a review of this research tradition, see Guiso et al. 2006).

The parasite-stress theory of values informs this research tradition by providing an ecological and evolutionary theory of the causes of values. Collectivist values—the values evoked by high parasite stress—create wealth disparity, educational disparity, and in-group-focused or parochial economic structure. The authoritarianism of collectivism legitimatizes the concentration of wealth and power in the hands of elites, which generates and ideologically validates wealth inequality (high Gini scores; see Chap. 10). Collectivism's ethnocentric and philopatric values generate local, even primarily family-based, production. The xenophobia component of collectivism reduces intergroup and between-region economic transactions. The neophobia of collectivism rewards conformity with and obedience of traditional ideas and discourages new ways of thinking and doing. This conservative propriety reduces the curiosity and creative activity needed for economic innovation as well as the willingness to consider or embrace economically progressive technologies. Xenophobia and neophobia limit not only between-group and inter-regional economic transactions, but also within-region diffusion of ideas and technologies. Non-transparency in governmental activities and widespread governmental corruption arise out of authoritarianism and in-group favoritism (e.g., nepotism and cronyism).

Individualistic values, however, have positive effects on economics across a region by increasing democracy and thereby reducing wealth and educational disparity, and by promoting transparency in government and widespread economic opportunity and networking across the region. The xenophilia and openness of individualism create a willingness to engage in intergroup sharing of goods and services, promote within-region diffusion, and reward curiosity and novel thinking. We documented above the strong positive cross-national association between individualism and GDP per capita.

Consistent with this framework of thinking are certain robust, empirical associations across countries discussed earlier in the book. We have emphasized that technological development, scientific development, diffusion of innovations, democracy, and individualism/liberalism go hand-in-hand. Or, said differently, that collectivism/conservatism is associated with autocracy and limited technology, science and diffusion (Chaps. 4 and 10; also Ferris 2010).

Another relevant pattern was documented by Lynn and Vanhanen (2002, 2006): the strong positive correlation between economic development or wealth and cognitive ability across countries. In studies we conducted with Chris Eppig, it was shown that parasite stress strongly and negatively correlates with cognitive ability across countries of the world, as well as across the states of the USA. (Eppig et al. 2010, 2011). Assuredly, low cognitive ability will limit the range of thinking and hence reduce economic, technological and scientific advances. The parasite-stress theory, however, suggests that the link between low IQ and the lag or absence of progress in economics and related areas of science and technology is the result not only of the negative influence of parasites on cognitive capacity, but also is caused by the conservative values associated with high parasite stress. Related to this, a robust relationship repeatedly reported is that IQ correlates positively with the personality factor of openness to new ideas and with liberalism (and negatively with conservatism) (Chap. 4). Hence, high parasite stress negatively affects economics by lowering IQ and through collectivism.

In sum, the parasite-stress theory applied to economics proposes that variable parasite stress affects regional variation in economic parameters through three causal and interrelated routes. One is that ill people are less productive than healthy people. Another is via the route of the various values that differ along the collectivism–individualism dimension. The third is through variable cognitive ability.

Below, we first discuss briefly our published studies of cognitive ability in relation to parasite stress. We then turn to our previously unpublished studies of the association between collectivism–individualism and IQ conducted cross-nationally and across the US states. After treating these topics, we turn to the topic of parasite stress and values in relation to within-region and between-region diffusion of ideas and technology. Finally, we examine the interrelationship between parasite stress, collectivism–individualism, and corruption versus transparency in government.

## 11.3 Eppig et al. (2010): International IQ Research

### *11.3.1 The Parasite-Stress Hypothesis of Cognitive Ability*

Lynn and Vanhanen's (2001, 2002, 2006) publication of quantitative data on the average national intelligence quotient (IQ) scores inspired empirical studies that attempted to explain the global distribution of variation in intelligence. The findings of these empirical studies provided potential confounds that, as explained below, were taken into account in Eppig et al.'s (2010) study of the relationship between IQ and parasite stress across the globe. In Eppig et al. (2010), a new hypothesis—the parasite-stress hypothesis of cognitive ability—was presented to explain the worldwide distribution of intelligence.

The brain is probably the most complex and costly organ in the human body. In human newborns, the brain demands 87 % of the body's metabolic budget; this declines to 44 % at age 5, 34 % at age 10, and about 25 % for adults (Holliday 1986). Presumably, if an individual cannot meet these energetic demands while the brain is growing and developing, the brain's growth and developmental stability will suffer. Lynn (1990, 1993) has argued that good nutrition is vital to high degrees of allocation to mental functioning during development, and has suggested that better nutrition may account for the Flynn effect: the large average increases in IQ over short periods of time as nations develop economically and technologically (Flynn 1987, and below). Lynn (1993) reviewed evidence showing that undernourished children have smaller brains and lower intelligence than sufficiently nourished children.

Parasitic infection affects the availability of energy to the body, and hence the brain, in multiple ways. Some parasites feed on the host's tissues—the tissue loss must be replaced at energetic cost to the host. Some parasites inhabit the intestinal tract or cause diarrhea, limiting the host's intake of otherwise available nutrients. Viruses use the host's cellular machinery and macromolecules to reproduce themselves, at the energetic expense of the host. Finally, the host must activate its classical immune system to fight off infection, at high energetic expense.

Given the high cost of energetic and material investment in the classical immune system and the high cost of the human brain, if exposed to infectious diseases in childhood, individuals are expected to experience detrimental effects to their brain development, and thus to their intelligence. Hence, investment in immune defense against parasites and investment in brain development are traded off. Parasites may negatively affect cognitive function by infecting the brain directly, but Eppig and we focused our discussion only on the large energetic costs of parasitism to the host, and the resultant trade-off between allocation to the classical immune system and allocation to cognitive capability.

Studies published before our research with Eppig reported a negative relationship between intestinal helminth ("worm") infection and cognitive ability (reviewed in Watkins and Pollitt 1997). Although explanations of this pattern have been proposed, none considered intestinal worms in the larger context of co-occurring infection by a broad range of parasites, nor has the energetic cost of host defense against infectious disease and its negative effects on brain development been considered fully. Other prior studies have shown relationships between helminth infection and economic and educational factors that are related to intelligence. For example, Bleakley (2007) studied the effects of control of hookworm in the southern USA during the early twentieth century, and found that areas where hookworm infections had been reduced greatly by health interventions that targeted hookworm had higher average incomes and educational achievement after treatment than areas that had not received the interventions. Venkataramani's (2010) findings from the malaria eradication program in Mexico are exemplary: children born in times of eradication showed higher IQ scores and increased success in skilled-labor occupations than did children born at other times; this effect was strongest in areas of highest pre-eradication incidences of malaria. A final example is provided by Jardin-Botelho et al. (2008), who found that Brazilian children infected with hookworm performed more poorly on cognitive tests than uninfected children, and that children infected with more than one species of intestinal helminth performed more poorly than children infected with only one.

The above-mentioned studies only address the effect of a small number of infectious diseases on cognitive ability. Eppig et al. (2010) extended this to a broader range of infectious diseases. From the parasite-stress hypothesis of cognitive ability, we predicted that average national intelligence will correlate significantly and negatively with infectious-disease adversity, and that infectious diseases will remain an important predictor of average national intelligence when other potentially confounding variables are controlled. As shown below, this prediction was supported strongly by Eppig et al. (2010).

### 11.3.2 Methods

Eppig et al. (2010) conducted analyses using three datasets on average IQ across countries. The results were very similar across the three data sets; hence, here we report the results from analyses with only one IQ data set, that of Lynn and

Vandanen (2006), comprised of a combination of collected and estimated IQ scores. (See Eppig et al. 2010 for details about the IQ scores and other measures mentioned in this section.)

*Infectious Disease DALY* was used as a measure of infectious disease adversity for each country. This measure combines years of life lost and years spent disabled owing to 28 representative and important human infectious diseases such that one DALY equals one healthy year of life lost from infectious-disease damage per 100,000 people (see Chap. 5). Although other cross-national measures of parasitic disease exist and are highly intercorrelated with the DALY measure, we argued in Eppig et al. (2010) that this measure was the best for our study of cross-national cognitive ability, because data exist for most countries of the world ($n=192$) and it is a reasonable measure of the physiological costs of infectious disease in terms of both mortality and morbidity.

As a measure of nutrient stress, DALY scores from nutritional deficiencies (WHO 2004) were used. These scores include mortality and healthy years lost (morbidity) owing to protein-energy malnutrition, iodine deficiency, vitamin A deficiency, and iron-deficiency anemia.

Potential confounds that earlier scholars had shown to be correlated with IQ were examined in Eppig et al. (2010). Average winter high temperatures were taken from Templer and Arikawa (2006), who showed that IQ and temperature are related negatively across regions. Several studies have found that IQ correlates positively with formal education across regions (Barber 2005 for review). We collected data on literacy rates from World Bank (2008, http://data.worldbank.org/topic/education). Enrollment in secondary school, completion of secondary school, and average years of education were taken from Barro and Lee (2001). Data on GDP per capita were taken from the World Factbook 2007. Kanazawa (2008) proposed that higher intelligence is especially functional in the evolutionarily recent environmental circumstances of higher latitudes, and that natural selection has crafted region-specific, genetically differentiated psychological adaptation for cognitive ability based on the degree of evolutionary recency or novelty. He showed that distance from central Africa—a measure meant to tap evolutionary novelty—correlated positively with IQ across regions. We calculated the distance from central Africa using Kanazawa's (2008) method in order to further examine his proposal. The variable wealth of nations as a potential confound was mentioned above.

Templer and Arikawa (2006) found a negative relationship between IQ and skin darkness. We did not, however, use skin darkness as a potential confound in our analyses for three reasons: (1) although evidence suggests that skin darkness is a measure of historical infectious disease intensity over evolutionary time, it is unclear as to which kinds of infectious diseases it is indicative of (see below); (2) Templer and Arikawa (2006) argued that the relationship between skin darkness and IQ is not causal; and (3) Templer and Arikawa (2006) did not sufficiently explain why the association between intelligence and skin darkness exists.

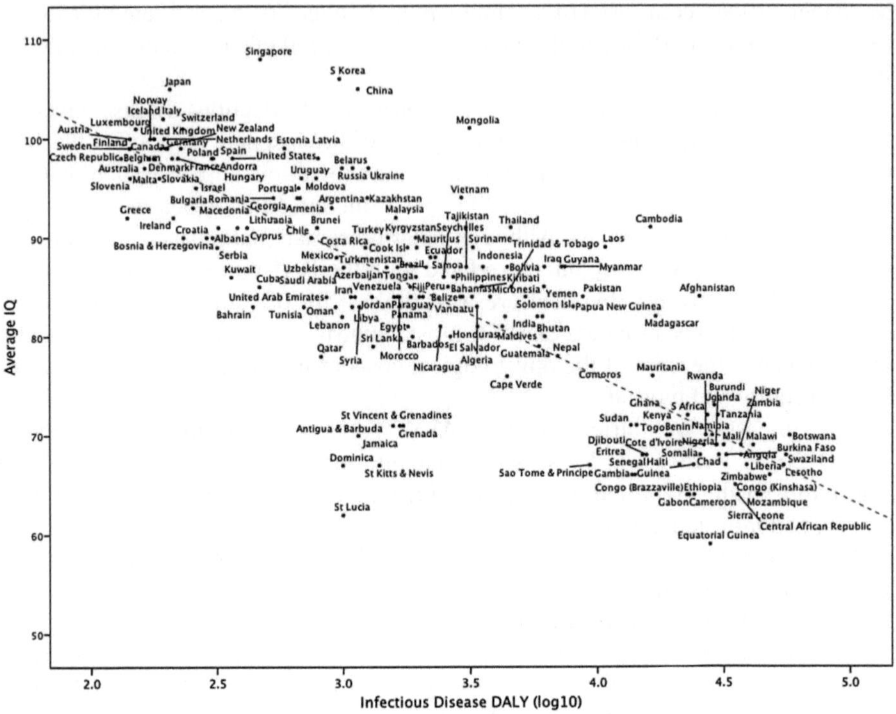


**Fig. 11.1** Log *Infectious Disease DALY* and average national IQ correlate at $r = -0.82$, $p < 0.0001$, $n = 182$ countries. The line is the regression line (results originally reported in Eppig et al. 2010)

## *11.3.3 Results*

The results in this section originally were presented in Eppig et al. (2010).

*Infectious Disease DALY* and average national IQ correlated negatively and strongly: $r = -0.82$, $p < 0.0001$, $n = 182$ (Fig. 11.1). Hence, higher parasite stress corresponds with lower cognitive ability. A hierarchical linear model was performed to determine whether this relationship is consistent across Murdock's six world regions; it is ($R^2 = 0.78$, $p = 0.0001$, $n = 184$). In separate analyses, in each of the six world regions, the significant global relationship between parasite adversity and IQ was repeated in five of the six regions; South America was the exception. DALY owing to nutritional deficiencies and IQ correlated highly at $r = -0.72$ ($p < 0.0001$, $n = 184$). *Infectious Disease DALY* and DALY owing to nutritional deficiencies correlated highly as well, at $r = 0.89$ ($p < 0.0001$, $n = 192$). However, the partial correlation between IQ and DALY from nutritional deficiencies with the effects of *Infectious Disease DALY* removed was near zero ($r = 0.028$; $p = 0.71$, $n = 184$), while the partial correlation between IQ and *Infectious Disease DALY* with the effects of DALY from nutritional deficiencies removed remained strong ($r = -0.56$; $p < 0.0001$, $n = 184$).

To select which, if any, education and wealth variables to include in a multiple regression analysis, partial correlations were performed independently between literacy, enrollment in secondary school, completion of secondary school, average years of education, and average national IQ, with the effects of infectious disease controlled. If a variable was no longer significant when the effects of infectious disease were removed, it was not included in the multiple regression. When the effects of *Infectious Disease DALY* were controlled, the correlation between IQ and literacy was $r=0.15$ ($p=0.09$, $n=113$); IQ and some secondary education was $r=0.09$ ($p=0.32$, $n=120$); IQ and completion of secondary education was $r=0.17$ ($p=0.08$, $n=110$); IQ and average years of education was $r=0.23$ ($p=0.008$, $n=127$); and IQ and GDP per capita was $r=0.05$ ($p=0.46$, $n=184$). The average years of education was the best predictor of IQ when the effects of infectious disease were controlled, so this education variable was used in multiple regressions. As such, it had the best chance of all the education variables of being significant in the multiple regression. Although GDP per capita was not a statistically significant predictor of IQ when the effects of infectious disease were removed, and the partial correlation coefficient was near zero (0.05), we included this variable as a control in some analyses at the request of a referee of the paper at the journal.

In a multiple regression, average national IQ was examined in relationship to infectious disease, winter high temperature, distance from sub-Saharan Africa, average years of education, and GDP per capita. Significant predictors in the regression were infectious disease, which had the predicted negative effect, and distance from Africa and winter high temperature, each of which had a positive effect; years of education was not significant. Infectious disease had a much larger effect than any other variable in the multiple regression. When GDP per capita was removed from the analysis, virtually identical patterns to those with GDP per capita included were found.

### *11.3.4 Discussion*

The negative relationship between cross-national indicators of infectious disease and IQ was strong. The correlation between *Infectious Disease DALY* and average national IQ was higher than that of any other potential confounding variable (the variables for which there was a previously proposed causal explanation in the scientific literature). The world regions analysis showed that the significant international pattern is repeated generally across all regions, but in separate region-specific analyses, only within five of the six regions. The world region in which this relationship was not significant was South America. This exception may be the result of the presence of several outlier countries. The group of conspicuous outliers in which IQ was much lower than expected in the worldwide trend (Fig. 11.1) are all Caribbean countries (St. Lucia, Dominica, St. Kitts and Nevis, Antigua and Barbuda, Grenada, St. Vincent and Grenadines, and Jamaica), which represent 4 of 23 nations in the South America analysis (St. Lucia, Dominica, Grenada, and St. Vincent and the

Grenadines). We hypothesized that, because these outliers are in the same geographical location, it is possible that local parasites that are not included in *Infectious Disease DALY* are causing these outliers.

Nutritional stress correlated strongly with average national IQ ($r = -0.72$), but this relationship was not significant when the effect of infectious disease was controlled. This result supports the suggested link between intelligence and nutrition, discussed above. Given the energetic cost of defense against infectious disease, individuals who are burdened with parasites may be more likely to be stressed nutritionally. Likewise, individuals who are suffering from nutritional stress may be less able to allocate to immunity.

Multiple regression showed that, of the variables infectious disease, temperature, evolutionary novelty and years of education, infectious disease was the best predictor of intelligence by a large margin. The effects of years of education were not significant, while temperature and evolutionary novelty seemed to have distinct predictive power beyond infectious disease. The independent effect of temperature may reflect the role of increased temperature in promoting the wellbeing of human parasitic diseases other than those captured in *Infectious Disease DALY*. Although the independent effect of distance from central Africa cannot be ruled out by the analysis, this effect is difficult to interpret due to doubt cast on the hypothesis underlying this variable. Wicherts et al. (2010) and Borsboom and Dolan (2006) criticized Kanazawa's hypothesis; for reasons they gave, we question the ability of linear distance from sub-Saharan Africa to measure evolutionary novelty, undermining the foundation of Kanazawa's hypothesis.

Hassall and Sherratt (2011) re-analyzed the data in Eppig et al. (2010), examining some additional potential confounding variables that they thought should be taken into account, but which had not been considered in Eppig et al. (2010). Their results were similar to our own, and they concluded that parasite stress probably is the best predictor of cross-national variation in cognitive ability. More recently, Daniele and Ostuni (2013) investigated the relationship between IQ and infectious disease across 138 countries while controlling for income, education and temperature; they reported a robust negative effect of infectious disease on cognitive ability.

### 11.3.5 Trade-Off Mechanisms

Eppig et al. (2010) suggested two mechanisms by which a trade-off occurs in the allocation of energy to immune function versus brain development and maintenance. First, during ontogeny, parasitic infection may intermittently cause the redirection of energy away from brain development. In this case, during periods of infection, the brain receives fewer energetic resources, but this allocation to brain capacity will return to pre-infection levels during healthy periods. During periods of infection, whatever aspects of the brain that are growing and developing will suffer reduced phenotypic quality. Second, exposure to infectious agents during ontogeny may cause a developmental pathway that permanently invests more energy into immune function at the expense of brain growth.

Eppig et al. (2010) also proposed a complementary hypothesis that may explain some of the effects of infectious disease on intelligence. As emphasized, a conditional developmental pathway may exist that invests more energy into the immune system at the expense of brain development. However, in an environment with a consistently high energetic cost associated with parasitic infection, selection would not favor the maintenance of conditionality in allocation. That is, if periods of health were rare, the conditional strategy of allocating more energy into brain development during periods of health would be lost evolutionarily. Thus, peoples living in areas of consistently high prevalence of infectious disease over evolutionary time may possess specific and genetically distinct adaptations that favor high obligatory investment in immune function at the expense of other metabolically expensive traits such as intelligence. For genetically distinct adaptations in intelligence to exist based on this principle, parasite levels must be quite consistent over evolutionary time. If this is not the case, then selection would maintain investment in the classical immune system and in the brain as a plastic or conditional trait.

Major increases in average intelligence across a few to several generations—the Flynn effect (Flynn 1987)—cannot be attributed to genetic differences caused by evolutionary processes, and therefore, indicate that conditional developmental causes are at work, at least in large part. Hence, it does not seem probable that region-specific genetically differentiated adaptations are the primary cause of the worldwide variation in intelligence. Fundamentally, the Flynn effect demands that any hypothesis regarding the worldwide variation and distribution of intelligence must be able to account for some factor that allows for large gains in average IQ over time spans too short to be attributed to evolution by natural selection. Eppig et al. (2010) hypothesized that reduced parasitic infection is the key factor. Moreover, as parasite stress declines in a society, individualism and democratization increase, which results in further reduction of parasite stress through the multiple, bidirectional causal pathways involving education, public health legislation, and infrastructure we have discussed.

### *11.3.6 Other Implications*

These findings about national IQ and infectious disease inform a number of other findings in the published literature. They suggest that the well-established heritable variation in intelligence may come from two sources: brain structure and immune system quality. Thus, two individuals may possess identical genes ontogenetically causing brain structure, but have different IQ owing to genetic differences in immune system quality reflecting their personal allocation of energy into brain development versus immunity.

Mackintosh (2001) presented comprehensive evidence that skin darkness and the associated cellular components (e.g., melanocytes) have an important role in defending against infectious disease. Moreover, Manning et al. (2003) found that, across regions of sub-Saharan Africa, rates of HIV infection were associated negatively with skin darkness; they attributed this relationship, in part, to lower infection rates of other parasites, especially bacteria and fungi that lead to tissue damage in the genital tract and hence increased likelihood of contracting HIV. Templer and Arikawa (2006) concluded that, despite the strong negative correlation between skin color and average national IQ, there must be an unknown mediating factor accounting for both, because there is no obvious reason for skin darkness to reduce IQ. Given the research linking skin color to infectious disease (Mackintosh 2001; Manning et al. 2003), coupled with the findings from Eppig et al. (2010) presented above, the unknown factor linking skin color and IQ may be infectious disease.

Several studies have shown a positive relationship between IQ and the bilateral symmetry of the body (e.g., Furlow et al. 1997; Prokosch et al. 2005; Bates 2007; Penke et al. 2009; but see also Johnson et al. 2008). IQ not only correlates positively with body symmetry, but the correlation increases as the quality of the IQ test increases (Prokosch et al. 2005). There is a large published literature indicating that body symmetry is a measure of developmental stability, an important component of which results from reduced contact with infectious disease (Thornhill and Møller 1997; Thornhill and Gangestad 2008). The findings of Eppig et al. (2010) suggest that IQ and body symmetry correlate because they are both affected negatively by exposure to high infectious disease. Individuals who are exposed to infectious disease may have many aspects of their body develop imperfectly, including the brain, negatively affecting both their body symmetry and cognitive ability. Indeed, research findings indicate that there is a positive relationship between body asymmetry and atypical brain asymmetries (Yeo et al. 2007).

### *11.3.7 Other Tests*

Longitudinal methods could be used to test further the parasite-stress hypothesis of cognitive ability. Children's IQ could be measured at an early age and re-measured later in life, while monitoring for infectious diseases throughout childhood. This would not only provide another test of the hypothesis, but could possibly determine the effects of individual infectious diseases on the ontogeny of cognitive development. Additionally, it could be determined which, if either, trade-off mechanism discussed above is responsible for the detrimental effects of infectious disease on intelligence. Both may operate, but with geographical differences based on the consistency of infectious disease over time. Moreover, as nations develop, they could be monitored for declining rates of parasitic infection to determine (1) whether this corresponds with elevated IQ and (2) whether any IQ gain observed is sufficient to account for the Flynn effect.

## 11.4 Eppig et al. (2011): US States IQ Research

This study examined IQ variation in relation to varying parasite stress across the states of the USA. It replicated the international study mentioned above, but did more than that as it examined the relationship between IQ and parasite stress on a much smaller scale of geography, one in which there is much less variation in IQ and parasite stress than on the international scale. This reduction in the range of variables increases the difficulty of finding significant results even when they exist. Also, the interstate analysis was useful because of the relative cultural uniformity across the USA compared to the entire world. This uniformity provides inherently more control of unknown cultural features that may be inadequately randomized in international analyses and possibly confound results. (See Chap. 2's discussion of randomization of confounds by the comparative method.)

### *11.4.1 Methods*

The variables used in analyses were as follows (see Eppig et al. 2011 for details). McDaniel's (2006) estimated average US state IQ scores were used; they are based on scores from the National Assessment of Educational Progress test (NAEP). We used *Parasite Stress USA*, which is described in Chap. 5. Two measures of education quality were used: (1) the student/teacher ratio and (2) the percentage of teachers in public schools teaching core classes who are "highly qualified." Data for (1) were from the 2008–2009 school year (National Center for Education Statistics 2009, http://nces.ed.gov/datatools/index.asp?DataToolSectionID=5). "Highly qualified" refers to teachers teaching core classes who are fully certified to teach in their state; these data were from the 2008–2009 school year (U.S. Department of Education 2009). It should be noted that, because the measure of states' IQ used was based on educational outcomes, the variables selected to measure education quality cannot also be based on educational outcomes. To this end, the qualifications of the teachers and the student/teacher ratios are effective ways to measure the quality of the educational opportunities to which students in each state have access. These two variables indicating educational quality correlated with each other only moderately ($r = -0.35$, $p = 0.012$, $n = 50$), so it was not appropriate to combine them. Three measures of state wealth were used: (1) median household income, (2) income per capita, and (3) gross state product. Median household income and income per capita were taken from the US Census Bureau from the year 2000 census. Gross state product is in millions of US dollars (from McDaniel 2006). The three measures of wealth (income per capita, gross state product, and median household income) correlated highly with one another and are all measures of state wealth. These variables were combined into a single wealth variable using unrotated principle component analysis. This component accounts for 87 % of the total variance among these three measures of wealth, and each measure of wealth loads on this component at 0.91 or higher. This constructed variable was referred to as "wealth."

In addition to economic and education variables, Eppig et al.'s (2010) cross-national study (presented above) examined geographical distance from deep-time evolutionary historical environments of humans (Kanazawa 2008). Since the majority of populations living in US states have not been occupying those locations over any extended length of evolutionary time, this variable was not used for the US analysis.

Temperature has been found to correlate negatively with average state IQ (Ryan et al. 2010). This variable was not included because Ryan et al. (2010) did not attempt to explain why this relationship exists; hence, temperature could not be treated as a potentially causal variable.

Other studies examining the distribution of IQ across the United States have controlled for race, as it has been documented to be associated with IQ (e.g., Ryan et al. 2010). The southern states in the USA have higher percentages of blacks. These states also have higher rates of infectious disease, due importantly to climate. Thus, controlling for race, the distribution of which closely follows the distribution of infectious disease, does not add meaningful information to the analysis. Indeed, across states, the percentage of a population who are black correlates much more strongly with infectious disease ($r=0.90$, $p<0.0001$, $n=50$, data from US Census Bureau 2000) than the percentage of blacks does with average IQ ($r=-0.51$, $p=0.0001$, $n=50$).

### 11.4.2 Results

The results in the section originally were presented in Eppig et al. (2011).

Average state IQ and parasite stress were highly correlated at $r=-0.67$ ($p<0.0001$, $n=50$; Fig. 11.2). Average IQ also correlated significantly with wealth ($r=0.32$, $p=0.03$, $n=50$), percentage of teachers highly qualified ($r=0.42$, $p=0.002$, $n=50$), and student/teacher ratio ($r=-0.31$, $p=0.03$, $n=50$).

A hierarchical regression was used to predict average state IQ using parasite stress, wealth, percentage of teachers highly qualified, and student/teacher ratio. Parasite stress (std. $\beta=-0.62$, $p<0.0001$), wealth (std. $\beta=0.30$, $p=0.0006$), percentage of teachers highly qualified (std. $\beta=0.29$, $p=0.0019$), and student/teacher ratio (std. $\beta=-0.22$, $p=0.015$) were all significant predictors of average state IQ. The whole model $R^2$ was 0.70 ($p<0.0001$). Of all the variables considered in the analysis, parasite stress had by far the greatest impact on cognitive ability.

### 11.4.3 Discussion

Across US states, there is a strong, negative association between infectious disease stress and average intelligence ($r=-0.67$, $p<0.0001$, $n=50$). This relationship remains robust and significant when economic and education variables are controlled for in a hierarchical regression. In this regression, infectious disease, wealth,

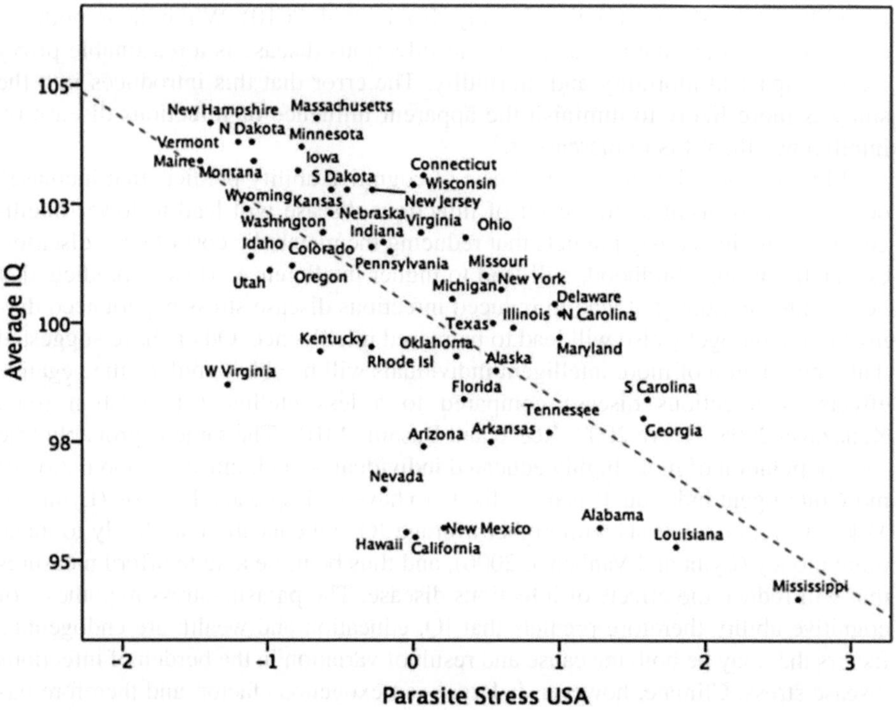


**Fig. 11.2** Relationship between average US state IQ and *Parasite Stress USA*. Average state IQ and parasite stress correlated at $r = -0.67$ ($p < 0.0001$, $n = 50$ states). The line is the regression line (results originally reported in Eppig et al. 2011)

percentage of teachers highly qualified, and student/teacher ratio were significant. Of these variables, infectious disease had more independent predictive power than wealth, percent teachers highly qualified, or student/teacher ratio.

In this analysis, infectious disease does not predict average IQ as well as it did in a similar analysis across nations, and education and economic variables have higher predictive power (Eppig et al. 2010). It is possible that this is an artifact of the way average IQ was measured across US states. Although the NAEP test, which was used to calculate average state IQ, is a valid measure of IQ (McDaniel 2006), it is likely influenced by education more than tests used cross-nationally designed to measure IQ more directly. It is also possible that the correlation between infectious disease and average state IQ ($r = -0.67$) is lower than the correlation between infectious disease and average national IQ ($r = -0.82$; Eppig et al. 2010) because there is a wider range of IQ and of parasite stress on the cross-national level than there is on the cross-state level within the USA. Despite this, of the variables examined, infectious disease is still a powerful predictor of average state IQ, and the best predictor of the variables we examined.

Some limitations exist in the US states study. The measure of infectious disease in the US study was based on number of cases, rather than on mortality and morbidity as in the cross-national study (Eppig et al. 2010). While the hypothesis concerns the latter, the prevalence of an infectious disease is a reasonable proxy for its impact in mortality and morbidity. The error that this introduces into the study is more likely to diminish the apparent influence of infectious disease on intelligence than it is to increase it.

While the parasite-stress hypothesis of cognitive ability predicts that increased developmental insult as the result of infectious disease will lead to lower intelligence, it simultaneously predicts that reducing the metabolic cost of such diseases, especially during childhood, will lead to higher intelligence. Thus, it predicts that health interventions that lead to reduced infectious disease stress experienced during human ontogeny also will lead to increased intelligence. Others have suggested that a population of more intelligent individuals will be able to reduce the negative effects of infectious disease compared to a less-intelligent population (e.g., Kanazawa 2008; Reeve 2009; Reeve and Basalik 2010). The same is probably true for a population of more highly educated individuals—and, indeed, a population of more intelligent individuals is more likely to have widespread education (Lynn and Mikk 2007). Populations with higher average IQ and education are likely to make more money (Lynn and Vanhanen 2006), and thus be more able to afford measures that will reduce the effects of infectious disease. The parasite-stress hypothesis of cognitive ability therefore predicts that IQ, education and wealth are endogenous factors that may be both the cause and result of variation in the burden of infectious disease stress. Climate, however, is largely an exogenous factor, and therefore has an independent effect on the geographical distribution of these diseases (Guernier et al. 2004; Dunn et al. 2010).

Gottfredson (1997) reviewed comprehensive evidence that IQ predicts job performance, especially in highly complex jobs; the complexity of occupations that one may succeed at; and the ability to solve problems encountered frequently in everyday life. IQ also may be involved in dealing effectively with the complexities of everyday social life. These patterns are consistent with the "social brain" hypothesis for the evolved social function of human-unique cognitive ability (e.g., Jolly 1966; Alexander 1989; Flinn et al. 2005; Dunbar and Shultz 2007; see also Chap. 5). These patterns also help account for the strong, positive relationship between cognitive ability and economic productivity across regions (Lynn and Vanhanen 2002; Lynn and Vanhanen 2006).

## 11.5 Conclusions from the Two Studies Combined

Across both countries and US states, IQ variation is strongly, negatively related to parasite stress. The relationship is not confounded by variables previously offered by researchers to account for the variation. Thus, the results supported the trade-off hypothesis for IQ variation. Infectious disease demands that human bodily resources are allocated to immune function and thereby less is available for allocation to brain function.

## 11.6 Values and IQ

Subsequent to our research with Eppig on IQ in relation to parasite stress (presented above), we examined IQ in relation to values across nation and US states. That research is presented in this section.

### *11.6.1 International*

As predicted by the parasite-stress theory of values, collectivism and IQ are strongly, negatively related across nations: *Strength of Family Ties* by IQ, $r=-0.66$, $p<0.0001$, $n=71$; the same significant pattern, but with smaller effect sizes, is seen across each of three additional measures of collectivism–individualism (average effect size ($r$)$=0.40$, all $p$'s$<0.004$). (Chap. 5 describes the *Strength of Family Ties* variable and the other cross-national collectivism–individualism variables.) Clearly, across countries, individualism corresponds to high IQ and collectivism to low IQ.

### *11.6.2 US States*

Collectivism and IQ also are strongly, negatively related across the 50 US states. *Strength of Family Ties USA* by IQ is $r=-0.64$, $p<0.0001$. *Collectivism* by IQ is $r=-0.65$, $p<0.0001$. (These two collectivism variables are described in Chap. 5.) Thus, the cross-national pattern of the negative relationship between collectivism and IQ is seen also at the interstate US scale. These findings add depth and clarity to understanding the earlier findings from several studies of Westerners showing that liberals (individualists) have higher IQs than conservatives (collectivists) (see Chap. 4).

## 11.7 GDP, IQ, and Collectivism

We predicted from the parasite-stress theory that collectivism would have an effect (negative) on economic productivity that is independent of the effect of IQ on GDP per capita (see above section, Parasite-stress Hypothesis of Economics). The prediction was supported in a multiple regression analysis. Cross-nationally, although the effect of IQ on GDP per capita is quite large, that of *Strength of Family Ties* on GDP per capita is much smaller and only marginally significant ($R^2=0.58$, $F=47.64$, $p<0.0001$, $n=71$; IQ std. $\beta=0.62$, $p<0.0001$; *Strength of Family Ties* std. $\beta=-0.20$, $p<0.06$). The *Strength of Family Ties* effect, however, is unequivocally statistically significant with a one-tail probability ($p=0.03$), which is appropriate for statistical inference in this case, because the negative direction of the association was predicted a priori. Thus, as predicted by the parasite-stress theory of values, collectivism itself has a significant (negative) influence on GDP per capita that is separate from the influence of IQ on GDP per capita.

Because of the central role of trust between interactants in promoting economic activity, Ball (2001) suggested that collectivism will sometimes promote economic productivity as a result of the trust shown by collectivists. We take issue with this, because the nature of trust is markedly different between collectivists and individualists. Collectivists have in-group trust and out-group distrust, whereas individualists are more trusting of in-group and out-group members (Chap. 4, Table 4.1a). Ball's hypothesis may apply to the localized economic productivity typical of collectivist in-groups, but certainly not to broad market economic productivity. Collectivist trust impedes out-group interactions and transactions that lead to societal economic productivity (see below). The strong, positive relationship between individualism and economic productivity cross-nationally discussed above shows that in fact collectivism impedes overall economic success of a region.

## 11.8 Parasite Stress Mediates the Relationship Between IQ and Economic Productivity

Lynn and Vanhanen (2002, 2006) argue that IQ variation is a cause of variation in economic productivity across nations. Supporting this, they show that the two variables are highly positively correlated. Our data revealed a very similar pattern. We found that the cross-national correlation between GDP per capita and IQ is $r=0.66$ ($p<0.0001$, $n=192$). The high, positive correlation across the world between these two variables, however, is actually the result of each variable's correlation with parasite stress. The strong relationship between the two variables falls to virtually zero when parasite stress is controlled statistically. This can be seen in a multiple regression analysis predicting GDP per capita from IQ and *Infectious Disease DALY*: $R^2=0.43$, $F=142.04$, $p<0.0001$, $n=192$; std. $\beta$ IQ $=-0.05$, $p=0.45$, std. $\beta$ *Infectious Disease DALY* $=-0.87$. This result shows that parasite stress is a powerful predictor of GDP per capita independently of IQ, and that IQ in itself has no predictive power beyond its relationship with parasite stress. This is another reason, beyond those mentioned early in this chapter, that scholarly economics needs the parasite-stress theory of values. The theory can account for why a key variable in economic research, IQ, is related to economic wellbeing.

## 11.9 Corruption and Value Systems

A large literature, reviewed by Ugur and Dasgupta (2011), documents the robust negative effect of governmental corruption on economic performance across countries. In this literature, corruption pertains to many types of abuses of power by governmental officials for personal gain.

Ball (2001) and various other scholars have commented about the negative economic influence of in-group favoritism that characterizes collectivism. Such bias involves favoritism (a) toward family members and (b) toward friends and other close associates with like values. The former bias is often referred to as nepotism and the latter as cronyism. These biases create ineffectiveness and incompetence in the operation of an economic unit as a result of hiring and rewarding in-group members irrespective of their competence and training. Here we examine this topic empirically by studying the interrelationship between parasite stress, collectivism–individualism, and governmental corruption.

Governmental corruption is measured by the Corruption Perceptions Index, which combines local people's perceptions and expert scholars' opinions of corruption in a nation across most of the countries of the world (http://www.transparency.org/policy_research/surveys_indices/cpi). It addresses corruption among public officials and politicians and their abuses of power for private gain. We averaged the Corruption Perceptions Index scores for each nation across the years 2005 and 2009 because these years provided the largest consistent samples of countries.

As predicted by the parasite-stress theory of values, governmental corruption is related positively to parasite stress and to collectivism (negatively with individualism) across countries. *Combined Parasite Stress* and corruption are related strongly: $r=0.65$, $p<0.0001$, $n=175$ countries. The higher the parasite stress in a nation, the more corrupt the government. Corruption and *Gelfand In-group Collectivism* are related strongly, too: $r=0.81$, $p<0.0001$, $n=57$. That is, more conservative politics are simultaneously more corrupt. Similarly, as predicted, *Suh Individualism* and corruption are related negatively and strongly ($r=-0.70$, $p<0.0001$, $n=55$), as are *Hofstede Individualism* and corruption ($r=-0.66$, $p<0.0001$, $n=67$). These relationships provide validity for the Corruption Perceptions Index because they show that the index relates in theoretically expected ways to collectivism and parasite stress. (*Combined Parasite Stress* and the values variables used in these analyses are described in Chap. 5.)

Political scientists often discuss the high corruption levels of autocratic governments (Vanhanen 2003 and literature reviewed in Ugur and Dasgupta 2011). It is unsurprising, then, that there is a strong, negative relationship across nations between democratization and corruption. As examples, the Freedom House scores (2008) (high scores equal low democratic freedom) and the Economist Intelligence Unit scores (2008)—democratization variables described in Chap. 10—each correlate strongly with the Corruption Perceptions Index: Freedom House, $r=0.72$, $p<0.0001$, $n=177$; Economist Intelligence Unit, $r=-0.76$, $p<0.0001$, $n=163$. Integrity and transparency in a government are largely the same things as the government's degree of democracy.

## 11.10 Innovation and Diffusion

We have discussed the robust relationship between the variable collectivism–individualism and the creation of knowledge and technologies. Compared to collectivist nations, individualistic nations exhibit higher rates of technological and scientific discovery, which strongly and favorably impacts the economic growth of nations (e.g., Gorodnichenko and Roland 2011; Taylor and Wilson 2012). In addition to the creation of useful knowledge and its impact on economies, economics scholars are interested in the diffusion of innovations within and across regions.

Greif (1994) discussed the difference between collectivist and individualist societies in social structure and network. Greif described collectivist cultures as "segregated" with each person economically and socially interacting primarily with in-group members (same region, ethnicity, and extended family). Greif mentioned, too, that, in such cultures, in-group members are highly involved in cooperation with in-group partners, with contract enforcement achieved through reputation within and surveillance by the in-group. Furthermore, in collectivist cultures non-cooperation is characteristic of the interactions between the members of different in-groups. In contrast, Greif emphasized that individualist cultures are "integrated"—economic and social transactions are among people from a variety of groups, and group boundaries are impermanent. Hence, people in individualist cultures are involved in widespread pure reciprocity—social interactions and exchanges with non-relatives are paramount (see Chap. 2). Under individualism, contract enforcement is accomplished by the legal court and other complex and economically costly social institutions. Greif proposed that these differences between collectivist and individualist cultures affect within- and between-region transactions and diffusion of technology and innovations, and thereby the economic wellbeing of a nation. Segregated economic units hinder such transactions and diffusion and thereby lower regional economic productivity, whereas integrated economic activity and associated widespread reciprocity promote the transactions and diffusion and thereby regional economic well-being.

In this section, we show that Greif's ideas are robust empirically. We show, too, that his topics are fundamentally part of the parasite-stress theory of sociality. First, we examine within-US diffusion of innovations in relation to collectivism and parasite stress. Then, from the same theoretical perspective, we turn to international innovation and diffusion.

### *11.10.1 Diffusion in the USA*

In Chap. 7, we treated the openness to new ideas and ways valued by individualists and the closed-mindedness to innovations of collectivists. This personality or value difference affects the spread of innovations and technology. The bounded, localized social network of collectivists retards diffusion, whereas the open and more cosmopolitan social networks of individualists promote diffusion.

We document the connection between values and diffusion of innovations below with data on the diffusion of hybrid corn, an agricultural innovation, as well as the diffusion of medical technologies and democratic laws across US states. We also mention a recent finding about consumer behavior at US supermarkets that supports this connection.

The planting of hybrid corn, in contrast to traditional corn, greatly increases the economic gain from corn farming. Griliches (1957) classic study of the adoption of hybrid corn across the USA provides data for the 31 corn-growing states on the date of "origin," the date, measured from 1940, when a corn-growing state first began to plant 10 % of its corn-farm acreage with hybrid corn.

The origin variable is comprised of features that reflect local values. One feature is when the idea of using hybrid corn was accepted by a state's agricultural establishment. Following acceptance, breeding at agricultural experiment stations for specific hybrids suitable for local state conditions was conducted. The duration of this experimental stage also is reflected in the origin datum of a state. Hence, origin has a component of openness to new ideas and a component of innovativeness at the experimental phase. Also, parasites of corn may vary regionally. In high parasite regions, the duration of the experimental phase could be lengthened if innovations were inadequate.

Griliches (1957) found regional differences in the timing of adoption of hybrid corn. The southeastern states were the laggards, with Alabama and Georgia being the last two states to adopt the new technology. According to the parasite-stress theory of values, this regional pattern is expected, because of the closed-mindedness and limited innovation that stems from collectivism and high parasite stress. This application of the parasite-stress theory of values is strongly supported. The relationship between origin and collectivism (Vandello and Cohen 1999) is $r=0.60$, $p=0.0004$, $n=31$; that between origin and *Parasite Stress USA* is $r=0.63$, $p=0.0002$, $n=31$. These patterns show that the more collectivist the state and the higher its parasite stress, the greater the delay in adoption of hybrid corn.

The low diffusion rates associated with high parasite stress and collectivism is not limited to hybrid corn. The pattern is also seen with medical technology and knowledge. Jencks et al. (2003) and Berwick (2003) have expressed worry about the variation in use of modern medical care across states of the USA. Jencks et al. (2003) provide data from 2000 to 2001 across US states for use of 22 evidence-based medical treatments for a range of major health problems (stroke, heart failure, diabetes, breast cancer, pneumonia, and so on) experienced by Medicare beneficiaries. The higher a state's score, the less use of the sum of the medical treatments. The range shown by the composite of the 22 treatments across the states is large. As expected from the parasite-stress theory of values, the use of the treatments is negatively related to a state's parasite stress and collectivism. *Parasite Stress USA* and treatment use show an $r$ of 0.66 ($p<0.0001$, $n=50$ states). Vandello and Cohen's collectivism and treatment use has an $r$ of 0.44 ($p=0.002$, $n=50$). High parasite-stress/collectivism states employ modern medical care less than low parasite-stress/collectivism states. This difference is not due to regional differences in innovation, because the treatments are documented in the scientific literature as evidence-based. The knowledge of the medical value of the treatments is available to all medical facilities regardless of state. Apparently, the difference is the result of regional variation in ways of thinking and attitude toward new and non-traditional ideas. The higher the parasite stress or collectivism, the less new ideas appeal, whether medical or agricultural improvements.

We mentioned in Chap. 10 that the states of the South were generally laggard in ratifying the 19th constitutional amendment allowing women to vote. This is not a surprising pattern given the greater gender inequality and collectivism in general in this region compared to other regions of the USA (Chap. 10). The movement of laws across the states of the USA has long been a major topic of research in political science and economics. With regard to laws, there are two issues of interest here: (a) the law's adoption date in a state and (b) its received level of monetary support by the state. In regard to (b), a law could be adopted by a state, but then its enforcement is not funded. The data available only address date of law adoption by a state. Here we treat the diffusion of two categories of laws, welfare laws (e.g., aid to families with dependent children) and civil rights laws (e.g., antidiscrimination in housing). The data we use are from Gray's (1973) review of the diffusion of these legal innovations across the 48 continental US states over the time period of 1883–1966; the higher the rank of a state, the later the state's adoption of the laws.

We found that all relationships are consistent with the parasite-stress theory of values. A later adoption of welfare laws and *Parasite Stress USA* are correlated positively with $r=0.41$, $p=0.004$, welfare laws and collectivism (based on Vandello and Cohen 1999) are also positively correlated with $r=0.35$, $p=0.02$. A later adoption of civil rights laws shows a positive relationship with collectivism of $r=0.35$, $p=0.01$, and with *Parasite Stress USA* of $r=0.33$, $p=0.03$. Hence, there are moderate and significant correlations between the adoption of each of the two types of laws examined and collectivism and parasite stress. At higher levels of parasite stress or collectivism, states have been slower in adopting laws that promote civil rights and welfare, two features of democracy. We predict that if data do emerge allowing tests of states' allocation of funding to these two categories of laws, the pattern found for initial adoption will be repeated. That is, as collectivism or parasite adversity increase, states will show increased reluctance to appropriate money to fund laws pertaining to civil rights and welfare. Note that all the topics of diffusion of innovations we have examined above are related to people's values, and equally so. According to the parasite-stress theory of values, the movement of hybrid corn or of medical innovations across states is the result of values as much as is diffusion of laws about antidiscrimination in housing or public facilities.

Diffusion of commercial products across regions is affected by the willingness of consumers to buy new or non-traditional products. The parasite-stress theory of values predicts that this willingness will be lower among conservative consumers than among liberal consumers. Indeed, the values of people affect the frequent choices they make at the supermarket in choosing laundry detergent, razor blades, milk, peanut butter, toilet paper, diapers, and other items. Conservatives buy based on valuing traditional brands and avoiding the perceived uncertainty of new products, whereas liberals are more open and willing to use novel products. Khan et al. (2013) documented these patterns in consumer behavior by investigating the brands purchased at supermarkets across counties of the USA. The database they used is weekly sales of thousands of products from about 2,000 stores in 135 supermarket chains over 6 years (2001–2006). Their data represent about 50 % of the population of the USA. They found that conservatism, measured as religiosity and Republican voting behavior, robustly predicted the favoritism of people in buying traditional product brands (as opposed to generic brands) and avoiding products newly introduced into stores.

### 11.10.2 Cross-National Technology Adoption

Above we analyzed the diffusion of medical and agricultural technologies across the states of the USA. We showed that the diffusion of both technologies across the states is related negatively to parasite stress and collectivism. Comin et al. (2008) provided data on diffusion of 100 technologies in 150 countries over the previous 200 years. The data are in the form of total number of technologies per country. Several categories of technology are included (e.g., medical, agriculture, finance, steel, telecommunications, transportation). Nearly one-half of the technologies Comin et al. consider are medical technologies (e.g., organ transplants, mastectomies, mammographs, magnetic resonance imaging). Given the over-representation of medical technologies, this data set allowed us to examine whether the medical-innovation-use patterns for US states are repeated across countries. They are. Total number of technologies was correlated negatively and significantly with *Combined Parasite Stress* ($r=-0.42, p<0.0001, n=143$ countries) and with *Strength of Family Ties* ($r=-0.43, p<0.0001, n=67$ countries). Although the total number of technologies across nations is more than just medical innovations, the medical technologies comprise a large part of the total. The findings indicate that countries with high parasite stress and associated collectivism, just like states with these characteristics, are the extreme laggards in adopting modern medical technologies. The findings also indicate that this is also the case for technology in general. Correspondingly, countries with high individualism are the most embracing of new technologies.

In a fascinating later investigation, Comin et al. (2010) assembled a dataset on the history of technology adoption across the world, and separately for the periods 1000 BC, 1 AD and 1500 AD. They emphasized their greater confidence in the 1500 AD technology data than in that of the earlier periods. Technology adoption was based on whether a particular technology was present in a time period, not how widespread it was used. Technology was tabulated across the six sectors of agriculture, transportation, industry, communication, writing, and military. The ancient locations of technology use were transformed into contemporary country locations. This provided data for more than 100 countries across each of the three periods. These data reflect both technological innovativeness and technology diffusion. The technologies tabulated are mechanisms that increase economic production and hence affected positively a region's wealth. Comin et al. (2010) also provide technology adoption measures for five sectors for 2000 AD (except for the military sector) and data on per capita income for 2002 across the countries. The 2000 AD data are for how widespread in a country a particular technology is used, not just a particular technology's presence or absence. The data set used by Comin et al. (2010) is at D. Comin's website (http://www.hbs.edu/faculty/Pages/profile.aspx?facId=438581&facInfo=res).

The regions of the world varied greatly in technology use in each of the ancient periods, just as they do today. The basic findings are reflected in the title of Comin et al.'s (2010) paper, "Was the Wealth of Nations Determined in 1000 BC?" They reported evidence of statistically significant persistence of technology adoption from 1000 BC? to 1 AD, and from 1 AD to 1500 AD across the countries analyzed. The technology use from 1500 AD to 2000 AD was the strongest relationship found between the periods. Moreover, income per capita in 2002 was positively, moderately correlated with the technology use at 1500 AD ($r=0.45$, $p<0.0001$, $n=108$ countries), and positively, strongly correlated with mean technology adoption in 2000 AD ($r=0.90$, $p<0.0001$, $n=122$ countries).

We examined some of Comin et al.'s (2010) findings in relation to the parasite-stress theory of values. Their variable "mean technology adoption," an average of all sectors, for 130 countries in 2000 AD shows a strong negative correlation of −0.74 ($p<0.0001$) with *Combined Parasite Stress*, a variable described in Chap. 5. At higher levels of parasite stress across countries, technology adoption was lower, a finding supporting the similar analysis of Comin et al. (2008) described above. When the sectors included in the 2000 AD data are considered separately, the correlations with *Combined Parasite Stress* and technology use all were negative and statistically significant (all $p$'s$<0.0001$, n's ranged from 117 to 130 countries): $r$'s ranged from −0.51 (transportation) to −0.82 (industry). Mean technology adoption for 1500 AD for 124 countries and *Combined Parasite Stress* were negatively and moderately related at $r=-0.41$. Hence, parasite adversity is a robust predictor of technology adoption. The higher the parasite adversity, the lower the adoption of technology on both the contemporary time scale, as well as at 1500 AD.

As discussed earlier in the book, parasite adversity tends to be somewhat stable across time in a region because parasite wellbeing is related to a region's climatic factors. This, we hypothesize, is a proximate cause of the temporal consistency of technology use that Comin et al. (2010) discovered. Their hypothesis for the temporal consistency is that early technology adoption makes subsequent technology use easier and more efficient. So once technology adoption gets started in a region, it just keeps on going through time. This may be true, but this notion is incomplete, as it does not explain why initial technology is invented and adopted in some regions but not others. Nor does it explain the maintenance of the openness and creativity needed to account for the historically consistent use of technology in some regions and not others. In essence, Comin et al.'s notion lacks a theory of values that can account for regional and temporal differences in the values that promote or retard technology use. According to the parasite-stress theory of values, high parasite stress evokes conservative values. Such values encourage parochial and traditional ways and means and discourage the innovative thought and reward for such thought necessary for technological discoveries and the maintenance of technology use in a region. Also, high parasite stress evokes avoidance of new ideas and ways and limits reciprocal transactions, and hence reduces the adoption of innovations that originate in out-groups or other regions. Finally, high parasite stress negatively affects cognitive ability, which is related in complex ways to innovativeness. In contrast, low parasite stress elevates cognitive ability and evokes liberal values that promote and reward creative thought and new ways and welcome out-group interactions and transactions.

The promise of the parasite-stress theory of values for understanding the history of technology invention and use across the world is indicated by the large body of evidence in our book supporting this theory of values. More specifically, the evidence in this and the previous chapter showing an important role of parasite adversity and associated values in diffusion and innovation of ideas indicates the scientific promise of the parasite-stress theory for understanding the history of technology use.

## 11.11 Knowledge Unity

We have provided a range of evidence indicating that some major features of scholarly economics fall into a subfield of evolutionary biology. More specifically, much of economics is a subdiscipline of the field of biological research that investigates the ecology and evolution of parasite–host interactions. Many of the phenomena of fundamental interest to economists from GDP, within- and between-region diffusion of innovation and technology, IQ, governmental transparency and efficiency versus corruption, autocracy–democracy, and so on appear to have their core, proximate causal basis in parasite stress and the values it causes. In terms of ultimate causation, these economics' topics reduce to the evolutionary historical selection that built the behavioral and classical immune systems of humans. Science can be very reductionistic, or, in Wilson's (1998) terminology and the vision of The Enlightenment, science at its best accomplishes consilience—it can reduce synthetically many fields of inquiry to a small number of core causal principles. The evidence we present in this book suggests that the parasite-stress theory is the synthesizer of many fields dealing with human affairs that have traditionally been studied as largely or entirely independent.

## 11.12 Economic Prosperity Has a Major Cost

It is often the opinion that economic success in a region is always a good thing because it creates an infrastructure of widespread individual, family, and community economic prosperity. As long as wealth inequality is harnessed during regional economic growth, these are certainly outcomes. A huge cost of economic prosperity, however, with or without democratic policy that moderates Gini, is the unbridled consumption of natural sources of energy. This energy is limited. Brown et al. (2011) have documented the near-perfect positive relationship between energy consumption per capita and GDP per capita across virtually all the countries of the planet. Most of the energy consumed is by the Western world. Continuing to ignore this cost of economic prosperity and democracy may have grave circumstances in the not too distant future. We return to this topic in more detail in Chap. 14.

## 11.13 Summary

We present the parasite-stress theory of economics. It argues that variable parasite stress across regions accounts for regional variation in economic productivity by three categories of proximate causes. One is that infectious diseases cause morbidity and hence reduce people's capability to produce. The second is that parasite stress evokes people's values, which, in turn, cause economic parameters of a region. As parasite stress increases, regions become increasingly collectivistic. Collectivism corresponds to regionally local economics, reduced innovativeness and interregional diffusion of innovations, political corruption, and autocracy. These effects stifle economic growth and productivity. Individualism corresponds to a willingness to transact with a wide diversity of people, creating market economies and interregional sharing of ideas and products, increased innovativeness, governmental transparency, and democracy. These effects promote economic prosperity. The third factor is the negative effect of infectious disease on cognitive ability. Reduced cognitive ability reduces innovativeness and thus economic wellbeing in a region.

Evidence supporting this framework is both diverse and copious. We discuss the established negative relationships between the important economic indicators, GDP per capita and Gini, and parasite stress and collectivism across the countries of the world. Studies confirming the negative relationship between the diffusion of various innovations and parasite stress and collectivism across countries and US states are presented. Impressive evidence shows that even the routine choices of people at supermarkets are consistent with the parasite-stress theory of values. We also discuss research indicating that parasite-stress variation across the globe affected wealth of regions as far back as 1500 AD. Cognitive ability is correlated negatively with parasite stress and collectivism both across countries and US states.

The chapter reveals the unity of knowledge about many major topics in economics that is accomplished by the parasite-stress theory of values. The implication is that this theory is a general hypothesis of economics. Accordingly, much of economic research is a subfield of the scientific study of the ecology and evolution of host–parasite relationships.

We mention, but will treat fully in Chap. 14, that economic prosperity has an ominous cost. It results in accelerated consumption of non-renewable energy.

## References

Alexander, R. D. (1989). Evolution of the human psyche. In *The Human Revolution* (eds. P. Mellars & C. Stringer), pp. 455–513. University of Edinburgh Press, Edinburgh, U.K.

Ball, R. (2001). Individualism, collectivism, and economic development. *Annals of the American Academy of Political and Social Science* 573: 57–84.

Barber, N. (2005). Educational and ecological correlates of IQ: A cross-national investigation. *Intelligence* 33: 273–284.

Barro, R. J., & Lee, J. W. (2001). International data on educational attainment: Updates and Implications. *Oxford Economic Papers* 53: 541–563.

Bates, T. C. (2007). Fluctuating asymmetry and intelligence. *Intelligence* 35: 41–46.

Berwick, D. M. (2003). Disseminating innovations in health care. *JAMA* 289: 1969–1975.

Bleakley, H. (2007). Disease and development: Evidence from hookworm eradication in the American South. *The Quarterly Journal of Economics* 122: 73–117.

Bonds, M. H., Keenan, D. C., Rohani, P. et al. (2010). Poverty trap formed by the ecology of infectious diseases. *Proceedings of the Royal Society B Biological Sciences* 277: 1185–1192.

Bonds, M. H., Dobson, A. P., & Keenan, D. C. (2012). Disease ecology, biodiversity, and the latitudinal gradient in income. *PLoS Biology* 10: e1001456.

Borsboom, D., & Dolan, C. V. (2006). Why *g* is not an adaptation: A comment on Kanazawa (2004). *Psychological Review* 113: 433–437.

Brown, J. H., Burnside, W. R., Davidson, A. D. et al. (2011). Energetic limits to economic growth. *Bioscience* 61: 19–26.

Comin, D. http://www.hbs.edu/faculty/Pages/profile.aspx?facId=438581&facInfo=res.

Comin, D., Hobijn, B., & Rovito, E. (2008). A new approach to measuring technology with an application to the shape of the diffusion curves. *Journal of Technology Transfer* 33: 187–207.

Comin, D., Easterly, W., & Gong, E. (2010). Was the wealth of nations determined in 1000 BC? *American Economic Journal: Macroeconomics* 2: 65–97.

Corruption Perceptions Index. (2005–2009). http://www.transparency.org/policy_research/surveys_indices/cpi.

Daniele, V. & Ostuni, N. (2013). The burden of disease and the IQ of nations. *Learning and Individual Differences* 28: 109–118.

Dunbar, R., & Shultz, S. (2007). Evolution in the social brain. *Science* 317: 1344–1347.

Dunn, R. R., Davies, T. J., Harris, N. C. et al. (2010). Global drivers of human pathogen richness and prevalence. *Proceedings of the Royal Society B* 27: 2587–2595.

Economist Intelligence Unit. (2008). http://www.eiu.com.

Eppig, C., Fincher, C. L., & Thornhill, R. (2010). Parasite prevalence and the worldwide distribution of cognitive ability. *Proceedings of the Royal Society of London B* 277: 3801–3808.

Eppig, C., Fincher, C. L., & Thornhill, R. (2011). Parasite prevalence and the distribution of intelligence among the states of the USA. *Intelligence* 39: 155–160.

Ferris, T. (2010). *The Science of Liberty: Democracy, Reason, and the Laws of Nature*. Harper, New York, NY.

Flinn, M. V., Geary, D. C., & Ward, C. V. (2005). Ecological dominance, social competition, and coalitionary arms races: Why humans evolved extraordinary intelligence. *Evolution and Human Behavior* 26: 10–46.

Flynn, J. R. (1987). Massive IQ gains in 14 nations: What IQ tests really measure. *Psychological Bulletin* 101: 171–191.

Freedom House. (2008). http://www.freedomhouse.org.

Furlow, F. B., Armijo-Prewitt, T., Gangestad, S. W. et al. (1997). Fluctuating asymmetry and psychometric intelligence. *Proceedings of the Royal Society B* 264: 823–829.

Gallup, J. L., & Sachs, J. D. (2001). The economic burden of malaria. *The American Journal of Tropical Medicine and Hygiene* 64: 85–96.

Gorodnichenko, Y., & Roland, G. (2011). Individualism, innovation, and long-run growth. *Proceedings of the National Academy of Sciences* 108: 21316–21319.

Gottfredson, L. S. (1997). Why *g* matters: The complexity of everyday life. *Intelligence* 24: 79–132.

Greif, A. (1994). Cultural beliefs and the organization of society: A historical and theoretical reflection on collectivist and individualist societies. *The Journal of Political Economy* 102: 912–950.

Griliches, Z. (1957). Hybrid corn: An exploration in the economics of technological change. *Econometrica* 25: 501–522.

Guernier, V., Hochberg, M. E., & Guégan, J. -F. (2004). Ecology drives the worldwide distribution of human diseases. *PLoS Biology* 2: 740–746.

Guiso, L., Sapienza, P., & Zingales, L. (2006). Does culture affect economic outcomes? *Journal of Economic Perspectives* 20: 23–48.
Hassall, C., & Sherratt, T. N. (2011). Statistical inference and spatial patterns in correlates of IQ. *Intelligence* 39: 303–310.
Holliday, M. A. (1986). Body composition and energy needs during growth. *Human Growth: A Comprehensive Treatise*, Vol. 2 (eds. F. Falkner & J. M. Tanner), pp. 101–117. Plenum, New York, NY.
Jardin-Botelho, A., Raff, S., Rodrigues, R. A. et al. (2008). Hookworm, *Ascaris lumbricoides* infection and polyparasit ism associated with poor cognitive performance in Brazilian schoolchildren. *Tropical Medicine and International Health* 13: 994–1004.
Jencks, S. F., Huff, E. D., & Cuerdon, T. (2003). Change in the quality of care delivered to Medicare beneficiaries, 1989–1999 to 2000–2001. *JAMA* 289: 305–312.
Johnson, W., Segal, N. L., & Bouchard Jr., T. J. (2008). Fluctuating asymmetry and general intelligence: No genetic or phenotypic association. *Intelligence* 36: 279–288.
Jolly, A. (1966). Social behavior and primate intelligence. *Science* 153: 501–506.
Kanazawa, S. (2008). Temperature and evolutionary novelty as forces behind the evolution of general intelligence. *Intelligence* 36: 99–108.
Landes, D. S. (1998). *The Wealth and Poverty of Nations*. W.W. Norton and Co., New York, NY.
Lynn, R. (1990). The role of nutrition in secular increases in intelligence. *Personality and Individual Differences* 11: 263–285.
Lynn, R. (1993). Nutrition and intelligence. In *Biological Approaches to the Study of Human Intelligence* (ed. P. A. Vernon), pp. 243–258. Ablex Publishing, Norwood, NJ.
Lynn, R., & Mikk, J. (2007). National differences in intelligence and educational attainment. *Intelligence* 35: 115–121.
Lynn, R., & Vanhanen, T. (2001). National IQ and economic development: A study of eighty-one nations. *Mankind Quarterly* 41: 415–435.
Lynn, R., & Vanhanen, T. (2002). *IQ and the Wealth of Nations*. Praeger, Westport, CT.
Lynn, R., & Vanhanen, T. (2006). *IQ and Global Inequality*. Washington Summit Publishers, Augusta, GA.
Mackintosh, J. A. (2001). The antimicrobial properties of malanocytes, melanosomes and melanin and the evolution of black skin. *Journal of Theoretical Biology* 211: 101–113.
Manning, J. T., Bundred, P. E., & Henzi, P. (2003). Melanin and HIV in sub-Saharan Africa. *Journal of Theoretical Biology* 223: 131–133.
McDaniel, M. A. (2006). Estimating state IQ: Measurement challenges and preliminary correlates. *Intelligence* 34: 607–619.
National Center for Education Statistics, 2009, http://nces.ed.gov/datatools/index.asp?DataToolSectionID=5.
Penke, L., Bates, T. C., Gow, A. J. et al. (2009). Symmetric faces are a sign of successful cognitive aging. *Evolution and Human Behavior* 30: 429–437.
Price-Smith, A. T. (2002). *The Health of Nations: Infectious Disease, Environmental Change, and Their Effects on National Security and Development*. MIT Press, Cambridge, MA.
Price-Smith, A. T. (2009). Contagion and Chaos: Disease, Ecology, and National Security in the Era of Globalization. MIT Press, Cambridge, MA.
Prokosch, M. D., Yeo, R. A., & Miller, G. F. (2005). Intelligence tests with higher *g*-loadings show higher correlations with body symmetry: Evidence for a general fitness factor mediated by developmental stability. *Intelligence* 33: 203–213.
Reeve, C. L. (2009). Expanding the *g*-nexus: Further evidence regarding the relations among national IQ, religiosity and national health outcomes. *Intelligence* 37: 495–505.
Reeve, C. L., & Basalik, D. (2010). Average state IQ, state wealth and racial composition as predictors of state health statistics: Partial support for '*g*' as a fundamental cause of health disparities. *Intelligence* 38: 282–289.
Ryan, J. J., Bartels, J. M., & Townsend, J. M. (2010). Associations between climate and IQ in the United States of America. *Psychological Reports* 107: 251–254.

Sachs, J., & Malaney, P. (2002). The economic and social burden of malaria. *Nature* 415: 680–685.

Smith, A. 1759 (2010). *The Theory of Moral Sentiments*. Penguin, New York, NY.

Smith, A. 1776 (2012). *An Inquiry into the Nature and Causes of the Wealth of Nations*. Simon and Brown, New York, NY.

Taylor, M. Z., & Wilson, S. (2012). Does culture still matter? The effects of individualism on national innovation rates. *Journal of Business Venturing* 27:234–247.

Templer, D. I., & Arikawa, H. (2006). Temperature, skin color, per capita income, and IQ: An international perspective. *Intelligence* 34: 121–139.

Thornhill, R. & Gangestad, S. W. (2008). *The Evolutionary Biology of Human Female Sexuality*. Oxford University Press, New York, NY.

Thornhill, R. & Møller, A. P. (1997). Developmental stability, disease and medicine. *Biological Review* 72: 497–548.

Ugur, M., & Dasgupta, N. (2011). Evidence on the economic growth impacts of corruption in low-income countries and beyond: A systematic review. EPPI-Centre, Social Science Research Unit, Institute of Education, University of London, London, UK.

U.S. Census Bureau, U.S. Department of Commerce. (2000). http://www.census.gov.

U.S. Department of Education. (2009). *State Contacts and Information.* http://www2.ed.gov/about/contacts/state/index.html.

Vandello, J. A., & Cohen, D. (1999). Patterns of individualism and collectivism across the United States. *Journal of Personality and Social Psychology* 77: 279–292.

Vanhanen, T. (2003). *Democratization: A Comparative Analysis of 170 Countries*. Routledge, New York.

Venkataramani, A. (2010). *Early Life Exposure to Malaria and Cognition and Skills in Adulthood: Evidence from Mexico*. Social Science Research Network, http://papers.ssrn.com/sol3/papers.cfm?abstract_id=1679164.

Watkins, W. E., & Pollitt, V. (1997). 'Stupidity or worms': Do intestinal worms impair mental performance? *Psychological Bulletin* 121: 171–191.

Wicherts, J. M., Borsboom, D., & Dolan, C. V. (2010). Why national IQs do not support evolutionary theories of intelligence. *Personality and Individual Differences* 48: 91–96.

Wilson, E. O. (1998). *Consilience: The Unity of Knowledge*. Knopf, New York, NY.

World Bank. (2008). http://data.worldbank.org/topic/education.

World Factbook. (2007). http://www.cia.gov.

World Factbook. (2008). http://www.cia.gov.

World Health Organization. (2004). *Global Burden of Disease: 2004 Update*. World Health Organization, Geneva, Switzerland.

Yeo, R. A., Gangestad, S. W., & Thoma, R. J. (2007). Developmental instability and individual variation in brain development: Implications for the origin of neurodevelopmental disorders. *Current Directions in Psychological Science* 16: 245–249.

# Chapter 12
# Wars, Revolutions and Coups, and the Absence of Peace Across the World

## 12.1 Introduction

This chapter addresses the application of the parasite-stress theory of values to cross-national incidences of within-country warfare and other types of political intergroup conflict within countries across the world. All the various types of such wars are investigated. We also treat the related topics of revolutions and coups. The methods and results used in some of the research on cross-national intergroup conflict described below are given in fuller detail in our two publications with Kenneth Letendre on intergroup conflict (Letendre et al. 2010, 2012); previously unpublished analyses are identified as such and described when they are introduced in this chapter. We also discuss a study that applied the parasite-stress theory of values to explain the regional diversity of college and university team sports across the states of the USA. First, we briefly discuss components of the parasite-stress theory of sociality relevant to intergroup conflict.

## 12.2 Relevant Aspects of the Parasite-Stress Theory

According to the parasite-stress theory of values, which was explained in detail in Chap. 3, temporally and spatially variable parasite stresses generated past selection that built species-typical, conditional psychological adaptations of humans that are functionally specialized for assessment of local parasite stress and for guiding the adoption and use of values (morals) pertaining to in-group and out-group behaviors that manage and avoid infectious diseases. Hence, parasite stresses generated the natural selection of individuals that caused the evolution of this conditional psychology in the first place (ultimate causation); and such stresses are the ancestral cues that cause that psychology's cognitive, emotional and behavioral manifestations within the lifetime of the individual (proximate causation).

R. Thornhill and C.L. Fincher, *The Parasite-Stress Theory of Values and Sociality: Infectious Disease, History and Human Values Worldwide*,
DOI 10.1007/978-3-319-08040-6_12, © Springer International Publishing Switzerland 2014

Host–parasite antagonistic coevolutionary races are variable and localized spatially across the range of a single human culture, yielding local coadaptation between hosts and their local parasites. This creates a situation in which contact and interaction with non-group members (out-groups) can be costly, because out-group members, relative to in-group members, may carry parasites to which in-group members are not adapted immunologically. This can involve different variants of single parasite species.

Therefore, people's core morality and associated social life arise to an important extent from assessments of contagion risk during ontogeny. Xenophobia—the avoidance of and antagonism toward out-groups—is an adaptation or evolved solution to the problem of being maladapted to the infectious diseases parasitizing out-groups. Ethnocentrism is a complementary, evolved solution to the fitness challenge imposed by parasite adversity: loyalty toward, dutiful assistance of, and interdependence with in-group members are defenses and insurance against the mortality and morbidity of local parasites that infect the in-group. Ethnocentrism is comprised of two parts: (a) nuclear- and extended-family nepotism, and (b) cooperation with in-group, non-family members with the same values and immunity. The greater the parasite stress in a region, the greater the ethnocentrism and xenophobia; likewise, the lower the pathogen prevalence, the lower the ethnocentrism and xenophobia. Low ethnocentrism is the value of prioritizing nuclear-family-focused nepotism and with limited extended-family interactions and in-group allegiance. Low xenophobia (=high xenophilia) is the value of attractiveness of out-group interactions and relations. Out-group interactions provide benefits to individuals of broader and more diverse social networks and intergroup alliances, but such benefits are expected to exceed costs when parasite stresses are reduced. Consequently, the parasite-stress theory of values proposes that parasites causally influence human values/morals pertaining to family life and to in-group and out-group feelings, motivations and behavior in general.

Moreover, the parasite-stress theory asserts that high infectious-disease intensity in a region leads to individuals with collectivist values/behaviors and, thus, emergent collectivist cultures, and that low levels of infectious diseases lead to individuals with individualistic values/behaviors and emergent individualistic cultures. The cross-regional relationship between a region's location on the collectivism–individualism values dimension and parasite adversity in the region provides strong support for this aspect of the theory: across the states of the USA and many countries of the world, high parasite stress corresponds to high collectivism, whereas low parasite stress corresponds to low collectivism, i.e., high individualism (Chap. 5). Collectivism (as opposed to individualism) is a value system of out-group devaluation; in-group support; conformity to in-group norms; closed-mindedness to new ideas and ways; and allegiance to traditional values, hierarchy and authority. The collectivist understands self as immersed in and interdependent upon in-group members, and places an emphasis on distinguishing in-group from out-group members. In contrast, the ideology of individualism recognizes the validity, safety and security of interactions with out-groups who have different norms and beliefs, and prioritizes openness to novelty, thus placing less importance on tradition, authority, and hierarchy. The individualist understands self as relatively independent of the in-group, and in-group and out-group boundaries are blurred and frequently change (Chap. 4).

Furthermore, pathogen stress and collectivism negatively relate to democratization across the countries of the world: high parasite stress and associated high collectivism correspond to low levels of democratization, i.e., to high autocracy. The interrelationship among collectivism, parasite stress, and democracy across countries is supportive of the parasite-stress theory of the causes of morality. Compared to individualistic countries, collectivist ones exhibit greater and more widespread poverty, inequality, morbidity, and mortality as a result of the reduced investment by the governing elites in public welfare, health, infrastructure, education, and other public goods and services. This reduced investment by elites stems from the collectivist ideology of devaluing out-group members, valuing in-group members, and a general acceptance of human inequality (Chaps. 10 and 11).

We apply these basic components of the parasite-stress theory of values to each of the various types of intra-nation warfare and conflict as we take them up for analysis below.

## 12.3 Civil War

### *12.3.1 Letendre et al. (2010)*

Civil wars are a type of intra-nation war waged between a government of a country and an armed, organized group(s) within the same country that seeks control of the government or a region, or seeks to change governmental policies in ways that best suit the non-government group's ideological preferences (e.g., Fearon and Laitin 2003). In our cross-national study of civil wars, conducted in collaboration with Kenneth Letendre (2010), it was hypothesized that the combination of increased (1) resource competition (due to widespread economic dearth and inequality), (2) ethnocentrism and (3) xenophobia, characteristics of collectivist societies, cause an increased frequency of civil war. Individualistic nations, in contrast, experiencing less severe resource competition (more equitable resource and political power distribution and higher gross domestic product (GDP) per capita), less ethnocentrism and less xenophobia (more xenophilia), are less prone to civil war. In individualistic countries, within-nation, escalated intergroup conflicts involving a national government versus an armed out-group are less likely to arise, and, when they do, are more likely to be reconciled diplomatically without war. In contrast, in collectivist countries, such intra-country conflicts are more likely to arise and escalate to civil war. This view is consistent with Hofstede's characterization of a "high risk of domestic intergroup conflict" in collectivist societies as a key difference from individualist societies (Hofstede 2001, p. 251).

Furthermore, the ideology of collectivism promotes within-nation regional factionalism and resultant fractionation based on strong and localized preferences for certain values coupled with xenophobic attitudes toward non-local values. This is seen empirically in encompassing form in the strong positive relation between parasite stress and the number of religions and languages across countries of the world (Fincher and Thornhill 2008a, b; treated in detail in Chap. 13). High parasite stress and the collectivist values it evokes lead to ideological and linguistic boundaries within single regions that can cause new cultures to arise.

To test this perspective on civil war, Letendre et al. (2010) employed two data sets on civil-war outbreaks across countries: Fearon and Laitin's (2003) data on outbreaks in 157 countries in the years 1945–1999, and Strand's (2006) data on outbreaks in 177 countries in the years 1946–2004. Strand's (2006) data include small civil wars resulting in at least 25 battle deaths in 1 year, as well as large civil wars. Fearon and Laitin (2003) data, based on the Correlates of War Intra-state War data set (Singer and Small 1994), tallied major civil wars—those killing at least 1,000, with a minimum yearly average of 100 dead, and at least 100 killed on both sides. Hence, data were analyzed for civil wars across a range of magnitude in terms of mortality. Letendre et al. (2010) used *Contemporary Parasite Severity* (see Chap. 5 for description) as a measure of cross-national parasite adversity.

The parasite-stress theory of values applied to civil war was supported (Letendre et al. 2010). The statistical analyses and associated empirical results indicated that parasite severity positively predicted the frequency of civil-war outbreaks across the globe; this was found in separate analyses for small civil wars with relatively low mortality, as well in large civil wars with high mortality. The effect sizes were moderate to strong. Several potential confounds suggested in publications of earlier researchers to cause civil war were included in analyses, but the positive relationship between parasite stress and civil war outbreaks remained robust when potential confounds were considered. The potential confounds examined were national GDP per capita, economic growth, population size, democratization, and political instability.

Hendrix and Gleditsch (2012) criticized the conclusion by Letendre et al. (2010) that support was found for the hypothesis that high parasite stress causes civil war onsets. These critics pointed out that the measure of parasite stress used was based on contemporary infectious diseases (in 2007), but the civil war onsets were earlier (from 1945 to 2004). Hence, they claimed that Letendre et al. (2010) could not conclude that the parasite stress occurring before the wars thereby caused the wars. Our reply to this criticism is of three parts. First, it was pointed out in Letendre et al. (2010) that the measure of *Contemporary Parasite Severity* used correlated strongly with historical parasite stress, and that climate in a region maintains through time the basic ecological requirements for a region's parasite abundance. Second, Hendrix and Gleditsch's criticism, however, motivated us to look again at the civil war onset patterns, this time in relation to *Historical Parasite Severity* (a variable described in Chap. 5). The result is basically the same as with *Contemporary Parasite Severity*: a moderate and statistically significant positive effect is found with the historical measure, $r=0.30$, $p=0.006$, $n=85$ countries. Third, we present below results analyzing *Historical Parasite Severity* in relation to within-country wars other than civil war, and find that the frequencies of onsets of these wars are significantly predictable from *Historical Parasite Severity*.

Hendrix and Gleditsch (2012) also argue that civil war causes infectious disease problems, but not vice versa. We agree that civil war causes increased parasite adversity. This is part of the parasite-stress hypothesis of civil war, as discussed later in this chapter (also in Letendre et al. 2010, 2012). We, of course, reason as well that civil war is caused by parasite stress. Causation is bidirectional in the parasite-stress theory of values applied to civil war.

Hendrix and Gleditsch, too, propose that civil war is caused by "poor government and public goods provision by the state …" (p. 166). Certainly, the parasite-stress theory of values predicts a strong correspondence between inept government, low governmental interest in public goods, and civil war onsets. This correspondence, according to the parasite-stress theory, is from the encompassing causes of high parasite stress and the collectivist values it evokes. Hendrix and Gleditsch ignore the role of human values or preferences altogether, and hence do not see the necessity of a validated theory of values in explaining human conflict. They ignore, too, the fact that the actors in intergroup conflict are evolved animals with psychological adaptations that cause all their decisions, including the decisions that determine these conflicts. As we have emphasized throughout our book, this fact is the most fundamental intellectual starting point for serious thought about the causes of people's behavior.

Letendre et al. (2010) reviewed prior literature advocating hypotheses of civil war based on environmental variables and the distribution and competition for resources. That review shows that the parasite-stress theory of sociality, as applied to civil wars, integrates many diverse findings and hypotheses reported in the traditional political-science literature on the incidence of civil war. These issues are treated near the end of this chapter.

### 12.3.2 Collectivism

Here we report analyses of collectivism–individualism, a variable not included in analyses in Letendre et al. (2010), as that paper looked only at the relationship between parasite adversity and civil war onsets. The parasite-stress theory of values predicts that civil-war frequencies across nations will not only be related positively to parasite stress, but also will show the same relationship to collectivism. This is the case. Fearon and Laitin's (2003) data show the following relationships with collectivism–individualism: with *Gelfand In-group Collectivism*, $r = 0.46$, $p = 0.0004$, $n = 56$; with *Suh Individualism*, $r = -0.46$, $p = 0.\ 0004$, $n = 55$. The relationship of civil war events with *Hostede Individualism* is similar, but shows a smaller effect ($r = -0.33$, $p < 0.01$, $n = 70$). All effect sizes are moderate in magnitude. (These measures of collectivism–individualism are described in Chap. 5.)

In sum, as predicted by the parasite-stress theory of values, civil wars are more frequent in countries with high parasite stress and collectivist values than in countries with low parasite stress and individualist values.

## 12.4 Other Intra-Nation Intergroup Conflicts

Civil war is not the only type of domestic/intra-nation intergroup conflict. There are additional types of these conflicts that the parasite-stress theory of values is expected to illuminate. Next, we explore extensions of the parasite-stress theory of values to frequencies of (1) non-state-government wars, hereafter "non-state wars," i.e., intergroup, within-country conflicts in which the federal government is not a combatant; (2) political coups; and (3) revolutions. As with civil war, all three of these additional intra-country conflicts derive from major differences in ideological preferences among groups within a nation. In non-state wars, organized groups, such as clans or tribes of ideological collectives, war against one another, and the national government is not a combatant. A coup (also called a coup d'état) occurs when a national government is suddenly usurped and replaced by a faction (often the military) of the same government. Revolutions, like coups, involve efforts to accomplish regime changes, but over longer periods of time and involving social transformation of the old government by a considerable segment of the society.

Letendre et al. (2012) hypothesized that these three types of intrastate conflicts arise, at least in part, from elevated out-group intolerance and devaluation, and in-group alliance and cooperation, and hence will be most frequent in nations with high parasite stress and related high collectivism. Specifically, for each of these three types of conflict, the parasite-stress theory of values predicts that parasite adversity and collectivism will correlate positively with the counts of events across countries, and that individualism will correlate negatively with the frequency of each of the three categories of conflicts. The analyses in Letendre et al. (2012) supported these ideas by showing that parasite stress was significantly related to each of the three types of intrastate conflicts when the same potential confounds mentioned above for analysis of civil wars in Letendre et al. (2010) were accounted for: GDP per capita, economic growth, population size, democratization, and political instability.

Letendre et al. (2012) also examined the application of the parasite-stress theory to a measure of peace, the Global Peace Index, across countries. The measure combines information about the presence or absence of internal and external war across many nations. The external war component of the peace measure allows the preliminary study of the parasite-stress theory's application to international warfare. From the parasite-stress theory, it is expected that, across countries, as parasite stress and collectivism decrease, peace will be more prevalent. Letendre et al. (2012) reported support for this hypothesis. That study showed in a path analysis that peace across countries increases as parasite stress declines.

In this chapter, we employ a fuller range of parasite-stress and collectivism–individualism variables in analyses of non-state wars, coups, revolutions and the

Global Peace Index than used in Letendre et al. (2012). Three measures of variation in parasite stress across countries are used here. They are positively and highly intercorrelated but assess somewhat different aspects of human infectious-disease stress. *Disease Richness* is the number of infectious diseases per country in a contemporary context. *Contemporary Parasite Severity* is the severity of a set of important human parasites in a current context. *Historical Parasite Severity* is the severity of a set of important human parasites in earlier decades back to the early 1900s. Hence, our measures of parasite stress cover the contemporary and the historical infectious-disease problems across countries (these disease variables are explained fully in Chap. 5). In Letendre et al. (2012) only *Contemporary Parasite Severity* was used to measure parasite adversity. The three collectivism measures we use in this chapter are also described in Chap. 5. Letendre et al. (2012) used *Strength of Family Ties* as the collectivism measure and only in an analysis with the Global Peace Index; this collectivism measure is also described in Chap. 5.

## 12.5 Non-State Wars

### *12.5.1 Methods*

As in Letendre et al. (2012), we used data on non-state war occurrences from the Uppsala Conflict Data Program (UCDP) WWW site; UCDP Non-State Conflict Dataset V.1.1, 2002–2005 at http://www.pcr.uu.se/research/UCDP/index.htm. This is a cross-national (255 countries) dataset with information about armed conflict onset between two organized groups within a country, neither of which is the government of the country, resulting in at least 25 battle-related deaths in a calendar year; both military and civilian deaths are counted as battle-related deaths. Hence, these non-state wars are a different type of conflict than civil wars, as the latter always involve the government of a state versus an organized warring group(s) within that state. This dataset lists 24 countries with at least one non-state war onset over the period of 2002–2005. There were a total of 125 such conflicts, and countries varied from 0 to 28 conflicts. These wars are escalated inter-ethnic or clan wars; examples are in Uganda, the Pokot clan versus the Sabiny clan; Syria, Arabs versus Kurds; Somalia, the Jareer subclan of the Hawiye clan versus the Jiddo subclan of the Digil clan. Our variable is the sum of the non-state war onsets per country over the period 2002–2005.

We use here two kinds of analyses to investigate non-state wars. First, countries were coded for presence (1) or absence (0) of a non-state war over the 2002–2005 period. This analysis was performed because of the large percentage of countries with zeros. For each of our hypothetical causal variables, the difference between the means of countries with non-state conflict present versus absent was tested by a *t*-test. The second kind of analysis was regression between the total number of non-state wars per country over 2002–2005 by the hypothetical causal variables. To reduce skew, the number of non-state wars was log-transformed.

### 12.5.2 Findings

As predicted by the parasite-stress theory of values, each of the three parasite-stress measures was associated with non-state war presence versus absence across countries: *Disease Richness*, $t=7.10$, $df=227$, $p<0.0001$, mean, std. dev., and $n$ for war present 220.50, 14.60 and 24, and for war absent 198.34, 13.53, 205; *Contemporary Parasite Severity*, $t=6.75$, $df=223$, $p<0.0001$, for war present 39.63, 4.74 and 24 and for war absent 30.53, 6.39, 201; *Historical Parasite Severity*, $t=6.35$, $df=91$, $p<0.0001$, for war present 0.70, 0.39, 13, and for war absent –0.12, 0.63, 80. In sum, high parasite stress corresponds to the presence of non-state wars, whereas low parasite stress corresponds to the absence of non-state wars.

Also as predicted, correlation analysis reveals significant positive covariation between each of the three measures of parasite stress and number of non-state wars across the countries. Each of the three relationships is highly significant by linear regression ($p<0.0001$) ($r$'s, 0.42–0.43) and is improved significantly by polynomial degree-2 analysis ($t$-ratio probability $\leq 0.005$). The polynomial analysis indicates that the number of non-state wars rises most rapidly at high parasite stress. Hence, parasite stress and number of non-state wars across countries are positively related with moderate effect sizes across the three measures of parasite stress.

As predicted also, collectivism is associated positively (individualism, negatively) with the number of non-state wars: *Gelfand In-group Collectivism*, $r=0.24$, $p=0.06$, $n=62$; *Hofstede Individualism*, $r=-0.27$, $p=0.02$, $n=70$; *Suh Individualism*, $r=-0.37$, $p=0.003$, $n=61$. Although the conventional level of statistical significance ($p=0.05$) is not reached in the relationship with *Gelfand In-group Collectivism*, this pattern is significant ($p=0.03$) with one-tailed probability, which is appropriate given the a priori prediction of the direction of the relationship. Effect sizes are small to moderate in magnitude.

## 12.6 Revolutions and Coups

### 12.6.1 Methods

As in Letendre et al. (2012), we used the Barro–Lee Dataset for a panel of 138 countries (Barro and Lee 1994). The Barro–Lee variable used was REVCOUP, which they define as "… [T]he number of revolutions and coups per year, averaged over the period 1960–1984." The source for these data is Banks (1979, updated). (This variable was used also in a recent cross-national analysis of political instability by Nettle et al. 2007.) The events appear to reflect the standard definitions of revolutions and coups as used in political science and described above.

### *12.6.2 Findings*

As predicted, countries with high parasite stress are more subject to civil political violence in the form of revolutions and coups than are countries with low parasite stress. The number of revolutions and coups correlated significantly and positively with each of the three parasite-stress measures: *Disease Richness*, $r = 0.23$, $p = 0.009$, $n = 131$; *Contemporary Parasite Severity*, $r = 0.39$, $p < 0.0001$, $n = 131$; *Historical Parasite Severity*, $r = 0.40$, $p < 0.0003$, $n = 77$. With the exception of the small effect size for *Disease Richness*, the effect sizes are intermediate in magnitude.

Collectivism covaries significantly across countries with the number of revolutions and coups in the direction predicted: *Gelfand In-group Collectivism*, $r = 0.37$, $p = 0.008$, $n = 50$; *Hofstede Individualism*, $r = -0.45$, $p < 0.0004$, $n = 58$; *Suh Individualism*, $r = -0.48$, $p = 0.0007$, $n = 46$. Hence, high collectivism (low individualism) corresponds to more frequent occurrences of revolutions and coups, whereas low collectivism (high individualism) corresponds to less frequent occurrences of these domestic conflicts. Effect sizes are intermediate in magnitude.

## 12.7 Peace

### *12.7.1 Methods*

As in Letendre et al. (2012), we used the Global Peace Index for 2008, collated and calculated by the Economist Intelligence Unit. The Index was available for 140 countries and is comprised of 24 qualitative and quantitative indicators, which combine factors pertaining to countries' relative peace status. The Index prioritizes measures of an absence of violent conflicts with neighboring countries and of internal wars. The 24 indicators include: political instability, relations with neighboring countries, the number of external and internal conflicts fought between 2000 and 2005, the number of deaths from both external and internal conflict, military expenditures, potential for terrorist acts, and homicide rate. The Index ranges from 1 to 5, where 1 is the most peaceful and 5 the least peaceful. Iceland in the most peaceful, with a score of 1.176; Iraq is the least peaceful, with a score of 3.514. Data and descriptions of ranking methods used are at http://www.visionofhumanity.org.

### *12.7.2 Findings*

Across the world, the relative peacefulness per country shows the patterns predicted by the parasite-stress theory of values. The lowest scores on the Global Peace Index correspond to relatively high peace, so we expected a positive relationship between parasite-stress and the Global Peace Index. The relationships of the Global Peace Index with each of the parasite-stress variables are: *Disease Richness*, $r=0.44$, $p<0.0001$, $n=140$; *Contemporary Parasite Severity*, $r=0.54$, $p<0.0001$, $n=140$; *Historical Parasite Severity*, $r=0.57$, $p<0.0001$, $n=89$. Thus, the higher the parasite stress, the lower the peacefulness. These effect sizes are intermediate to strong in magnitude.

Moreover, the predicted patterns were seen with the three values' measures and the Global Peace Index: *Gelfand In-group Collectivism*, $r=0.51$, $p<0.0001$, $n=56$; *Hofstede Individualism*, $r=-0.43$, $p<0.0003$, $n=66$; *Suh Individualism*, $r=-0.49$, $p<0.0002$, $n=54$. Effect sizes were intermediate to strong. Thus, across nations, we found that the higher the individualism (or the lower the collectivism), the higher the peacefulness in a country.

## 12.8 Discussion and Conclusions

The overall findings are that, across the countries of the world, the parasite-stress theory of values provides an empirically fruitful theory for major types of intergroup, within-nation political conflicts. As predicted, the number of events of civil wars, non-state wars (i.e., clan, tribal, and ethnic wars), and political revolutions and coups covaried with parasite stress and collectivism (and, hence, individualism); the predicted relationships also were seen with a measure of peacefulness, the absence of internal and external conflict. Specifically, in countries with higher levels of parasite stress and collectivism, there were a larger number of civil-war onsets, non-state war onsets, and revolutions and coups. As well, in countries with higher levels of parasite stress and collectivism, peacefulness, as measured by the Global Peace Index, was lower. Our analyses of the Global Peace Index across countries allow preliminary examination of the parasite-stress theory in relation to international political conflicts because such conflicts are a component of this index. All results indicate that the parasite-stress theory is a useful way to understand major political conflicts of all types.

We now turn to a more detailed discussion of the parasite-stress theory of values as applied to intergroup political conflicts in order to further clarify this application. Then we turn to some additional considerations arising from the empirical findings reported above.

## 12.9 Parasite Stress and Civil Conflict: Further Clarifications

The parasite-stress theory of values proposes that civil political conflicts can be understood as follows. They are caused by behavioral and psychological features functionally designed for (i.e., directly selected in the context of) intragroup embeddedness, cohesion and cooperation (i.e., in-group assortative sociality) critical in defense against coercion and aggression by out-groups and in offensive coercion and aggression against out-groups. In this regard, the psychology of collectivism is causal, because it is designed (a) for distinguishing group boundaries through collective adherence to shared in-group values and norms and, hence, for identifying out-groups, and (b) when combined with xenophobia, for producing negative feelings (dislike, disgust) toward out-groups. Collectivist emotions and behavior are designed, too, for investment in, support of, and loyalty toward in-group members comprised of extended family and other group members with the same values/morals. Collectivism is an interdependency on and a high valuation of in-group members with a simultaneous devaluation and avoidance of out-group members in conjunction with xenophobia.

Much of collectivism is appropriately cast as in-group cooperation. Therefore, collectivism is the basis of success in both defensive and offensive out-group hostility. The degree of cooperation achieved among members of a warring group—whether a raiding party of relatively egalitarian hunter–gatherers or a highly hierarchical army—is recognized widely as critical for effectiveness in warring (e.g., see Buss 2004 discussion of warfare). Coalitional aggression against out-groups is pursued almost exclusively by men (Wrangham and Peterson 1996). However, the people (including women) not participating directly in warring importantly provide moral support and associated assistance, which, like the amity among members of the warring coalition, is promoted by collectivist ideology.

There is increasing evidence of condition-dependent psychological adaptation in men that is functionally designed for war. Its information-processing capacities include an assessment of benefits from war in the form of access to women and other resources, as well as an assessment of coalitional support and strength of own versus enemy group (Duntley and Buss 2008). This adaptation may have been directly sexually selected in the context of men's competition for women and the status and related resources that can give access to multiple sexual partners (Low 1993; Wrangham and Peterson 1996; Buss 2004). In the parasite-stress theory as applied to war, the war adaptation interacts with the psychological adaptation for adopting and using human values such as xenophobia and in-group allegiance (collectivism) to result in the decision that war is the appropriate means for dealing with intergroup conflict. Hence, warfare is caused partly by war adaptation in men and partly by collectivist values.

Furthermore, the parasite-stress theory proposes that the psychology of collectivism and of war give rise to a major political conflict when the perceived benefits of intergroup conflict exceed its high costs. One benefit of the pursuit of conflict is access to resources in the event of a victory over the out-group. Another benefit is the exclusion of the out-group from the region and, in some cases, out-group extermination. Although warfare may expose warriors to the risk of contracting new diseases from the enemy during combat, warfare may reduce importantly future intergroup contact and interaction. According to the parasite-stress theory, during human evolutionary history, this future reduction of intergroup contact and interaction provides inclusive fitness benefits greater than the cost of contracting infectious diseases during combat. Hence, in this view, the xenophobia that motivated intergroup aggression had, as its net effect, an avoidance of problems resulting from infectious diseases.

In the context of infectious disease, past selection created a condition-dependent moral psychology—the psychological adaptations that manifest in human cognitions and behavior as collectivism and associated xenophobia and ethnocentrism, or as individualism. The moral psychology is designed to incorporate values during development (ontogeny) by learning socially those values well suited to local parasite prevalence. We have discussed some possible ontogenetic ancestral cues that may guide historically adaptive construction of individuals' moral repertoires (Chap. 3). High parasite stress causes a willingness to accept the costs of intergroup conflict, whereas low parasite stress builds pacifism and other positivism toward out-groups. The parasite-stress theory, then, may explain much of the variation in the values affecting within-nation conflicts as well as international conflicts across the globe.

In the parasite-stress theory, the following are proximate causes of political conflict, as well as its antipole, pacifism/absence of such conflict: the war psychological adaptation, the moral psychological adaptation, collectivism–individualism, the psychology that assesses local parasite stress, and the ontogenetic events involved in the production of all this phenotypic machinery. Of course, the ontogeny includes the important role of social learning of values within and across generations, which gives rise to what some researchers call "cultural evolution," referring to changes in the frequencies of ideas, values and related behavior (Richerson and Boyd 1998; also see Chap. 2). The ontogeny of the social-learning machinery, like the ontogeny of all phenotypic features, is causally dependent on genes as a partial proximate cause.

As explained earlier in the book, we use the concept of "cause" in its typical, scientific sense: that, without which, an effect will not occur. Each proximate cause listed above is necessary, but insufficient alone, to generate political conflict. Each is a partial cause; again, using the standard conception of cause in science.

By definition, proximate causes are those that act to generate an effect within the lifetime of the organism. Each piece of machinery comprising the above list of proximate causes is the product of evolutionary historical causation, i.e., ultimate causation. We have treated only the selection history of this machinery and ignored phylogenetic ultimate causation, a distinct and complementary causal framework that addresses the location on the Tree of Life where traits first appeared in the history of life (Chap. 2). According to the parasite-stress theory, the selection that built all the proximate causes (listed above) was direct selection in the context of parasite stress, or in the case of the war adaptation, direct sexual selection for condition-dependent warring behavior.

In regard to testing, the parasite-stress theory of political conflicts predicts (i.e., requires for its support) that the frequency of political conflicts across countries will show a positive correlation with parasite stress and collectivism (and a negative correlation with individualism). If these patterns are not seen, the theory is false; the findings to date reported herein and in Letendre et al. (2010) and Letendre et al. (2012) support the theory.

The civil-conflict literature is voluminous, especially with regard to civil war (partial reviews in Alesina et al. 1996; Hegre and Sambanis 2006; Nettle et al. 2007; Abadie and Gardeazabal 2008; Sosis and Alcorta 2008). This literature proposes various causes for these conflicts, as well as numerous tests of these causes. Often, the factors of population size, GDP per capita, Gini (wealth inequality), time since last conflict (in the case of civil war), inconsistent democratic institutions, political instability, war-prone and undemocratic neighboring countries, ethnic diversity, and a low rate of economic growth are considered to be basic causes of such conflicts. Also, typically in research looking at one or a few of these variables that predict conflict, some of the other variables are considered confounds and, hence, statistically controlled.

In the parasite-stress theory of values, however, all the variables just mentioned are effects of the same underlying cause—parasite stress. Even increased population size, which is correlated positively with the frequency of within-country conflicts (e.g., Hegre and Sambanis 2006), may be an effect of parasite stress in many parts of the globe, because infectious disease is correlated positively with birth rate cross-nationally (Guégan et al. 2001). This correlation, we argue, is the result, in part, of moderate parasite stress acting as an intrinsic mortality factor promoting a high reproductive rate associated with collectivist extended family nepotism and the result, in part, of extreme parasite stress as an extrinsic mortality factor promoting even higher reproduction (fast-track life history strategy) (Chap. 14.6). Separately, we have treated in detail how GDP per capita, economic growth, and democratization are predicted consequences of the parasite-stress theory: upon relative emancipation from infectious diseases, peoples' values become more individualistic or liberalized, which results in greater economic productivity and investment in public goods and services and the welfare of out-groups in general (Chaps. 10 and 11). The relationship between ethnic or cultural diversity and parasite stress is treated in Chap. 13. Furthermore, pertaining to the variable "war-prone neighbors," we propose that frequently there is a spatial autocorrelation in domestic-conflict events among countries in a geographical region, because there are regional differences in the ecological conditions (e.g., rainfall and temperature) affecting parasite stress (Chap. 3). We propose, too, that the variable "time since last civil conflict" is an important effect of the parasite-stress theory. Hence, the parasite-stress theory cannot be tested appropriately by controlling statistically these variables. For example, to control for GDP per capita and/or democratization in an analysis of, say, parasite stress and non-state wars would reduce the ability to detect the predicted relationships, because GDP per capita and democratization are consequences and causes of parasite stress: low GDP per capita and low democracy derive from high parasite stress (and associated collectivist values) and feedback to increase parasite stress, and high GDP per capita and high democracy result from low parasite stress (and associated individualistic values) and feedback to reduce parasite stress. (On the bidirectional relationship between parasite adversity and values, see Chap. 10.) Although some of these variables were statistically controlled in analyses in Letendre et al. (2010) and (2012) it is important to realize that the statistics obtained with such controls are hard to interpret, given that the controls used are effects and in some cases causes of parasite stress and values. (See also the discussion of the partialling fallacy in Chap. 5.)

Hegre and Sambanis (2006) point out that published analyses of civil conflicts across countries are highly variable in specifications of relevant statistical-control variables. They then say this is because "[W]e do not know the true model" (p. 513). The models of human activity that are most general and useful for scientific discovery are those based in the evolutionary science of human functional design (examples are the empirical cornucopia from Hamilton's model of nepotism; Trivers' model of direct reciprocal altruism; Alexander's model of indirect reciprocity and reputation; see Chap. 2). Hence, the most encompassing and useful models of human political conflict will be those based in human mental functional design resulting from an evolutionary history of selection for inclusive fitness maximization; hypotheses ignoring evolved mental adaptations are of limited scientific value. In the long tradition of political conflict research, there is no generally accepted model, because the research has not been inspired by evolutionary theory. This chapter is an attempt to identify a general model of intergroup conflict that is inclusive of all the values and their effects that arise under high parasite stress as well as under emancipation from parasite stress. These values interact with men's psychological war adaptation.

The parasite-stress theory does not suggest that there is evolved adaptation that functions specifically in the context of one or more of the various types of political conflicts we have addressed in this chapter. Hence, there is no adaptation functionally designed for civil war per se, or for coups. Instead, the theory implies that these conflicts are manifestations of moral psychological adaptation designed for historically adaptive in- and out-group relations coinciding with the level of local parasite stress.

Moreover, the theory does not imply or require that these conflicts are adaptive currently. For example, it does not predict that civil war, on average, has a net benefit in promoting inclusive fitness of the warriors. From the theory, the adaptive value is in terms of the moral psychological adaptation, and solely in evolutionary historical environmental settings that caused its evolution by direct selection. Modern human environments often differ greatly from the evolutionary historical settings that were responsible ultimately for the effective selection of human traits. Each of the types of conflicts we have treated may be currently adaptive or maladaptive at the individual level, depending on the circumstances.

As mentioned above, the parasite-stress theory of intergroup conflict is compatible with the proposals in the scientific literature that men's sexually selected pursuits of high mate number affect positively men's decisions to engage in coalitional aggression (e.g., Low 1993; Wrangham and Peterson 1996; Buss 2004). Low (1990) reported that, across traditional societies, polygynous marriage systems are more frequent in geographical regions of high parasite stress than in regions of low parasite stress (see Chap. 6). She also found that, in traditional societies, wife-capture from neighboring groups by warring men is most frequent under high parasite stress. These findings support Low's hypothesis that high parasite stress intensifies sexual selection on males (i.e., increases the variance among men in access to mates with whom children are produced). In this case, the sexual selection intensity stems from parasites generating high phenotypic and associated genetic variance in male quality that is visible to females during mate choice. Hence, parasite stress, through its effect of enhancing polygyny and associated limitations on men's access to mates, may generate a net benefit of intergroup aggression to obtain out-group mates.

Klavina et al.'s (2011) recent study supports our thinking that collectivism is a cause of intergroup conflict and war. In Chap. 6, we discussed the evidence in that study indicating that collectivist men, compared to individualistic men, are more concerned about out-group men taking their mates. This concern of collectivist men is part of their prejudice against out-group men and may contribute to the relationship between collectivism and intergroup coalitional aggression as well as interpersonal aggression.

We emphasize that the parasite-stress hypothesis of intergroup political conflict is consistent with an important role for nepotistic coalitions in warring decisions (e.g., Low 1993). Nepotistic adaptation is central to collectivism and hence, as we have explained, to cooperation in warring. Also, as we have stressed, collectivism is more than nepotism: it includes in-group assortative favoritism toward others with like values but who are not genetic relatives. Both of these aspects of collectivism, according to the parasite-stress theory of values and its empirical tests (Chap. 5), are related causally and positively to parasite stress.

## 12.10 Limitation of Our Findings on Intra-Nation Conflicts

Our treatment of international war, in relation to the parasite-stress theory, is quite preliminary, as we could not separate the Global Peace Index components pertaining to intra-nation conflict versus inter-nation war.

## 12.11 Prospects for Eliminating Civil Conflicts

Letendre et al. (2010) discuss evidence that, since about the end of World War II, civil wars have killed six times as many soldiers as international wars. If, over this period of time, intra-nation wars other than civil wars are added, the number of warrior deaths from civil conflicts in general would be even more in excess of the number resulting from international wars. Of course, warrior deaths from these conflicts are only part of the mortality they cause. Often in civil conflicts of all types, noncombatants suffer high mortality as well (Ghobarah et al. 2003). We assume that many people would agree on moral grounds that less civil conflict is conducive to a better world than is more civil conflict. The solution to attaining this moral goal offered by the findings in this chapter is to reduce parasite stress across the world. According to the parasite-stress theory of sociality, foreign aid in the forms of sanitation infrastructure, medical assistance, health education and other means of reducing parasite adversity would reduce the incidence of civil conflicts in the geographical areas of high current conflicts by shifting values toward more liberalism while simultaneously creating democratic institutions.

With K. Letendre, we have proposed a disease trap, by which nations in regions with high levels of infectious disease become trapped in self-reinforcing poverty (Letendre et al. 2010; also see Bonds et al. 2010, 2012). As we explained in Chap. 11, infectious disease depresses economic development through its negative effects on human capital. Additionally, as we have discussed, the xenophobia evoked by high intensity of infectious disease further diminishes the willingness of people to invest in public goods and services that are shared across groups, such as economic and health infrastructure, municipal clean-water sources and sanitation systems; xenophobia also diminishes the willingness to engage in transactions with neighboring groups for useful ideas, goods and technologies. Moreover, the violent conflict that erupts among impoverished collectivist groups who are unwilling to seek cooperative solutions when inter-group competition arises further compounds the infectious disease problem as violent conflict causes death, disability, and disease beyond those killed directly in conflict (Ghobarah et al. 2003).

Other researchers have recommended international aid targeted at building economic institutions in conflict-ridden regions (e.g., Elbadawi and Sambanis 2000). Considering the effects infectious diseases have on societies, we instead recommend international aid be targeted at the control and elimination of these diseases. Based on our findings in this chapter and in Letendre et al. (2010, 2012), we advocate that this sort of targeted aid has the greatest potential to get directly to the root cause of poverty and civil conflict, and to disrupt the infectious disease trap that locks billions of people into poverty and civil conflict. Dunn et al. (2010) found that governmental investment in public health significantly and negatively affects the number of cases of human parasitic diseases; thus, there is evidence that such directed investment in public health can be effective.

We also stress that it is in countries with a high intensity of infectious disease that foreign aid directed at economic development may be most frequently misappropriated for the personal benefit of corrupt government officials. In Chap. 11, we showed that parasite stress correlates strongly and positively with governmental corruption cross-nationally. It is widely recognized by scholars that foreign aid directed to corrupt governments is largely wasted (e.g., Burnside and Dollar 2004; Easterly and Pfutze 2008); yet, because the adversity of infectious disease causes both poverty and the establishment of autocratic and corrupt regimes, foreign aid directed at alleviating this poverty is necessarily directed toward countries where it is most likely to be misappropriated or otherwise squandered by corrupt governments. We suggest, too, that direct economic aid may be more readily misappropriated, whereas aid directed at diminishing the intensity of infectious disease, such as delivery of vaccines or the construction of sewage-treatment facilities and municipal water systems, may be more likely to provide the intended benefit to the people of these countries.

## 12.12 The American Civil War

As discussed in Chap. 1, the American Civil War (1861–1865) has received much attention from scholars. There we claimed that our analysis of civil conflicts would elevate understanding of this war by placing it in the context of a general hypothesis of civil conflict as provided by the parasite-stress theory of values. At the very minimum, our findings tell scholars of this war that they should study the parasite-stress theory of values and its diverse empirical discoveries, including those pertaining to civil conflicts. Most generally, evidence indicates that the American Civil War, like other such wars wherever and whenever they occur, was caused ultimately by evolution by selection that favored psychological features that defend against infectious disease and was caused proximately by the collectivism that high parasite stress evokes. That is why the high parasite region of the USA was antagonistic toward the North and its ways and values and then seceded from the United States forming the Confederate States of America that fought for its independence from the United States. The clash of different regional values is now understandable as arising from the difference in parasite stress between the North and South and the region-specific values evoked as a result. This provides clarity as to why slavery was a major political agenda in the war. Inequality of people is a part of collectivist morality whereas equality is individualist morality.

The parasite-stress theory also explains other topics on which there is frequent speculation among researchers interested in the American Civil War. The reason the South lost the war was because of collectivist values and associated high parasite stress. The American Civil War was an early industrial war. It involved railroads, telegraph, ships, sophisticated weaponry, and other technology. Collectivist values limited the South's technology, because of the reduced innovations and openness to new technologies associated with these values. The more innovative and technologically advanced North was sure to win from the beginning. And why did the South continue to fight long after it was clear that defeat and surrender were inevitable? Collectivism's dutiful rigidity toward the values and goals of the in-group provides a fundamental answer that is corroborated by the diverse evidence discussed throughout this book.

## 12.13 Team-Sport Competition in the USA

Team sports are based on intergroup conflict. The broad applicability of the parasite-stress theory of intergroup conflict is seen not only in the range of such conflicts discussed above, but also in its ability to predict the diversity of sports teams in a region. Dan Colman, a doctoral student at The University of New Mexico, first saw the relevance of the parasite-stress theory for study of the regional diversity of team sports (D. Colman, unpublished manuscript, May 7, 2011). Team athletic programs from children's sports through high school, college and university to the professional level emphasize the in-group's mission of winning over that of individual team member's success. Training of these athletes at all levels prioritizes the group's unity for the success of the collective. Individualism is strongly discouraged. The training promotes collectivist values of embeddedness in the team as the means of success in intergroup battles or games. Team supporters or fans join the team players to influence and celebrate wins by their favorite team. Team sports arise from collectivist values and, as we have shown, collectivist values are regionally local in scope. Collectivism involves parochial embeddedness in a local collective and its boundary from outlander groups. This same kind of thinking led to our research with Kenneth Letendre on intergroup warfare, discussed above.

Given these considerations, sports team involvement, support and diversity are anticipated to covary positively with collectivism (negatively with individualism) and parasite stress across certain regions. Dan Colman examined the diversity component of this hypothesis as applied to collegiate football, basketball and baseball teams (Division 1 and 2 teams) across the states of the USA in 2010. (Data are at the National Collegiate Athletic Association's data and statistics web site: http://www.ncaa.org/wps/wcm/connect/public/NCAA/Resources/Stats/). As predicted, he found, across the 48 continental states, that the total number of such teams per state shows a significant, positive relationship with Vandello and Cohen's (1999) collectivism ($r=0.44$, $p=0.002$) and *Parasite Stress USA* ($r=0.47$, $p=0.0007$). (These measures of collectivism and parasite-stress are described in Chap. 5.) As expected, the number of these teams per state correlates strongly and positively with state population size. (Data for the year 2000 are at http://www.census.gov/popest/) Both collectivism and *Parasite Stress USA*, however, show a significant relationship with the number of the teams per state when population size per state is controlled statistically (partial $r$ for collectivism$=0.29$, $p=0.05$; for *Parasite Stress USA*, $r=0.41$, $p=0.004$). The evidence discussed indicates that collectivism and parasite stress promote local in-group ethos and boundary, which yield more team-sport teams as each of these variables increases.

## 12.14 Summary

This chapter documents the applicability of the parasite-stress theory of values to the frequencies of the major types of within-nation intergroup conflict across contemporary countries: civil wars, non-state wars (intrastate wars in which warring groups do not include the government of the state), and coups and revolutions. Collectivist values of people promote interdependence with, and loyalty toward, in-group members (ethnocentrism) and goals but antagonism toward out-group members (xenophobia) and goals. Host–parasite antagonistic coevolutionary races produce variation among regions in the specificity of immune defenses and of parasites. According to the parasite-stress theory, the collectivist values of ethnocentrism and xenophobia are defenses against novel infectious diseases harbored in out-groups and to which local people are not adapted. From this, we proposed that high parasite stresses and associated collectivist values, then, promote all the major types of within-region civil conflict. As predicted, based on this, the frequency of civil wars, non-state wars, and coups and revolutions are associated positively with parasite stress and collectivism across countries of the world; peacefulness shows the predicted negative relationships with parasite stress and collectivism. These findings indicate that occurrences of civil conflicts would be reduced by reducing parasite stress and associated collectivist values.

The parasite-stress theory of values provides a general causal model of intergroup conflict. The American Civil War is revisited in light of this general model. We show, too, that the parasite-stress theory of values applies to coalitional conflict as seen in team sports. We discuss how the parasite-stress theory of intergroup conflict relates to other hypotheses for coalitional aggression.

## References


Abadie, A., & Gardeazabal, J. (2008). Terrorism and the world economy. *European Economic Review* 52: 1–27.

Alesina, A., Ozler, S., Roubini, N. et al. (1996). Political instability and economic growth. *Journal of Economic Growth* 1: 189–211.

Banks, A. S. (1979, updated). Cross-national Time Series Data Archive, Center for Social Analysis, State University of New York at Binghamton. Sept. 1979, updated.

Barro, R. J., & Lee, J. W. (1994). Data set for a panel of 138 countries. http://www.nber.org/pub/barro.lee/.

Bonds, M. H., Dobson, A. P., & Keenan, D. C. (2012). Disease ecology, biodiversity, and the latitudinal gradient in income. *PLoS Biology* 10: e1001456.

Bonds, M. H., Keenan, D. C., Rohani, P. et al. (2010). Poverty trap formed by the ecology of infectious diseases. *Proceedings of the Royal Society B* 277: 1185–1192.

Burnside, C., & Dollar, D. (2004). Aid, policies, and growth: Reply. *American Economic Review* 94: 781–784.

Buss, D. M. (2004). *Evolutionary Psychology: The New Science of the Mind*, 2nd ed. Allyn and Bacon, Boston, MA.

Dunn, R. R., Davies, T. J., Harris, N. C. et al. (2010). Global drivers of human pathogen richness and prevalence. *Proceedings of the Royal Society B* 27: 2587–2595.

Duntley, J. H., & Buss, D. M. (2008). The origins of homicide. In *Evolutionary Forensic Psychology: Darwinian Foundations of Crime and Law* (eds. J. Duntley & T. Shackelford), pp. 41–64. Oxford University Press, New York, NY.

Easterly, W., & Pfutze, T. (2008). Where does the money go? Best and worst practices in foreign aid. *Journal of Economic Perspectives* 22: 29–52.

Elbadawi, I., & Sambanis, N. (2000). Why are there so many civil wars in Africa? Understanding and preventing violent conflict. *Journal of African Economies* 9: 244–269.

Fearon, J. D., & Laitin, D. D. (2003). Ethnicity, insurgency, and civil war. *The American Political Science Review* 97: 75–90.

Fincher, C. L., & Thornhill, R. (2008a). A parasite-driven wedge: Infectious diseases may explain language and other biodiversity. *Oikos* 117: 1289–1297.

Fincher, C. L., & Thornhill, R. (2008b). Assortative sociality, limited dispersal, infectious disease and the genesis of the global pattern of religion diversity. *Proceedings of the Royal Society of London, Biological Sciences* 275: 2587–2594.

Ghobarah, H. A., Huth, P., & Russett, B. (2003). Civil wars maim and kill people—long after the shooting stops. *American Political Science Review* 97: 189–202.

Global Peace Index. (2008). http://www.visionofhumanity.org/gpi/results/rankings.php

Guégan, J. -F., Thomas, F., Hochberg, M. E. et al. (2001). Disease diversity and human fertility. *Evolution* 55: 1308–1314.

Hegre, H., & Sambanis, N. (2006). Sensitivity analysis of empirical results on civil war onset. *Journal of Conflict Resolution* 50: 508–535.

Hendrix, C. S., & Gleditsch, K. S. (2012). Civil war: Is it all about disease and xenophobia? A comment on Letendre, Fincher and Thornhill. *Biological Reviews* 87: 163–167.

Hofstede, G. (2001). *Culture's Consequences: Comparing Values, Behaviors, Institutions, and Organizations Across Nations*, 2nd ed. Sage Publications, Thousand Oaks, CA.

Klavina, L., Buunk, A. P., & Pollet, T. V. (2011). Out-group mating threat and disease threat increase implicit negative attitudes toward the out-group among men. *Frontiers in Psychology* 2: 1–8.

Letendre, K., Fincher, C. L., & Thornhill, R. (2010). Does infectious disease cause global variation in the frequency of intrastate armed conflict and civil war? *Biological Reviews* 85: 669–683.

Letendre, K., Fincher, C. L., & Thornhill, R. (2012). Infectious disease, collectivism, and warfare. In *The Oxford Handbook on Evolutionary Perspectives on Violence, Homicide, and Warfare* (eds. T. Shackelford & V. Weekes-Shackelford), pp. 351–371. Oxford University Press, New York, NY.

Low, B. S. (1990). Marriage systems and pathogen stress in human societies. *American Zoologist* 30: 325–339.

Low, B. S. (1993). An evolutionary perspective on war. In *Behavior, Culture and Conflict in World Politics* (eds. W. Zimmerman and H. K. Jacobson), pp. 13–55. University of Michigan Press, Ann Arbor, MI.

National Collegiate Athletic Association, http://www.ncaa.org/wps/wcm/connect/public/NCAA/Resources/Stats/

Nettle, D., Grace, J. B., Choisy, M. et al. (2007). Cultural diversity, economic development and societal instability. *PLoS* 9: 1–5.

Richerson, P. J., & Boyd, R. (1998). The Evolution of Human Ultra-sociality. In *Indoctrinability, Ideology, and Warfare: Evolutionary Perspectives* (eds. I. Eibl-Eibisfeldt & F. Salter), pp. 71–95. Berghahn Books, New York, NY.

Singer, J. D., & Small, M. H. (1994). *Correlates of War Project: International and Civil War Data, 1816–1992*. Data file, Inter-University Consortium for Political and Social Research, Ann Arbor, MI.

Sosis, R., & Alcorta, C. (2008). Militants and martyrs: Evolutionary perspectives on religion and terrorism. In *Natural security: A Darwinian Approach to a Dangerous World.* (eds. R. Sagarin and T. Taylor), pp. 105–24. University of California Press, Berkeley, CA.

Strand, H. (2006). Onset of armed conflict: A new list for the period 1946–2004, with applications. In Strand, H., *Reassessing the Civil Democratic Peace*, Ph.D. dissertation, Department of Political Science, University of Oslo and Centre of the Study of Civil War, PRIO.

Uppsala Conflict Data Program (UCDP) www site; UCDP Non-State Conflict Dataset V.1.1, 2002–2005. http://www.pcr.uu.se/research/UCDP/index.htm.

U.S. Census. (2000). http://www.census.gov/popest/.

Wrangham, R., & Peterson, D. (1996). *Demonic Males*. Houghton Mifflin, Boston, MA.

# Chapter 13
# Biodiversity and the Parasite-Driven Wedge

## 13.1 Introduction

Biodiversity is comprised of two types of variation. One type is the variation in trait expression across geographical space. In previous chapters, we have documented patterns of geographical variation in human cultural traits. There is variation, for example, across human societies in religious commitment, inbreeding behavior, collectivism, homicide, autocracy, and the importance attributed to physical attractiveness; and in each case, the trait variation shows a robust positive relationship with infectious-disease stress. These discoveries, as well as many others we discuss in this book, indicate that the parasite-stress theory of sociality is a general theory for understanding variation in cultural traits. The second type of biodiversity is the geographical variation in numbers or kinds. As examples, the number of species of songbirds and beetles vary geographically, as does the numbers of languages and religions.

The second type of biodiversity is the type usually labeled biodiversity by biologists and other researchers. We examine in this chapter the diversity of kinds, hereafter referred to as biological diversity or biodiversity. The evidence we present reveals that the parasite-stress theory of sociality informs many aspects of this biodiversity.

Parasite diversity itself is a typical aspect of biodiversity by showing the pattern of greater numbers of kinds at low than at high latitudes (Guernier et al. 2004; Dunn et al. 2010). It is well established, too, that biological diversity in number of species declines in many taxa with increasing latitude (Hillebrand 2004). This same pattern of more kinds at low than at high latitude is seen also in the case of human indigenous societies (Cashdan 2001). The latitudinal gradient in biodiversity has attracted a great deal of interest from biologists and geographers. Various hypotheses have been proposed to explain this pattern with no consensus on causal mechanisms (Gaston 2000; Schemske 2002; Willig et al. 2003; Mittelbach et al. 2007). Because the pattern is so widespread, some scholars appeal to a single causal mechanism (e.g., MacArthur and Connell 1966), while others like Gaston (2000) argue there is no reason to suppose that a single process must explain this general pattern for all or most taxa.

R. Thornhill and C.L. Fincher, *The Parasite-Stress Theory of Values and Sociality: Infectious Disease, History and Human Values Worldwide*,
DOI 10.1007/978-3-319-08040-6_13, © Springer International Publishing Switzerland 2014

We have hypothesized that an important context causing variation in biodiversity across the globe is the magnitude of infectious diseases that organisms have faced and continue to face throughout the world. We proposed that the behavioral avoidance and management of infectious disease can result in reproductively isolated populations and culturally isolated human groups (e.g., by language or values) and thereby provide the necessary conditions for the genesis of new species as well as the genesis of new languages, religions, and other kinds of cultural differences. Accordingly, the higher the parasite prevalence in a region, the more frequent the parasite-based processes of speciation (the origin of new species) and ethnogenesis (the origin of new human cultures) will occur, explaining the latitudinal biodiversity gradient (Fincher and Thornhill 2008a, b). At least for human parasites, which are well known relative to infectious diseases of other species, parasite diversity of kinds (Guernier et al. 2004; Dunn et al. 2010; Jones et al. 2008) as well as parasite severity (number of disease cases) (Fincher et al. 2008; Fincher and Thornhill 2012) show robust patterns of greater numbers/severity at low than at high latitudes. Although knowledge of the parasitism of host species other than humans is quite limited, where data do exist, the negative relationship between infectious disease and latitude typically is seen (birds: Calvete 2003; Møller et al. 2004; nonhuman primates: Nunn et al. 2003; Nunn et al. 2005). Latitudinal variation corresponds to climatic variation and parasitic diseases thrive to a greater extent in ecological conditions that are warm and moist compared to other climates (Guernier et al. 2004; Dunn et al. 2010).

Below, first, we describe the hypothesis we have proposed for how infectious disease creates biodiversity. In our two papers in 2008 cited just above, we referred to this process as the parasite-driven-wedge model of biodiversity. We give here a more expanded and detailed model than that in our 2008 papers. Then, we apply and test the model using two major types of human cultural biodiversity across the countries of the world: language and religion number. After that, we examine political factionalism in the Old South and the origin of caste social systems as potential outcomes of the parasite-driven-wedge model. Finally, we discuss the broader implication of the model for speciation and biodiversity in general.

## 13.2 The Parasite-Driven-Wedge Model

We hypothesize that the wedge mechanism works as follows in a species or cultural group residing in a region of parasite adversity to generate discontinuity/segmentation and divergence, and hence new kinds:

### *13.2.1 Step 1: The Ancestral Culture/Species*

Initially, the species or the culture has a homogeneous phenotype and immunity across its geographical range. Individuals interact, mate and reproduce throughout the geographical range without bias. Parasites invade the species or the people of the culture.

### 13.2.2 Step 2: Localized Host–Parasite Races

Over time, parasite–host coevolutionary races become spatially distinct and localized across the range of the species or the culture. This spatial variation is the result of the localized emergence of new parasites and the evolution of locally adaptive classical immunity in the host. The new parasites involved may be species that are new to a locale or varieties of already present species. (See Chap. 3 for evidence and further discussion of geographical localization of host–parasite coevolutionary races.)

### 13.2.3 Step 3: Spatially Variable Immunity

At this point in the process, there is spatial variation across the species or the culture in individuals' ability to meet immune challenges when encountering infected conspecific individuals or habitat or other environmental features having contagion risk. Individuals with alleles that contribute to classical immunity against local parasitic diseases are favored by natural selection. Antipathogen values (preferences) and associated behaviors that arise will increase in frequency because they are adaptive. These values and behaviors function in (a) avoidance of conspecifics that are infected or potentially infected with novel parasites to which local classical immunity is reduced; (b) developing strong and interdependent social ties with local conspecifics (and hence immunologically locally adapted individuals), which provide social investment and protection against the effects of parasitic infection; and (c) restricting movements to the local habitat (philopatry). The resultant in-group assortative sociality—by way of its components of xenophobia (avoidance of non-local conspecifics), social favoritism toward and embeddedness with local conspecifics (in-group members), and limited dispersal—defends against contact with novel parasites harbored in non-local conspecifics (out-groups) and manages and reduces the negative effects of infectious diseases within the local group. These components of in-group assortative sociality are aspects of behavioral immunity (Chap. 3). In the case of humans, behavioral immunity in the form of cultural behaviors that promote parasite avoidance and management are originated and adopted widely. These cultural behaviors—namely, xenophobia, parochial social contact and interaction, and philopatry—serve the same functions as in (a)–(c) above.

There are two types of preferences or values present here. One is social preference for individuals who manifest the locally adaptive antipathogen values and behaviors. Social interactions are biased toward these individuals because they have classical and behavioral immunity to local parasites and hence present less contagion risk than occurs in interactions with behaviorally dissimilar individuals. This in-group social preference refers to the preferential alliance and transaction with similar and local others, including mating and other social contact (e.g., nepotism, reciprocity, cooperative hunting, cooperative breeding). In humans, this includes behaviors of discriminative affiliation based on the presence of similar normative behavior (norms), styles of adornment, religious and other values, dialects and other language use, or other cultural traits that distinguish local from non-local people.

Thus, in humans, there are multiple categories of potential phenotypic markers of kin and other in-group affiliates and simultaneously markers of boundary between in-group and foreignness/out-group. The second type of preference involved is for the acquisition by individuals of values and behaviors that are locally adaptive against infectious disease. In humans, this involves discriminative enculturation. Earlier in the book, we discussed human psychological adaptations for the strategic adoption of values and other cultural items that promote inclusive fitness (Chap. 2). In animal species in which the behavioral immune system is not comprised of culturally acquired behavior, only the first type of preference will be seen. In people, both preferences will be important for achieving defense against local parasites.

Behavioral immunity in humans overlaps with human personality traits as we documented in Chap. 7. We propose that, in nonhuman animals, behavioral immunity traits correspond to some of the individual and group differences behavioral ecologists have labeled personality traits or behavioral syndromes. Thus, in nonhuman animals, the two personality traits referred to as "dispersal-prone" and "adventurous" correspond functionally to personality features of low philopatry and openness to experiences in people. And similarly, the nonhuman animal personality traits "sedentary" and "shy" correspond to high philopatry and introversion in people. As suggested in Chap. 7, the parasite-stress theory of values may allow the synthesis of the study of human personality with that of nonhuman personality into a single research field that is predictive of personality across all animal species based on degree of infectious disease adversity (for reviews of research on nonhuman animal personality, see Barber and Dingemanse 2010; Sih et al. 2012; Wolf and Weissing 2012). If these comparative connections are accurate then it is appropriate and informative to discuss values and behaviors of nonhuman animals along a dimension of collectivism–individualism or conservatism–liberalism.

### *13.2.4 Step 4: Subdivision of the Ancestral Range*

In-group assortative sociality, including philopatry, subdivides the originally uniform, unstructured species or culture and generates a wedge that pushes the segments apart. This subpopulation structuring, along with reproduction in the natal region, increases genetic relatedness among members of localized in-groups. In turn, the increased genetic similarity promotes effective natural selection for kin altruism (nepotism), because increased relatedness of interactants raises the reproductive gains (inclusive reproductive success) from local altruism (Hamilton 1964). In the case of humans, the collectivist values of in-group embeddedness, xenophobia, and philopatry generate subpopulation structuring that subdivides an ancestral culture into subunits based on genetic similarity. In contrast, individualism reduces subpopulation structuring by mixing a cultural group through dispersal, out-group social interactions including mating, and non-local reproduction.

Best et al. (2011) have clarified through mathematical modeling the coevolution of host–parasite dynamics in spatially segmented populations. Their ideas apply to the parasite-driven wedge as follows. Increased relatedness among locals arising from assortative sociality and local reproduction not only promotes the evolution of nepotism, but also promotes the rapid spread of new mutant alleles that confer resistance to local parasites. Given the relatedness of local conspecifics, when such a mutant first arises, its bearers likely will be in the close proximity of other individuals who also have the mutant. Then, in-group assortative social favoritism and altruism and local reproduction will magnify the natural selection for the mutant, providing rapid and widespread immunity in the in-group against local parasites. As alleles for resistance to local parasites increase in frequency in the local group as a result of this natural selection, the parasite severity (number of cases of the disease) also declines. In turn, as Best et al. (2011) emphasized, declining parasite severity coincides with reduced host-to-host transmission rate of the parasite. And reduced transmission rate of a parasite is an ecological context in which natural selection favors lower virulence in the parasite (Ewald 1994). Hence, the combination of subpopulation structure and concomitant increased genetic relatedness of in-group members, in-group assortative sociality, and effective selection for both local parasite resistance in hosts and low virulence in parasites synergistically produce an encompassing defense against local contagion. The benefits of in-group assortativeness of coping with contagion extend beyond the values of xenophobia and ethnocentrism. Its long-term benefits, including those of philopatry, are the reduction of the prevalence *and* virulence of parasitic diseases that are present in the local group.

Limited dispersal in the form of philopatry refers to behaviors that reduce movements away from the natal location. In areas of high pathogen adversity, compared to areas of low pathogen adversity, high philopatry will be the optimal habitat preference, because of the correspondent increase in association with immunologically similar individuals and decreased contact with more distant, and differently parasitized, conspecifics and their habitats. High philopatry under high infectious-disease stress is optimal for people as well as other organisms.

On the one hand, philopatry is assortative sociality, because high philopatry restricts social interactions, including altruism and mating, to the natal locale. In the parasite-driven-wedge model, however, discriminative local contact with conspecifics is promoted by psychological adaptations designed for associating with these individuals and not others. On the other hand, philopatry (and its opposite, dispersal), requires different adaptations that function specifically to either restrict and localize movement or facilitate dispersal.

In Chap. 5 we showed, using a large sample of indigenous societies, that parasite stress is negatively related to a measure of dispersal (home range size) of the people comprising a society. Thus, home range size and related dispersal across these many different cultures decreases as parasite stress increases. This finding is consistent with the biogeographical pattern that, generally, indigenous human societal range size is reduced in the tropics compared to temperate areas (Cashdan 2001). (Also see Chaps. 5 and 14 for additional evidence that human philopatry is a defense against parasites.)

The same latitudinal pattern is observed in the geographical range sizes of species and is referred to by ecologists as Rapoport's rule (Stevens 1989). We have argued that range reduction in low-latitude traditional human groups and other host species results from the adaptive response of limited dispersal in pathogen-rich tropical areas (Fincher and Thornhill 2008b). Related comparative evidence is the finding that, across marine fish species and species in certain marine invertebrate taxa, the presence of adaptations for egg and larval distance dispersal are correlated positively with latitude (Bradbury et al. 2008). Given the negative covariation of parasite adversity and latitude, Bradbury et al.'s finding indicates effective selection for dispersal under low parasite stress and for philopatry under high parasite stress.

Freeland (1979), Møller et al. (1993), and Loehle (1995) discussed how limited dispersal may reduce exposure to infectious diseases, and also argued for the importance of territoriality and restricted home-range behavior—forms of limited dispersal—as adaptations for reducing contact with dissimilar conspecifics that may carry novel diseases unlike those carried by local conspecifics. Moreover, we mentioned in Chap. 5 that territoriality in nonhuman species may function, in part, in a way similar to human xenophobia by reducing contact with out-group individuals carrying novel parasites.

Freeland (1979) also provided evidence consistent with the thesis that reduced dispersal is an adaptive response to high local parasite levels. In a study of rain forest-dwelling versus savannah-dwelling primates, he showed that savannah baboons, with characteristically higher rates of inter-group movement of individuals, shared protozoan faunas, while rain forest primates, with lower rates of inter-group movements of individuals, had more unique protozoan faunas. Freeland argued that because of increased parasite adversity in rain forest primates, out-group contact was too costly to promote more frequent exchange of individuals. Meanwhile, because of the lower parasite adversity for the savannah primates, greater interchange of individuals and associated out-group contact had lower costs.

### *13.2.5 Step 5: Divergence*

As we envision the wedge mechanism, as individuals adaptively contact local conspecifics preferentially (as they show a contact bias in altruism, associative behavior, mating) and adaptively limit dispersal, fractionation/segmentation within the original or ancestral distribution of a species or culture continues. Simultaneously, divergence between the segments in the traits involved in the wedge ensues. The fractions or segments diverge, in part, because selection favors local genetic immunity as well as traits of behavioral immunity that promote local social affiliation, avoidance of non-local conspecifics, and limit dispersal. In humans, divergence results also from preferential adoption of socially learned or cultural values that promote in-group embeddedness, out-group avoidance and limit dispersal. Moreover, divergence is further increased because philopatry and the other components of in-group assortative sociality reduce and may eliminate gene flow or cultural-item flow among the differentiated segments.

The parasite-driven wedge at the border of two diverging and incipient species or cultures may sometimes exhibit reinforcement—a greater difference in traits at the contact areas than at non-contact areas—at least for a period of time, in the development of the wedge. Behaviors comprising behavioral immunity will be favored by selection and, in the case of humans, by discriminative choice of values and other cultural elements that reduce the likelihood of intergroup contact including mating. Out-group mating, compared to in-group mating, would lead to the production of off-spring with a reduced genetic defense against the parasites of the parental groups (i.e., hybrid fitness-disadvantage). Such processes will be stronger at the wedge than elsewhere in the distribution of each diverging group. That is, assortative sociality behaviors of xenophobia and in-group sociality and embeddedness, as well as limited dispersal, will be reinforced at the wedge as a result of selection favoring individuals that avoid mating and mating attempts with non-local individuals.

### *13.2.6 Step 6: Coupling of Immunity Traits*

At this stage, the four traits of (a) local genetic immunity, (b) preference for the local habitat (philopatry), (c) social preference for in-group members with local behaviors, and (d) avoidance of out-group conspecifics will become coupled genetically within the individuals comprising a segment, as a result of multiple processes. First, each of the four traits corresponds to high inclusive fitness of individuals and hence becomes increasingly represented in descendant generations as a result of natural selection, and additionally, in humans, as a result of discriminative social learning. Second, the coupling itself, like each of its four components, is defensive against contagion and will be favored by selection, giving rise to a coadapted gene complex that defends against local parasites. Mating is non-random; specifically, mating and reproduction are local and hence produce inbreeding. Inbreeding behavior itself may have the selective advantage of coupling coadapted alleles that defend against local infectious diseases. In regard to humans, in Chap. 6 we presented evidence that frequency of consanguineous marriages is related positively to parasite stress both across contemporary countries of the world and indigenous societies. The genetic isolation of the segments and, in the case of humans, the cultural isolation, combined with the coupling of the traits and the positive selection for local genetic disease resistance, local social interactions, and philopatric behavior, further increase differentiation and divergence of the segments.

The coupling of traits just described leads us to hypothesize that other complementary processes act to increase intergroup divergence at the parasite-driven wedge. In species in which cultural behavior is not a part or major part of behavioral immunity, the coupling of local genetic disease resistance, in-group assortative mating and other sociality, and philopatric behavior will include their genetic correlation arising from linkage per se (if the traits are affected by pleiotropic genes on the same chromosome) or much more commonly from linkage disequilibrium, i.e., the non-random association in gametes of alleles of unlinked loci (i.e., genes on different chromosomes). This genetic correlation of traits promotes rapid divergence between the differentiating segments because the correlation means that the traits acting adaptively in the wedge are *mutually reinforcing*. The genetic linkage of in-group sociality, philopatry, and local genetic immunity creates a context in which all the traits involved in the linkage are simultaneously favored when any one is favored. Moreover, this genetic linkage gives rise to an additional force of selection favoring assortative sociality, because the behavioral expression of assortative sociality favors itself—it is *self-reinforcing*. For instance, a preference of biasing altruism or mating toward an in-group member with local behavioral immunity is an action toward an individual who possesses the same preference, and whose reproductive success is enhanced by the altruism received or by the in-group mating.

This stage of the parasite-driven wedge has some similarity with Fisher's runaway sexual selection hypothesis in which female mate preference for a male display trait favors itself because of the genetic correlation between the female preference and the male trait preferred (Fisher 1930). Fisher's hypothesis involves a sexually dimorphic trait such as tail length in a bird where female preference for tail length and male tail length are heritable (show underlying genetic variation among individuals). Females preferring males with a given tail length produce sons with such tails and daughters with a preference for such tails. This yields the genetic correlation of male trait and female preference for the male trait based on linkage disequilibrium. If females preferring a particular tail trait predominate (say, many females prefer the longest tails), sexual selection by female choice can lead rapidly to highly exaggerated male tail length (Lande 1981; Kirkpatrick 1982). This sexual selection mechanism when involving geographically variable and localized female mate preferences may be a cause of the multiplication of species (Lande 1981, 1982; West-Eberhard 1983). The Fisherian sexual selection process, in the versions mathematically modeled by Lande (1981) and Kirkpatrick (1982), involves female preference that is neutral to selection while the male trait is directly sexually selected by female choice. In this case, the female preference evolves as a genetically correlated by-product trait along with the directly selected (by female preference) male trait. In the parasite-driven-wedge model, each of the genetically correlated traits is under positive direct selection and in the same direction of providing adaptive solutions to infectious disease avoidance and management. Assortative sociality is self-reinforcing and collectively the genetically correlated traits are mutually reinforcing. Hence, the parasite-driven-wedge model may lead to rapid and extensive divergence across a population that was originally uniform.

In the literature on speciation, linkage disequilibrium is established as an important genetic mechanism contributing to the separation of incipient species. It is also emphasized in the same literature that genetic recombination from meiosis in the presence of gene flow between incipient species can reduce linkage disequilibrium and thereby erode the potential for completed speciation (Felsenstein 1981; Rice 1984). In the parasite-driven-wedge scenario, however, the linkage and each of its component traits are under strong positive selection as described above, and in-group assortative sociality and especially out-group avoidance, hybrid disadvantage and philopatry eliminate the gene flow that can make recombination a factor that retards divergence and the completion of speciation. The parasite-driven-wedge model creates effective spatial isolation of, by way of borders between, incipient cultures or species simultaneously with their divergence.

Trait coupling, we argue, may play an important role not only in creating new species but also in creating new cultures. Consider a human society living in a setting of high parasite adversity and associated localized parasite–host antagonistic coevolution. Suppose discriminating in-group members from other people can be achieved by a local cultural innovation, say, a new religious belief or dialect. Both acquiring the new cultural item and preferring those who show it in their behavior are adaptive against novel parasites in out-groups because the cultural item is not available to and used by the out-group people who may possess novel parasites; therefore, the cultural item distinguishes the in- versus out-group. Individuals who adopt the new cultural trait and prefer its presence in others with whom social interactions occur have more descendants than individuals who do not and the predominant descendants socially learn the adaptive cultural item and in-group social bias. With time, both the cultural trait and the preference for others with it become common among members of the in-group. Within individuals, the cultural item will be linked with the preference for others with it, as well as with any present genetic resistance against local parasites. Simultaneously, cultural ideas that promote philopatry, as well as additional cultural items that focus social investment on in-group members, arise and become common because they are adaptive against local infectious disease. Genetic immunity and all the cultural traits of assortative sociality become linked within individuals and hence mutually reinforce each other. In addition, when an individual expresses a social preference of investment in another person with a cultural trait of local behavioral immunity, the preference reinforces itself, because the preferred individual has the same preference.

Similarly, given the localization of host–parasite races, an adjacent group is undergoing the same processes, but involving its group-unique cultural innovations and alleles for defense against local parasites. In the adjacent group, xenophobia, local habitat preference and in-group favoritism are linked to each other and to local genetic immunity, giving rise to the mutual reinforcement and self-reinforcement described above. In time, cultural isolation results and contiguous groups diverge to the status of two distinct cultures. Thus, cultural divergence among incipient cultures involves the mutual reinforcement and self-reinforcement aspects of the parasite-driven-wedge model discussed above as important in divergence of incipient species.

We suggest here some new labels that identify the processes just described of mutual reinforcement and self-reinforcement of cultural traits. We call the process of the linkage of cultural items within individuals "cultural linkage disequilibrium." We define it as the non-random association of cultural items or preferences or values within the minds of individual people. We label the self-reinforcing elements of cultural assortative sociality "cultural self-reinforcement." Boyd and Richerson (1985) recognized that a cultural trait and the socially learned preference for others with the same trait can become coupled within individuals. They explored in mathematical models this coupling's positive effect on the rapid spread and divergence of cultural traits. McElreath et al. (2003) also recognized this coupling's role in the divergence of ethnic cultural markers. These earlier researchers, however, did not connect their ideas to a hypothesis like the parasite-driven-wedge model.

Empirical evaluation of cultural linkage disequilibrium is straightforward. Consider a human society with a cultural repertoire consisting of the following: prefer local mates, prefer local habitat, dislike and avoid strangers (out-group people). These three values may be possessed exclusively by three different groups of the society in one extreme. In another extreme, all three may be within the minds of all societal members. The degree to which the three items correspond to the cognitive preferences within individuals versus between individuals or groups is the magnitude of cultural linkage disequilibrium.

### 13.2.7 Step 7: The Wedge Causes New Parapatric Kinds

In overview, in ecological settings of parasite adversity, the parasite-driven wedge causes the divergence of segments of an ancestral species or culture. The localized host–parasite races that are the basis of the wedge can endure. In the case of humans, the parasite-driven divergence, by the processes we have proposed, can bring about new adjacent cultures that arise side-by-side—that is, arise parapatrically. In other species, this divergence can lead to reproductive isolation of the groups—that is, yield new species—that originated contiguously. Biologists refer to the origin of new species from contiguous, initially interbreeding segments of an ancestral population as parapatric speciation. (See Thornhill and Fincher 2013 for further discussion of the wedge model applied to parapatric speciation.)

Thus, the parasite-driven wedge consists of behaviors (philopatry, xenophobia, and in-group transaction and embeddedness), the coupling of these immunity-providing behaviors with each other and with genetic immunity, mutual reinforcement by selection of the traits involved, and self-reinforcement of assortative sociality. The wedge subdivides an ancestral species or culture and thereby pushes segments apart, creating new species and cultures. The ubiquity of parasites occupying all living organisms implies that this is a common driver of the origin of new kinds. Furthermore, the parasite-driven-wedge model provides a context for the fractionation and divergence within a species or culture in the total absence of segmentation created by mountains or other geographical barriers. Therefore, the model forms a basis for widespread parapatric divergence leading to new subspecies, species, and cultures.

### 13.2.8 Step 8: Parasites Also Speciate

Moreover, the localized antagonistic coevolutionary races between hosts and parasites along with localized behavioral immunity will drive population divergence among the parasites themselves, generating novel kinds of infectious disease.

Hence, the parasite-driven-wedge model gives rise to parapatric diversification of parasites as well as their hosts. Also, such races guarantee through mutation new solutions in parasites that circumvent host defenses, which fuels the continuation of the localization of antagonistic coevolutionary interaction between hosts and their parasites. It is the simultaneous parapatric diversification of parasites and their hosts that can account for the diversity of both covarying with latitude. Low latitudes provide the ideal, basic ecological conditions for parasites, but this does not explain their greater diversity there. The wedge model does.

### *13.2.9 General Prediction*

The higher the parasite adversity experienced by a species or cultural group, the more opportunity for spatial variation in parasite–host coevolutionary races and the resulting components of the parasite-driven-wedge model in the geographical range of a species or culture. The frequency, duration, and intensity of the steps discussed above will covary positively with the infectious disease adversity. That is, relatively high parasite adversity yields more localized parasitic disease problems across geographical ranges of hosts and hence increased parasite-driven intergroup divergence by the parasite-driven-wedge mechanism. Locally high parasite adversity simultaneously maintains (a) the high cost of out-group contact that would arise from inter-group interactions or distance dispersal and (b) the large benefit of in-group social interactions and philopatry. According to the parasite-driven-wedge model, biological diversity is predicted to be highest where parasite adversity is highest. The parasite-driven-wedge processes also take place in areas or low parasite diversity; however, other diversification processes may predominate.

## 13.3 The Parasite-Driven Wedge Creates Ecological Diversification

We emphasize that the parasite-driven-wedge model is an ecological model of diversification in that divergence is caused by adaptation to local ecological challenges. It differs from the many other ecological models of diversification most fundamentally in the combination of features involved. In the wedge model, there is preference for and adaptation to local conditions of habitat, mating and other social milieu, and immune challenges, and local immune challenges are met by both classical and behavioral immunity. The wedge process automatically generates the coexistence of new species and cultures in a region because it creates new habitat, social and immunological niche dimensions at the same time as it fragments, isolates and diversifies an ancestral species or culture.

The interaction between the origin of new species and the need for ecological diversification among species in order for the species to attain long-term coexistence is commonly discussed in the literature. (A recent review is Weissing et al. 2011.) Some authors have argued that speciation by sexual selection lacks the ability to create this necessary ecological divergence, and hence cannot provide long-term coexistence. The argument is as follows. Spatial variation in sexual selection diversifies mate choice and male traits affected by mate choice, but may result in species with identical or similar ecological niche requirements; thus they are unable to persist in the face of between-species competition and competitive exclusion. Speciation by sexual selection generates reproductive isolation of species by diversifying pre-mating isolating mechanisms, but cannot produce long-term coexistence. In contrast, ecological speciation diversifies ecological niches of incipient species, but generally lacks inherent processes that produce behavioral isolation of species. As just mentioned, the parasite-driven-wedge model is an ecological model of speciation. Niche diversification arises in preference for local habitat and social allies, including mates, and in local immunity. Mate preference for locals makes the wedge model a sexual selection model too, and specifically a good-genes sexual selection model, given the preference for mates resistant to local parasites and adapted to the in-group social setting. Hence, the parasite-driven-wedge model has some overlap with the combined ecological–sexual selection model proposed by Weissing et al. (2011). It differs from their model in the features that account for divergence of the incipient species and their long-term coexistence.

We emphasize that the parasite-driven-wedge model is not merely a new sexual selection model of diversification. The wedge is comprised of the behaviors of local habitat preference, social preference for locals including local mates, and behavioral avoidance of non-locals.

We now discuss our model's ability to explain the geographical distribution of two forms of biological diversity: human language and religion diversity. Although we focus this empirical analysis on a single species, it gives the advantage in exploring group diversification within a restricted and hence controlled phenotypic landscape. Nevertheless, humans are found worldwide in various ecologies, providing a backdrop for studying, in a general way, the ecology of diversification. And, far more is known about parasites of humans than the infectious diseases of other species. After the treatment of human cultural diversity, we discuss briefly cultural diversity in nonhuman animals, and thereafter return to topics of speciation in relation to parasite stress.

## 13.4 Language

Language diversity across regions has received attention from researchers employing ecological and evolutionary thinking (e.g., Mace and Pagel 1995; Nettle 1999a; Harmon 2002; Sutherland 2003; Maffi 2005). Mace and Pagel (1995) demonstrated a latitudinal gradient in the number of indigenous languages on the North American continent: a higher diversity of languages at low than high latitudes. Sutherland (2003) showed that the number of languages is related negatively to latitude worldwide. Nettle (1999a) provided a hypothesis for the distribution of languages across the globe, arguing that an environmental variable causing language diversity is the ecological risk from climatic factors that people face. He measured this risk using the average growing season. He suggested that areas with long growing seasons provide conditions that allow for self-reliance on local resources within a particular ethnic group. In contrast, people in areas with short growing seasons must participate in contact and transactions with other ethnic groups to buffer the effects of the ecological risk from a harsh climate. Under these conditions, many languages emerge and persist in areas with long growing seasons, while few languages emerge from areas with short growing seasons and associated ecological insecurity, because under short growing seasons individuals must be able to speak with many different people across an expansive area. Therefore, in Nettle's hypothesis, the high ecological risk of short growing seasons is a homogenizing mechanism across human groups, leading to low language diversity. However, Sutherland's (2003) research questioned Nettle's hypothesis; Sutherland reported that mean growing season was not an important predictor of language diversity across the world.

We provide evidence below supportive of the hypothesis that a significant mechanism underlying the latitudinal diversity gradient in languages is the parasite-driven wedge elaborated above. See Fincher and Thornhill (2008a) for more details of the methods and analyses of human *Disease Richness* (number of infectious diseases) and language diversity discussed below.

### 13.4.1 Methods

First, we examined the correlation between contemporary measures of human *Disease Richness* and human language richness across most countries of the world (the global analysis). Then, we evaluated the same correlation in each of six world regions and three longitudinal geographical bands.

We acquired, during August 2007, *Disease Richness* for all contemporary countries worldwide listed in the database, Global Infectious Disease and Epidemiology Network (GIDEON; www.gideononline.com) (see Chap. 5 for a discussion of this database). Our tally includes all human infectious diseases listed in GIDEON for a country. The average *Disease Richness* score per country was $201 \pm 15$ ($M \pm$ SD; $n = 229$ countries).

We acquired in 2007 the language richness values for each country from Ethnologue (www.ethnologue.com), a resource commonly used in the study of human language diversity (e.g., Harmon 1996; Nettle 1999a; Sutherland 2003). We confined our analyses to the number of living indigenous languages per country reported in *Ethnologue* ($38 \pm 94$, $M \pm$ SD; $n = 225$ countries), because this is the category of languages that originate within a region and hence the type to which our model of parapatric differentiation of languages (the parasite-driven wedge) applies. Because the language richness scores were highly skewed, we used log-transformed values for analysis.

In our analysis, we examined the influence of potentially confounding factors. Given that latitude is negatively related to language richness (Sutherland 2003) and the general importance of latitude to the traditional study of biodiversity in general (Hillebrand 2004), we included absolute latitude measured at the midpoint of each country as a control variable. Because country-level wealth might affect language diversity within a country (Nettle 2000), we entered Gross Domestic Product (GDP) per capita as a potential confounder. A larger population (irrespective of country land area) may have more languages than a smaller population; therefore, we controlled the effect of human population size per country. Moreover, some countries may have more languages within their borders simply because they are geographically larger countries; therefore, we controlled a country's land area. Data for these potential confounders came from the *World Factbook* 2007 (www.cia.gov). GDP per capita, population size, and land area were all log-transformed prior to analysis.

Furthermore, it is possible that the diversity of languages across the world is caused by different patterns of human settlement on the continents (e.g., more recent colonization; Nettle 1999b) or by different patterns of conquest (Diamond 1998). We accounted for this potential confound using three methods. First, we explored the predicted positive correlation between *Disease Richness* and language richness in each of six world culture areas. Second, we considered the correlation between *Disease Richness* and language richness using the averages for both variables from the six world regions. Third, we compared the pattern of the correlation among three longitudinal geographical bands.

For the world culture areas analysis, we divided the countries of the world into Murdock's (1949) six world regions: North America, South America, West Eurasia, East Eurasia, Africa, and Insular Pacific. Primarily, the countries were easy to separate into these world regions; however, the division between West Eurasia and East Eurasia was less distinct. Russia, Kazakhstan, Tajikistan, Kyrgystan, Pakistan, and India were placed in the East Eurasia category; Uzbekistan, Afghanistan, Turkmenistan, and Iran were placed in the West Eurasia category.

Similar to the procedure in Collard and Foley's (2002) analysis of the global distribution of cultures, we analyzed the relationship between *Disease Richness* and language richness across three longitudinal geographical bands that have been subjected to different histories of conquest and colonization (McNeill 1981; Diamond 1998; Collard and Foley 2002). The American band incorporates those countries and territories from 180°W to 30°W; the Europe–Africa band incorporates the countries and territories from 30°W to 60°E, including Iran and Iceland; the Asia–Australia band incorporates the countries and territories from 60°E to 180°W, including Kazakhstan, Russia, Turkmenistan, and Uzbekistan. The bands provide three distinct longitudinal bands.

### *13.4.2 Results*

In the global analysis, *Disease Richness* was positively and highly correlated with living indigenous language richness ($r=0.78$, $p<0.0001$, $n=221$; Fig. 13.1). The correlations between language richness, *Disease Richness* and the control

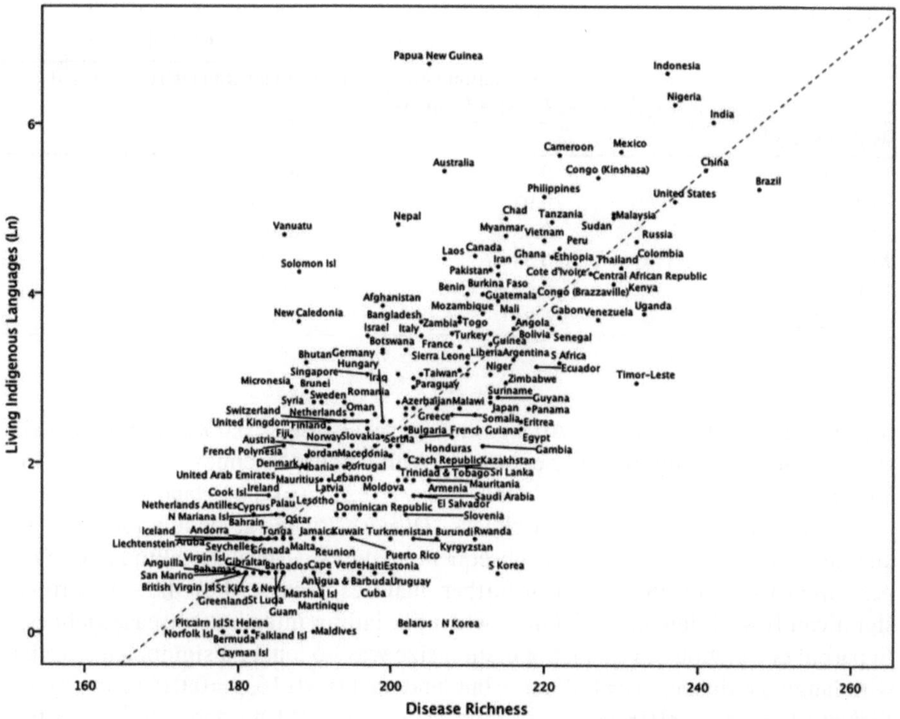


**Fig. 13.1** The Pearson correlation between *Disease Richness* and the natural log of indigenous living languages per country found across the world ($r = 0.78$, $p < 0.0001$, $n = 221$ countries). The line is the regression line (results originally reported in Fincher and Thornhill 2008a)

**Table 13.1** Pearson zero-order correlations between living indigenous language richness, infectious-disease richness, and control variables for the countries of the world; $n$ = number of countries in analysis (results originally reported in Fincher and Thornhill 2008a)

| | Correlation with indigenous language richness (Ln) | | |
|---|---|---|---|
| Variable | *r* | *p* | *N* |
| *Disease Richness* | 0.78 | <0.0001 | 221 |
| Absolute latitude | –0.21 | 0.0014 | 220 |
| GDP per capita (Ln) | –0.27 | <0.0001 | 218 |
| Population size (Ln) | 0.69 | <0.0001 | 220 |
| Land area (Ln) | 0.70 | <0.0001 | 220 |

variables are presented in Table 13.1. All of the control variables were correlated significantly with language richness—absolute latitude (–0.21), GDP per capita (–0.27), population size (0.69), land area (0.70)—thus, each of these variables was considered in partial correlations between *Disease Richness* and language richness.

The correlation between *Disease Richness* and language richness was 0.77 when partialling (controlling) absolute latitude ($n = 217$), 0.75 when partialling GDP per capita ($n = 215$), 0.54 when partialling population size ($n = 217$), and 0.54 when

**Table 13.2** Pearson zero-order correlations between living indigenous language richness and *Disease Richness* separated by the six world regions; $n$ = number of countries in the analysis

| | Correlation between indigenous language richness and *Disease Richness* | | |
|---|---|---|---|
| World region | $r$ | $p$ | $N$ |
| North America | 0.94 | <0.0001 | 22 |
| South America | 0.93 | <0.0001 | 28 |
| Africa | 0.83 | <0.0001 | 55 |
| West Eurasia | 0.58 | <0.0001 | 63 |
| East Eurasia | 0.62 | 0.0023 | 22 |
| Insular Pacific | 0.75 | <0.0001 | 31 |

partialling land area ($n=217$). All significance levels were $<0.0001$. Population size and land area were significantly correlated with language richness when partialling *Disease Richness* ($n=217$; population size: $r=0.21$, $p=0.001$; land area: $r=0.26$, $p<0.001$). Therefore, language richness, *Disease Richness*, population size, and land area were all considered in subsequent analyses. (Absolute latitude and GDP per capita were not considered in further analyses because they did not correlate significantly with language richness when partialling infectious-disease richness.) In partial correlation ($n=217$), population size was no longer a significant correlate with language richness ($r=0.05$, ns), but land area ($r=0.16$, $p=0.016$) and *Disease Richness* ($r=0.51$, $p<0.001$) were significant correlates of language richness. We then regressed language richness on both *Disease Richness* and land area. This multiple regression was significant ($R^2=0.63$, $F_{2,214}=185.60$, $p<0.0001$) with both *Disease Richness* and land area contributing distinct effects, but with *Disease Richness* showing a much larger effect (standardized betas: *Disease Richness* = 0.60, land area = 0.25, both $p$'s $<0.0001$).

We discovered after the publication of Fincher and Thornhill (2008a, b) that there were some misclassifications of the countries into the world regions. We published the corrected classification as part of the electronic appendices associated with Fincher and Thornhill (2012a). The results we report here for the world regional and geographical band analyses use the corrected classifications. The findings are qualitatively similar to that published in Fincher and Thornhill (2008a).

For each of the six world regions, the correlations between language richness and *Disease Richness* were all positive and significant ($r$'s 0.58–0.94; all $p$'s $\leq 0.002$; Table 13.2). The correlation between language richness and *Disease Richness* using the means from the world regions was 0.79 ($p=0.06$, $n=6$). And, in all three longitudinal bands, language richness was correlated positively to *Disease Richness* (American: $r=0.92$, $p<0.0001$, $n=60$; Europe–Africa: $r=0.81$, $p<0.0001$, $n=115$; Asia–Australia: $r=0.60$, $p<0.0001$, $n=46$).

We conducted partial correlations of language richness, *Disease Richness*, and land area in all six world regions and three longitudinal bands. Infectious-disease richness and language richness were correlated positively when partialling the effect of land area in five of the six world regions and all three bands (for the six world regions, $r$'s $=0.29$–$0.85$; for the three bands, $r$'s $=0.29$–$0.76$; Table 13.3). The correlation

**Table 13.3** Partial correlations and *p* values between indigenous language richness and *Disease Richness* when partialling the effect of land area in the six world regions and three longitudinal bands; $n$ = number of countries in the analysis (results originally reported in Fincher and Thornhill 2008a)

| | Partial correlation between indigenous language richness (Ln) and *Disease Richness*, partialling the effect of land area (Ln) | | |
|---|---|---|---|
| World Region | *r* | *p* | *N* |
| North America | 0.85 | <0.0001 | 21 |
| South America | 0.76 | <0.0001 | 26 |
| Africa | 0.70 | <0.0001 | 54 |
| West Eurasia | 0.29 | 0.022 | 63 |
| East Eurasia | 0.33 | 0.144 | 22 |
| Insular Pacific | 0.38 | 0.038 | 31 |
| *Longitudinal band* | | | |
| American | 0.76 | < 0.0001 | 57 |
| Europe–Africa | 0.66 | < 0.0001 | 114 |
| Asia–Australia | 0.29 | 0.053 | 46 |

was not significant in the world region East Eurasia; however, a sign test on the direction (+or –) of each correlation shows that the pattern of a positive correlation across the world regions is not spurious ($0.5^6 = 0.016$).

### 13.4.3 Parasite Severity

At the time we conducted our initial cross-national research on language diversity in relation to infectious diseases discussed above, we did not yet have data compiled on parasite severity per se (the number of cases of the diseases). Hence, in Fincher and Thornhill (2008a) we investigated only parasite richness as our measure of parasite adversity across countries. Here, we add the analysis of language richness in relation to two measures of parasite severity described in Chap. 5. The strong relationship between language diversity and parasite richness is repeated with the parasite-severity measures: *Contemporary Parasite Severity*, $r = 0.63$, $p < 0.0001$, $n = 218$; *Combined Parasite Stress*, $r = 0.56$, $p < 0.0001$, $n = 192$. However, each measure of parasite severity shows a weaker relationship with language diversity than does *Disease Richness*. Later in this chapter we discuss further the two measures of parasite adversity, infectious-disease richness and parasite severity, and their somewhat variable relationships to cultural diversity.

### 13.4.4 Conclusion

Language diversity across the globe is highly, positively related to human parasite richness. The various potential confounds considered do not alter this conclusion. Language diversity also is strongly, positively correlated with infectious-disease severity across the world. Hence, the distribution of language number across the globe supports the parasite-driven-wedge model for the genesis of biodiversity. We will discuss the implications of these findings for increasing the understanding of biodiversity in general after we treat geographical variation in religion diversity.

## 13.5 Religion

In this section, we briefly present the methods and findings from our paper Fincher and Thornhill (2008b), which showed that the contemporary global pattern of religion diversity correlates with infectious-disease adversity in ways that support the parasite-driven-wedge model of diversification.

### *13.5.1 Methods*

We were interested in whether there is a positive correlation between religion number and human infectious-disease problems across the countries of the world. To this end, we correlated the number of religions per contemporary country (=religion richness) with pathogen problems measured in two complementary ways. We used tallies of the total number of religions per country ($n=219$ countries), which includes the number of major religions and ethnoreligions, from Barrett et al.'s (2001) *World Christian Encyclopedia*. Barrett et al. (2001) is a highly regarded source among religion scholars (Grim and Finke 2006). A religion is defined by Barrett et al. (2001) as "a grouping of persons with beliefs about God or gods, and defined by its adherent loyalty to it, by their acceptance of it as unique and superior to all other religions, and by its relative autonomy." The average religion richness per country was $30.8 \pm 69.4$ ($M \pm \mathrm{SD}$) and ranged from 3 to 643. The number of religions was log transformed prior to analysis. One of our measures of parasite adversity was *Disease Richness*, as used above in our analysis of languages; our second measure was *Contemporary Parasite Severity*, described in Chap. 5.

We included in our analyses potentially confounding variables. At the time of our research on religion diversity, there had been very little research attempting to address the topic, and, therefore, there were no established paradigms for explaining religion diversity. However, given the general importance of latitude to the study of human infectious diseases and biodiversity (Guernier et al. 2004; Hillebrand 2004), we considered as a control variable the absolute latitude measured at the midpoint of each country. We also included as a potential confound the effect of human population size, as well as a country's land area. Autocratic political systems can restrict the ability to form religions and worship freely. Potentially, then, a country's governmental system could influence religion number in the country. Hence, we included as a potential confound Vanhanen's (2003) Index of Democratization (described in Chap. 10); higher values indicate greater democracy while lower values indicate greater autocracy.

**Table 13.4** Pearson zero-order correlations and *p* values between religion richness, *Disease Richness*, *Contemporary Parasite Severity*, and control variables; $n$ = number of countries in the analysis (results originally reported in Fincher and Thornhill 2008b)

| | Correlation with the natural log of religion richness | | |
|---|---|---|---|
| | *r* | *p* | *n* |
| *Disease Richness* | 0.75 | <0.0001 | 214 |
| *Contemporary Parasite Severity* | 0.62 | <0.0001 | 211 |
| Absolute latitude | −0.37 | <0.0001 | 218 |
| Population size (Ln) | 0.59 | <0.0001 | 214 |
| Land area (Ln) | 0.58 | <0.0001 | 214 |
| Democracy Index | −0.14 | 0.08 | 167 |
| GDP per capita (Ln) | −0.33 | <0.0001 | 213 |
| Gini | 0.38 | <0.0001 | 122 |

The secularization hypothesis for explaining religiosity (religious commitment; see Chap. 9) argues that a religion's vitality (commitment to and participation in a religion by adherents) decreases under increasing economic development, because individuals have less "need" for religion and its benefits when economic resources are favorable. An extension of the secularization hypothesis to explain religion diversity in a region would be that, under conditions of increasing economic development, religion diversity should decrease due to fewer adherents, while reduced economic conditions would increase religion diversity. Thus, a negative relationship between economic development and religion diversity is anticipated by the extended secularization hypothesis. To deal with this potential confound, we used GDP per capita and Gini (wealth inequity) as economic indicators.

Data for these potential confounders, except for Vanhanen's (2003) Index of Democratization (see Chap. 10), came from the *World Factbook* 2007 (http://www.cia.gov). GDP per capita, population size, and land area were log-transformed.

The diversity of religions across the world might be influenced by different patterns of human settlement on the continents or different patterns of conquest (as discussed above for language diversity). We accounted for this effect using two methods. First, we explored the predicted positive correlation between parasite richness and parasite stress and the number of religions in each of Murdock's six world culture areas described above. Second, to further account for different histories of conquest and colonization, we compared the pattern of the correlations between parasite richness and parasite stress and religion richness within three longitudinal geographical bands: American, Europe-Africa, and Asia-Australia (see above).

### 13.5.2 Results

The correlations between religion richness and the explanatory or control variables are presented in Table 13.4. Religion richness was related positively and strongly to *Disease Richness* ($r = 0.75$, $p < 0.0001$, $n = 214$; Fig. 13.2) and *Contemporary*

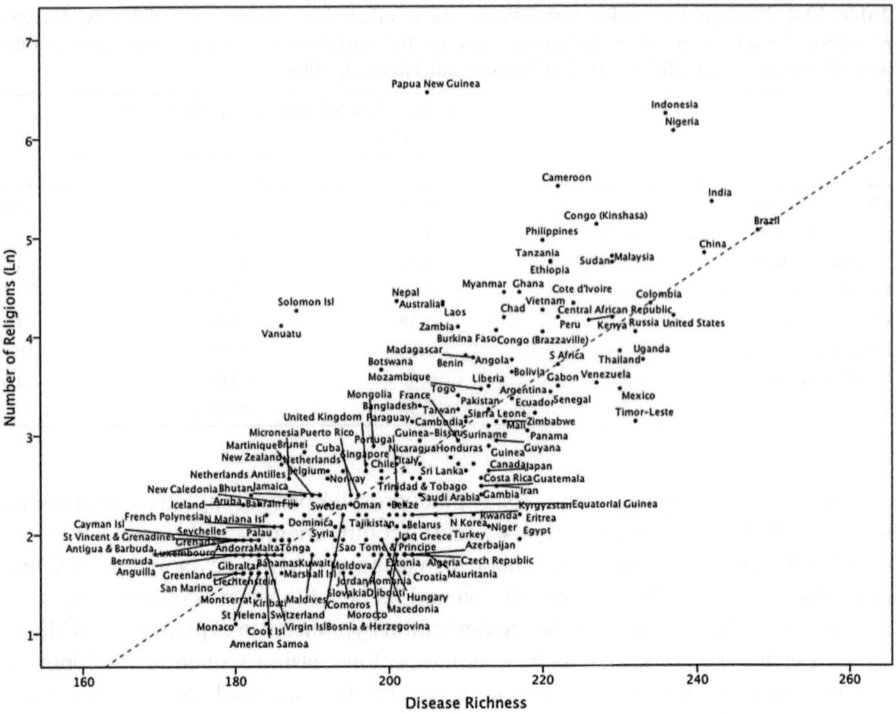


**Fig. 13.2** The Pearson correlation between human parasite richness and the natural log of religion richness across the countries of the world ($r=0.75$, $p<0.0001$, $n=214$). The line is the regression line (results originally reported in Fincher and Thornhill 2008b)

*Parasite Severity* ($r=0.62$, $p<0.0001$, $n=211$). Religion richness also was related significantly to population size ($r=0.59$), GDP per capita ($r=-0.33$), Gini ($r=0.38$), absolute latitude ($r=-0.37$), and land area ($r=0.58$). Democracy ($r=-0.14$, $p=0.08$) was close enough to conventional statistical significance for retention in subsequent analyses.

We conducted partial correlations between religion richness and parasite richness or parasite stress while controlling each of the potentially confounding variables (Table 13.5). The correlation between religion richness and *Disease Richness* was robust to the effects of the confounding variables ($r$'s 0.59–0.73; Table 13.5); as well, the correlation between religion richness and Co*ntemporary Parasite Severity* was robust to the confounding variables ($r$'s 0.45–0.58; Table 13.5). Only absolute latitude was significantly related to religion richness while partialling the effects of *Disease Richness* ($r=-0.27$, $p<0.0001$, $n=214$). Thus, we regressed religion richness on *Disease Richness* and absolute latitude. This multiple regression was significant ($R^2=0.59$, $F_{2,211}=156.72$, $p<0.0001$) with both *Disease Richness* and absolute latitude contributing distinct effects (standardized betas: *Disease*

**Table 13.5** Partial correlations between religion richness and *Disease Richness* or *Contemporary Parasite Stress* while partialling the effects of potentially confounding variables. All *p*'s are <0.0001; number of countries is in parentheses following each partial correlation coefficient (results originally reported in Fincher and Thornhill 2008b)

| | Correlation between *Disease Richness* and religion richness | Correlation between *Contemporary Parasite Severity* and religion richness |
|---|---|---|
| *Variable partialled* | | |
| Absolute latitude | 0.73 (214) | 0.54 (211) |
| Population size (Ln) | 0.59 (210) | 0.47 (207) |
| Land area (Ln) | 0.60 (210) | 0.46 (207) |
| Democracy | 0.72 (167) | 0.58 (167) |
| GDP per capita (Ln) | 0.72 (209) | 0.56 (206) |
| Gini | 0.68 (121) | 0.45 (121) |

*Richness* $=0.71$, absolute latitude $=-0.19$, both *p*'s $<0.0001$). Population size, land area, and democracy were related significantly to religion richness when partialling the effects of *Contemporary Parasite Severity* (respective *r*'s and *p*'s: $r=0.37$, $p<0.0001$; $r=0.34$, $p<0.0001$; $r=-0.21$, $p=0.006$). Thus, we regressed religion richness on *Contemporary Parasite Severity*, population size, land area, and democracy. This multiple regression was significant ($R^2=0.44$, $F_{4,162}=33.49$, $p<0.0001$). Only *Contemporary Parasite Severity* and population size contributed significant, distinct effects to the regression (standardized betas: *Contemporary Parasite Severity* $=0.50$, $p<0.0001$; population size $=0.24$, $p=0.006$; land area $=0.14$, $p=0.11$; democracy $=-0.12$, $p=0.08$).

As described in Sect. 13.4.2, we discovered after the publication of Fincher and Thornhill (2008a, b) that there were some misclassifications of the countries into the world regions. We published the corrected classification as part of the electronic appendices associated with Fincher and Thornhill (2012a). The results we report here for the world regional and geographical band analyses use the corrected classifications. The findings are qualitatively similar to that published in Fincher and Thornhill (2008b).

In all six world regions, religion richness was related positively to *Disease Richness* (and significantly so) (*r*'s 0.26–0.96; Table 13.6) and all three longitudinal bands (American: $r=0.93$, $n=56$; Asia–Australia: $r=0.65$, $n=45$; Europe–Africa: $r=0.78$, $n=113$; all *p*'s $<0.0001$). In all six world regions, religion richness was related positively to *Contemporary Parasite Severity* (*r*'s 0.15–0.83; Table 13.6); however, the correlation in West Eurasia was not significant. A sign test on the direction (+or –) of each correlation shows that the pattern of a positive correlation across the world regions is not spurious ($0.5^6=0.016$). In all three bands, the correlation between religion richness and *Contemporary Parasite Severity* was significantly positive (American: $r=0.78$, $n=54$; Asia–Australia: $r=0$ .63, $n=45$; Europe–Africa: $r=0.70$, $n=112$; all *p*'s $<0.0001$).

**Table 13.6** Pearson zero-order correlations and *p* values between religion richness and *Disease Richness* or *Contemporary Parasite Severity* separated by the six world regions; $n$ = number of countries in the analysis (results originally reported in Fincher and Thornhill 2008b)

| | Disease Richness | | | Parasite stress | | |
|---|---|---|---|---|---|---|
| World Region | *r* | *p* | *n* | *r* | *p* | *n* |
| North America | 0.96 | <0.0001 | 22 | 0.49 | 0.02 | 21 |
| South America | 0.91 | <0.0001 | 27 | 0.83 | <0.0001 | 27 |
| Africa | 0.79 | <0.0001 | 55 | 0.62 | <0.0001 | 54 |
| West Eurasia | 0.26 | 0.04 | 61 | 0.15 | 0.25 | 61 |
| East Eurasia | 0.72 | <0.0002 | 22 | 0.70 | <0.0003 | 22 |
| Insular Pacific | 0.75 | <0.0001 | 27 | 0.72 | <0.0001 | 26 |

### *13.5.3 Conclusions and Discussion*

As predicted, across the countries of the world, we found that religion diversity is the highest where the number of human infectious diseases is also the highest, and the lowest where human parasite diversity is also the lowest. In addition, we found that religion diversity corresponds positively to human parasite stress as measured by the number of cases of the diseases. Our analyses indicated that infectious disease is a more important driver of religion diversity than economic conditions, population size, land area, and the governmental system of countries. Likewise, the effects of different histories of colonization and conquest did not confound our general patterns of finding more religions where there are also more infectious diseases. Moreover, the patterns are observed in the six world culture areas and in three distinct geographical longitudinal bands of nations.

These findings support the parasite-driven-wedge hypothesis of ethnogenesis as applied to the origin of religions. Specifically, religions emerge because intergroup cultural boundaries form in response to the spatial variation and localization of infectious-disease problems. These boundaries are generated by the behavioral immune system's adaptations of assortative sociality and limited dispersal/philopatry.

Earlier evolutionary models of religious behavior provided important insights into religiosity/religious commitment, but did not offer an explanation for why religion diversity varies spatially across the globe. Basically, these models are of two types. One type emphasizes that religiosity, given its high costs, fundamentally functions in honest signaling of in-group commitment and thereby in avoidance of free-riders/cheaters and in in-group coordination and solidarity (e.g., Richerson and Boyd 1998; Irons 2001; Wilson 2002; Sosis 2003). The other type of model purports that religiosity arises from incidental effects of psychological adaptations for purposes other than religiosity (e.g., Kirkpatrick 1999; Boyer 2002; Atran 2002; Dawkins 2006). A more complete understanding of how religiosity relates to religion diversity, however, must include why anti-cheating psychology and the need for in-group coordination and solidarity lead to greater genesis of religions in the tropics compared with temperate regions or why incidental effects generate different religion diversity in the tropics versus temperate regions. The earlier models of religiosity did not do this.

Evidence presented in earlier chapters strongly supported the hypothesis that the importance of in-group coordination and embeddedness and out-group avoidance is actually the manifestation of the importance of in-group assortative sociality and associated limited dispersal for defense against infectious diseases. Although religiosity establishes a social marker and signals in-group boundary, alliance and allegiance, at the most fundamental level, religiosity appears to function for the avoidance and management of infectious disease (Chap. 9). Said succinctly, religiosity is part of the human behavioral immune system. This addition to the earlier models of religiosity mentioned above allows understanding of how religiosity manifests as different numbers of religions in different areas of the world. Across regions, the greater the threat of infectious diseases, the greater the in-group commitment and embeddedness and out-group dislike—and hence the higher the religiosity. And the higher the religiosity in a region, the more the parasite-driven wedge acts to diversify the kinds of religion in the region. Our analysis of religion diversity across the world and our findings about religious commitment throughout the world (discussed in Chap. 9) are mutually reinforcing.

## 13.6 Studies' Findings and Their Relationship to Other Models of Cultural Diversity

Across the countries of the world, we found that both language and religion diversity—two major components of cultural diversity—are the highest where the number of human infectious diseases is also the highest and the lowest where human parasite diversity is also the lowest. In addition, we found that both types of cultural diversity correspond positively to parasite stress as measured by the number of cases of the diseases. Our analyses indicate that infectious disease is a more important driver of both of the types of diversity than a range of other potentially confounding factors included in our analyses. Moreover, the patterns are observed in the six world areas and in three distinct longitudinal bands of nations. The findings support the parasite-driven-wedge hypothesis of parapatric ethnogenesis as applied to the origin of languages and religions.

We found that infectious-disease richness is correlated more strongly with language richness and religion richness than is parasite severity. This, we propose, is because infectious-disease richness often better represents the history of antagonistic coevolutionary arms races across the world than does parasite severity. Each infectious disease, irrespective of its prevalence among hosts, has the potential, over time, to generate localized and spatially variable coevolutionary races and thereby render out-group contact very costly and xenophobia and in-group favoritism and philopatry beneficial.

A common argument for the current distribution of human cultures and their languages is that the pattern is due to regional differences in colonization and conquest and the culture carried along (e.g., McNeill 1976; Diamond 1998; Nettle 1999a; Cashdan 2001). The common, positive correlation between language richness and *Disease Richness* found in each of the six world regions and each of the three longitudinal bands speaks strongly against this as the primary factor for explaining differences in the distribution of languages and religions across the world.

Language or religion diversity is a component of ethnic diversity. Cashdan (2001) provided findings supportive of the parasite-driven-wedge model of the generation of cultural diversity. Using a different measure of infectious-disease burden than we do, which taps the relative severity of a select set of parasites (based on Low 1994) for each society, Cashdan showed that, for 186 indigenous human societies (the Standard Cross-cultural Sample), ethnic diversity was related positively to pathogen stress. To explain this finding, Cashdan (2001) suggested pathogen severity was a limiting factor in the successful dispersal of humans to different areas of the world. She provided examples (mostly from McNeill 1976) indicating that variable levels of immunity to parasites among conquerors and the conquered, in some cases, facilitated invasion of new territory; whereas, in other cases, differences in immunity staved off invasion by conquerors. Cashdan (2001) did not suggest as we have that spatial differences in infectious diseases and the consequent differences in the evocation of behavioral immunity provided the necessary conditions for populations or cultures to fractionate and diverge.

Collard and Foley (2002) and Cashdan (2001) showed that human culture diversity declines with increasing latitude. This finding, along with the already mentioned negative correlations between latitude and human parasite richness (Guernier et al. 2004; Dunn et al. 2010) as well as human pathogen severity (e.g., Cashdan 2001; Fincher and Thornhill 2012), are also generally supportive of the parasite-driven-wedge model.

Earlier in this chapter Nettle's argument for the role of risk and uncertainty from climatic factors in the genesis of languages was discussed. Similar to Nettle's hypothesis, Cashdan (2001) and Collard and Foley (2002) argued for the role of ecological uncertainty in the distribution of human groups. The parasite-driven-wedge model agrees completely with the basic thesis that there will be greater intergroup contact (via more dispersal and reduced xenophobia and collectivism in general) in areas with high climatic risk (high latitudes). We have hypothesized that local levels of pathogen adversity determine to a large extent whether and what forms of intergroup interactions are adaptive and will take place (Fincher and Thornhill 2008a, b; Thornhill et al. 2009). Consequently, we predicted that human groups in areas of high pathogen severity will be characterized by little exchange of goods, norms, and other cultural items. Moreover, widespread networks of exchange and other transactions are expected to primarily persist in areas of low parasite severity. We expect widespread social networks in high pathogen severity areas to be relatively uncommon and impermanent. We provide a range of evidence for these ideas in Chap. 11, which deals with diffusion of innovations among regions as a consequence of parasite stress and associated value systems.

The parasite-driven-wedge model and its supporting evidence provide significant context for two other models of cultural divergence in the literature: Nettle's (1999a, c) social-marker model of language divergence and McElreath et al.'s (2003) model of human ethnic markers. Nettle (1999a, c) argued that languages serve as markers (rather, targets) for individuals to direct reciprocal exchange (see also Nettle and Dunbar 1997). Similarly, Fitch (2004) suggested language dialects serve as markers of kinship and thereby as the basis of nepotistic altruism. Importantly, Nettle (1999a) recognized the empirical inability to explain language diversity based solely on language divergence as a result of geographical barriers. He argued that through the action of the social selection of language (i.e., "[the language] learner does not just pick up all the language activity going on around him or her, but instead homes in specifically on that of a target group. …"; Nettle 1999a, p. 29), languages can emerge and remain distinct without geographical barriers (i.e., they can arise parapatrically). We suggest that while this is accurate, language as a social marker is for the direction of assortative in-group sociality to defend against and manage pathogens, not solely for reciprocal exchange (and not solely for kin recognition and discriminative nepotism as in Fitch's hypothesis).

McElreath et al. (2003) argued that, in humans, normative social behaviors are markers that create cultural boundaries through in-group coordination and cooperation. They argued this relationship between marker and normative behavior provides reliable in-group identification and thereby gives a resolution to the free-rider/cheater problem inherent in any social system that involves altruism. As explained earlier in our book, social norms, as markers used in assortative sociality, probably most generally function in the context of avoiding and managing parasites (see also Schaller and Duncan 2007). McElreath et al.'s model does not offer an explanation for the worldwide distribution of cultural diversity as we do. Furthermore, we predicted in Fincher and Thornhill (2008b) that the spatial variation in pathogen adversity will correspond to regional distribution of adherence to societal norms and values, with greater adherence in areas of high pathogen severity and lower adherence in areas of low pathogen severity. This prediction has been strongly supported empirically by the cross-national research of Murray et al. (2011).

Both models (Nettle and McElreath et al.) indicate the ability of assortative sociality to generate divergence without geographical isolation. While not denying the validity of ethnogenesis resulting from divergence of cultures under geographical separation, we propose that the dominant mode of culture divergence and ethnogenesis in high parasitic-stress regions (e.g., low latitudes) has been parasite-driven, parapatric ethnogenesis.

## 13.7 Political Factionalism

The parasite-driven-wedge model creates cultural diversity by dividing and factionalizing groups based, in part, on amity within each group and enmity between groups. Each group understands itself as having ideology and goals that are distinct from those of neighboring groups. Hence, the parasite-driven-wedge model may apply to the diversity of political groupings in regions within nations. Collectivism and parasite adversity are predicted to fractionate and factionalize political groups within such regions. A well-studied case supporting this was the highly politically factionalized Old South of the Reconstruction Era (1863–1877) that continued subsequently in the region for almost 100 years.

In a remarkable book published in 1949 about the political structure in this period of each of the Old South states—the 11 states that seceded from the USA to form the Confederacy—V.O. Key documented fully the multi-factional nature of the internal politics in these states. Key concluded that, "[t]he South really has no political parties" (p. 16). As Key discussed, many white southerners expressed affiliation with the Democratic Party, or the Southern Democratic Party. But, as Key demonstrates and writes, such parties are "merely a holding-company for a congeries of transient squabbling factions … that fail by far to meet the standards of permanence and cohesiveness that characterize the political party" (p. 16). Southern political structure was based on parochial factions led by a charismatic leader or oligarchy; governance in the South did not rely on political competition between political parties, but, again citing Key (p. 16), "on fortuitous groupings of individuals, on spectacular demagogues odd enough to command the attention of considerable numbers of voters, on men who have become persons of political consequence in their own little bailiwicks, and on other types of leaders whose methods to attract electoral attention serve as substitutes for leadership of a party organization." In the period spanning Reconstruction through the 1960s, the Old South's political factionalism resembled that of other highly collectivist regions such as contemporary Somalia or Afghanistan. The parasite-stress theory provides an encompassing understanding of the otherwise incomprehensible political complexity and chaos of the localized political factionalism that was, as Key emphasized in his book, the Old South's fundamental political and social problem, as well as the gradual recession of this problem—and its associated authoritarianism, classism and apartheid—as health improvements reduced the region's parasite adversity and associated collectivism (see Chap. 10).

## 13.8 Caste Social Systems

Van den Berghe (1981) provided a fascinating, brief review of the scholarly literature on human caste systems, which are based on high social stratification and rigid interclass boundaries. These social systems exhibit cultural diversification arising sympatrically in precisely the same region. (Below we treat sympatric speciation by the parasite-driven wedge.) He documents the lack of consensus among scholars in their efforts to define and explain caste social systems. The parasite-driven diversification model applied to sympatric cultural diversification offers a unique interpretation of caste systems and one supported by the body of evidence in our book.

Caste systems are identified, in part, by an abundance of rules about socially acceptable interactions and conduct. The hierarchy of caste systems is strictly enforced by such rules, and status is typically hereditary and enduring throughout life. A person is born to a class role and this is permanent. According to the ideology of caste systems, the castes or classes vary in their degree of morality, competence, responsibility, sound judgment, and humanness. People know and accept their station in society, and those of low classes have a perceived need to be guided by the ideologically designated wiser and superior humans of higher class. Obedience and deference to hierarchical norms and rules and assumed elite superiority are basic features of high authoritarianism (Chap. 4). A second category of rules in caste societies focuses on creating and maintaining boundary between castes. Who can and cannot be touched, married, addressed in conversation, or otherwise associated with provide rigid fractionation of the society. The Black Codes laws and the Jim Crow laws and norms in the Old South and the similar social rules and regulations until fairly recently in South Africa are on the same dimension of cultural preferences that guide social life, intergroup segregation, and disenfranchisement of lower castes in the caste systems of India and other regions. Both the authoritarianism and the rigid social boundaries reflect highly conservative values and hence prejudice. At a more encompassing and basic proximate level, these features, we propose, are caused by high parasite adversity and the behavioral immune system.

The hypothesis that the castes of India arose out of cultural responses to infectious disease is not original with us, but was proposed by McNeill (1998). We add to this the hypothesis that the degree of caste social strictness and hierarchical boundary will covary positively with degree of conservatism and thus parasite adversity across times and places. Consistent with McNeill's hypothesis is the finding that the castes of India have some caste-specific infectious diseases and classical immunity (Pitchappan 2002). We discussed this in Chap. 3 as an example of the localization of host–parasite coevolutionary races. This finding also reveals that distinct and major cultural differences can emerge, stabilize, and persist for long periods of time when there are different host–parasite coevolutionary races in the same region.

Caste diversification in an area involves caste-specific behaviors of hierarchy and contagion avoidance that arise sympatrically, not parapatrically. Of course, the privileged people of caste systems often isolate the unprivileged people in a separate region. This is seen with the US government's Indian Removal Act that placed Native Americans in specific, bounded reservations spatially separated from whites. It is seen, too, with the Nazi concentration camps and similar instances of the spatial isolation of people deemed subhuman by highly conservative ideologues in their genocidal efforts. Also in this category was the use of Black Codes laws and later Jim Crow laws by white elites to restrict, isolate and marginalize residence locations of blacks throughout the Old South; these residence restrictions continued into the 1960s. Yet despite the various concerted efforts of the socially privileged groups in caste societies to create parapatric separation from their respective untouchables, cultural divergence within a caste system is largely sympatric in origin.

The evidence in our book indicates that caste systems, like all prejudicial institutions and norms, can only be eliminated by the emancipation of the people from infectious diseases through sanitation, healthcare and other public health interventions. Efforts to democratize caste systems without this will fail. As a well-documented example, Reconstruction after the American Civil War failed in its goal of bringing democracy and racial equity to the Old South. Instead, during this period, apartheid was maintained or strengthened in the Old South (see Chap. 10).

## 13.9 Cultural Diversity in Nonhuman Animals

Additionally, while our analyses of cultural diversification above are limited to human cultures, we propose that the basic mechanisms of the parasite-driven wedge can lead to the genesis and maintenance of cultures in other animals in which socially learned mating and other social behavior occurs. For example, culturally acquired song dialects in birds involve assortative mating within spatially distinct dialects of the species-typical song. Apparently, immunological barriers can generate the adaptive value of within-group (-dialect) preference (see MacDougall-Schackleton et al. 2002). The prediction from this is that bird dialects will show more diversity of kinds in high-bird-parasite regions than in low-bird-parasite regions. This prediction of the parasite-driven-wedge model could be tested indirectly by using latitude as a surrogate for bird parasite richness/severity.

## 13.10 On the Origin of Species

### *13.10.1 The Mystery of Mysteries*

According to the parasite-driven-wedge model, localized parasite–host coevolutionary races favor and result in limited dispersal (philopatry), in-group favoritism in social interactions, and out-group avoidance. We have hypothesized that, together, these features of behavioral immunity provide not only an engine creating new cultures, but an engine of speciation as well (Fincher and Thornhill 2008a; Thornhill and Fincher 2013). Earlier in this chapter, we added to the basic parasite-driven-wedge model by emphasizing the linkage between the traits of behavioral immunity and genetic immunity. This linkage results in a synergistic force of evolutionary change: all the traits are mutually reinforcing and the traits of in-group social preference are self-reinforcing. As an engine of species' creation, the model offers a new solution to Charles Darwin's "mystery of mysteries," the label he mentioned in the opening paragraph of his famous book, *On the Origin of Species*, first published in 1859, for the puzzle of the causes of speciation.

As with cultural diversity, species diversity also correlates negatively with latitude. This geographical pattern of species was apparently first recognized by the biologist Alfred R. Wallace, the co-discoverer with Charles Darwin of evolution by natural selection. We have proposed that there are more species in the tropics, because parasites are better able to thrive in tropical environments and thus parasite adversity and the processes of the parasite-driven wedge are most salient in the tropics.

Parasites, in general, are featured in other models of speciation proposed to explain the geographical distribution of species numbers. One popular theme is that parasites are strong selective agents on hosts (similar to predators as selective agents on prey) and hence may be important in generating the evolution of species diversity (e.g., Haldane 1949; Price et al. 1986; Summers et al. 2003). We are not in disagreement with this view, but emphasize that the parasite-driven-wedge model casts parasites in a more encompassing and specific role as depicted in the model's sequential events presented earlier in this chapter (Sect. 13.2).

### *13.10.2 Speciation Processes*

Darwin was an advocate of all three basic theories of species formation recognized in biology: allopatric speciation, parapatric speciation, and sympatric speciation. He thought each of the models made a contribution to the origin of species and the diversity of species across the planet (Coyne and Orr 2004). In this section, we treat allopatric speciation, parapatric speciation, and the parasite-driven-wedge model as a type of parapatric speciation, after which we discuss sympatric speciation and its relationship to diversification by parasite stress. We will argue that the parasite-driven-wedge is not only involved in parapatric speciation, but also in sympatric speciation.

Allopatric speciation begins with the division of the range of a species into segments by a distinct geographical barrier, extensive unsuitable habitat, or by colonization of a separate range by a subset of the species. This first step is followed by evolutionary divergence of the segments. The genetic and phenotypic divergence in allopatry (in separated geographical areas) leads to a lowering or an absence of reproductive compatibility between the segments of the species. If the segments subsequently overlap in their ranges, they may be reproductively isolated as a by-product of their divergence in allopatry. Or, they may have partial reproductively compatibility with hybrid disadvantage, leading to direct selection for reinforcement of trait differences that reduce hybridization. The status of distinct species is achieved when complete or near complete reproductive isolation occurs. Biologists generally have assumed that allopatric speciation was the predominant mode of speciation (Endler 1977; Otte and Ender 1989; Coyne and Orr 2004).

Parapatric speciation is thought by most biologists to be a relatively infrequent mechanism of species' formation compared to allopatric speciation (Endler 1977). For instance, parapatric speciation typically receives only brief discussion in college and university textbooks for courses in evolution, whereas allopatric speciation receives detailed coverage. The etymology of allopatric is separate (allo) and patry (land), referring to the geographical separation of the population segments that diverge to become distinct species. The etymology of parapatric also is descriptive of the concept it labels: para (beside) and patry (land). In this case, species arise side-by-side/contiguously and without subdivision of the ancestral population by a geographical barrier.

The essence of the parapatric theory of speciation is as follows. Initially, the original range of a species has an ecological gradient—say, one side of the range is dry and the other side is wet. In each of the two sides, there is natural selection for the ability to cope with the local climate. This leads to an evolutionary divergence between the two portions of the species. With sufficient divergence between the two sides, hybrids produced by inter-mating between parents adapted to their locally distinct environments have reduced reproductive success compared to offspring produced by non-hybridizing parents. This hybrid disadvantage gives rise to Darwinian selection favoring individuals that prefer the local habitat and prefer locals as mates (Fisher 1930; Endler 1977).

The parasite-driven-wedge model is a special case of parapatric diversification. In this case, the ecological gradient across the range of a species is the result of different parasite species or different genotypes of a single parasite species. There is selection for the ability to cope with the local parasites, which leads to evolutionary divergence in adaptations for local genetic immunity and behavioral immunity. Hybrids produced by inter-mating between individuals adapted to the different infectious-disease environments have reduced reproductive success compared to offspring produced by non-hybridizing individuals. The Darwinian selection for local immunity and avoidance of hybridization favors parochialism. Specifically, as explained earlier in this chapter, direct selection favors individuals that (1) have immunity to local parasites, (2) can distinguish local from non-local individuals, (3) prefer locals as mates and other social allies, (4) avoid non-locals, and (5) are philopatric. As traits (1)–(5) increase in frequency across generations, they become combined within individuals. Also, the linkage itself is directly selected, including selection for the compatibility of its components. The traits of local classical immunity, and local habitat and in-group social preference are in genetic linkage disequilibrium. As a result of the linkage, when any one of the traits is favorably selected, all the other linked traits are favored simultaneously. The linkage also results in the self-reinforcement of social preferences for local individuals: the individuals preferred and benefitted because they are preferred to have the same preferences as those that prefer them. Given the strength of selection typically attributed to infectious diseases, the hybrid reproductive disadvantage is considerable and the selection pressures for (1)–(5)—i.e., selection for locally adaptive classical and behavioral immunity—and their linkage are strong. Distinct species that have arisen side-by-side without geographical separation or barriers are the final result of the parasite-driven-wedge processes.

Certainly, we are not claiming that the allopatric mechanism is unimportant in the origin of new species. Our proposal is that a dominant mode of speciation in high parasitic stress regions (e.g., low latitudes) has been parasite-driven parapatric speciation.

Evolutionary biologists often remark that, with current data, it is difficult to distinguish species that originated allopatrically from those that arose parapatrically. The current data referred to are the patterns seen in phylogenetic trees, geographical ranges of closely related species, and the nature of hybrid zones (see various chapters in Otte and Ender 1989). These data sources are almost always ambiguous with regard to whether speciation was by allopatric or parapatric means. There are, however, some cases that many biologists accept as resulting from parapatric speciation (Endler 1977).

We suggest three unambiguous tests for distinguishing allopatric speciation from parasite-driven parapatric speciation. First, when hybrids of closely related species can be obtained, if the parasite-driven wedge has resulted in speciation, these hybrids will be inferior in fitness only or primarily because of their lowered fitness in the contexts of classical immunity, local habitat preference and assortative social behavior rather than in a wide range of fitness components that would be affected by divergence in geographical isolation. In contrast, allopatric speciation is expected to generate hybrid disadvantage across all fitness components with more-or-less equal likelihood.

In Chap. 3, we mentioned some studies showing that a hybrid disadvantage among closely related species is attributable to reduced classical immunity in the hybrids. This is not sufficient evidence to prove parapatric speciation, but it is consistent with it. The evidence required by our distinguishing criterion is that when parasite-driven parapatric speciation has given rise to closely related species, their hybridization will lead to hybrid descendants with relatively more fitness loss from the parasites of either parent species than from the many environmental hostile forces that can negatively impact fitness, but are not specific to the parasite-driven-wedge model. We know of no current data that could be used to assess this method of distinguishing parasite-driven parapatric from allopatric speciation. Hence, new research is required that can discriminate between the predictions of the wedge model and predictions from allopatric speciation.

A second way to distinguish between parasite-driven parapatric speciation and allopatric speciation is by adaptationism. This method of studying functional design is the only way to empirically identify and document historical selection pressures that produced adaptations (Chap. 2). Closely related species that emerged from the parasite-driven-wedge model will show classical-immunity and behavioral-immunity adaptations that are species-specific. The behavioral-immunity adaptations involved will exhibit functional design for in-group assortative sociality, out-group avoidance and philopatry. Such species will differ primarily in these regards. Closely related species produced by allopatric events, however, will show a wider range of trait differences that may not include any divergent adaptations of classical or behavioral immunity. The designs of adaptations that function in avoidance of social contact with out-groups are empirically distinguishable from the many by-product traits, including premating behaviors, that may reproductively isolate closely related species that speciated in allopatry.

A third way to separate parasite-driven parapatric speciation from allopatric speciation is by determination of the genetic structure among traits. Allopatric speciation makes no specific predictions about genetic correlation of traits affecting reproductive isolation. In contrast, the parasite-driven-wedge model makes specific predictions about the genetic intercorrelation of immunity to local parasites, local habitat preference, and in-group assortative social behaviors. Of the three ways we have suggested for distinguishing parasite-driven speciation from allopatric speciation, the study of genetic covariances among behavioral immunity traits and classical immunity is probably the most demanding of time and resources (for methods, see Falconer and Mackay 1996).

Next we treat some earlier research relevant to the parasite-driven-wedge model of parapatric speciation.

### 13.10.3 Other Research Relevant to the Wedge Model of Parapatric Speciation

The parasite-driven-wedge model of speciation adds significantly to Hochberg et al.'s (2003) model of socially mediated speciation. Hochberg et al.'s (2003) social-marker-based model showed that premating reproductive isolation can emerge without geographical barriers through the action of directing altruism to similarly marked individuals and in-group assortative mating. While we consider Hochberg et al.'s speciation model to be a useful starting point, it does not explain why the rates of speciation have differed dramatically worldwide. Furthermore, it gives no context for the presence of relevant markers in the first place; their existence is assumed. Missing from their model is the incorporation of the worldwide spatial variation in parasite adversity and the resultant variation in assortative sociality and philopatry. Our addition to this model is that the importance of in-group markers for discriminating appropriate social partners, including mates, and preferring in-group ones, will increase as parasite stress increases. This addition allows Hochberg et al.'s model to be consistent with the copious data documenting the latitudinal gradient in species diversity.

We consider our model of parasite-mediated diversification of species to be an important synthetic link between previous research showing that species richness is related positively to warmer temperatures, greater precipitation, amount of tropical forestation, ecological productivity, ecological specialization, limited dispersal, or physiological specialization (Reed and Fleagle 1995; Allen et al. 2002; Hawkins et al. 2003; Belmaker et al. 2012; Salisbury et al. 2012). All these factors are components of the parasite-driven-wedge model of speciation.

Others have recognized the importance of limited dispersal in generating population divergence. Belliure et al. (2000) provide evidence that, in British birds, species characterized by limited dispersal also had higher numbers of associated subspecies indicative of relatively greater population diversification. Since subspeciation can precede speciation, these findings indicate that limited dispersal increases the rate of speciation.

Researchers have found positive correlations between parasite diversity and clade diversity that support the hypothesis that speciation is caused by the parasite-driven-wedge, but they attributed the correlations to different causes. Clade diversity refers to the number of species in a clade—a group of species descended from the same common ancestral species. Krasnov et al. (2004) found that small-mammal species richness was correlated positively with the species richness of their fleas. In their words (Krasnov et al. 2004, p. 1861), "This positive correlation suggests that diversification of parasites is a response to diversification of hosts." Nunn et al. (2004) used similar reasoning to explain their finding that the more speciose a primate clade, the more speciose the clade's parasites. These findings and ideas, while generally supportive of our model, do present a logical challenge because we could interpret our findings on diversity similarly. We reiterate, however, the alternative framework that we propose: the presence of parasite adversity generates selection for mechanisms to prefer in-group, immunologically similar conspecifics and avoid dissimilar ones, and limit dispersal. While this discriminative social engagement generates population divergence among hosts, it also generates population divergence among the hosts' parasites. Simultaneously, parasites and hosts are involved in antagonistic, coevolutionary races that can lead to further divergence within host and parasite populations. Short of clade extinction, this positive feedback is unending in its effect on diversification and has presumably been so since the time in life's history when the first infective agent attacked a host.

As we have mentioned, because of the selective advantage of limited dispersal under severe parasite levels, we have suggested that the general finding that species' geographical ranges are smaller in the tropics than in temperate areas (Rapoport's rule; Stevens 1989) derives from the greater pathogen severity found at low latitudes. Thus, dispersal variation in response to local pathogen severity provides a new potential context for Rapoport's rule, an important ecological pattern. This new context is attractive as a reasonable explanation because it is consistent with the wide variety of evidence we present in this chapter for the role of parasites in causing biodiversity.

Parasites play a role in the distribution of species through their effects on range expansion by host species. For example, parasites are known to facilitate range expansion by serving as biological weapons that invading species or groups carry along, effectively eliminating resident competitors (Freeland 1983; Diamond 1998). Hosts also have been able to invade successfully new territories by leaving their parasites behind (Mitchell and Power 2003; Torchin et al. 2003). Prior research has also implicated parasites as important moderators of species coexistence and community composition (see, e.g., Freeland 1983; Pagel et al. 1991). While certainly important for understanding the distribution of species, parasites in these roles, in contrast to their role in the parasite-driven-wedge model, do not provide the types of general mechanisms that explain the latitudinal species-diversity gradient.

The parasite-driven-wedge model is also different from the Theory of Pest Pressure (Gillett 1962; Janzen 1970; Connell 1971; Gilbert 2005). This hypothesis is based on the idea that high density-dependent mortality due to parasites at low latitudes creates a condition whereby no species can be exceptionally common. Thus, high levels of tropical diversity arise and persist because many more species can coexist in low latitude areas than in temperate areas. The Theory of Pest Pressure might be consistent with the data we present: more parasites, more diversity. However, research has shown that similar levels of density-dependent mortality are experienced in temperate areas (Hille Ris Lambers et al. 2002), reducing the likelihood that density-dependent mortality due to infectious disease is a general explanation for the latitudinal biodiversity gradient.

Future research can best distinguish between the parasite-driven diversification model and the Theory of Pest Pressure by searching for adaptations predicted uniquely by the parasite-driven-wedge model. The parasite-driven-wedge model predicts that the psychology of in-group assortative social preference and philopatry will differ in accord with parasite intensity levels. The Theory of Pest Pressure makes no such predictions.

The genetic distance between nearby populations should be greater in areas of high parasite severity. This, we pose, results from the history of local reproduction and strong selection for in-group assortative social interactions, philopatry, and avoiding out-group contact in these areas. In low parasite areas, out-group mating and dispersal will have been less costly and may manifest oftentimes as genetically more similar adjacent populations. This prediction about genetic distance of nearby populations is supported by the study by Martin and McKay (2004) of the genetic divergence of populations in 62 vertebrate taxa. They reported that the genetic divergence is related negatively to latitude (and hence is related positively to parasite adversity). Also, these patterns across vertebrate taxa are consistent with the positive relationship between dispersal and latitude that was discussed earlier in this chapter. High parasite adversity leads to host populations becoming geographically subdivided and isolated genetically.

Similarly, spatial differences in parasite prevalence might produce patterning in hybrids and hybrid zones. Hybrid matings provide avenues for out-group parasites to travel between host populations. Hence, the wedge model predicts that hybrids among species will be more common in low parasite severity regions than in regions of high parasite adversity. Moreover, because increased contact with novel parasites is a cost of hybrid mating, it is predicted that the extent of hybrid zones will vary across parasitic gradients, with wider zones found in low parasite adversity areas and narrow zones found in high parasite adversity areas.

A current debate centers on whether the transmission of culture (including language) proceeds primarily through vertical transmission (between generations, e.g., parent to offspring) or horizontal transmission (society to society), and whether a bifurcating phylogeny and associated analytical techniques is the best method for studying human cultural history and diversification (Borgerhoff Mulder et al. 2006; Collard et al. 2006). Vertical transmission results in branching cultural phylogenies (strictly bifurcating), while horizontal transmission results in blending cultural phylogenies (reticulate). The spatial variation of parasite intensities may help resolve this debate. Cultures that emerge from high-parasite-adversity regions are expected to produce predominantly bifurcating phylogenies representing significant philopatry, in-group assortative sociality and associated inter-group divergence and isolation, while cultures that come from low-parasite-adversity regions may be best represented by reticulate, blending phylogenies resulting from lower levels of philopatry and assortative sociality and increased exchange of ideas and values with out-groups.

We add that phylogenies for other species also may contain histories of blending and branching in accordance with variation in parasite intensities. We predict that the phylogenies of species that emerge at high latitudes (low parasite intensity) will appear more reticulate than phylogenies of species that emerge at low latitudes (high parasite intensity), which will appear more bifurcate. In sum, we propose that spatial variation in parasite adversity has major implications for understanding the phylogenetic histories of both cultures and species.

### 13.10.4 Parasite-Driven Sympatric Speciation

Hochberg et al.'s socially mediated speciation model, discussed above, involves diversification within an ancestral population resulting from preferences for local social makers of in-group and associated assortative mating. Their model is consistent with processes of both parapatric and sympatric speciation.

Sympatric speciation is speciation in the same locality without any geographical separation. It does not involve species arising adjacently across the range of an ancestral species, as with parapatric speciation. Instead a population's segments become ecologically isolated and divergent through local adaptation to ecological differences that are intermixed in the same locale (e.g., different host-plant species) (Diehl and Bush 1989; Tauber and Tauber 1989).

A possible case of this is the diversification of *Rhagoletis* flies in North America. According to Diehl and Bush (1989), the following historical events caused speciation in *Rhagoletis*. Originally, a species of endemic *Rhagoletis* parasitized hawthorn as its host for egg laying and larval development. After the apple, a near relative of hawthorn, was introduced into North America, some *Rhagoletis* successfully colonized apple as a host. As a result of natal host-plant fidelity for mating and egg laying and disruptive selection favoring adaptation to the natal host plant but not the other host, there was a divergence between the original *Rhagoletis* stock and the apple *Rhagoletis*, to the point of near reproductive isolation. Note that natal host fidelity is a type of philopatry and local habitat preference that leads to assortative mating by natal host type or habitat.

Research by Blais et al. (2007) is supportive of Hochberg et al. (2003) and our suggested addition to it (discussed above). Blais et al. (2007) reported that apparent sympatric speciation of a pair of African cichlid fish species was related to assortative mating based on adaptive divergence in immunity genes. These immunity genes function as a defense against infectious disease and are purportedly important social markers affecting mating preference. Here, mate choice and evolutionary divergence are based on the same factor: immunity.

Tauber and Tauber (1989) presented evidence indicating that sympatric speciation may occur in certain phytophagous insects other than *Rhagoletis*. We suggest that parasite-driven diversification in sympatry is a frequent process of speciation in phytophagous insects, especially in high-parasite-adversity regions, such as low latitudes. When a portion of an ancestral species shifts to a new host in the same local, the parasite-driven-wedge mechanism may result in speciation if the new host presents novel infectious disease challenges. Such challenges are indicated by the evidence that parasite severity of the phytophagous insects that exploit multiple host types is variable across the types (Price et al. 1980; Jaenike 1990; Feder 1995). Furthermore, a basic theoretical feature expected to promote sympatric speciation (and parapatric speciation) is a mating preference for similar individuals that is based on the same criterion as that causing the population fractions to diverge evolutionarily. This is a fundamental feature of the parasite-driven-wedge model of diversification.

High parasite stress favors classical and behavioral immunity adaptations that defend against local parasites, including local habitat preference and in-group favoritism in mating. These traits link through genetic linkage disequilibrium and are mutually reinforced by selection. Hence, local mate preference is linked with local habitat and host preference. The adaptive value of this combination, as well as each of its components, is its resistance to local parasites. Also, the wedge model includes the self-reinforcement of assortative mating preference. As a result of the trait linkages involved in the wedge, when individuals use a particular host type and have a mating preference for others that do the same, the mating preference reinforces itself because preferred mates also have the same mating preference.

It is an onerous task to demonstrate that sympatric speciation accounts for the speciation history of closely related species (see Coyne and Orr's 2004 review and critical discussion). We suggest tests for parasite-mediated sympatric speciation analogous to the three suggested above for parasite-driven parapatric speciation. These were as follows: (a) the hybrids of closely related species or incipient species will show hybrid inferiority predominantly in behavioral and classical immunity; (b) closely related species will differ primarily in terms of behavioral and classical immunity adaptations; and (c) such species will exhibit genetic correlations in traits specified by the parasite-driven-wedge model (see above). Such results would question the allopatric process for the speciation. If these results are found, the task then is to separate the two speciation processes that can yield them, either parapatric or sympatric speciation. Evidence from range patterns, ecology, and other natural history information may allow separation of the two processes in each case.

Earlier in this chapter we discussed cultural diversification within highly stratified human social systems. There we proposed that the parasite-driven wedge creates diversification in behavioral and classical immunity in a sympatric context and thereby causes caste societies.

## 13.11 Summary

The parasite-stress theory of sociality includes a theory of biodiversity: the parasite-driven-wedge model. Regionally localized coevolutionary races between parasites and their hosts result in three anti-pathogen behaviors: preference for in-group affiliation and interaction, out-group avoidance (xenophobia), and limited dispersal (philopatry). The first two of these traits, social favoritism toward in-group members and out-group avoidance, comprise in-group assortative sociality. All three behaviors comprise behavioral immunity. They become linked within individuals through genetic linkage disequilibrium. In the case of human cultural behavioral immunity, within-individual linkage of behavioral immunity traits results in what we refer to as cultural linkage disequilibrium. Linkage by either process also includes correlation or linkage with genetic immunity to local parasites. These linked traits are mutually reinforcing in that as any one increases in frequency due to its adaptiveness, the others do as well. Also, preference for in-group members with behavioral immune values and behavior is self-reinforcing because the in-group members preferred and favored have the same preference.

These events create a wedge that gives rise to intergroup boundaries that effectively fractionate, locally isolate, and diversify the original range of a culture or a species, leading to the genesis of two or more discrete groups from one. The higher the parasite stress in a region, the greater the frequency and intensity of these processes of biodiversity genesis. The parasite-driven-wedge model, then, provides a parapatric (side-by-side) diversification mechanism. We do not deny an important role for vicariance events (isolation of segments of an ancestral population by geographical barriers) in the genesis of biodiversity, but argue that parapatric processes will be relatively predominant in regions of high parasite stress, leading to the high diversity of species and cultures in the regions. This fractionation of host populations through the parasite-driven wedge also parapatrically diversifies parasites, leading to even greater geographical localization of host–parasite races.

The parasite-stress model of parapatric divergence may be a general theory of biodiversity, as suggested by five main lines of evidence: (1) the cross-taxa negative relationships between species richness and, in humans, cultural richness, and latitude; (2) the positive relationships between human cultural diversity (in terms of endemic languages and religions) and parasite stress; (3) the negative relationship between home range size and parasite stress in traditional human societies; (4) the positive relationship between species' range size and latitude (Rapoport's rule); and (5) the negative relationship between the genetic differentiation between populations (and thus limited allele flow) and latitude in vertebrate taxa. Other patterns we discuss also suggest support.

Parasite-driven divergence may lead to sympatric speciation, especially at low latitudes. Methods are discussed for distinguishing parapatric and sympatric parasite-driven diversification from allopatric diversification during speciation.

Parasite-driven cultural diversity that arises in sympatry may account for caste and related highly hierarchical social systems characterized by conservatism and its associated authoritarianism and prejudice and rigid ideological boundaries between strata. Such social systems can be made less authoritarian and prejudicial by emancipating the people involved in them from parasite adversity.

## References

Allen, A. P., Brown, J. H., & Gillooly, J. F. (2002). Global biodiversity, biochemical kinetics, and the energetic-equivalence rule. *Science* 297: 1545-1548.

Atran, S. (2002). *In Gods We Trust: The Evolutionary Landscape of Religion*. Oxford University Press USA, New York, NY.

Barber, I., & Dingemanse, N. J. (2010). Parasitism and the evolutionary ecology of animal personality. *Philosophical Transactions of the Royal Society B–Biological Sciences* 365: 4077-4088.

Barrett, D. B., Kurian, G. T., & Johnson, T. M., Eds. (2001). *World Christian Encyclopedia: A Comparative Survey of Churches and Religions in the Modern World, Volume 1: The World by Countries: Religionists, Churches, Ministries*, 2nd ed. Oxford University Press, Oxford, U.K.

Belliure, J., Sorci, G., Møller, A. P. et al. (2000). Dispersal distances predict subspecies richness in birds. *Journal of Evolutionary Biology* 13: 480-487.

Belmaker, J., Sekercioglu, C. H., & Jetz, W. (2012). Global patterns of specialization and coexistence in bird assemblages. *Journal of Biogeography* 39: 193-203.

Best, A., Webb, S., White, A. et al. (2011). Host resistance and coevolution in spatially structured populations. *Proceedings of the Royal Society B* 278: 2216-2222.

Blais, J., Rico, C., van Oosterhout, C. et al. (2007). MHC adaptive divergence between closely related and sympatric African cichlids. *PloS ONE* 2: e734. doi:10.1371/journal.pone.0000734

Borgerhoff Mulder, M., Nunn, C. L., & Towner, M. C. (2006). Cultural macroevolution and the transmission of traits. *Evolutionary Anthropology* 15: 52-64.

Boyd, R. & Richerson, P. J. (1985). *Culture and the Evolutionary Process*. Chicago University Press, Chicago, IL.

Boyer, P. (2002). *Religion Explained: The Evolutionary Origins of Religions Thought*. Basic Books, New York, NY.

Bradbury, I. R., Laurel, B., Snelgrove, P. V. R., Bentzen, P., & Campana, S. E. (2008). Global patterns in marine dispersal estimates: The influence of geography, taxonomic category and life history. *Proceedings of the Royal Society B* 275: 1803-1809.

Calvete, C. (2003). Correlates of helminth community in the red-legged partridge (*Alectoris rufa* L.) in Spain. *Journal of Parasitology* 89: 445–451.

Cashdan, E. (2001). Ethnic diversity and its environmental determinants: Effects on climate, pathogens, and habitat diversity. *American Anthropology* 103: 968-991.

Collard, I. F., & Foley, R. A. (2002). Latitudinal patterns and environmental determinants of recent human cultural diversity: Do humans follow biogeographical rules? *Evolutionary Ecology Research* 4: 371-383.

Collard, M., Shennan, S. J., & Tehrani, J. J. (2006). Branching, blending, and the evolution of cultural similarities and differences among human populations. *Evolution and Human Behavior* 27: 169-184.

Connell, J. H. (1971). On the role of natural enemies in preventing competitive exclusion in some marine animals and in rain forest trees. In *Dynamics of Numbers in Populations* (Proceedings of the Advanced Study Institute, Osterbeek 1970), pp. 298-312. Centre for Agricultural Publication and Documentation, Wageningen, The Netherlands.

Coyne, J. A., & Orr, H. A. 2004. *Speciation*. Sinauer Assoc., New York, NY.

Dawkins, R. (2006). *The God Delusion*. Houghton Mifflin, Boston, MA.

Diamond, J. (1998). *Guns, Germs and Steel: The Fates of Human Societies*. W. W. Norton and Co., New York, NY.

Diehl, S. R., & Bush, G. L. (1989). The role of habitat preference in adaptation and speciation. In *Speciation and Its Consequences* (eds. D. Otte & J. A. Endler), pp. 345-365. Sinauer Associates, Inc., Sunderland, MA.

Dunn, R. R., Davies, T. J., Harris, N. C. et al. (2010). Global drivers of human pathogen richness and prevalence. *Proceedings of the Royal Society B* 27: 2587-2595.

Endler, J. A. (1977). *Geographic Variation, Speciation, and Clines*. Princeton University Press, Princeton, NJ.

*Ethnologue*, www.ethnologue.com

Ewald, P. W. (1994). *Evolution of Infectious Disease*. Oxford University Press, New York, NY.

Falconer, D. S., & Mackay, T. F. C. (1996). *Introduction to Quantitative Genetics*, fourth edition. Benjamin Cummings, New York, NY.

Feder, J. L. (1995). The effects of parasitoids on sympatric host races of *Rhagoletis pomonella* (Diptera: Tephridae). *Ecology* 76: 801-813.

Felsenstein, J. (1981). Skepticism towards Santa Rosalia, or why are there so few kinds of animals? *Evolution* 35: 124-138.

Fincher, C. L., & Thornhill, R. (2008a). A parasite-driven wedge: Infectious diseases may explain language and other biodiversity. *Oikos* 117: 1289-1297.

Fincher, C. L., & Thornhill, R. (2008b). Assortative sociality, limited dispersal, infectious disease and the genesis of the global pattern of religion diversity. *Proceedings of the Royal Society of London, Biological Sciences* 275: 2587-2594.

Fincher, C. L., & Thornhill, R. (2012). Parasite-stress promotes in-group assortative sociality: The cases of strong family ties and heightened religiosity. *Behavioral and Brain Sciences* 35: 61-79.

Fincher, C. L., Thornhill, R., Murray, D. R. et al. (2008). Pathogen prevalence predicts human cross-cultural variability in individualism/collectivism. *Proceedings of the Royal Society of London Biological Sciences* 275: 1279-1285.

Fisher, R. A. (1930). *The Genetic Theory of Natural Selection.* Claredon Press, Oxford, U.K.

Fitch, W. T. (2004). Kin selection and "mother tongues": A neglected component in language evolution. In *Evolution of Communication Systems: A Comparative Approach* (eds. D. K. Oller & U. Griebel), pp. 275-296. MIT Press, Cambridge, MA.

Freeland, W. J. (1979). Primate social groups as biological islands. *Ecology* 60: 719-728.

Freeland, W. J. (1983). Parasites and the coexistence of animal host species. *American Naturalist* 121: 223-236.

Gaston, K. J. (2000). Global patterns in biodiversity. *Nature* 405: 220-227.

Gilbert, G. S. (2005). Dimensions of plant disease in tropical forests. *Biotic Interactions in the Tropics* (eds. D. Burslem, M. Pinard, & S. Hartley), pp. 141-164. Cambridge University Press, Cambridge, U.K.

Gillett, J. B. (1962). Pest pressure, an underestimated factor in evolution. *Systematics Association Publication Number* 4: 37-46.

Grim, B. J., & Finke, R. (2006). International religion indexes: Government regulation, government favoritism, and social regulation of religion. *Interdisciplinary Journal of Research on Religion* 2: 1-40.

Guernier, V., Hochberg, M. E., & Guégan, J. -F. (2004). Ecology drives the worldwide distribution of human diseases. *PLoS Biology* 2: 740-746.

Haldane, J. B. S. (1949). Disease and evolution. *Ricerca Scientifica Suppl. A* 19: 68-76.

Hamilton, W. D. (1964). The genetical evolution of social behaviour, I & II. *Journal of Theoretical Biology* 7: 1-52.

Harmon, D. (1996). Saving nature's legacy: Protecting and restoring biodiversity. *Society and Natural Resources* 9: 435-437.

Harmon, D. (2002). *In Light of Our Differences: How Diversity in Nature and Culture Makes Us Human*. Smithsonian Institution Press, Washington D C, USA.

Hawkins, B. A., Field, R., Cornell, H. V. et al. (2003). Energy, water, and broad-scale geographic patterns of species richness. *Ecology* 84: 3105-3117.

Hille Ris Lambers, J., Clark, J. S., & Beckage, B. (2002). Density-dependent mortality and the latitudinal gradient in species diversity. *Nature* 417: 732-735.

Hillebrand, H. (2004). On the generality of the latitudinal diversity gradient. *American Naturalist* 163: 192-211.

Hochberg, M. E., Sinervo, B., & Brown, S. P. (2003). Socially mediated speciation. *Evolution* 57: 154-158.

Irons, W. (2001). Religion as a hard-to-fake sign of commitment. *Evolution and the Capacity for Commitment* (ed. R. M. Nesse), pp.292-309. Russell Sage Foundation, New York, NY.

Jaenike, J. (1990). Host specialization in phytophagous insects. *Annual Review of Ecology and Systematics* 6: 365-397.

Janzen, D. H. (1970). Herbivores and the number of tree species in tropical forests. *American Naturalist* 104: 501-528.

Jones, K. E., Patel, N. G., Levy, M. A. et al. (2008). Global trends in emerging infectious diseases. *Nature* 451: 990-993.

Kirkpatrick, L. A. (1999). Toward an evolutionary psychology of religion and personality. *Journal of Personality* 67: 921–952.

Kirkpatrick. M. (1982). Sexual selection and the evolution of female choice. *Evolution* 36: 1-12.

Krasnov, B. R., Shenbrot, G. I., Khokhlova, I. S. et al. (2004). Relationship between host diversity and parasite diversity: Flea assemblages on small mammals. *Journal of Biogeography* 31: 1857–1866.

Lande, R. (1981). Models of speciation by sexual selection on polygenic traits. *Proceedings of the National Academy USA* 78: 3721-3725.

Loehle, C. (1995). Social barriers to pathogen transmission in wild animal populations. *Ecology* 76: 326-335.

Low, B. S. (1994). Pathogen severity cross-culturally. *World Cultures* 8: 24-34.

MacArthur, R. H., & Connell, J. H. (1966). *The Biology of Populations*. John Wiley and Sons, Inc., New York, NY.
MacDougall-Schackleton, E. A., Derryberry, E. P., & Hahn, T. P. (2002). Nonlocal male mountain white-crowned sparrows have lower paternity and higher parasite loads than males singing local dialect. *Behavioral Ecology* 13: 682-689.
Mace, R., & Pagel, M. (1995). A latitudinal gradient in the density of human languages in North America. *Proceedings of the Royal Society of London B* 261: 117-121.
Maffi, L. (2005). Linguistic, cultural, and biological diversity. *Annual Review of Anthropology* 29: 599-617.
Martin, P. R., & McKay, J. K. (2004). Latitudinal variation in genetic divergence of populations and the potential for future speciation. *Evolution* 58: 938-945.
McElreath, R., Boyd, R., & Richerson, P. J. (2003). Shared norms and the evolution of ethnic markers. *Current Anthropology* 44: 122-129.
McNeill, W. H. (1976). *Plagues and Peoples*. Anchor Books, Harpswell, ME.
McNeill, W. H. (1981). Migration patterns and infection in traditional societies. In *Changing Disease Patterns and Human Behavior* (eds. N. F. Stanley & R. A. Joske), pp. 27-36. Academic Press, Salt Lake, UT.
McNeill, W. H. (1998). *Plagues and Peoples*. Anchor, Harpswell, ME.
Mitchell, C. E. & Power, A. G. (2003). Release of invasive plants from fungal and viral pathogens. *Nature* 421: 625-627.
Mittelbach, G. G., Schemske, D. W., Cornell, H. V. et al. (2007). Evolution and the latitudinal diversity gradient: Speciation, extinction and biogeography. *Ecology Letters* 10: 315-331.
Møller, A. P., Dufva, R., & Allander, K. (1993). Parasites and the evolution of host social behaviour. *Advances in the Study of Behavior* 22: 65-102.
Møller, A. P., Martin-Vivaldi, M., & Soler, J. J. (2004). Parasitism, host immune defense and dispersal. *Journal of Evolutionary Biology* 17: 603-612.
Murdock, G. P. (1949). *Social Structure*. MacMillan, New York, NY.
Murray, D. R., Trudeau, R., & Schaller, M. (2011). On the origins of cultural differences in conformity: Four tests of the pathogen prevalence hypothesis. *Personality and Social Psychology Bulletin* 37: 318-329.
Nettle, D. (1999a). *Linguistic Diversity*. Oxford University Press, Oxford, NY.
Nettle, D. (1999b). Linguistic diversity of the Americas can be reconciled with a recent colonization. *Proceedings of the National Academy of Science USA* 96: 3325-3329.
Nettle, D. (1999c). Language variation and the evolution of societies. In *The Evolution of Culture* (eds. R. Dunbar et al.), pp. 214-227. Rutgers University Press, Rutgers, NJ.
Nettle, D. (2000). Linguistic fragmentation and the wealth of nations: The Fishman-Pool hypothesis re-examined. *Economic Development and Cultural Change* 48: 335-348.
Nettle, D., & Dunbar, R. I. M. (1997). Social markers and the evolution of reciprocal exchange. *Current Anthropology* 38: 93-98.
Nunn, C. L., Altizer, S., Jones, K. E. et al. (2003). Comparative tests of parasite species richness in primates. *American Naturalist* 162: 597-614.
Nunn, C. L, Altizer, S., Sechrest, W. et al. (2004). Parasites and the evolutionary diversification of primate clades. *American Naturalist* 164: S90-S103.
Nunn, C. L., Altizer, S. M., Sechrest, W. et al. (2005). Latitudinal gradients of parasite species richness in primates. *Diversity and Distributions* 413: 249-256.
Otte, D., & Ender, J. A., Eds. (1989). *Speciation and Its Consequences*. Sinauer Associates, Sunderland, MA.
Pagel, M. D., May, R. M., & Collie, A. R. (1991). Ecological aspects of the geographical distribution and diversity of mammalian species. *American Naturalist* 137: 791-815.
Pitchappan, R. M. (2002). Castes, migration, immunogenetics and infectious diseases in south India. *Community Genetics* 5: 157-161.
Price, P. W., Bouton, C. E., Gross, P. et al. (1980). Interactions among three trophic levels: influence of plants on interactions between insect herbivores and natural enemies. *Annual Review of Ecology and Systematics* 11: 41-65.

Price, P. W., Westoby, M., Rice, B. et al. (1986). Parasite mediation in ecological interactions. *Annual Review of Ecology Evolution and Systematics* 17: 487-05.

Reed, K. E., & Fleagle, J. G. (1995). Geographic and climatic control of primate diversity. *Proceedings of the National Academy of Sciences USA* 92: 7874-7876.

Rice, W. R. (1984). Disruptive selection on habitat preference and the evolution of reproductive isolation: A simulation study. *Evolution* 38: 1251-1260.

Richerson, P. J., & Boyd, R. (1998). The Evolution of Human Ultra-sociality. In *Indoctrinability, Ideology, and Warfare: Evolutionary Perspectives* (eds. I. Eibl-Eibisfeldt & F. Salter), pp. 71-95. Berghahn Books, New York, NY.

Salisbury, C. L., Seddon, N., Cooney, C. R. et al. (2012). The latitudinal gradient in dispersal constraints: Ecological specialization drives diversification in tropical birds. *Ecology Letters* 15: 847-855.

Schaller, M., & Duncan, L. (2007). The behavioral immune system: Its evolution and social psychological implications. In *Evolution and the Social Mind: Evolutionary Psychology and Social Cognition* (eds. J. P. Forges, M. G. Haselton & W. Von Hippel), pp. 293-307. Psychology Press, New York, NY.

Schemske, D. W. (2002). Ecology and evolutionary perspectives on the origins of tropical diversity. In *Foundations of Tropical Forest Biology: Classical Papers with Commentaries* (eds. R. L. Chazdon & T. C. Whitmore), pp. 163-173. University of Chicago Press, Chicago, IL.

Sih, A., Cote, J., Evans, M., Fogarty, S. et al. (2012). Ecological implications of behavioural syndromes. *Ecological Letters* 15: 278-289.

Sosis, R. (2003). Why aren't we all Hutterities? Costly signaling theory and religious behavior. *Human Nature-An Interdisciplinary Biosocial Perspective* 14: 91-127.

Stevens, G. C. (1989). The latitudinal gradient in geographical range: How so many species coexist in the tropics. *American Naturalist* 133: 240-256.

Summers, K., McKeon, S., Sellars, J. et al. (2003). Parasitic exploitation as an engine of diversity. *Biological Reviews* 78: 639-675.

Sutherland, W. J. (2003). Parallel extinction risk and global distribution of languages and species. *Nature* 423: 276-279.

Tauber, C. A., & Tauber, M. J. (1989). Sympatric speciation in insects: Perception and perspective. In *Speciation and Its Consequences* (eds. D. Otte & J. A. Endler), pp. 307-344. Sinauer Associates, Inc., Sunderland, MA.

Thornhill, R., & Fincher, C. L. (2013). The parasite-driven-wedge model of parapatric speciation. *Journal of Zoology* 291: 23-33.

Thornhill, R., Fincher, C. L., & Aran, D. (2009). Parasites, democratization, and the liberalization of values across contemporary countries. *Biological Reviews* 84: 113-131.

Torchin, M. E., Lafferty, K. D., Dobson, A. P. et al. (2003). Introduced species and their missing parasites. *Nature* 421: 628-630.

Van den Berghe, P. L. (1981). *The Ethnic Phenomenon*. Elsevier, New York, NY.

Vanhanen, T. (2003). *Democratization: A Comparative Analysis of 170 Countries*. Routledge, New York, NY.

Weissing, F. J., Edelaar, P., & van Doorn, G. S. (2011). Adaptive speciation theory: A conceptual review. *Behavioral Ecology and Sociobiology* 65: 461-480.

West-Eberhard, M. J. (1983). Sexual selection, social competition, and speciation. *The Quarterly Review of Biology* 58: 155-183.

Willig, M. R., Kaufman, D. M., & Stevens, R. D. (2003). Latitudinal gradients of biodiversity: Pattern, process, scale, and synthesis. *Annual Review of Ecology Evolution and Systematics* 34: 273-309.

Wilson, D. S. (2002). *Darwin's Cathedral: Evolution, Religion, and the Nature of Society*. University of Chicago Press, Chicago, IL.

Wolf, M., & Weissing, F. J. (2012). Animal personalities: Consequences for ecology and evolution. *Trends in Ecology and Evolution* 27: 452-461.

World Factbook. (2007). http://www.cia.gov

# Chapter 14
# Reflections, Criticisms, and Future Research

## 14.1 Introduction

In the first part of this final chapter of our book, we reflect on the implications of the findings presented in previous chapters for advancing scientific discovery and synthesis as well as humanitarian and democratic goals. In previous chapters, we addressed certain criticisms of the parasite-stress theory of values/sociality and also pointed out some future research directions suggested by the theory. In this chapter, we answer some additional criticisms of the theory and discuss some additional research directions.

Before going to these topics, we state briefly here our overall conclusions supported by the evidence presented in previous chapters. That evidence indicates that both a wide span of human affairs and major aspects of human cultural diversity can be understood in light of variable parasite stress and the range of value systems evoked by variable parasite stress. The same evidence supports the hypothesis that people have psychological adaptations that function to adopt values dependent upon local infectious-disease adversity. Evidence we have presented also indicates that the parasite-stress theory of sociality informs other topics in ecology and evolutionary biology such as variable family organization and speciation processes and biological diversity in general in nonhuman animals.

## 14.2 Creativity and the Temporary Limit of Science

> The grand aim of all science is to cover the greatest number of empirical facts by logical deduction from the smallest number of hypotheses or axioms.—Albert Einstein, quoted in *Life Magazine*, January 9, 1950.

Our overall endeavor in the preceding chapters was directed primarily at building a synthesis or unity, anchored in the parasite-stress theory of sociality, of many long-standing areas of scholarship that traditionally have been investigated separately

R. Thornhill and C.L. Fincher, *The Parasite-Stress Theory of Values and Sociality: Infectious Disease, History and Human Values Worldwide*,
DOI 10.1007/978-3-319-08040-6_14, © Springer International Publishing Switzerland 2014

and independently. Among the diverse scholarly areas we have treated are biodiversity, enculturation, economics, ethnogenesis, interpersonal psychology, family life in people and other animals, governmental systems, human conflict and violence, morals, personality, political science, religiosity, and speciation. This kind of unifying effort was first discussed and highly valued during the Scientific Revolution and the Enlightenment, and reflects Francis Bacon's claim that the explanatory scope of the natural world is small—diverse and seemingly discrete natural phenomena can be unified and understood deeply by a small number of causes. Synthetic theoretical and related empirical unification of scholarly disciplines is the grand aim of science, as pointed out by Albert Einstein in the quote above. Such synthetic effort begins with ideas about how various areas of scholarship might be more fundamentally illuminated by integrating them under shared causes. Then one begins to read broadly in those areas to evaluate if the ideas seem to have any credibility at all to make it worth the time-consuming and sometimes monetarily expensive task of data collection, hypothesis testing, and analysis. Does there appear to be a connectedness across the relevant variables in ways that make sense if one's ideas about synthetic causation have scientific merit? If so, further thinking and testing are warranted. Scholars work by the opportunity-cost principle of effort allocation. This principle is a central part of life-history theory in biology and refers to the trade-off that effort put into one endeavor cannot be used in another. In evolutionary analysis, the alternative endeavors and their contingent use are considered by psychological mechanisms that evaluate the endeavors in terms of their effects on inclusive fitness in evolutionarily ancestral settings. Synthetic ideas themselves are not time consuming or otherwise expensive to generate, but these ideas, the ones that in the end actually lead to the discovery of fundamental causes of broad interrelated patterns in nature, are very rare. It is this rarity that limits scientific progress toward the theoretical and empirical fusion of scholarly disciplines.

We are fascinated by why these big ideas are so rare. Consider, for example, how long it took before a biologist, Charles Darwin, arrived at the big idea that life's history on our planet was an evolutionary history, which is the most synthetic idea in biology. And it was more than 100 years later when William Hamilton proposed the fundamental ideas about inclusive fitness that have revolutionized biologists' understanding of fitness and adaptation. Scientists get stuck in traditional and popular views in their field. This is seen in every branch of evolutionary biology. George Williams was noted for his ability to see the stagnation of intellectual activity in evolutionary biology that traditional thought generates and then motivate and direct biologists toward more progressive thinking and research. Williams' critiques of thinking about various evolutionary topics, especially adaptation, group-level natural selection, and phylogeny, were published primarily in his 1966 and 1992 books. Many other examples of how traditional thinking and the authoritarian acceptance of the status quo stagnate scientific discovery could be mentioned. Thomas Kuhn wrote a popular book, *The Structure of Scientific Revolutions* (2012), about this phenomenon; he documented that the sciences stagnate until some folks come along and point to new and more productive research directions, and he called these better research directions "paradigm shifts."

But to say that authoritarian acceptance of traditional ideas limits scientific discovery is just a description of the pattern all scientists are aware of occurring in their own field. The challenge is to explain the pattern. What are the causes of the typical stagnation of research and correspondent rarity of fundamental ideas in science? What is the reason that most graduate students in doctoral programs cannot be simply taught to think in terms of ideas that yield consilience, to use Edward Wilson's term for large-scale scientific synthesis based on a small number of core causes? Or asked differently, why do these graduate students' mentors, study committees, and institutions not value highly, and indeed require, consilience in doctoral dissertations in order to grant the doctoral degree? We submit that the question of the causation of the rarity of unifying ideas is the most important and urgent unsolved scientific problem facing scholars. Such rarity sets all limits on discoveries and thus on knowledge of nature as well as on solving the short-term and long-term challenges that face humans now and in the future.

Our hypothesis for the rarity of big ideas in science arises from some central themes of this book. In this book, we spent some time investigating intelligence variation across regions and its relation to parasite stress, innovation, and values. Although intelligence must figure somewhat into the answer to the question we pose, it is not the encompassing answer. Although psychometric cognitive ability (IQ) correlates positively and substantially with creativity, the relationship is not unity, and appears to be reduced when IQ scores exceed 120 (Sternberg 2005). More fundamentally for our question is the matter of the rarity of synthetic creative thought. Creativity is a kind of originality. Anyone can be original and erect thoughts that no other person has ever had. For example, let us say as we sit in our office writing this, we have an original idea; a very silly and useless idea, but an original one: it is that a person of 5 ft 7.5 in. in height with quadri-colored hair will knock on the office door in 10 s. Creativity requires more than original thoughts. Creativity is about useful novel thinking—novel ideas that yield "fruit." The fruit varies depending on the endeavor. A creative artist generates new art that works because it pleases certain aesthetic adaptations in human brains. A creative businessperson generates a new idea that makes money. A creative scientist generates a new idea that leads to discovery of the natural world. It is creative originality that is rare, and the bigger or more encompassing the realms addressed by creative ideas, the rarer they are. This rarity, we hypothesize, is explained by the parasite-stress theory of values.

As discussed in earlier chapters of this book, across regions, scientific and technological advances are related to values and hence to parasite adversity. As collectivism increases across regions, science and technology correspondingly decline. It is liberalism that promotes scientific and technological advances. This, we have argued, arises, in part, from the established pattern of openness in the thinking of liberals/individualists compared to the closed-mindedness and dogmatic traditionalism of conservatives. In part, too, the scientific advances arise from the anti-authoritarianism that characterizes liberals. Scientific creativity requires the questioning of earlier and established opinions. It also requires the personal opinion that one can do better than those who came before. Hence, an ingredient of importance, and one fundamental to anti-authoritarianism, is the intellectual independence of the individual, again an established correlate of liberalism. Conservatives think in terms of the traditional opinions of their collective—they have an interdependent cognition, which is the antithesis of the creativity that can lead to scientific discovery (Chap. 4). Our hypothesis proposes that some combination of openness, anti-authoritarianism, and the conception of self as both independent and effective produces the big ideas to which consilience refers.

Empirical implications of our hypothesis include the following. The winning of Nobel and other major international prizes in science is predicted to correlate positively with individualism and negatively with parasite adversity across regions. Also, we predict that competitive federal grants in the USA, such as those given by the National Institutes of Health and National Science Foundation, will show these same patterns across states of the USA. Also registered patents will show the same patterns across regions. Finally, the so-called "impact scores" of scientists, which reflect the degree to which one's published ideas and findings are considered and cited by other scientists and thereby influence future research direction, will show a positive relationship with individualism and a negative one with parasite adversity across regions. In general, scientific ideas and findings with high impact address more fundamental issues of causation and discovery than scientific ideas and findings with low or no impact.

There is evidence supporting our hypothesis that emancipation from parasites and associated individualism are important in promoting innovation and scientific and technological advances across regions. Murray (2014) has studied per capita Nobel prizes of all types through 2012 in relation to parasite adversity across the 72 countries in which at least one such prize has been awarded. The relationship is negative and strong ($r = -0.75$). The result was similar when only science Nobel awards were analyzed (that is, Nobel Peace and Literature prizes were excluded). Murray (2014) reported robust, negative relationships across countries between each of four additional measures of scientific and technological innovation per capita and parasite adversity.

Taylor and Wilson (2012) also have provided findings supporting our hypothesis. These scholars first review earlier studies that support the link between inventiveness and individualism across countries. They then examine for 62 countries the relationship between collectivism–individualism and per capita engineering and scientific technology patents that are influential (as measured by impact/widespread use) and scientific publications that are influential (as measured by citation impact). They report strong, positive relationships between individualism (negative with collectivism) and technological and scientific progress across nations. According to our hypothesis, the associations with values found by Taylor and Wilson are caused by variable parasite stress across countries, which evoke the values differences leading to regional variation in innovation. Supporting this thinking, Murray (2014) found a strong, negative relationship between total patents per capita and parasite stress across countries.

Cappell (2009) has provided additional evidence that supports our hypothesis. That research reported more than a 100-fold higher per capita rate of authorship of articles published in medical science journals from relatively democratic countries compared to their relatively undemocratic neighboring countries. For example, Israel was compared with the rest of the Middle East, South Korea versus North Korea, and Taiwan versus mainland China. In the same paper, Cappell proposed the hypothesis that the freedom and liberties of democracy promote intellectual creativity and scientific medical research. This is overlapping with our hypothesis that individualism and associated emancipation from parasite adversity are the encompassing cause. If our hypothesis continues to be supported, the practical implication is that infectious-disease control across the world will advance knowledge in multiple ways. It will create more people with interests in knowledge of how the world works, or said differently, how nature actually is. It also will produce more people who are more than just interested and thus want to make their own discoveries about nature. And it will increase the number of people who quest to identify and answer the big questions.

Corner bookstores sell self-help books that claim to have useful advice to liberal-mined parents about how to make their children creative and free-thinking (e.g., McGowan et al. 2009). There are even product lines of children's playthings that claim to enhance creativity in children who use them. In conservative settings of child-raising, children are reprimanded oftentimes for novel or creative thinking. In the conservative culture of the Old South in which Thornhill was raised, he was reminded frequently that "curiosity killed the cat," meaning he should not think and act outside of traditional norms of thought and behavior. He also was said to have "book sense, but not common sense," meaning he did not conform to the ideological path of tradition.

Our suggestion as to how to promote curiosity and creativity in children is based on the diverse and substantive evidence supporting the parasite-stress theory of values. We hypothesize that children's curiosity about and passion to know nature is dependent importantly, if not primarily, on freedom from infectious disease during ontogeny. Of course, a child's genetic resistance to parasites matters, too, in causing her or his values. However, in an ontogenetic environment free of parasites, alleles that contribute to lack of immunity in an ontogenetic setting containing parasites cannot yield manifest infectious disease in people. As explained early in our book, because of the way ontogeny works, genes alone are completely impotent in making phenotypic features, those of immunity or otherwise.

As also mentioned earlier, we have observed that students' awareness that their values guide their thinking about everything is critical for opening their minds and promoting new ways of thinking. It is useful for students and other people to know this, as well as the proximate and ultimate causes of their values and associated biases in thinking, in order to reach their full, creative potential.

Early in the book, we traced the history of our thinking about parasites and sociality. There we list a number of scientists whose earlier ideas sparked our thinking and research about these topics. This book could not have been written without the creative work of scholars like those and others we cite whose work in many cases provides the basis for our studies herein. We cast a huge net with the parasite-stress theory across the traditional fields of scholarship dealing with human affairs and some other topics such as biodiversity and family life outside the human species.

In so doing, we claim that there are errors or, in many cases, incompleteness in earlier work. It is, of course, our job as scientists to try to accomplish more than our predecessors. We hope others will identify the errors and incompleteness in our book and build more accurate and encompassing knowledge. Good ideas set the limit of scientific discovery, but that limit is temporary and is surpassed as others see nature more deeply in terms of consilience.

## 14.3 There Is No Such Thing as a Free Lunch

In this book, we have discussed and supported with many bodies of evidence our view that emancipation from parasites is a key factor in the emancipation of people from prejudice, poverty, oppression, domestic conflict, and violence. Accordingly, cleaning up the parasites in a region will liberalize the values of the people in the region and result in increased out-group tolerance and amity and reduced authoritarianism, classism, and sexism. Hence, relief from infectious disease does more than improve health, increase longevity, and reduce child and overall morbidity and mortality rates—such relief produces people with an egalitarian priority. Those interested in making a more humane planet should study the ideas and findings reported in this book. We have emphasized that the evidence we have presented in itself does not identify a more democratic planet as a morally correct goal; instead, the evidence provides the way to get there if people desire such a world. Yet, the same evidence points the way to make the Old South rise again or to turn the whole planet into only ultra-conservative, totalitarian governmental polities. Those undemocratic outcomes can be accomplished simply by promoting the welfare of human infectious diseases across the world.

The folkloric phrase that is the title of this section refers to the fact that there is a cost to everything. If something seems cost-free, it is because it has not been examined carefully to reveal the hidden costs. Thus, we can ask, what are the costs of emancipation from parasites and the associated liberalization of values and economic productivity? Similarly, what are the costs of increased democracy? Here we seek an encompassing answer. We are not satisfied with considering only the monetary cost of improved healthcare, sanitation, and vector control. Even though this cost is substantial, it is small compared to the expense of not investing in the reduction of parasite adversity. For example, if we are correct scientifically, emancipation from parasites will reduce civil warfare. But, the monetary cost of rebuilding infrastructure after a single, average-size civil war could be enough, we guess, to provide and sustain modern healthcare and sanitation across a developing country for generations.

The biggest cost of emancipation from infectious diseases and the prejudicial values the diseases evoke is increased energy consumption. Energy is limited. Our colleagues James Brown, Bill Burnside and others (Brown et al. 2011) only recently have documented empirically this very real cost of democracy. They address this matter as the cost of economic growth and productivity, measured as Gross Domestic

Product (GDP) per capita, in using up nonrenewable energy; they apparently did not see its connection to democracy and values and hence to the parasite-stress theory of values. Here, we discuss their results and show how their results connect to the parasite-stress theory.

Revisiting the strong, positive relationship between energy use per capita and GDP per capita reported by Brown et al. (2011), we find a similar pattern in data for 2008. The $r$ is 0.88, $p<0.0001$, $n=206$ countries. GDP per capita is logged values from *World Factbook* 2008. Energy use is the logged total energy consumed per capita for 2008, the most recent year available at US Energy Information Administration (www.eia.gov). Democratization, measured as Vanhanen's Resource Distribution (see Chap. 10), also correlates strongly and positively with energy use across countries; $r=0.74$, $p<0.0001$, $n=169$. *Combined Parasite Stress* (see Chap. 5) shows a strong, negative relationship with energy consumption across the world: $r=-0.76$, $p<0.0001$, $n=185$. Hence, as economic productivity and democracy increase across the planet, energy use simultaneously increases. But, at higher levels of parasite stress, people use less energy. These effects are large. Hence, a major cost of widespread economic wellbeing or democracy is copious energy consumption. We have discussed the strong, negative relationship of parasite stress and GDP per capita. For the variables used here, the $r=-0.74$, $p<0.0001$, $n=192$. *Combined Parasite Stress* and energy consumption, however, show a significant negative relationship, even when GDP per capita in controlled statistically: partial $r=-0.31$, $p<0.0001$. Parasite stress reduces energy consumption, even when energy consumed by economic production is not a contributing factor. On a related note, collectivism reduces energy consumption, whereas individualism increases energy use. *In-group Assortativeness*, a measure of collectivism that includes religiosity (see Chap. 9), shows a strong, negative relationship with energy use across countries ($r=-0.64$, $p<0.0001$, $n=66$). Other measures of collectivism–individualism discussed in Chap. 5 show similar patterns in relation to energy use across the world.

An important question is whether the innovations, research and technological development needed to accomplish the production and availability of alternative energies, as well as feasible means of reducing per capita use of nonrenewable energy, will keep up sufficiently with economic growth and development. It seems clear however from the evidence in this book that a future focus in policy on infectious-disease reduction is the primary path worthy of consideration and implementation to cope with the issue of energy limitation. Such policy would evoke widespread innovation, as well as liberal egalitarianism with a priority of profound concern about the welfare of our planet, and of people in general across the globe and into distant future generations. Moreover, the widespread liberal values so evoked would promote, support and finance the high-level educational infrastructure and the scientific and engineering creativity and research that will be required to achieve a future with levels of energy availability that could maintain economic development and wellbeing.

Another cost of emancipating people from infectious diseases is an increase in the rate of some autoimmune diseases. The hygiene hypothesis, proposed by Strachan (1989), states that low exposure to infectious organisms during child development produces a hyperactive classical immune system response that damages tissue and manifests as asthma, allergies, or other chronic autoimmune disorders. When levels of infectious disease exposure during children's ontogeny are low enough to be evolutionarily novel, as in many circumstances in the modern West, the classical immune system overreacts in order to locate and defend against elusive, but actually absent, infectious agents. Although the proximate mechanisms involved are incompletely understood, there is mounting evidence that the hygiene hypothesis is correct (e.g., Peters et al. 2006; Vogel et al. 2008; Gangal and Chowgule 2009; Ege et al. 2011; Rook 2012; Fox et al. 2013).

A closely related idea is the "old friends hypothesis" (Rook and Brunet 2002). It, like the hygiene hypothesis, involves evolutionary novelty, but in reduced contact with commensal and mutualistic organisms of human bodies, not reduced contact with parasites. The old friends hypothesis argues that these nonpathogenic microbes, with which humans have had a long evolutionary historical intimacy, cue the classical immune system to not activate in their presence. In the absence or great reduction of the friendly microbes resulting from Western hygiene and sanitation, the classical immune system fails to acquire the necessary experiences with the microbes that allow it to distinguish friendly microbes from pathogenic ones, which can lead to certain immune-system diseases. Evidence for the old friends hypothesis is accumulating rapidly (see reviews in Raison et al. 2010; Rook 2012).

In sum, the emancipation of people from infectious diseases has both benefits and costs. The benefits are increased longevity, living with reduced illness (morbidity reduction), enhanced democracy/equalitarianism, increased intelligence, and scientific and technological progress. The costs, however, are major. The most ominous of these is the exhaustion of nonrenewable energy for the future that coincides with technological and economic productivity. This is one of many reasons that the parasite-stress theory of values is relevant to human affairs and everyday life.

## 14.4 Criticisms of the Parasite-Stress Theory of Sociality

The next several sections of this chapter deal with some criticisms of the parasite-stress theory and of certain empirical results claimed to support the theory. We are fortunate to have critics who take time away from their research and personal lives to study our work and seek to improve it with their criticisms. Science is self-correcting because, in the evolutionary perspective, scientists like all people are designed by past Darwinian selection to succeed socially. Hence, scientific social competition includes publishing opinions about the errors or incompleteness in the work of other scientists. The criticisms we address below pertain to our papers on parasite stress and social life. Recently, we received a diversity of criticisms in the context of our paper published in *Behavioral and Brain Sciences* (Fincher and Thornhill 2012a). That paper dealt with the relationships, both across nations and states of the USA, among the variables parasite stress, family values, and religiosity, topics we treated in detail in Chaps. 5 and 9. The paper served as the "target article" for commentators to comment about and criticize. On our target article, there were 21 published commentaries by a total of 38 authors from across a wide range of scholarly disciplines. As authors of the target article, we then had the opportunity to reply to the commentators in a follow-up article (Fincher and Thornhill 2012b), published in the same issue of *Behavioral and Brain Sciences* as the target article and commentaries. Although the target article was specifically about family values and religiosity in relation to parasite stress, some commentators offered criticisms of the parasite-stress theory in general or its findings other than those of religiosity and family life. Below we discuss the criticisms of commentators and answer them basically as we did in Fincher and Thornhill (2012b) but add additional comments. Also, some of the *Behavioral and Brain Sciences* commentators provided interesting suggestions and hypotheses about extending the parasite-stress theory to new arenas; our discussion of these in Fincher and Thornhill (2012b) is included in the section on future research directions near the end of this chapter.

## 14.5 Alternative Hypotheses

Several commentators on our *Behavioral and Brain Sciences* target paper offered alternative models they thought explain better certain findings that we interpreted as supportive of the parasite-stress theory of values. We address these suggested alternatives in this section.

Van de Vilert and Postmes (2012) argued in their commentary that climatic stress is a salient causal feature for the development of cross-national cultural differences, especially when accounting for the wealth of a country. Van de Vliert's earlier work (2009) presented a hypothesis proposing that climatic stress is met by a compensating cultural response tempered by the average wealth of citizens within a country (i.e., under harsh climatic conditions, citizens from wealthy countries have different options than those from poor countries). Van de Vliert (2009) presented a measure of climatic harshness that indexes the sum of absolute temperature deviations from 22 °C for the average lowest and highest temperatures in the hottest and coldest months for a country; he called this the *Total Index*. The *Total Index* was used in Van de Vliert and Postmes' analyses that they presented in their commentary. They suggested that harsh climatic temperatures are more demanding of resources and that people in richer versus poorer countries will meet the demands differently: people from poor countries will rely on their in-group affiliates, while people from rich countries will see the demands as challenges and this will increase their individualism.

We examined the association between their measure of climatic stress (*Total Index* data collected from Van de Vliert 2009) and two life-history measures that pertain to human inclusive fitness, and found their measure to be lacking in ecological validity. We correlated their *Total Index* of climate harshness with child mortality under 5 years of age (variable was logged, and represents the average for data from the years 1990, 1995, 2000, and 2005 collected from data.worldbank.org) and found the correlation was −0.40 ($p < 0.0001$, $n = 188$ countries). We correlated their *Total Index* with the life expectancy at birth for both sexes (variable was logged, and represents the average for data from the years 1960 to 2008 collected from data.worldbank.org) and found the correlation was 0.39 ($p < 0.0001$, $n = 186$). Thus, their measure of climatic harshness actually corresponds to significantly increased lifespan and reduced mortality of young children, just the opposite of what is expected if their index measures ecological harshness. In large contrast, one of the focal measures of parasite stress used in our target article, nonzoonotic parasite severity (described in Chap. 5), correlated $r = 0.75$ ($p < 0.0001$, $n = 191$) with under-five mortality and −0.76 with life expectancy ($p < 0.0001$, $n = 198$). Given these findings, we consider their analyses that involve the *Total Index* and our analyses that involve measures of parasite stress as totally incomparable.

Moreover, in their commentary, Van de Vliert and Postmes use climate stress and average wealth to predict a cross-national measure of in-group favoritism, *Societal Collectivism*, a variable developed by Van de Vliert (2011). In this analysis, they showed that parasite stress was not a significant factor predicting *Societal Collectivism*. Given the problem we just documented with their measure of climatic harshness, we used the same type of approach as in Chap. 5 for exploring non-parasite-stress causation. We determined the residual life expectancy from regressing lifespan expectancy for both sexes combined on nonzoonotic parasite prevalence and correlated those residuals with Van de Vliert's measure of *Societal Collectivism*. (In Chap. 5, we conducted similar analyses of residuals.) We found a nonsignificant negative correlation ($r = -0.13$, $p = 0.18$, $n = 119$ countries), indicating that variation in lifespan when independent of the effects of parasite-stress was not associated reliably with *Societal Collectivism*.

Van de Vliert and Postmes' climate-stress model also is weakened because they do not explain why a person seeks the assistance of in-group members versus out-group members under times of stress. Why is it that when poor and under stress, individuals turn to in-group instead of out-group members who may offer many forms of assistance unattainable within the in-group? The parasite-stress theory offers an explanation for this. As we have emphasized, out-group interactions can provide many benefits to individuals, but such benefits apparently do not exceed costs of infectious-disease encounters under high parasite stress.

Van de Vliert and Postmes treat international wealth variation as a given aspect of the ecological setting, but they do not attempt to explain wealth variation itself. As documented in Chap. 11, the parasite-stress theory of values has an inherent theoretical basis that informs economics. Consequently, wealth variation arises, in large part, due to variation in parasite stress. Economic productivity is affected negatively through parasite-mediated reductions in intelligence (Eppig et al. 2010, 2011) and in health (Gallup and Sachs 2001; Price-Smith 2002; Sachs and Malaney 2002). Moreover, the various values that differ along the collectivism–individualism dimension are evoked by variable parasite stress and affect economic productivity. Collectivism retards economic development because it is associated with parochial economics, sometimes even restricted to the extended family, and closed-mindedness to new technologies and other innovations. Individualism, in contrast, has positive

**Table 14.1** Results from multiple regression of Unified Democracy Scores on climatic harshness, average wealth and *Combined Parasite Stress*

| Predictor | $b$ ($\beta$) | $T$ |
|---|---|---|
| Climatic harshness | −0.004 (−0.12) | $-1.78^{ns}$ |
| Ln GDP per capita (average for 1960–2008) | 0.128 (0.23) | $2.41^{*}$ |
| Climate * GDP per capita | 0.009 (0.30) | $5.02^{**}$ |
| *Combined Parasite Stress* | −0.109 (−0.37) | $-3.74^{**}$ |
| $R^2$ | 0.38 | |
| $F_{4,178}$ | $28.11^{**}$ | |

Regression coefficients are unstandardized estimates with standardized $\beta$ in parentheses (results originally reported in Fincher and Thornhill 2012a; reprinted with permission)
*Note*: *Ln* natural-log transformation, *ns* is not significant, $^{*}p<0.05$; $^{**}p<0.001$

effects on economic productivity by increasing democracy and thereby reducing health, wealth and educational disparities and enhancing openness, economic opportunity and networking across a region.

Van de Vliert and Postmes criticized aspects of our published research on international governmental systems not presented in our target article, but presented earlier in Thornhill et al. (2009) and Thornhill et al. (2010). (This research is described in detail in Chap. 10.) In Chap. 10, we provide considerable evidence that variation in democratization, both cross-nationally and across indigenous human societies, arises from variation in parasite stress and associated evoked value systems. Specifically, democratization is related negatively to parasite stress and collectivism. Taking the cross-national Unified Democracy Scores (Pemstein et al. 2010) for 2008, discussed and analyzed by Van de Vliert and Postmes in their commentary, we found—as predicted by the parasite-stress theory of values—that these scores were correlated strongly ($r=-0.49$, $p<0.0001$, $n=189$) with *Combined Parasite Stress* (a variable described in Chap. 5). Van de Vliert and Postmes analyzed the ability of climatic harshness and average wealth and their interaction in predicting democracy versus autocracy in comparison to our measures of parasite stress (described in Chap. 5), nonzoonotic parasite prevalence, zoonotic parasite prevalence, and their interaction. Using a hierarchical regression where they entered climatic harshness (*Total Index*) and average wealth before entering parasite stress, Van de Vliert and Postmes reported that parasite stress was relatively inconsequential for the explanation of cross-national democracy variation. We repeated the analysis they presented in their commentary using *Combined Parasite Stress*, one of our primary pathogen measures in our target article. We also used multiple regression rather than hierarchical regression and found that parasite stress was the largest contributor in terms of standardized beta to the cross-national variation in democratization (Table 14.1; see Fincher and Thornhill 2012a for more details about this analysis).

We evaluated too Van de Vliert and Postmes' basic hypothesis that individuals existing in harsh conditions, but with access to wealth, behave differently (and produce different culture) than individuals without access to wealth. We tested this idea using an ecologically valid measure of ecological harshness, parasite stress. We conducted a new series of multiple regressions using *Combined Parasite Stress*, GDP per capita

**Table 14.2** Results of multiple regression analyses considering parasite-stress, average wealth and their interaction for predicting aspects of religiosity and strength of family ties

| | In-group assortativeness | Religious participation and value | Proportion of believers | Strength of family ties | Societal collectivism (Van de Vliert 2011) |
|---|---|---|---|---|---|
| GDP per capita (GDP per cap) | −0.25* | −0.16 | −0.15 | −0.34** | −0.68*** |
| *Combined Parasite Stress* (PS) | 0.62*** | 0.62*** | 0.51*** | 0.53*** | −0.05 |
| GDP/cap * PS | 0.22* | 0.08 | 0.28*** | 0.30** | 0.25** |
| $R^2$ | 0.58 | 0.51 | 0.49 | 0.56 | 0.46 |
| *F* | 28.3 | 29.4 | 41.9 | 27.6 | 32.1 |
| *N* | 65 | 89 | 135 | 69 | 118 |

The rows for the predictor variables contain the standardized beta coefficients. All regressions are statistically significant at $p<0.0001$ (results originally reported in Fincher and Thornhill 2012a; reprinted with permission)

***$p<0.001$, **$p<0.01$, *$p<0.05$. Otherwise, the coefficients are not statistically significant

(average from 1960 to 2008, logged), and the interaction between the two for predicting a selection of the dependent variables we used in the target article, as well as the variable, *Societal Collectivism*, presented and analyzed in the commentary by Van de Vliert and Postmes. (In the target article, we presented multiple regression analyses of the independent effects of *Combined Parasite Stress* and GDP per capita and found that parasite stress remained a significant predictor of our dependent variables, even when controlling the effects of average wealth.) The model of Van de Vliert and Postmes predicts a significant interaction between wealth and ecological harshness. Specifically, their model predicts that, in conditions of ecological harshness and increased wealth, there will be increased individualism, and in conditions of ecological harshness and low wealth, there will be increased in-group assortativeness/collectivism.

The results are presented in Table 14.2. For the variables *In-Group Assortativeness*, *Proportion of Believers*, *Strength of Family Ties*, and *Societal Collectivism* (but not *Religious Participation and Value*), we found a significant interaction between average wealth and ecological harshness, but not the interaction predicted by the hypothesis of Van de Vliert and Postmes. (Our variables used here are described in Chaps. 5 or 9 depending on the variable.) The interaction plots revealed that, for the countries with high average wealth, as parasite stress increases, so does the *Strength of Family Ties*, *Proportion of Believers*, *In-Group Assortativeness*, and *Societal Collectivism*. Thus, the interaction was not consistent with that expected by Van de Vliert and Postmes. For the countries with low average wealth, the change due to increasing parasite stress is minimal (slightly negative or slightly positive) for the *Strength of Family Ties*, *Proportion of Believers*, and *In-Group Assortativeness*, but is strongly negative for *Societal Collectivism*. Hence, the patterns we found were not what was expected or were in the wrong direction. Overall, these findings did not support the hypothesis presented by Van de Vliert and Postmes. Moreover, *Combined Parasite Stress* had the largest effect for all dependent variables except for *Societal Collectivism*. In the case of predicting *Societal Collectivism*, the effect of *Combined Parasite Stress* was small and nonsignificant. However, there was a significant interaction between *Combined Parasite Stress* and GDP per capita, suggesting an important role for the variation due to parasite stress in explaining this measure of collectivism. We expected greater concordance between the findings of the analyses involving our measure of *In-Group Assortativeness* and *Societal Collectivism* considering they are supposed to be measuring the same cultural features. The two measures *Societal Collectivism* and *In-Group Assortativeness* were correlated positively ($r=0.54$, $p<0.0001$, $n=65$), but not as highly as we expected. Overall, our findings suggest that the variable *Societal Collectivism* needs evaluation and refinement in light of the parasite-stress theory of values.

Paul's (2012) criticism in his commentary was that our hypothesis about religiosity and parasite stress in the target article is actually a component of a larger socioeconomic dysfunctionality theory that explains the negative correlation between religiosity and wealth on the premise that as conditions become more benign, then people need the benefits of religion less and thus religiosity declines (atheism increases, for example). We discussed this model in the target article calling it the "conditions-of-living" model (see Chap. 9). The "conditions-of-living" model (also called the socioeconomic dysfunctionality model (Paul 2009), the "uncertainty hypothesis" (e.g., Barber 2011), and the "deprivation theory" (Solt et al. 2011)) is incomplete, because it does not offer an explanation for why people do not turn to out-groups in harsh ecological settings. As mentioned above, the parasite stress theory explains this. Specifically, under harsh conditions (which are those where parasite stress is high), the cost of out-group contact can be relatively high (because of the potential for contacting new infectious diseases), meaning that the benefits of widespread out-group contact may not outweigh its costs. The outcome is functional avoidance of out-group members under high parasite-stress conditions.

We have long been aware of the intense interest that many researchers maintain for the effects of wealth as the explanation of all-things-cultural. Therefore, in the analyses included in the target article, we demonstrated that all of the dependent variables for both the cross-national and US interstate comparisons were explained by parasite stress significantly and positively, even when removing the effects of wealth resources and wealth inequality. In many cases, parasite stress was the only significant predictor; in others, parasite stress had the largest effect. In only four of 16 regressions, wealth had a larger effect size than parasite stress, but not by much (e.g., −0.41 versus 0.34). And in these cases, parasite stress was still a significant factor in the predicted direction. Nevertheless, we reported in our article that responded to critics (Fincher and Thornhill 2012b) a different analysis of wealth resources and the independent effects of parasite stress for a representative selection of the dependent variables we analyzed in the target article.

We compared through a multiple-regression analysis the relative effects of wealth inequality (measured with the Gini index in net household income from the Standardized World Income Inequality Database, SWIID, Solt 2009; higher values indicate greater inequality) and *Combined Parasite Stress* for explaining our dependent variables in the target article. Our prediction was that parasite stress would

**Table 14.3** Results of multiple-regression analyses using two model specifications: (1) both the Gini index and *Combined Parasite Stress* are considered independent predictors, and (2) includes the addition of the interaction between the Gini index and *Combined Parasite Stress*

| | In-group assortativeness | | Religious participation and value | | Proportion of believers | | Strength of family ties | |
|---|---|---|---|---|---|---|---|---|
| | 1 | 2 | 1 | 2 | 1 | 2 | 1 | 2 |
| Gini index (*G*) | 0.11 | 0.19 | 0.05 | 0.08 | 0.22** | 0.25** | 0.17 | 0.27* |
| *Combined Parasite Stress* (PS) | 0.64*** | 0.65*** | 0.67*** | 0.68*** | 0.54*** | 0.46*** | 0.53*** | 0.55*** |
| *G* * PS | – | −0.21* | – | −0.10 | – | −0.24*** | – | −0.29** |
| $R^2$ | 0.51 | 0.55 | 0.50 | 0.51 | 0.48 | 0.53 | 0.42 | 0.49 |
| F | 32.3 | 24.7 | 42.1 | 28.7 | 56.6 | 46.6 | 23.6 | 20.6 |
| *N* | 64 | 64 | 88 | 88 | 128 | 128 | 68 | 68 |

The rows for the predictor variables contain the standardized beta coefficients. All regressions are statistically significant at $p<0.0001$ (results originally reported in Fincher and Thornhill 2012a; reprinted with permission)

***$p<0.001$, **$p<0.01$, *$p<0.05$. Otherwise, the coefficients are not statistically significant

have a unique predictive effect in spite of any predictive effects attributable to wealth inequality. We tested two models: one that considered wealth inequality (Gini) and parasite stress (*Combined Parasite Stress*) as independent predictors (model 1) and a second model that included the interaction between these two variables (model 2). The results of the analyses are presented in Table 14.3. For all dependent variables, using model 1, parasite stress was a significant, positive predictor, whereas wealth inequality was a significant predictor in only one case, the *Proportion of Believers*. Using model 2, parasite stress was a significant, positive predictor for all dependent variables. Wealth inequality was a positive predictor for the *Proportion of Believers* and *Strength of Family Ties*. The interaction between wealth inequality and parasite stress was a significant predictor for three of the four dependent variables, *In-Group Assortativeness*, *Proportion of Believers*, and *Strength of Family Ties*. In all regressions, either model 1 or 2, parasite stress had the largest effect size (for additional analytical detail, see Fincher and Thornhill 2012b). Taken together, these results indicate that although wealth inequality is a significant contributor to the variation in religious affiliation and strength of family ties, parasite stress is a more general contributor to the variation in religiosity and strength of family ties.

In a separate study, Barber (2011) published a cross-national analysis of religiosity testing the uncertainty hypothesis (discussed above) using a measure of Gini, a measure of parasite adversity from one of our previous publications (Fincher and Thornhill 2008b), and a few other factors such as living in a Communist country as predictors of the variable we called *Proportion of Believers* (this variable is described in Chap. 9). Barber's conclusion supported the uncertainty hypothesis. However, as stated above, the uncertainty hypothesis is incomplete, because of the lack of consideration of noncontact with out-groups under harsh, uncertain conditions.

Both Rees (2009) and Solt et al. (2011) also have examined the influence of wealth inequality in explaining different aspects of religiosity. We suggest that the best hypothesis will include wealth inequality variation itself arising from the causal effects of parasite-stress variation (see Chap. 11).

Currie and Mace (2012) in their commentary suggested that we should think more about causes of religiosity other than parasite adversity. They seem to favor latitude or GDP, a measure of international wealth, but do not state a hypothesis. We explained in the target article that economic indicators and latitude are not variables that are independent of parasite stress, and hence any analysis that includes these two variables and parasite stress will be difficult to interpret. Keeping in mind that a theory determines what variables to control and what variables are simply effects of a theory's causal variable(s), Currie and Mace fail to provide a theory as to why their suggested alternatives are actually alternatives to the parasite-stress theory of values. As we stressed earlier in this book, economic indicators and climate variables (related to latitude) are part of the parasite-stress theory, not alternatives to it. Admittedly, in our analyses in the target article, we controlled for some variables that are not independent of the parasite-stress variables. We did, however, qualify our relevant control analyses by emphasizing that the results should be viewed in light of the causal covariation between parasite stress, economics and climatic factors. (See also the discussion of the partialling fallacy in Chap. 5.)

Vigil and Coulombe (2012) in their commentary argue that within-region assortative social behavior is best explained by Vigil's (2009) socio-relational model rather than the parasite-stress theory of sociality. Vigil's model addresses variation in people's emotional expression of cues of social interest/disinterest while functioning in different kinds of social networks. First, the socio-relational model, Vigil and Coulombe claim, is consistent with the finding that happiness is higher at low latitudes (high parasite stress) than at high latitudes. But this, Vigil and Coulombe argue, is the opposite of prediction from the parasite-stress theory of values, because happiness solicits new social partners and therefore carries risk of out-group contact and associated contagion. On the other hand, sadness, they argue, obtains in-group support, and thus, according to the parasite-stress model, should be greatest at low latitudes, not high latitudes.

We suggest the following approach to better study variation in happiness or worry across regions and individuals within regions. The psychometric (questionnaire) procedures should be modified to determine what makes one happy or worry-free. The parasite-stress theory of values predicts that collectivists will reply that the harmony of their connections with extended family and other long-term in-group members will be paramount—the more in-group harmony, the more happiness and less worry. And, the theory predicts that individualists will respond positively based on harmony and success in a network of people outside the extended family and ideologically similar in-group members. Similarly, it is expected that collectivists will tie self-esteem less to personal success and more to in-group success whereas individualists will do the opposite. As we discussed in Chap. 4, Gelfand et al. (2004) provided evidence from research that supports the predictions from the parasite-stress theory, at least pertaining the self-esteem.

Atran (2012) in his commentary argues that historical events and social structures account better than the parasite-stress theory of values for the democracy and secularism in the West. Cultural history, however, is never an alternative to ecological and ultimate causal frameworks for understanding culture. And, as explained in Chap. 1, cultural inertia is inaccurate as a theory of enculturation. Chapter 10 discusses a hypothesis, based on the parasite-stress theory of values, for the rise of the earliest democracies and the related liberal value system encompassed in the Enlightenment, both of which are Western phenomena. The explanatory potential of this hypothesis is its consistency with a range of evidence supporting the parasite-stress theory's application to the various components of democratization. In the target article, we present a wide range of evidence supporting the application of the parasite-stress theory of values to religiosity (and hence secularism) (see Chap. 9).

## 14.6 Other Criticisms

In the same commentary, Atran points out that political scientists have documented that democratization reliably corresponds to an expansion of the middle class. As we have discussed, this is the result of the redistribution of wealth and power from the exclusive hands of elites to the people at large and arises from the liberalization of values that comprise democratization. Hence, Atran does not provide an alternative to the parasite-stress theory of values applied to democratization. The creation and expansion of the middle class is a definition or description of democratization, not an explanation. Our research with colleagues has attempted to explain the proximate and ultimate causal bases of democratic and autocratic values (see Chap. 10).

Grotuss's (2012) commentary raises the question of how parasite stress may impact indirectly societal structure, a topic we did not discuss in the target article. We did discuss this, however, in an earlier paper on parasite stress, values and governmental systems. Therein we proposed a bidirectional, proximate causal feedback between parasite stress, economic factors, and liberalization of values (Thornhill et al. 2009; see Chap. 10). As parasite stress declines in a region and peoples' values change to bring about widespread economic and other wellbeing in the region, the changes will further reduce parasite stress through increases in widespread nutrition, sanitation, vector control, and improved general living conditions, and access to medical care and educational information. These humanitarian advances cycle back to reduce mortality and morbidity from parasites. Thus, as parasite stress declines, democratization factors increase, which, in turn, further reduce parasitic disease. The opposite also holds: as parasite stress increases or maintains high levels, the values of prejudice, inequality and authoritarianism that arise further magnify the morbidity and mortality from infectious disease. Hence, a society's general level of democracy/nondemocracy arises from parasite stress and feeds back and affects the society's level of parasite stress.

Chang et al. (2012) in their commentary suggest that Islam and Christianity did not arise in extremely high parasite-stress areas. This is true and deserves more research, because of the strong, positive relationship that we have documented between parasite stress and religion number (and hence the genesis of new religions) across the countries of the world. We have argued that, in high parasite-stress areas, high in-group assortative sociality, specifically its components of ethnocentrism, xenophobia and philopatry, fractionate an original culture's range and thereby give rise to new religions. This is supported not only by the positive relationship between parasite stress and religion number across countries, but also by the cross-national, positive relationship between parasite stress and language number (Fincher and Thornhill 2008a, b; Chap. 13). Cashdan's (2001) finding that parasite stress positively predicts the number of ethnic groups also supports the parasite-stress theory's application to the causes of ethnogenesis. Perhaps, the fact that Islam and Christianity arose in areas that are moderate in parasite-stress contributed to their spread across regions through the ability to amass resources (enhanced economic productivity) and with members more apt to disperse (low philopatry) than members of religions that arose from high parasite-stress regions.

In his commentary, Atran (2012) states that "the most expansive and successful religions aimed to include as many genetic strangers as possible" as a counter-argument to our claim that religious groups use their unique supernatural belief systems in order to heighten costs of participation and distance themselves from out-groups (see Chap. 9). This is considered by Atran to be a "most problematic" feature of our arguments regarding religiosity. It may be that the most successful religions that Atran is referring to (presumably, measured by the number of adherents) are also the most expansive, and that they may be both successful and expansive because of their origin in regions with moderate parasite stress rather than in regions with high parasite stress. These so-called successful, expansive religions are infrequent. There are thousands of other religions, and many smaller religions (as judged by the number of adherents) could be considered successful by other measures such as longevity or isolation-ability. For those religions in high parasite-stress regions, a long period without the introduction of an infectious-disease epidemic could be a resounding success. By this reckoning, the presence of large, expansive religions is not contrary to our hypothesis regarding parasite stress and religiosity. In fact, the parasite-stress model could be used to explore why some religions are expansive and others are not. The assumption that all religions should be expansive, and therefore that a religion's success should be measured through historical expansion, is inaccurate.

In their commentary, Wall and Shackelford (2012) suggested that the USA is more religious than can be accounted for by measures of infectious diseases. This, they propose, is the result of the USA's high immigration rates that create a hyper-activation of assortative sociality including religiosity in US citizens to the point that the extraordinarily high assortative sociality mismatches the actual parasite stress.

Evidence does not support the claim that the USA is a unique positive outlier from the cross-national pattern of association between religiosity and parasite stress. Figure 14.1 is a cross-national plot of *Religious Participation and Value* regressed on *Combined Parasite Stress*. The US datum is a positive residual but not as great as Malta and Jordan, which are the two largest positive residuals. Figure 14.2 shows

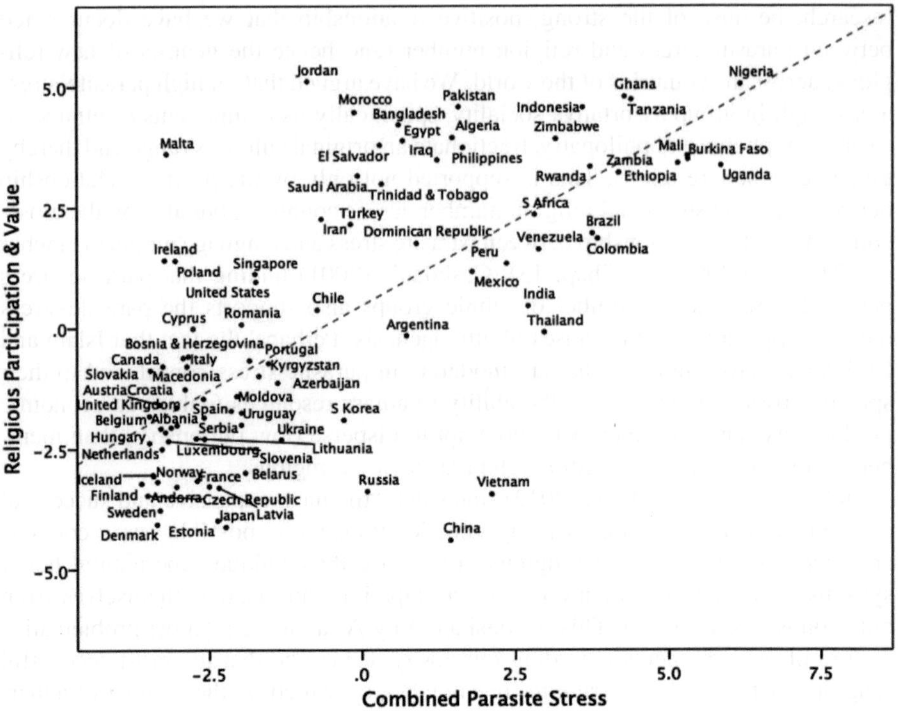


**Fig. 14.1** *Religious Participation and Value* regressed on *Combined Parasite Stress* (results originally reported in Fincher and Thornhill 2012b; reprinted with permission)

the *Proportion of Believers* regressed on *Combined Parasite Stress*. Both linear and quadratic relationships are shown. The US datum is nearly on the line in this case.

Paul (2009) suggested in his earlier study of religiosity and dysfunctional societies that the best cross-cultural sample for examining religiosity is the "prosperous democracies," because it avoids extra confounding factors like "former communist country" or poor data quality for less-developed countries. We explored the question of US religiosity uniqueness by focusing on the 17 prosperous democracies included in his sample. They are primarily Western (e.g., Norway, Spain), but also include Japan. We discovered in the process of constructing this book that there were errors in how the software plotted the figures we produced for Fincher and Thornhill 2012b pertaining to these 17 democracies. Thus, we have redone the investigation and present the correct figures and discussion here. Figure 14.3 shows a plot of *Religious Participation and Value* regressed on *Combined Parasite Stress* across these 17 democracies. The USA is a positive residual, but Ireland is even further from the line. And, we plotted the *Proportion of Believers* on *Combined Parasite Stress* for the 17 prosperous democracies (Fig. 14.4). Also, in this case, the US datum is above the line, but to a lesser extent than some of the other prosperous democracies. Examining both figures, one can see that the USA does stand out among the prosperous democracies, but this

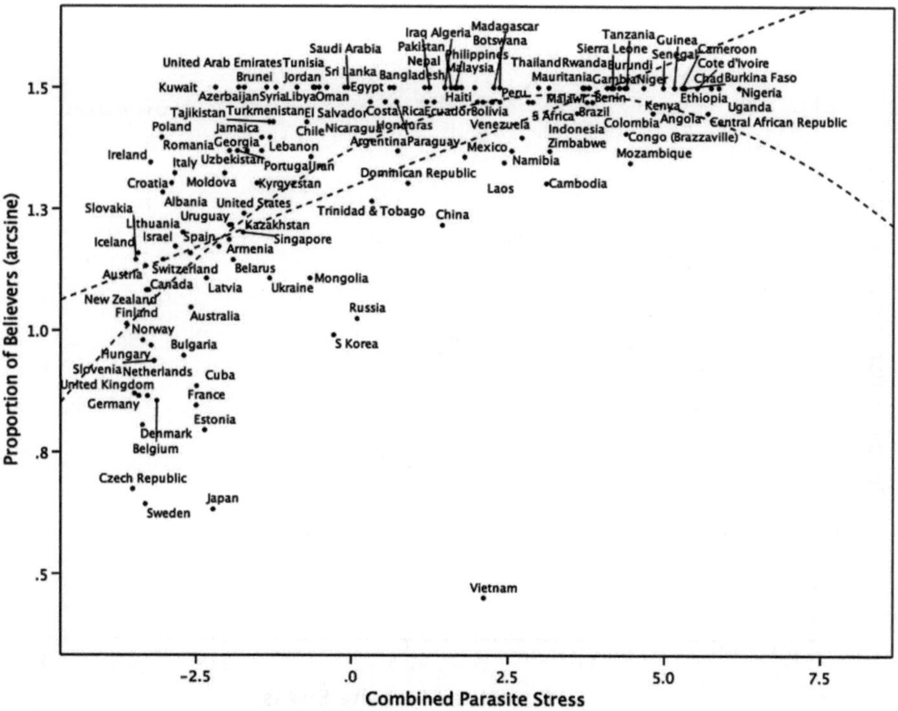


**Fig. 14.2** *Proportion of Believers* regressed on *Combined Parasite Stress*. Both linear and quadratic relationships are shown (results originally reported in Fincher and Thornhill 2012b; reprinted with permission)

is because of its high level of parasite stress in comparison to the other prosperous democracies, not because of its religiosity.

There is much evidence for the hyperactivity of behavioral immunity that Wall and Shackelford emphasize. It is likely that hyper-vigilant parasite detection and avoidance is a psychological adaptation. The penalty for error in detecting parasite presence can be literally grave. Thus, selection has favored the hypersensitivity of this detection and deduction bias of significant parasite threat. Said differently, humans are designed by past natural selection to adaptively accept many false-positive cues of potential infectious-disease presence in the environment. As we mentioned earlier in this book, hyperactivation of xenophobia is responsible for human prejudice against people who deviate from the normal range of phenotypes in weight (over- or underweight) or behavior (e.g., the physically or mentally challenged); this also may explain prejudice against the elderly and groups with minority sexual orientations. Most of these prejudices have been tied empirically to the parasite-stress theory of values by showing their relationship to perceived vulnerability to disease or disgust sensitivity, and/or their immediate activation by parasite-relevant cues, or their specificity in avoidance of physical contact (versus nonphysical contact) (Park et al. 2003; Park et al. 2007; Duncan and Schaller 2009; Terrizzi et al. 2010; Schaller and Park 2011; Park et al. 2013).

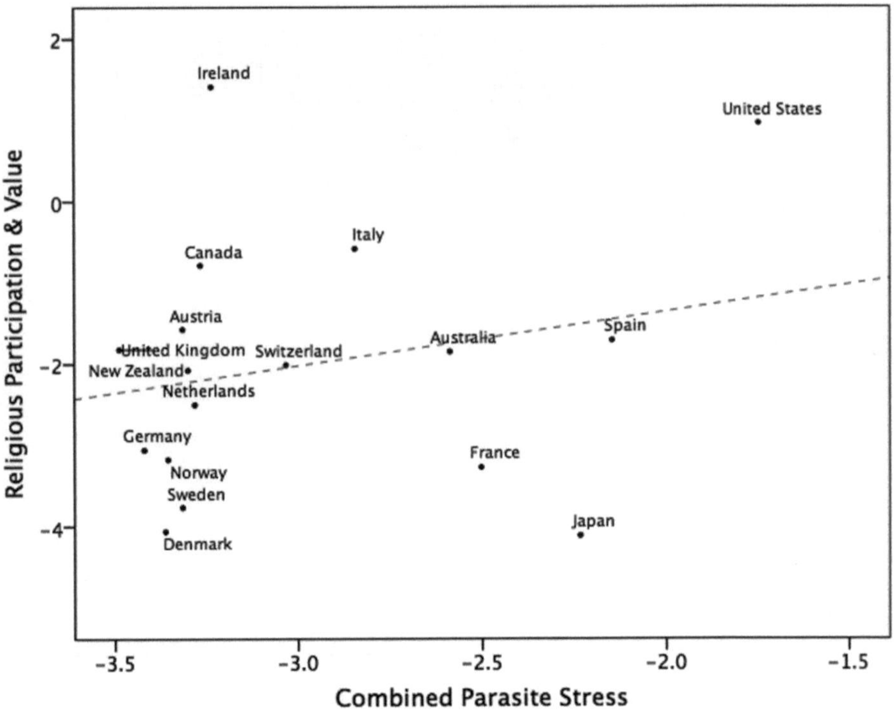


**Fig. 14.3** *Religious Participation and Value* regressed on *Combined Parasite Stress* for the 17 prosperous democracies used in Paul (2009)

Figueredo et al. (2012) in their commentary suggest that xenophobia and ethnocentrism may be by-products of adaptation and not, as we argue, adaptations that function in defense against parasites. They propose that these components of in-group assortativeness are the result of reduced impulse control associated with the adaptation for energetic trade-off between allocation to cognitive ability (IQ) and allocation to classical immune defense. With Chris Eppig, we have published findings showing strong negative correlations between IQ and parasite stress across nations and the US states supportive of this trade-off (see Chap. 11). Hence, Figueredo et al. conjecture that high parasite-stress reduces allocation to cognitive development and thereby reduces impulse control, and the lower impulse control in turn manifests as certain collectivist values related to religiosity.

We maintain that important aspects of xenophobia and ethnocentrism are adaptations that function in defense against parasites. First, increasing evidence indicates these two cultural features are allocations to immunity due to evolved design and function in behavioral immunity (see Chaps. 4 and 5). Second, these cultural features have high costs and thus would have been eliminated or greatly reduced in prevalence by selection unless they were adaptive (ancestrally). Hence, it is unlikely that these features are incidental effects. The two features also occur widely in nonhuman vertebrates

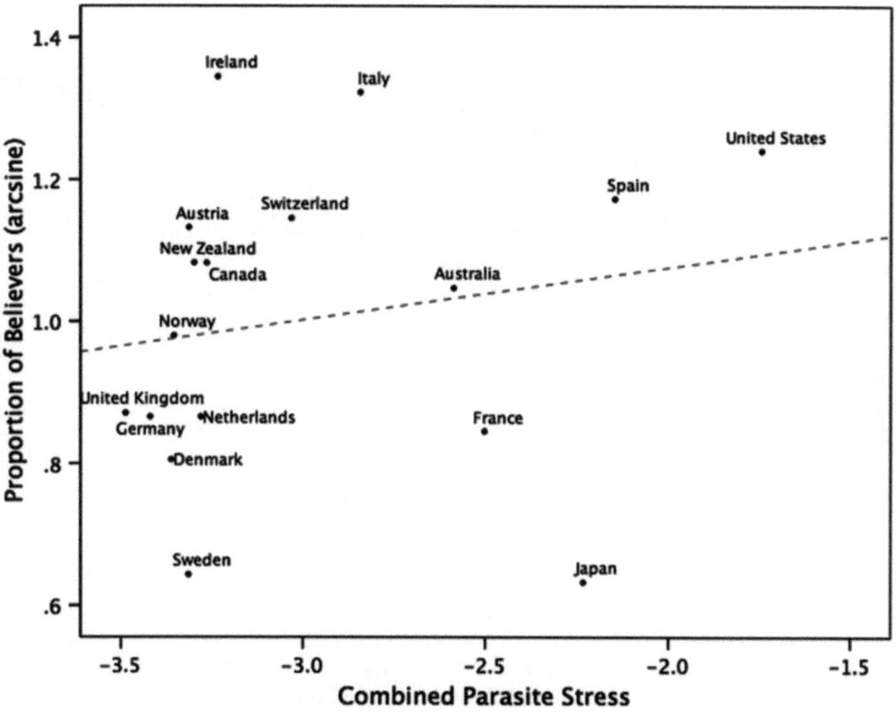


**Fig. 14.4** *Proportion of Believers* regressed on *Combined Parasite Stress* for the 17 prosperous democracies used in Paul (2009)

(e.g., Freeland 1976; also see Chap. 5), which implies that the incidental-effect hypothesis would need to account for the comparative evidence.

We hypothesize that reduced impulse control is best framed as a design component of a fast life history—an ancestrally adaptive feature that motivates high risk acceptance in order to acquire immediately available resources under extrinsic mortality (for additional discussion, see Thornhill and Palmer 2004).

Figueredo et al. in their commentary also argue that factors other than parasites may generate extrinsic mortality and lead to the fast life-history strategy of early reproduction. We agree. Our effort in the target article was to suggest that parasite stress is a source of extrinsic mortality that has not been appreciated fully by life-history researchers (with the exception of Quinlan 2007). As we hypothesized in Chap. 5, when extremely high parasite adversity yields extrinsic mortality in humans, the in-group investment and embeddedness of assortative sociality is not defensive against it and should decline. Hence, this hypothesis predicts that the relationship between parasite stress and collectivism will be curvilinear, such that in conditions of extreme parasite stress, collectivism declines. Our initial test of this (Chap. 5) across the world regions is quite preliminary, but does suggest support of the predicted curvilinear pattern. Africa is exceptional in that, compared to other world regions, the relationship between collectivism and parasite stress is negative

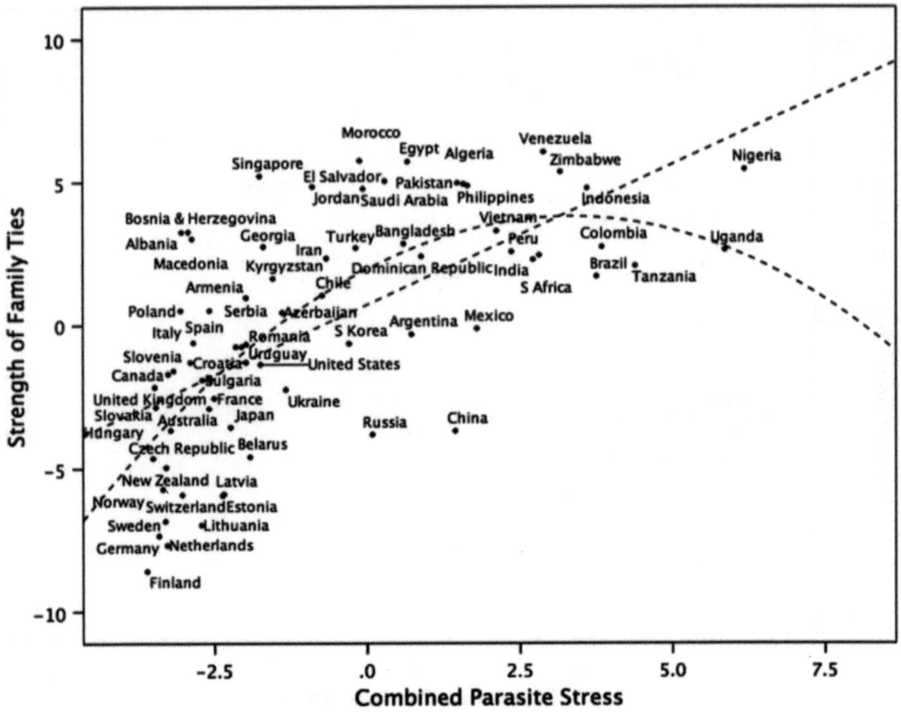


**Fig. 14.5** *Strength of Family Ties* regressed on *Parasite Stress* across nations. Both the linear and quadratic relationships are shown; see text for statistics ($n = 69$ countries)

rather than positive. This supported our hypothesis that high extrinsic mortality, as from the high parasite stress in Africa, leads to reduced collectivism, as predicted by life-history theory.

We provided a stronger test in our article responding to critics (Fincher and Thornhill 2012b)—the cross-national test that Figueredo et al. requested in their commentary. The cross-national relationship between *Strength of Family Ties* and *Combined Parasite Stress* is improved significantly by the quadratic model with an $r^2$ of 0.40 for the linear and 0.47 for the quadratic ($p = 0.003$, $n = 69$ countries) (Fig. 14.5). Thus, the improved model fit is supportive of the hypothesis that high levels of parasite stress can become an extrinsic mortality factor and reduce nepotistic and other in-group social investment. We return to the connection between life-history theory and the parasite-stress theory near the end of this section.

Figueredo et al. ask also about the magnitude of the correlation between collectivism and conservatism and individualism and liberalism. The large overlap of these variables is documented in Chaps. 4 and 5. Conservatism and liberalism correspond closely with collectivism and individualism, respectively.

Schaller and Murray (2012), Beit-Hallahmi (2012), and other commentators mentioned the need for more research on the proximate psychological mechanisms involved in the acquisition of culture that cause the range of values from high collectivism to high individualism across individuals and regions. We agree completely, but emphasize that this aspect of the parasite-stress theory's foundation is far from a total black box. We have summarized earlier in this book a range of research studies that confirm specific predictions from the parasite-stress theory about ontogenetic, perceptual, affective, cognitive, and behavioral mechanisms that defend against the negative fitness effects of infectious diseases (see Chaps. 3 and 4).

Currie and Mace (2012) in their commentary criticize the target article because its new empirical analyses do not provide evidence for "a cognitive mechanism that is sensitive to parasite stress and causes people to exhibit more in-group favoritism ...." Certainly they are correct. Our analyses in the target article assumed such mechanisms (and some other mechanisms focused on perception, affect, and motivation), a reasonable assumption given the research on such mechanisms discussed in the target article (and mentioned in the previous paragraph). Also, the cross-regional patterns we document in the target article and throughout this book are consistent with such mechanisms in people's heads. These patterns must arise from information processing and deduction by individuals' brains.

In their commentary, Vigil and Coulombe (2012) see an inconsistency with the parasite-stress model from literature indicating that, in some Western samples, people of high religiosity and conservatism have larger social networks than less religious, more liberal people. Certainly, the evidence is mixed on this pattern, as Gelfand et al. (2004) concluded the opposite, at least with regard to collectivists versus individualists: collectivists have smaller groups and more intimate and durable relations among members than individualists. The pattern may be sample-dependent.

The prediction about social networks from the parasite-stress theory, however, is more about the nature of social relations between the two types of ideologues rather than social network size. The hypothesis that collectivism compared to individualism is characterized by tight social networks, more cohesive and cooperative friendship groups, in-versus out-group distinctiveness, more permanence of group membership, and more intensity or intimacy of social interactions is supported by a range of studies reviewed in Chap. 4. Hence, it does appear that individualists have more superficial and less durable relationships and with a wider variety of people than collectivists. This pattern is as predicted by the parasite-stress theory of values in light of the established positive covariation between collectivism and parasite stress (Fincher et al. 2008; Thornhill et al. 2010; Chap. 5).

In their commentary, Vigil and Coulombe (2012) also raise the interesting issue of sex differences in values. They wondered if some of our findings about religiosity or family ties in the target article are only specific to one sex and not the other. There is some evidence that females may allocate more to classical immune function than males (see the references in Vigil and Coulombe (2012)). Also, there is considerable evidence that women are more disgust sensitive than men (Curtis et al. 2011; Fleischman 2014). In our article responding to commentators (Fincher and Thornhill 2012b) we returned to the data in the target article for the cross-national variables *Strength of Family Ties* (described in Chap. 5) and *Religious Participation and Value* (see Chap. 9) and computed sex-specific values. The Cronbach's $\alpha$ for the *Strength of Family Ties* for males was 0.87 and 0.83 for females; the Cronbach's $\alpha$ for

*Religious Participation and Value* for males was 0.95 and 0.91 for females. For both variables, the correlation between the sex-specific value score and *Combined Parasite Stress* was identical: *Strength of Family Ties*: male $r=0.63$, female $r=0.63$; $n=69$ for both correlations; *Religious Participation and Value*: male $r=0.70$, female $r=0.70$; $n=89$ for both correlations. Thus, we found no sex differences in the relationship between these dimensions of values and parasite stress. Nevertheless, given that other researchers have found significant sex differences in values related to differences behavioral immune system activation (Terrizzi et al. 2014), further research on sex differences in values due to the effects of parasite stress certainly is warranted.

Furthermore, Vigil and Coulombe suggest in their commentary that certain sex differences in values can be explained by an evolutionary history of male-biased philopatry and conversely female-biased dispersal from the natal locale. We explored this in Fincher and Thornhill (2012b) across contemporary countries by focusing on the question of whether a respondent lived at home with his or her parents (an item of the variable *Strength of Family Ties*). For this item, we found a significant male-bias in philopatry. On average, a greater proportion of males reported living at home with parents (male mean [$M$]=0.32; female $M=0.25$; $t_{93}=-10.5$, $p<0.0001$). However, the positive correlation between the proportion of those that lived at home with their parents and *Combined Parasite Stress* for the sexes was not significantly different (male $r=0.46$; female $r=0.52$; $z=-0.52$, $p=0.6031$; $n=90$ countries for both). Thus, while there is a significant male-biased philopatry, it does not necessarily lead to a sex-difference in the relationships between parasite stress and values such as *Strength of Family Ties* or *Religious Participation and Value*.

Currie and Mace (2012) in their commentary state their view that, because of shared cultural history, countries often are not statistically independent. They do not mention, however, the alternative view that countries and other regions, and cultures themselves, are independent, even in the case of recently shared cultural history. Early in this book, we discussed our view of enculturation that implies cultural independence: humans are designed by a history of evolution by selection to acquire contingently the cultural items that provide solutions to local ecological, including social, problems. Also, we cited various researchers who view enculturation in ways similar to our view, based on various findings. The claim of cultural nonindependence is inconsistent with the evolution of human cultural capacity—the set of psychological adaptations that are designed functionally for adaptive (ancestrally) enculturation of individuals. This capacity was favored by selection because it promoted inclusive fitness; automatically, arbitrarily or maladaptively learning cultural elements was always selected against. Thus, the values that people adopt during ontogeny in one region are independent of the values adopted in another region, even adjacent regions (cultures), and regardless of the degree of cultural item flow between the regions or the historical connections between cultures involved (see Chap. 2).

Our analyses in this book within world area regions and US census regions are not for dealing with issues on nonindependence of countries or US states. Our use of subregions is an exercise that can identify if any particular subregion does not fit the general global analysis. The value of this was found with collectivism and parasite stress in Africa. As mentioned above, the relationship of these two variables was negative in Africa, whereas it was positive in other world regions. This supported our hypothesis that high extrinsic mortality, as from the high parasite stress in Africa, leads to reduced collectivism, as predicted by life-history theory (see above).

Currie and Mace do their own regional subdivision of the globe in their commentary. The world regions identified by Murdock and used by us reflect a legacy of research validating the subdivision. At the least, Currie and Mace should validate their novel subdivision. In the absence of such validation, it is difficult to interpret their regional analyses.

Certain referees of some of our earlier papers on parasite stress in relation to cultural diversity, other than Currie and Mace, have advanced also the criticism of non-independence of cultures. That is why we spent time on the topic of the independence of cultures in Chap. 2. It is a common error to think that variation among cultures cannot be studied unless their historical connectedness is somehow corrected or controlled. Interdependent mindedness (sometimes called holistic cognition), in contrast to independent mindedness (analytic cognition), leads, we hypothesize, to this criticism. The independence perspective is supported theoretically by the theory that evolution by selection for maximum individual inclusive fitness has led to human cultural capacity—the psychological adaptations by which individuals acquire the cultural items, including values, that allow coping with local ecological adversity. The independence perspective is supported empirically by the repeated ability of the parasite stress theory of values to predict successfully human cultural diversity across nations, US states, and indigenous societies, as reported in this book.

The commentary by de Barra and Curtis (2012) questions an assumption that they perceive is part of the parasite-stress theory of sociality: "A critical assumption of Fincher and Thornhill's thesis is that pathogens carried by out-groups … will be more dangerous than those of one's own family and community. (p. 85)" Actually, we do not make this assumption. In the target article, we discussed that behavioral immunity has two design features as a result of direct Darwinian selection for them: protection against novel parasites harbored in out-groups to which individuals in the in-group are not adapted and managing the negative effects of parasites within the in-group. Accordingly, xenophobia and limited dispersal are the adaptations for out-group contagion avoidance. In-group embeddedness, including family ties and religiosity, functions for managing present parasites within the groups. If out-group contagion were "more dangerous" than in-group contagion, one would expect out-group psychological and behavioral defenses to be better (or only) designed for pathogen defense than in-group defenses. This is not an empirically apparent pattern, as seen in this book. For example, in the case of religiosity, our results indicate that it is well suited both for in-group embeddedness and out-group boundary formation (Chap. 9). Furthermore, it may be enough that contracting infectious diseases from out-groups, even if they are not "more dangerous" than what already plagues someone's in-group could still represent a significant cost against reproductive success that is best to avoid.

Both in-group and out-group parasites can present the host with novelty to which it is not immune. In-group parasites do so during the co-evolutionary races with a host—new features arise in the parasite that circumvent evolved host defense. Out-group parasites can be as dangerous as in-group parasites because the spatially localized host–parasite races may not equip the host with immunity to new out-group parasites.

de Barra and Curtis (2012) in their commentary also discuss examples of parasites that have more success in out-groups of hosts than in in-groups. This is an interesting scenario because when a parasite can achieve high reproductive success by invading individuals across a cultural boundary, selection is strong on the parasite to invade the adjacent group. This, in turn, will promote the adoption of cultural behaviors of xenophobia and more restricted dispersal (philopatry) in the culture being invaded by the parasite.

Grotuss's (2012) commentary mentions that the published research so far inspired by the parasite-stress theory of sociality has not addressed the matter of psychological and behavioral components to reduce contact with parasites present within the in-group. We agree with Freeland (1976) and Kurzban and Leary (2001) in predicting that stigmatization and prejudice resulting in marginalization, isolation, ostracism, and periods of quarantining are adaptations for this purpose. If this is true, these features are components of the behavioral immune system.

Figueredo et al. (2012) in their commentary question the combination of philopatry, ethnocentrism, and xenophobia into the variable we refer to as in-group assortative sociality. They point out that across published research studies ethnocentrism and xenophobia show a range of positive, negative, or no correlations, depending upon the society or sample investigated. We recognized this in the target article and proposed circumstances under which xenophobia and ethnocentrism would not be positively correlated, and by extension might even show negative or no correlation in a region. As well, there is good reason to expect that in some areas and under certain conditions such as high levels of parasite stress that these elements of in-group assortative sociality will be strongly intercorrelated but in areas of low pathogen stress there may be more variation in the intercorrelation among people of these values (see the discussion of cultural linkage disequilibrium in Chap. 13 (Sect. 13.2.6)).

Cashdan's commentary (2012) also discussed our combination of values. Cashdan's commentary presents a summary of results found by her and M. Steele based on cross-cultural data on values from the Standard Cross-cultural Sample of indigenous human societies (Cashdan and Steele 2010; Cashdan and Steele 2013). As discussed in Chap. 5, they found that, across these societies, parasite stress significantly predicts negatively the peoples' residential mobility among communities, a measure of restriction of movement and related philopatry. They also found that parasite stress positively predicts degree of xenophobia. Cashdan concludes that these two findings support the parasite-stress theory's prediction that philopatry and xenophobia are features of assortative sociality that reduce contact with other groups and their habitats in high parasite-stress situations. However, in follow-up analysis reported in a more recent paper (Cashdan and Steele 2013), using a different measure of parasite stress, the xenophobia result was not repeated. Also, Cashdan and Steele (2013) found no evidence that the variable "ethnic loyalty" across these societies corresponded to variation in infectious disease stress. Cashdan and Steele hypothesized that ethnic loyalty may defend against various ecological stresses in addition to parasite stress. Certainly, we have no criticism of this hypothesis.

An empirical question, then, in light of the criticisms of Figueredo et al. and Cashdan, is whether people have adaptation which functions to promote in-group ties and support under the threat or presence of infectious agents per se. The parasite-stress theory of values implies that people will have such adaptation and to both nonzoonotic and zoonotic disease cues, because the adversity of either category of diseases can make local in-group embeddedness adaptive. The experimental paradigm used, for example, by Mortensen et al. (2010) or Schaller et al. (2010) involving presentation of parasite-salient cues to research participants (see Chap. 3) could explore this in detail and simultaneously include tests of cues of ecological stressors other than parasite stress. The parasite-stress theory of values predicts that people will show a shift to increased in-group alliance and investment upon perceiving contagion risk, and that the shift will be greater in people with high perceived vulnerability to disease or disgust sensitivity than people with low perceived vulnerability to infectious agents or disgust. The already published findings of Park et al. (2003), Faulkner et al. (2004), Park et al. (2007), Navarrete and Fessler (2006), and Navarrete et al. (2007) provide preliminary support for these predictions.

The commentary by Vigil and Coulombe (2012) claims that disgust sensitivity to pathogens is not higher in conservatives than in liberals, citing Tybur et al. (2010). Actually, multiple other studies have found that conservatives have greater disgust than liberals (Inbar et al. 2009, 2012; Terrizzi et al. 2010; Terrizzi et al. 2013). Indeed, this pattern was found across 121 different countries and was statistically significant in each of ten geographic regions of the world (Inbar et al. 2012). In addition, Inbar et al. (2012) reported from the US sample that disgust sensitivity predicted voting preferences and actual voting in the 2008 US presidential election: high disgust individuals and US states favored John McCain (conservative), whereas their low disgust counterparts favored Barack Obama (liberal). When the various research studies showing the positive relationship between disgust and conservatism are combined with the large literature on the functional design of disgust for pathogen avoidance (Oaten et al. 2009; Curtis et al. 2011), the hypothesis that disgust will be greater in conservatives (collectivists) than in liberals (individualists) is strongly supported.

In a recent paper, Hackman and Hruschka (2013) criticized our published research on the cultural variables homicide, child maltreatment, religiosity, and family ties across US states. We published analyses of these variables in relation to parasite stress in Thornhill and Fincher (2011) (homicide and child maltreatment, see Chap. 8) and Fincher and Thornhill (2012a) (family ties and religiosity, see Chaps. 5 and 9, respectively). Hackman and Hruschka (2013) provide two criticisms and conclude from their analyses that the findings reported in our two papers are more consistent with their interpretation of life-history theory than with the parasite-stress theory of values. (Our responses to Hackman and Hruschka were published in Thornhill and Fincher (2013) and are given below.)

One of their criticisms is that the positive relationships of the cultural variables with parasite adversity, measured as *Parasite Stress USA* (see Chap. 5) reported in our papers are driven significantly by sexually transmitted diseases (STDs), but not by non-STDS. Hackman and Hruschka (2013) show that STDs comprise about three-quarters of the infectious disease cases overall and about 90 % of the cases in one-half of the years in the time-span of the infectious-disease data we compiled (from 1993 to 2007) in constructing *Parasite Stress USA*. They argue that STDs and non-STDs should each covary significantly with the cultural variables if the parasite-stress theory is correct, but only STDs should show the pattern if life-history theory is paramount, because only STDs are the result of fast-track life history, reflecting incautious mating decisions by risk-prone people. They go on to show that STDs are positively associated with the cultural variables, while non-STDs are not.

Our response to this criticism is that the parasite-stress theory does not distinguish STDs as a special or exceptional case. Importantly, human STDs are nonzoonotic infectious diseases and hence are expected to be especially robust predictors of the cultural variables, as we have argued and shown empirically throughout this book. Given the huge over-representation of STDs in the US parasite stress measure, elimination of STDs from this measure of parasite adversity removes a substantial proportion of the nonzoonotic parasite stress and, therefore, creates a poor measure of parasite adversity and one inconsistent with the parasite-stress theory. Hence, it is difficult to interpret null findings when the analyses involve a measure of parasite adversity that eliminates most of the theoretically relevant disease stress.

Hackman and Hruschka (2013) do not take issue with our findings from international analyses of parasite adversity in relation to the cultural variables, which support the parasite-stress theory and are reported along with the US analyses in Thornhill and Fincher (2011) and Fincher and Thornhill (2012a) (see the book chapters listed above). The predominance of STDs in parasite adversity measures may apply widely, especially in countries in the West.

Hackman and Hruschka's (2013) second criticism is their claim that the parasite-stress theory has to account for the cultural variables even after statistically controlling for fast-track life-history factors. For example, they argue that adult-on-adult homicides are caused by a fast life-history strategy. We agree that this is a partial cause, but only briefly mentioned this in Thornhill and Fincher (2011). In that paper, we focused on collectivism as the value set that is a cause of the homicides (see Chap. 8).

Previously in this chapter, we discussed and empirically supported the prediction from the parasite-stress theory that when infectious disease adversity reaches the status of an extrinsic mortality factor, then fast life history will be the predominate value system, and not the collectivist value system. As Hackman and Hruschka (2013) correctly point out, only intrinsic mortality factors are controllable by individuals; extrinsic mortality factors cannot be offset by people's nepotism or other social investments. Hence, collectivist investment is not a defense for coping with extrinsic mortality factors. Hackman and Hruschka (2013) use rates of teenage mothers per state as a measure of fast life history. They statistically control this variable, which reduces or eliminates the covariation between the cultural variables and parasite adversity. We propose, however, that teen-mother rates are a mixed variable in terms of causation, and the mix depends on whether collectivism is effective or not for offsetting offspring mortality across the range of parasite adversity. As parasite stress increases, but remains below the tipping point of extrinsic mortality, birth rate increases, because the extended family support of collectivism cues a condition suitable for high reproductive rate by women. The importance of extended family support in enhancing women's reproductive rate is supported by multiple studies (Gibson and Mace 2005; Sear and Mace 2008; Meehan et al. 2013), and this effect is a prediction (i.e., an empirical necessity) of the inclusive fitness hypothesis for human nepotism. But when parasites achieve the high adversity corresponding to extrinsic mortality, collectivism declines strategically and fast life history of many births and low investment per child becomes optimal. It is only in the ecological setting of extrinsic mortality from parasites (and/or other factors resulting in extrinsic mortality) that the life-history theory prediction that people will adopt fast life-history values applies. Hence, contrary to Hackman and Hruschka (2013), teen birth rates are not a straightforward and unambiguous measure of fast life history. It is only such a measure after the transition from intrinsic to extrinsic mortality in a region. Statistically controlling teen birth rate in analysis of parasite stress in relation to cultural variables produces results that are hard to interpret and can be misleading.

Hackman and Hruschka (2013) use the percent of a state's population that is African-American ("percent blacks") as a measure of extrinsic mortality risk in the state. They summarize a range of evidence showing that blacks have higher rates of mortality than other racial categories. They then statistically control for percent blacks and show that parasite stress no longer significantly predicts the cultural variables. We have two criticisms of their analytical approach. First, we question the use of percent blacks as a suitable proxy for extrinsic mortality. Second, the very high correlation between percent blacks and parasite stress across states makes percent blacks an unsuitable variable to remove in the analysis. With regard to our first criticism, African-Americans in the USA are expected to be high in collectivism in general, because they often face high healthcare disparity and impoverished conditions that promote infectious diseases. The relationship between percent blacks and *Parasite Stress USA* is huge, a correlation of 0.90 across the 50 states (Eppig et al. (2011) reports an $r$ of 0.92, which includes the District of Columbia for an $n$ of 51. See that paper for the source of data for the percent blacks across the US regions. Hackman and Hruschka (2013) summarize additional evidence of relatively high infectious disease rates among blacks in the USA). As we have explained, collectivism defends against parasite stress as long as the level of stress is intrinsic mortality. Hence, percent blacks is a combination of collectivist as well as fast life-history strategies and not a suitable or unambiguous measure of extrinsic mortality. Our second criticism arises from the strong correlation between percent blacks and parasite stress just mentioned. In our research with Chris Eppig on IQ variation across states of the USA (see Chap. 11), we pointed out that, given the near perfect correlation between percent blacks and parasite stress, it is not appropriate to control percent blacks in analysis of IQ variation and parasite stress. (Some earlier researchers had controlled percent blacks in their analyses of IQ variation across states. In Eppig et al. (2011) reasons other than statistical for not controlling percent blacks also are discussed.) The huge correlation between percent blacks and parasite stress implies that percent blacks is a proxy measure of parasite stress. Hence, controlling for percent blacks, as Hackman and Hruschka (2013) did, is to control for infectious disease variation, i.e., it eliminates infectious disease as an independent variable in analysis.

Hackman and Hruschka (2013) describe what they perceive as a "puzzle" arising from their analyses. Gladden et al. (2009) provide evidence that religiosity is a component of long-term social investment characteristic of a slow life-history strategy. (We discuss Gladden et al.'s study further in Chap. 9.) Hackman and Hruschka (2013), however, feel that their own results are contrary to this pattern, because religiosity increased among people they assumed were uniformly fast in life history (teen mothers and African Americans). The apparent contradiction across the two studies is resolved by the parasite-stress theory—specifically, by recognition that both teen mothers and African-Americans, in part, are collectivists, and religiosity is a central component of collectivism (see Chap. 9).

It is likely that infectious diseases are a cause of human life-history allocations and trade-offs on both the ecological and evolutionary time scales. Hackman and Hruschka (2013) treat life-history theory and the parasite-stress theory as alternative evolutionary and ecological theories of cultural variables. We have proposed their complementariness. Infectious diseases are causes of both intrinsic and extrinsic mortality, depending on the degree of adversity. Infectious diseases are a (possibly the) leading cause of natural selection on contemporary humans; genes associated with immunity are evolutionary hotspots (see Chap. 3). Furthermore, the magnificently designed classical and behavioral human immune systems are documentation of the importance of infectious diseases as agents of natural selection in human evolutionary history.

Shrira et al. (2013) have criticized some of our findings on adult-on-adult homicide across the states of the USA discussed in Chap. 8 and first presented in Thornhill and Fincher (2011). Shrira et al. calculated a new parasite-adversity measure for each state based on only the eight most frequent infectious diseases. Their measure correlates highly with ours, which is based on a larger set of diseases reported from each state ($r = 0.87$). When Shrira et al. controlled for various variables they felt were confounds, their "family homicide" variable is marginally significantly related to parasite adversity as measured by their method and insignificant as measured by ours. However, their family-homicide variable is not the same as the romantic-partner homicide variables used in Thornhill and Fincher (2011; and see Chap. 8). Their variable was all murders in which the perpetrator and victim were family members of any sort (spouse, child, parent, sibling, etc.). Therefore, their findings and ours are not comparable. Shrira et al., however, offer some interesting new ideas about parasite stress and crime in general that we discuss in the section below on additional future research.

It is of interest to examine other recent papers that address the parasite-stress theory of values as applied to human behavior. Three found results supporting the theory. The fourth reported some findings that were interpreted by its authors as inconsistent with the theory and, based on these findings, criticized the theory in general. We contrast these four studies here in this section on criticisms.

Varnum (2012) researched so-called nonconformist voting in presidential elections across the 50 US states for each of the years from 1968 to 2008. Basically, the USA has a two-party political system, and typically, votes for third-party presidential candidates do not exceed 10 % of the voters. Third-party votes, then, derive from voters with nonconformist values. We have reviewed evidence that conformity is related positively to collectivism and parasite adversity (Chaps. 4 and 5). From the parasite-stress theory, Varnum (2012) predicted that the percentage of third-party votes (nonconformist votes) across states will relate negatively to parasite stress in each of the years, with the exception of 1968 when George Wallace ran as an independent third-party candidate with a political strategy based on racial segregation and white supremacy. Varnum's empirical findings supported the theory. Across the states, third-party voting was related negatively to parasite stress across all years, except in 1968, where third-party voting was related positively to parasite stress. Presumably, the support for Wallace in 1968 reversed the typical pattern because his conservative supporters felt that Wallace, if elected, would restore the tradition of racist political policy in the USA.

A second paper supporting the parasite-stress theory of values investigated trust across approximately 100 nations (Le 2013). Based on the parasite-stress theory and using a measure of extensive trust in people in general, and hence trust in out-group members, from the World Values Survey, Le (2013) predicted and found a negative relationship between trust and parasite adversity. As we have discussed, collectivists tend to trust in-group members and distrust out-group members, whereas individualists are more broadly trusting and hence affiliate with a diversity of people (Chaps. 4 and 5). In the same paper, Le also reported cross-national, positive relationships between parasite stress and the variables of religion diversity, language diversity, ethnic diversity, and governmental corruption, and negative relationships between parasite stress and economic well-being and latitude. These findings replicate a range of findings we have treated earlier in this book. Le (p. 17) concludes that "… [infectious] disease burden turns out to be the most robust predictor … and a robust predictor of the development of societies."

The third study was conducted by Varnum (in press). It examined, across the states of the USA, the relationship between aspects of collectivism–individualism and *Parasite Stress USA*, a measure of parasite adversity (described in Chap. 5). A robust, negative relationship was found between parasite stress and an index of social capital that taps people's involvement in nonlocal social networking and organizations involving contacts with strangers. Also, generalized trust—trust outside the in-group—was related negatively to parasite stress. Hence, as expected from the parasite-stress theory, across states of the USA, social involvement and trust beyond the in-group are least in collectivist/high parasite stress states and maximum in individualist/low parasite-stress states.

The fourth study was by Hruschka and Henrich (2013). It concludes that, across nations, parasite stress does not predict governmental effectiveness and in-group social preferences as robustly as some researchers, including us, have assumed. They used several measures of governmental effectiveness and in-group preferences. Governmental effectiveness, for example, was measured by the quality of government-provided public services in a country. In-group preference was measured, for example, by *Hofstede Individualism* (the converse of in-group preference) and our measure *Strength of Family Ties.* (These two collectivism–individualism variables are described in Chap. 5.) Hruschka and Henrich's (2013) hypothesis was that when governments are ineffective in providing for the many needs of people, the people then turn to the in-group for their needs, but when governments provide the basic needs, people do not engage, or engage less, in ethnocentrism. As we have mentioned earlier in this chapter and in Chap. 9, those proposing this hypothesis and related hypotheses, which we have combined and labeled the conditions-of-living model, fail to address why people will prefer in-group under many harsh circumstance rather than ally with out-groups, given that out-groups potentially offer substantial benefits (ideas, support, resources) that can alleviate adversity. Also, Hruschka and Henrich (2013) just accept as a given that governments vary in effectiveness and do not recognize that explaining this variation is a scientific challenge in itself. As we have explained throughout this book, the parasite-stress theory of values offers an explanation and considerable evidence as to why people seek in-group membership under parasite adversity per se (rather than out-group membership) and why governments vary in providing basic needs. Hruschka and Henrich (2013) showed that governmental effectiveness is related negatively to in-group preference/collectivism, which replicates the multiple studies we discussed in Chaps. 10 and 11 on topics (democratization, governmental transparency) corresponding closely with their variable of governmental effectiveness. Also, they reported that parasite stress predicts positively in-group favoritism, which replicates the extensive research on parasite stress and collectivism and religiosity presented in Chaps. 5 and 9. They then introduce various control variables into the analyses of the relationship between parasite stress and in-group preference and find that parasite stress has either limited or no effect, depending on the particular controls used. Moreover, they believe that shared cultural factors among countries create nonindependence in country-level data and hence they control for shared religious ideology and world region in their analyses. We have mentioned earlier in this chapter and explained in detail in Chap. 2 why such controls are inappropriate and can give misleading empirical results. Furthermore, by controlling the effects of world region they effectively controlled the effects of parasite stress and made it impossible to find any further contribution of parasite stress variation to the model because world regions differ significantly in their levels of parasite stress (e.g., Africa has a significantly higher level of parasite stress compared to the other world regions, see Sect. 5.14.1). Given these problems, Hruschka and Henrich's (2013) results involving the various controls they use are not pertinent to the parasite-stress theory of values. (See also the discussion of the partialling fallacy in Chap. 5.)

In summary, the parasite-stress theory has been evaluated and criticized by a range of researchers across numerous scholarly disciplines and is continually being refined. The theory is far-reaching and thus open to criticisms from the many research fields that study human affairs and various other topics related to the parasite-stress theory. To date, the theory seems not to be threatened by criticisms. We hope other scholars will have an interest in the ideas and findings we present in this book and identify areas of the research that need clarification, or that are erroneous.

We conclude this section on criticisms of the parasite-stress theory of sociality with a criticism that has not yet been made, but we believe will be forthcoming.

According to the parasite-stress theory of values, people are parasitically modified animals as a result of their interactions with parasites in evolutionary ancestral generations as well as during the lifetime of individuals. The evolutionary ancestral interaction gave rise to the natural selection that crafted the classical and behavioral immune systems of humans as well as the functional integration and synergy of these two systems in dealing with pathogens. The interactions with parasites and infectious-disease related cues during the lifetime of individual humans proximately cause the values and feelings that serve to conditionally optimize personal sociality for infectious-disease levels locally. This view of humans as parasitically modified animals is supported strongly by evidence in this book.

For completeness, however, we must consider a causally distinct/alternative hypothesis for the aspects of human sociality upon which we have focused in this book: host manipulation by parasites. According to the alternative perspective, people are parasitically modified to promote the fitness of parasites rather than their own fitness. Critics of the parasite-stress theory have not yet raised this alternative, but we expect it will be raised when biologists discuss the parasite-stress theory more widely.

In numerous parasite–host associations, the parasite manipulates the host's psychology, behavior and/or external morphology as an evolved adaptation to increase transmission to new hosts (Moore 2002). This is in the reproductive interest of the parasite, but maladaptive for the host. Effective host manipulation is seen in a wide range of parasite types from viruses and bacteria to fungi and "worms," and in a wide range of host taxa. The nature of host-manipulation adaptations of parasites depends upon whether the parasite involved has a direct or an indirect life cycle. In the case of a parasite with an indirect life cycle, the parasite's host-manipulation adaptation functions to increase the intermediate host's probability of being eaten by a species of predator that is the definitive host of the parasite—that is, the host in which sexual reproduction of the parasite occurs. In the case of direct life cycle parasites (those without an intermediate host), the host-manipulation adaptation functions to transmit the parasite to susceptible members of the host species. Humans are not the host of any parasite that is transmitted by predation of humans to a definitive host. However, the host-manipulation strategy of parasites to increase direct transmission among members of a host species is a possibility for human parasites. A recent study by Rode et al. (2013) reported that brine shrimp increase temporary grouping behavior (referred to as swarming) when parasitized, which promotes transmission of the parasite involved among group members. This begs the question: Is the in-group assortative social preference of conservative people (ethnocentrism, xenophobia, and philopatry) a parasite-manipulation adaptation that functions to promote transmission to new local hosts?

There are good reasons to reject the hypothesis that conservative values of people are simply a parasite's strategy to invade new human hosts. The parasite manipulation hypothesis requires that the in-group sociality will be maladaptive for hosts and show functional design for parasite transmission (Poulin 2010). This is negated in the human case by the diverse and copious evidence in this book indicating that the values comprising in-group sociality are effective means of avoiding and managing infectious diseases and thereby are functional for human hosts in high parasite-adversity regions. Furthermore, as explained in Chap. 13, the in-group values and associated local reproduction promote the evolution of genetic immunity, which reduces host-to-host transmission of a parasite in an in-group; the reduced transmission, in turn, selects for reduced parasite virulence. Nevertheless, future research may find that some aspects of human psychology are manipulable by parasites (Lindova et al. 2011).

Although, we favor this same kind of thinking applied to relevant behaviors across nonhuman animal species in high parasite stress regions, we recognize that the parasite-manipulation hypothesis may apply to some behaviors outside humans. Research on the sociality of nonhuman animals would benefit from keeping the parasite-manipulation hypothesis in mind as an alternative along with the parasite-stress theory of sociality.

## 14.7 Additional Future Research

In previous chapters, we suggested some future research directions as we analyzed topics. Also, in the section just above, we pointed to certain additional future research areas. In this section, we discuss some other research directions suggested by commentaries on our *Behavioral and Brain Sciences* target article (Fincher and Thornhill 2012a).

Uskul (2012) in her commentary calls for more experimentation in the future. We agree, and pointed out in the target article that additional experimental testing of the parasite-stress theory of values is important. We suggested, too, in the target article some lab experiments that would complement the recent experimental research using infectious-disease salient stimuli as in studies by Schaller et al. (2010) and Mortensen et al. (2010). (These two studies are first described in Chap. 3.) Other experimental suggestions we made in the target article are field experiments, which have the advantage of more inherent ecological validity compared to lab experiments. A combination of lab and field experiments and naturalistic observations potentially can best address the specific nature of mediating psychological processes of enculturation that result in the relationship between parasite stress and cultural patterns. We emphasize though that this combination approach in research is only one test-ground for the parasite theory of sociality; comparative research is equally important (see Chap. 2).

Uskul (2012) wonders in her commentary how the parasite-stress theory might apply to certain well-established social and behavior science research programs other than the focal areas treated in the target article (collectivism–individualism, family life, and religiosity). She mentions established research programs such as the rugged individualism that underlies the "frontier spirit," residential mobility patterns of Westerners, the creation and diffusion of innovations, cognitive or reasoning styles, and the nature of units of economic productivity.

The frontier spirit, so important in the immigration history of the USA and the Hokkaido region of Japan (Kitayama et al. 2006), we suggest, is caused by psychological traits enculturated by low parasite stress and characterizing individualism: independent self, self-efficacy, dispersal proneness, openness to new experiences, and associated willingness to engage the adventure of the frontier (see Chaps. 4 and 5). Hence, the frontier spirit is the antithesis of high philopatry or remaining in or near the natal region throughout life.

We hypothesize that the empirical connection that Oishi's (2010) earlier research has found between patterns of residential mobility of people in Western societies and the people's values of self-identity (independent versus interdependent) and group affiliation arise from variation in valuing philopatry versus dispersal. Residential mobility is defined by Oishi (2010) as the number of residential moves for an individual or the percentage having moved residence recently for a neighborhood. There is evidence that dispersal patterns are caused by different ontogenetic experiences with infectious disease and from the evoked values of collectivism or individualism (Chap. 5). This evidence includes reduced residential movement between communities by individuals in indigenous societies and reduced interstate residential movement by people in the USA in settings of high parasite adversity compared to settings of low pathogen adversity. As we have argued, dispersal has benefits, but also costs in terms of exposure to novel parasites, and is expected therefore to be characteristic of individualistic people and low parasite-stress conditions (see Chap. 5). To further test our hypothesis for residential movement behavior of people, one might measure perceived vulnerability to disease and/or disgust sensitivity in relation to history of movement frequency and distance moved. We predict people who have a high perceived vulnerability to disease/disgust sensitivity will be more philopatric than those scoring low. The component of the parasite-stress theory of values pertaining to the psychology and behavior of dispersal also could be examined experimentally by showing people parasite-salient pictures and measuring their immediate value changes pertaining to dispersal.

The two cognitive styles, holistic and analytical reasoning, have been tied to collectivism–individualism by prior researchers (e.g., Nisbett et al. 2001; Uskul et al. 2008; Chap. 4). Holistic reasoning is the interdependent thinking that prioritizes the in-group's wellbeing, harmony, and goals. According to the parasite-stress theory of values, holistic reasoning is part of in-group embeddedness, and hence is predicted to be characteristic of relatively high parasite-stress regions and individual ontogenies. In contrast, analytical cognition is intellectual autonomy that prioritizes personal achievement rather than the achievement of in-group goals. Analytical reasoning is described also as a thinking mode that dissects the whole into causal parts that then give a comprehensive explanatory picture of the whole. According to the parasite-stress theory, analytical cognition is optimal when parasite stress is reduced. Under low parasite stress, innovativeness and openness are normative and rewarded, and there is less need to construct and maintain strong and permanent in-group affiliations that function to offset the negative reproductive consequences from parasites. We propose that the experimental exposure of individuals to parasite-salient cues will shift their cognition to more holistic styles of reasoning. Also, we predict that individuals with high perceived-vulnerability-to-disease scores or conservatism scores will exhibit more holistic reasoning.

Uskul et al. (2008) proposed in earlier research that certain subsistence ecologies such as farming promote interdependent cognitive styles. We suggest that the degree of parasite adversity associated with different subsistence ecologies will explain the cognitive styles involved. Some economists are interested in why the unit of economic productivity varies across the world. Regions vary in the degree to which economic units are in-groups, in extreme just the extended family, versus large social networks or markets. Alesina and Giuliano (2010) have provided evidence that collectivism correlates positively (and individualism negatively) with the degree to which economic productivity derives from in-group production. According to the parasite-stress theory, this variation arises from variable parasite-stress across regions as well as the values that are known to covary with it (Chap. 11). Hence, Uskul et al.'s (2008) proposal, and the evidence they present for it across subsistence types, may be part of the more general patterns of parasite-influenced economic patterns and associated cognitions.

Regarding the question raised by Uskul in her commentary of the origin and dissemination of innovations, earlier in this book we proposed a connection with variable parasite-stress and corresponding evoked values. Individualistic values promote and reward intellectual independence and therefore novel thinking and doing, as well as openness to new ideas and experiences. Collectivist values promote and reward adherence to traditional norms and ways of thinking, as well as closed-mindedness to the new and different. Hence collectivism is predicted to reduce diffusion of innovation and individualism to increase it. See Chap. 11 for various types of evidence supporting this prediction, ranging from within US diffusion of innovations to inter-regional transfers of technology.

Navarrete (2012) in his commentary suggests some interesting extensions of the parasite-stress theory of sociality that could be studied in future research. He notes that this theory proposes an important role for assortative sociality in increasing the inclusive fitness of in-group members afflicted with parasites. From this, he proposed the hypothesis that there will be more norms and values specifically focused on extended family and local-religious in-group healthcare in high than in low parasite regions. This might be tested using data from indigenous and/or contemporary human societies. Navarrete also proposes the interesting hypothesis that the human behavioral immune system may contain psychological adaptations that function to cause individuals to adopt healing behaviors during infection and while mending the bodily damage caused by parasites. Certainly, people in both modern and traditional societies use medicinal defenses against parasites, but do people have in addition some of the other "sickness" behaviors shown by various nonhuman animals and discussed by Hart (1988, 2011)?

Chang et al. (2012) in their commentary argue that in traditional societies, matrilocal residence, compared to patrilocal residence, provides strong family ties and extended family support and hence, according to the parasite-stress theory of sociality, is expected to be more typical in high parasite stress regions than in low parasite stress regions. Consistent with this idea, they provide evidence, using data from the Ethnographic Atlas, that matrilocal residence is more frequent in world regions of high than low parasite stress. They also suggest that the cultural practice of matrilocality disfavors individualism and modernity in general. The interplay between parasite stress, residence pattern, and collectivism–individualism may be a fruitful direction of future research.

The commentary by Vandello and Hettinger (2012) provides a novel perspective on the connection between the parasite-stress theory of sociality and the aspect of the culture of honor involving female purity (religious, sexual, moral, and hygienic) as a female marriage strategy. They document in their commentary a strong cross-national positive relationship between a new variable they created, "emphasis on female purity" (relative to male purity), and parasite stress. Hence, the ideology of female purity and its associated signaled conformity with traditional and conservative feminine roles becomes increasingly salient as infectious diseases increase across countries. They interpret this pattern as follows. It is well established that the marriage of women can involve marrying-up the social ladder, and such marriages can be highly valued by and beneficial to both the bride and her family in cultures of honor, because of such cultures' high stratification of resources and social influence. The ideology of female purity increases the likelihood that a female can marry-up, because the purity signals the female's freedom from contamination by parasites as well as contamination from the ideologies of out-groups that can involve exposure to novel parasites. This increases her marketability, they argue, especially in regions of high parasite stress.

We suggest a complementary hypothesis for the positive relationship of female purity with parasite stress. Across regions, parasite stress is correlated positively with the importance of good looks in mate selection, as shown in Chap. 6. Good looks are signals of phenotypic and genetic quality (Thornhill and Gangestad 2008). Because many women may marry a man who provides nongenetic material benefits, but lacks high genetic quality, female extra-pair mating may occur. The threat of extra-pair copulation to male paternity may be greater in high parasite regions, because of the value women put on good-genes (good looks) in such areas. Accordingly, female purity may be a competitive and honest female signal of likely faithfulness in a romantic relationship and is especially valued by resourceful and hence desirable long-term male partners in high parasite regions (also see Chap. 8).

Grotuss's (2012) commentary raises the issue of the role of mutualistic microbes of humans in shaping in-group assortative sociality. We treated this topic in our early paper on parasite stress in relation to the large variation in number of religions across countries of the world (Fincher and Thornhill 2008b). In that paper, we proposed that in-group assortative interactions and philopatry increase inclusive fitness in two ways: (1) avoiding and managing parasites and (2) acquisition and maintenance of an individual host's mutualistic (and commensalistic) microbial community—their microbiome.

Humans begin acquiring their nonpathological microbiota at birth, but the development and maintenance of this community occurs over the lifetime. Several benefits to the survival and reproductive success of individuals are provided by these symbionts. These include provision of metabolic by-products that can be used as fuels (e.g., butyrate). Also such microbes act as a defense system through immunoregulation and competition with pathogens preventing the pathogen's colonization and infectivity (reviewed in Dethlefsen et al. 2007). The immunoregulatory capacities of symbionts are fundamental to both the hygiene hypothesis and the old-friends hypothesis discussed earlier in this chapter. Montiel-Castro et al. (2013) brought together much evidence on the interactions between the human microbiota, gut, and brain—a system that seems to affect many aspects of human functioning including social behavior.

Owing to the localized coevolutionary races between hosts and parasites, and because in high-parasite regions these races occur in smaller areas within relatively behaviorally isolated populations, it is probable that humans living in high-parasite areas will experience greater specificity and local adaptation in their mutualistic/commensalistic communities. Interacting with out-group members has the potential to disrupt these communities as well as lead to the acquisition of novel pathogens. Hence, we hypothesized that both parasitic and mutualistic/commensalistic interactions may be driving the assortative social life and limited dispersal that influence human social behavior (Fincher and Thornhill 2008b). Beneficent symbionts may similarly play a major role in creating the sociality of species other than humans. Indeed, this is an argument also made by Lombardo (2008) at about the same time as our writing on this.

Grotuss's (2012) commentary also points out the elegant design of the behavioral immune system implied by the combination of findings reported in our target article that nonzoonotic human parasites impact human values and behavior more than zoonotics and the separate findings on the important role of mutualistic and commensalistic microbes in human health and fitness. The most general implications are that the human psychological adaptations proposed by the parasite-stress theory are responsible not only for a person's ontogenetic acquisition of values, based on experiences with infectious-disease cues, but also are functionally designed to identify and differentially respond to parasite-presence versus parasite-absence in one's environment and/or self, the nature of a present parasite (nonzoonotic versus zoonotic), and the presence of local beneficial microbes in one's social environment and/or self. This implies, too, that the classical immune system is far more sophisticated than traditionally thought. Recognizing self versus nonself is not enough, as immunologists now recognize. Non-self can include beneficial microbes, which should not be destroyed by the classical immune system. Similarly, regarding the behavioral immune system, the beneficial microbes harbored in conspecifics or other features of the environment should not be avoided.

The commentaries of Grotuss (2012) and Swartwout et al. (2012) point out that some religious and other cultural practices such as scarification and proselytizing increase the spread of disease or likelihood of infection. We argued in the target article that religiosity functions to create (a) a cultural boundary between in-group and out-group that reduces contact with novel out-group parasites and (b) a reliable, embedded social network that defends against parasites within the group (see Chap. 9). Hence, religious practices that increase exposure to contagion are very interesting cases. As Swartout et al. emphasized, the most honest signals of commitment and embeddedness—those that cannot be faked by low commitment individuals—sometimes involve compromising the signaler's immune system. They mention scarification and other bloody rituals, extreme physical exertion, fasting, and ingesting poisons. We hypothesize that proselytizing may be another example of this, and serves as an unfakable signal of one's commitment to in-group values, given its high costs in terms of contagion risk. These extreme displays of commitment are best framed in the context of trade-offs where benefits from social embeddedness exceed high costs from contagion and other personal risks.

We stress that proselytizing is potentially an important area for future research into the breadth of the parasite-stress theory of sociality as applied to religiosity. We understand the honest-signal theory of biology to imply that there will be competition among signalers to use those signals that most honestly define the communicated information (in-group commitment and boundary in the case of religiosity). Optimal signals of in-group commitment in high parasite-stress regions sometimes may be those that confront the most feared ecological feature—infectious disease.

One area of research that needs much investment of effort is the development of individual differences in behavioral immune system reactivity. For example, almost nothing is known about the antecedents of individual differences in the perception of disease vulnerability. There has been a single published study on the developmental ecology of infectious disease experiences and adult mate preferences (de Barra et al. 2013). The development of disgust has received more research effort (Stevenson et al. 2009; Stevenson et al. 2010), but there is still a great deal more to learn. Discovering more about the development of the behavioral immune system is important for not only learning more about what's happening at the level of individuals but also for understanding how cultural differences emerge from parasite–host interactions (Clay et al. 2012; Schaller and Murray 2011).

Shrira et al. (2013) have proposed some ideas, derived from the parasite-stress theory of values, that may have promise for understanding rates of certain crimes. They first point out that high parasite stress evokes the pathogen-defensive cultural behaviors of xenophobia and ethnocentrism. They then hypothesize that the values of in-group favoritism and out-group dislike and disrespect basic to these cultural behaviors proximately cause increased property and violent crime against members of out-groups. As explained in Chap. 12 on warfare, we have applied an analog of this thinking to between-group aggression (warfare). Shrira et al.'s focus is on interpersonal criminal acts rather than intergroup aggression. In support of their hypothesis, their analyses across the states of the USA show that parasite stress predicts rates of property crimes (larceny, motor vehicle theft, burglary) and violent crimes (aggravated assault, homicide, rape, robbery) more strongly than a range of other independent variables previously shown by researchers to covary with these crimes. They also report that parasite stress is a strong predictor of homicides perpetrated on strangers (hence, potentially on members of out-groups). Shrira et al.'s interesting findings point to the need for more research on property and violent crimes in which offender–victim relationships are known.

The application of the parasite-stress theory of values to human violence suggests another future research direction. Pinker (2011) has provided evidence that violent behavior has declined steadily from ancient times to the present. His work focuses on the West and includes a range of violent behaviors that show the pattern of historical decline, including various types of homicides, intergroup violence (warfare), and wife abuse. In Chap. 8 we provided evidence, from both analyses among nations and among states of the USA, that contemporary rates of interpersonal homicide show robust, positive relationships with parasite adversity. Further evidence for this same association across contemporary US states was provided by Shrira et al.'s (2013) research discussed in the previous paragraph. Also in Chap. 8, using contemporary data, we show the positive association between rates of adult-on-adult interpersonal homicides and conservative values across regions. In Chap. 12 we presented analyses of the frequencies of the various types of within-nation wars across the globe in recent decades, the results of which reveal that both parasite adversity and collectivism covary with the frequencies of onsets of these wars. Overall, then, the empirical findings in Chaps. 8 and 12 and Shrira et al. (2013) lead to the strong conclusion that, at least on a contemporary time scale, rates of violence are quite consistent with the parasite-stress theory of values. This leads us to hypothesize that the patterns of historical decline in violence treated by Pinker (2011) are also effects of a reduction in parasite adversity and the increase in liberalization of values associated with this reduction.

We hope historians will take an interest in our hypothesis and test it using historical records of violence rates and parasite adversity. In general, evidence appears to support our hypothesis. Pinker (2011) suggested that the decline in violence in the West over time was caused by an increase in humanitarian values, empathy toward strangers, widespread trade markets and commerce, recognition of women as fully human, and the popularity of analytical cognition (reasoning). To this list, Martin Daly (2011), in his book review of Pinker (2011), added another variable: a steady increase in democratization and related decline in wealth disparity. Certainly, the cultural trends suggested by Pinker and the trend toward democratic governance and wealth redistribution suggested by Daly actually happened and coincide with the reduction of violence across eras in the West. As our book has documented, all these changes in values are the predicted effects of the increased emancipation of people from infectious diseases. The history of parasite-adversity decline in the West is documented after the beginning of the twentieth century (see Chap. 10), but earlier records of sanitation improvements and other indicators of health-related improvements could be examined from historical records earlier than the twentieth century in the West. Future research on violence across time in the West should consider Gregory Hanlon's (2013) critique of certain conclusions by Pinker (2011).

In closing this section on future research, we predict that in the near future the parasite-stress theory of values will become a foundational research paradigm of the scholarly discipline of Darwinian or evolutionary medicine: the understanding of human health and disease in terms of evolutionary theory (Nesse and Williams 1996). Darwinian medicine's traditional domain of study is expanded greatly by knowledge of how the values associated with behavioral immunity are evoked by infectious diseases and then affect the health of those with the values as well as the health of those impacted by others' values. We have shown that conservative values defend against infectious diseases, but when such values are widespread in a society, they yield autocratic governance and endorse disenfranchisement of, and associated reduced health in, many people in the society. In contrast, liberal values of social equity extend healthcare and welfare in general widely across a society. Also, we have provided evidence that conservative values constrain science within a region and the diffusion of new ideas and technologies, including medical technology that saves lives, into the region; liberal values promote science and medical advancement. These findings tie Darwinian medicine to major topics in political science and economics.

## 14.8 Summary

The overall goal of the book is to create a synthesis or unity, based on the parasite-stress theory of values/sociality, of many topics that traditionally have been viewed and studied as distinct. The book presents the utility of the parasite-stress theory for unification of areas of research and knowledge ranging from parasitology, immunology, moral systems, civil conflict, governmental systems, family life, sexual behavior, dispersal patterns, economics, personality, violence, religious commitment, biodiversity, and so on. The book supports the claim of the scientific revolution that the realm of explanation is small—diverse and seemingly unconnected parts of nature can be unified by a few shared and basic causes. The evidence provided in our book supports the synthetic conclusion that evolutionary and ecological encounters with infectious diseases cause values and these values affect behaviors that account for the range of topics we consider. We invite others to clarify and expand the unity we have only started to lay out.

We ask, why are synthetic scientific ideas so rare? We hypothesize that this rarity is explained by the parasite-stress theory. Our hypothesis proposes that the combination of the features of openness, anti-authoritarianism, and conception of self as independent agent can produce the big ideas that unify knowledge. These features are most prevalent in the heads of people who go through ontogeny in low parasite-stress regions or are relatively less affected by parasites as a result of personal immunity. We hypothesize that emancipation of people from infectious diseases not only will reduce mortality and morbidity and increase liberalism and associated egalitarianism, but also will increase the frequency of scientifically encompassing ideas. We provide evidence in support of this hypothesis and make some predictions that would further test this perspective on scientific creativity.

We address the benefits and costs of emancipating people from infectious diseases. The benefits are increased longevity, reduced illness (morbidity reduction), increased democratization, increased intelligence, and scientific and technological innovation and progress. Costs are increased autoimmune disease associated with evolutionarily novel low levels of infectious, commensal and mutualistic organisms encountered by children during their development and the exhaustion of non-renewable energy for the future coinciding with technological and economic productivity.

We discuss and respond to a range of criticisms of the parasite-stress theory of values or its claimed empirical support. We conclude that the criticisms to date do not falsify the theory, moderate its application to any of the topics it purports to explain, or question the empirical support of the theory.

Numerous future research directions and associated hypotheses are presented. Colleagues have suggested many of these in their published comments on the parasite-stress theory.

## References


Alesina, A., & Giuliano, P. (2010). The power of the family. *Journal of Economic Growth* 15: 93–125.

Atran, S. (2012). Parasite stress is not so critical to the history of religions or major modern group formations. *Behavioral and Brain Sciences* 35: 79–80.

Barber, N. (2011). A cross-national test of the uncertainty hypothesis of religious belief. *Cross-Cultural Research* 45: 318–333.

Beit-Hallahmi, B. (2012). Connecting biological concepts and religious behavior. *Behavioral and Brain Sciences* 35: 80–81.

Brown, J. H., Burnside, W. R., Davidson, A. D. et al. (2011). Energetic limits to economic growth. *Bioscience* 61: 19–26.

Cappell, M. S. (2009). A more than one-hundred-fold higher per capita rate of authorship of five democratic nations versus their relatively undemocratic neighboring nations among 6,437 articles in 14 medical journals: Does democracy and civil liberties promote intellectual creativity and medical research? *Digestive Diseases and Sciences* 54: 1609–1620.

Cashdan, E. (2001). Ethnic diversity and its environmental determinants: Effects on climate, pathogens, and habitat diversity. *American Anthropology* 103: 968–991.

Cashdan, E., & Steele, M. (2010). Do pathogens and other risks promote ethnocentrism, xenophobia, and ethnic diversity? Unpublished paper presented at the Annual Meeting of the American Anthropological Association, New Orleans. November 17–21, 2010.

Cashdan, E., & Steele, M. (2013). Pathogen prevalence, group bias, and collectivism in the standard cross-cultural sample. *Human Nature* 24: 59–75.

Chang, L., Lu, H. J., & Wu, B. P. (2012). Pathogens promote matrilocal family ties and the copying of foreign religions. *Behavioral and Brain Sciences* 35: 82–83.

Clay, R., Terrizzi Jr., J. A., & Shook, N. J. (2012). Individual differences in the behavioral immune system and the emergence of cultural systems. *Journal of Social Psychology* 43: 174–184.

Currie, T. E., & Mace, R. (2012). Analyses do not support the parasite-stress theory of human sociality. *Behavioral and Brain Sciences* 35: 83–85.

Curtis, V., de Barra, M., & Aunger, H. (2011). Disgust as an adaptive system for disease avoidance behaviour. *Philosophical Transactions of the Royal Society B* 366: 389–401.

Daly, M. (2011). A farewell to arms. *Nature* 478:453–454.

de Barra, M., & Curtis, V. (2012). Are the pathogens of out-groups really more dangerous? *Behavioral and Brain Sciences* 35: 85–86.

de Barra, M., DeBruine, L., Jones, B. et al. (2013). Illness in childhood predicts face preferences in adulthood. *Evolution and Human Behavior* 43: 384–389.

Dethlefsen, L., McFall-Ngai, M., & Relman, D. A. (2007). An ecological and evolutionary perspective on human-microbe mutualism and disease. *Nature* 449: 811–818.

Duncan, L. A., & Schaller, M. (2009). Prejudicial attitudes toward older adults may be exaggerated when people feel vulnerable to infectious disease: Evidence and implications. *Analyses of Social Issues and Public Policy* 9: 97–115.

Ege, M. J., Mayer, M., Normand, A. C. et al. (2011). Exposure to environmental microorganisms and childhood asthma. *New England Journal of Medicine* 364: 701–709.

Eppig, C., Fincher, C. L., & Thornhill, R. (2010). Parasite prevalence and the worldwide distribution of cognitive ability. *Proceedings of the Royal Society of London B* 277: 3801–3808.

Eppig, C., Fincher, C. L., & Thornhill, R. (2011). Parasite prevalence and the distribution of intelligence among the states of the USA. *Intelligence* 39: 155–160.

Faulkner, J., Schaller, M., Park, J. H. et al. (2004). Evolved disease-avoidance mechanisms and contemporary xenophobic attitudes. *Group Processes and Intergroup Relations* 7: 333–353.

Figueredo, A. J., Gladden, P. R., & Black, C. J. (2012). Parasite stress, ethnocentrism, and life history strategy. *Behavioral and Brain Sciences* 35: 87–88.

Fincher, C. L., & Thornhill, R. (2008a). A parasite-driven wedge: Infectious diseases may explain language and other biodiversity. *Oikos* 117: 1289–1297.

Fincher, C. L., & Thornhill, R. (2008b). Assortative sociality, limited dispersal, infectious disease and the genesis of the global pattern of religion diversity. *Proceedings of the Royal Society of London, Biological Sciences* 275: 2587–2594.

Fincher, C. L., & Thornhill, R. (2012a). Parasite-stress promotes in-group assortative sociality: The cases of strong family ties and heightened religiosity. *Behavioral and Brain Sciences* 35: 61–79.

Fincher, C. L., & Thornhill, R. (2012b). The parasite-stress theory may be a general theory of culture and sociality Response. *Behavioral and Brain Sciences* 35: 99–119.

Fincher, C. L., Thornhill, R., Murray, D. R. et al. (2008). Pathogen prevalence predicts human cross-cultural variability in individualism/collectivism. *Proceedings of the Royal Society of London Biological Sciences* 275: 1279–1285.

Fleischman, D. S. (2014). Women's disgust adaptations. In *Evolutionary Perspectives on Human Sexual Psychology and Behavior* (eds. V.A. Weekes-Shackelford & T.K. Shackelford), pp. 277–296. Springer, New York, NY.

Fox, M., Knapp, L.A., Andrews, P.W. et al. (2013). Hygiene and the world distribution of Alzheimer's disease: Epidemiological evidence for a relationship between microbial environment and age-adjusted disease burden. *Evolution, Medicine, and Public Health*: 173–186. doi: 10.1093/emph/eot015.

Freeland, W. J. (1976). Pathogens and the evolution of primate sociality. *Biotropica* 8: 12–24.

Gallup, J. L., & Sachs, J. D. (2001). The economic burden of malaria. *The American Journal of Tropical Medicine and Hygiene* 64: 85–96.

Gangal, C., & Chowgule, R. (2009). Infections in early life and susceptibility to allergic diseases: Relevance of hygiene hypothesis. *Current Science* 96: 784–793.

Gelfand, M. J., Bhawuk, D. P. S., Nishii, L. H., et al. (2004). Individualism and collectivism. In *Culture, Leadership, and Organizations: The GLOBE Study of 62 Societies* (eds. R. J. House, P. J. Hanges, M. Javidan, et al.), pp. 437–512. Sage Publications, Thousand Oaks, CA.

Gibson, M. A., & Mace, R. (2005). Helpful grandmothers in rural Ethiopia: A study of the effect of kin on child survival and growth. *Evolution and Human Behavior* 26: 469–482.

Gladden, P. R., Welch, J., Figueredo, A. J. et al. (2009). Moral intuitions and religiosity as spuriously correlated life history traits. *Journal of Evolutionary Psychology* 7: 167–184.

Grotuss, J. (2012). The evolution and development of human social systems requires more than parasite-stress avoidance explanation. *Behavioral and Brain Sciences* 35: 88–89.

Hackman, J., & Hruschka, D. (2013). Fast life histories, not pathogens, account for state-level variation in homicide, child maltreatment, and family ties in the U.S. *Evolution and Human Behavior* 34: 118–124.

Hanlon, G. (2013). The decline of violence in the West: From cultural to post-cultural history. *English Historical Review* 128: 367–400.

Hart, B. L. (1988). Biological basis of the behavior of sick animals. *Neuroscience and Biobehavioral Reviews* 12: 123–137.

Hart, B. L. (2011). Behavioural defences in animals against pathogens and parasites: Parallels with the pillars of medicine in humans. *Philosophical Transactions of the Royal Society B* 366: 3406–3417.

Hruschka, D. J., & Henrich, J. (2013). Institutions, parasites and the persistence of in-group preferences. *PLoS One* 8: e63642. doi: 10.1371/journal.pone.063642.

Inbar, Y., Pizarro, D. A., & Bloom, P. (2009). Conservatives are more easily disgusted than liberals. *Cognition and Emotion* 23: 714–725.

Inbar, Y., Pizarro, D. A., Iyer, R. et al. (2012). Disgust sensitivity, political conservatism, and voting. *Social Psychological and Personality Science* 5: 537–544.

Kitayama, S., Ishii, K., Tmada, T. et al. (2006). Voluntary settlement and the spirit of independence: Evidence from Japan's "Northern Frontier." *Journal of Personality and Social Psychology* 91: 369–384.

Kuhn, T. S. (2012). *The Structure of Scientific Revolutions*, 50th Anniversary Ed., University of Chicago Press, Chicago, IL.

Kurzban, R., & Leary, M. R. (2001). Evolutionary origins of stigmatization: The functions of social exclusion. *Psychological Bulletin* 127: 187–208.

Le, S. H. T. (2013). Societal trust and geography. *Cross-Cultural Research.* doi: 10.1177/1069397113485330

Lindova, J., Priplatova, L., & Flegr, J. (2011). Higher extraversion and lower conscientiousness in humans infected with Toxoplasma. *European Journal of Personality* 26: 285–291.

Lombardo, M, P. (2008). Access to mutualistic endosymbiotic microbes: An underappreciated benefit of group living. *Behavioral Ecology and Sociobiology* 62: 479–497.

McGowan, D., Matsumura, M., Metskas, A. et al. (2009). *Raising Freethinkers: A Practical Guide for Parenting Beyond Belief.* AMACOM, New York, NY.

Meehan, C. L., Quinlan, R., & Malcom, C. D. (2013). Cooperative breeding and maternal energy expenditure among Aka foragers. *American Journal of Human Biology* 25: 42–57.

Montiel-Castro, A. J., González-Cervantes, R. M., Bravo-Ruiseco, G. et al. (2013). The microbiota-gut-brain axis: Neurobehavioral correlates, health and sociality. *Frontiers in Integrative Neuroscience* 7: 1–16.

Moore, J. (2002). Parasites and the Behavior of Animals. Oxford University Press, New York, NY.

Mortensen, C. R., Becker, D. V., Ackerman, J. M. et al. (2010). Infection breeds reticence: The effects of disease salience on self-perceptions of personality and behavioral avoidance tendencies. *Psychological Science* 21: 440–447.

Murray, D. R. (2014). Direct and indirect implications of pathogen prevalence for scientific and technological innovation. *Journal of Cross-Cultural Psychology* 45: 971–985.

Navarrete, C. D. (2012). Coping with germs and people: Investigating the link between pathogen threat and human social cognition. *Behavioral and Brain Sciences* 35: 89–90.

Navarrete, C. D., & Fessler, D. M. T. (2006). Disease avoidance and ethnocentrism: The effects of disease vulnerability and disgust sensitivity on intergroup attitudes. *Evolution and Human Behavior* 27: 270–282.

Navarrete, C. D., Fessler, D. M. T., & Eng, S. J. (2007). Elevated ethnocentrism in the first trimester of pregnancy. *Evolution and Human Behavior* 28: 60–65.

Nesse, R. M., & Williams, G. C. (1996). *Why We Get Sick: The New Science of Darwinian Medicine.* Vintage Books, New York, NY.

Nisbett, R. E., Peng, K., Choi, I. et al. (2001). Culture and systems of thought: Holistic vs. analytic cognition. *Psychological Review* 108: 291–310.

Oaten, M., Stevenson, R. J. & Case, T. I. (2009). Disgust as a disease-avoidance mechanism. *Psychological Bulletin* 135: 303–321.

Oishi, S. (2010). The psychology of residential mobility: Implications for the self, social relationships, and well-being. *Perspectives on Psychological Science* 5: 5–21.

Park, J. H., Faulkner, J., & Schaller, M. (2003). Evolved disease avoidance processes and contemporary anti-social behavior: Prejudicial attitudes and avoidance of people with disabilities. *Journal of Nonverbal Behavior* 27: 65–87.

Park, J. H., Schaller, M., & Crandall, C. S. (2007). Pathogen-avoidance mechanisms and the stigmatization of obese people. *Evolution and Human Behavior* 28: 410–414.

Park, J. H., Van Leeuwen, F., & Ypapanti, C. (2013). Disease-avoidance processes and stigmatization: Cues of substandard health arouse heightened discomfort with physical contact. *Journal of Social Psychology* 153: 212–228.

Paul, G. S. (2009). The chronic dependence of popular religiosity upon dysfunctional psychosociological conditions. *Evolutionary Psychology* 7: 398-441.

Paul, G. S. (2012). High illness loads (physical and social) do not always force high levels of mass religiosity. *Behavioral and Brain Sciences* 35: 90.

Pemstein, D., Meserve, S. A., & Melton, J. (2010). Democratic compromise: A latent variable analysis of ten measures of regime type. *Political Analysis* 18: 426–449.

Peters, M., Kauth, M., Schwarze, J. et al. (2006). Inhalation of stable dust extract prevents allergen induced airway inflammation and hyperresponsiveness. *Thorax* 61: 134–139.

Pinker, S. (2011). *The Better Angels of our Nature: The Decline of Violence in History and its Causes*. Allen Lane, London, U.K.

Poulin, R. (2010). Parasite manipulation of host behavior: An update and frequently asked questions. *Advances in the Study of Behavior* 41: 151–186.

Price-Smith, A. T. (2002). *The Health of Nations: Infectious Disease, Environmental Change, and Their Effects on National Security and Development*. MIT Press, Cambridge, MA.

Quinlan, R. J. (2007). Human parental effort and environmental risk. *Proceedings of the Royal Society B* 274: 121–125.

Raison, C. L., Lowry, C. A., & Rook, G. A. W. (2010). Inflammation, sanitation, and consternation. *Arch Gen Psychiatry* 67: 1211–1224.

Rees, T. J. (2009). Is personal insecurity a cause of cross-national differences in the intensity of religious belief. *Journal of Religion and Society* 11: 1–24.

Rode, N. O., Lievens, E. J. P., Flaven, E. et al. (2013). Why join groups? Lessons from parasite-manipulated Artemia. *Ecology Letters* 16: 493–501.

Rook, G. A. (2012). Hygiene hypothesis and autoimmune diseases. *Clinical Reviews in Allergy and Immunology* 42: 5–15.

Rook, G. A., & Brunet, L. R. (2002). Give us this day our daily germs. *Biologist* 49: 145–149.

Sachs, J., & Malaney, P. (2002). The economic and social burden of malaria. *Nature* 415: 680–685.

Schaller, M., & Murray, D. R. (2011). Infectious disease and the creation of culture. In *Advances in Culture and Psychology* (eds. M. Gelfand, C.-y. Chiu, & Y.-y. Hong), pp. 99–151). Oxford University Press, New York, NY.

Schaller, M., & Murray, D. R. (2012). Mechanisms by which parasites influence cultures, and why they matter. *Behavioral and Brain Sciences* 35: 91–92.

Schaller, M., & Park, J. H. (2011). The behavioral immune system (and why it matters). *Current Directions in Psychological Science* 20: 99–103.

Schaller, M., Miller, G. E., Gervais, W. M. et al. (2010). Mere visual perception of other people's disease symptoms facilitates a more aggressive immune response. *Psychological Science* 21: 649–652.

Sear, R., & Mace, R. (2008). Who keeps children alive? A review of the effects of kin on child survival. *Evolution and Human Behavior* 29: 1–18.

Shrira, I., Wisman, A., & Webster, G. (2013). Guns, germs, and stealing: Exploring the link between infectious disease and crime. *Evolutionary Psychology* 11: 270–287.

Solt, F. (2009). Standardizing the World Income Inequality Database. *Social Science Quarterly* 90: 232–342.

Solt, F., Habel, P., & Grant, J. T. (2011). Economic inequality, relative power, and religiosity. *Social Science Quarterly* 92: 448–465.

Sternberg, R. J. (2005). *Handbook of Creativity*. Cambridge University Press, New York, NY.

Stevenson, R. J., Case, T. I., & Oaten, M. J. (2009). Frequency and recency of infection and their relationship with disgust and contamination sensitivity. *Evolution and Human Behavior* 30: 363–368.

Stevenson, R. J., Oaten, M. J., Case, T. I. et al. (2010). Children's response to adult disgust elicitors: Development and acquisition. *Developmental Psychology* 46: 165–177.

Strachan, D. P. (1989). Hay-fever, hygiene, and household size. *British Medical Journal* 12: 1259–1260.

Swartwout, P., Purzycki, B. G., & Sosis, R. (2012). Form and function in religious signaling under pathogen stress. *Behavioral and Brain Sciences* 35: 92–93.

Taylor, M. Z., & Wilson, S. (2012). Does culture still matter? The effects of individualism on national innovation rates. *Journal of Business Venturing* 27:234–247.

Terrizzi Jr., J. A., Shook, N. J., & Ventis, W. L. (2010). Disgust: A predictor of social conservatism and prejudicial attitudes toward homosexuals. *Personality and Individual Differences* 49: 587–592.

Terrizzi Jr., J. A., Shook, N. J., & McDaniel, M. A. (2013). The behavioral immune system and social conservatism: A meta-analysis. *Evolution and Human Behavior* 34: 99–108.
Terrizzi, J. A., Jr., Clay, R., & Shook, N. J. (2014). Does the behavioral immune system prepare females to be religiously conservative and collectivistic? *Personality and Social Psychology Bulletin* 40: 189–202.
Thornhill, R., & Fincher, C. L. (2011). Parasite stress promotes homicide and child maltreatment. *Philosophical Transactions of the Royal Society: Biological Sciences* 366: 3466–3477.
Thornhill, R. & Fincher, C.L. (2013). Commentary on Hackman, J. And Hruschka, D. (2013). Fast life histories, not pathogens, account for state-level variation in homicide, child maltreatment, and family ties in the U.S. *Evolution and Human Behavior* 34: 118–124.
Thornhill, R., Fincher, C. L., & Aran, D. (2009). Parasites, democratization, and the liberalization of values across contemporary countries. *Biological Reviews* 84: 113–131.
Thornhill, R., Fincher, C. L., Murray, D. R. et al. (2010). Zoonotic and non-zoonotic diseases in relation to human personality and societal values: Support for the parasite-stress model. *Evolutionary Psychology* 8: 151–169.
Thornhill, R., & Gangestad, S. W. (2008). *The Evolutionary Biology of Human Female Sexuality*. Oxford University Press, New York, NY.
Thornhill, R., & Palmer, C. T. (2004). Evolutionary life history perspective on rape. In *Evolutionary Psychology, Public Policy and Personal Decisions* (eds. C. Crawford & C. Salmon), pp. 249–274. Lawrence Erlbaum, Mahwah, NJ.
Tybur, J. M., Merriman, L. A., Caldwell Hooper, A. E. et al. (2010). Extending the behavioral immune system to political psychology: Are political conservatism and disgust sensitivity really related? *Evolutionary Psychology* 8: 599–616.
U.S. Energy Information Administration, http://www.eia.gov
Uskul, A. K. (2012). Rethinking innovative design to further test parasite-stress theory. *Behavioral and Brain Sciences* 35: 93–94.
Uskul, A. K., Kitayama, S., & Nibett, R. E. (2008). Ecocultural basis of cognition: Farmers and fisherman are more holistic than herders. *Proceedings of the National Academy of Sciences* 105: 8552–8556.
Van de Vliert, E. (2009). *Climate, Affluence, and Culture.* Cambridge University Press, New York, NY.
Van de Vliert, E. (2011). Bullying the media: Cultural and climato-economic readings of press repression versus press freedom. *Applied Psychology: An International Review* 60: 354–376.
Van de Vilert, E., & Postmes, T. (2012). Climato-economic livability predicts societal collectivism and political autocracy better than parasitic stress does. *Behavioral and Brain Sciences* 35: 94–95.
Vandello, J. A., & Hettinger, V. E. (2012). Parasite-stress, cultures of honor, and the emergence of gender bias in purity norms. *Behavioral and Brain Science* 35: 95–96.
Varnum, M. E. W. (2012). Frontiers, germs, and nonconformist voting. *Journal of Cross-Cultural Psychology* 44: 832–837.
Varnum, M. E. W. (In press). Sources of regional variation in social capital in the U.S. *Evolutionary Behavioral Sciences.*
Vigil, J. M. (2009). A socio-relational framework of sex differences in the expression of emotion. *Behavioral and Brain Sciences* 32: 375–428.
Vigil, J. M., & Coulombe, P. (2012). Intra-regional assortative sociality may be better explained by social network dynamics rather than pathogen risk avoidance. *Behavioral and Brain Sciences* 35: 96–97.
Vogel, K., Blumer, N., Korthals, M. et al. (2008). Animal shed *Bacillus licheniformis* spores possess allergy-protective as well as inflammatory properties. *Journal of Allergy and Clinical Immunology* 122: 307–312.
Wall, J. N., & Shackelford, T. K. (2012). Immigration, parasitic infection, and United States religiosity. *Behavioral and Brain Sciences* 35: 97–98.

# Index

A
Acceptance of uncertainty, 179
Accommodation, 10–11, 17, 22, 51, 97
Acent, 73, 100
Adaptationism, 4–5, 7, 11–14, 17, 26, 28–31, 33–43, 45, 46, 49–50, 52–54, 60–62, 64, 66–70, 75, 83, 95, 96, 145, 148–150, 174, 180–181, 187, 197–198, 202, 210, 217, 226, 271, 285, 289, 292–293, 296, 297, 303, 309, 313, 333, 334, 337, 343, 344, 346, 347, 356–358, 363, 374, 382, 383, 385, 387, 388, 395–397, 413, 414, 418–421, 427, 430, 432
Aesthetics, 11, 21–54, 397
Africa, 3, 43, 62, 139, 147, 173, 181, 185, 190, 217, 258, 294, 296, 309, 311, 312, 366, 368, 371, 373, 379, 387, 415, 416, 419, 423, 424, 426
African cichlid, 387
Age of Reason, 22
Alexander, R.D., 33
Allopatric speciation, 70, 381–383, 388, 389
Altruism, 29–33, 51, 61, 63, 69, 113, 114, 139, 146, 148, 153, 155, 239, 240, 346, 356–358, 360, 377, 384
Altruistic punishment, 32, 33
13th Amendment, 294
19th Amendment, 289, 324
American Civil War, 2, 5, 6, 17, 173, 185, 269, 349, 351, 379
Analytical reasoning, 96, 429
Anthropology, 3, 4, 23, 24, 43, 45, 47, 67, 85, 119, 171, 172, 183, 189, 283
Anti-authoritarian, 269, 271, 285, 291, 296, 298, 397, 398, 435
Antibiotics, 140, 149, 221, 231, 256, 292, 298
Anti-Enlightenment tradition, 53, 54
Ants, 33, 93–95, 146
Apartheid, 294, 378, 379
Arbitrary learning, 36
Assortative sociality, 15, 62–66, 75, 76, 141, 146, 147, 237, 239, 243, 245, 250, 254, 257–260, 343, 355–362, 374, 375, 377, 383–386, 388, 407, 411, 415, 420, 427, 430, 431
Authoritarianism, 1, 10, 15, 22, 40, 52, 85–87, 94, 95, 98, 100, 105, 106, 138, 140, 153, 210, 214, 217, 222, 266, 268–272, 282, 283, 285, 286, 291, 293–296, 298, 306, 334, 335, 378, 379, 389, 396–398, 400, 410, 435
Autocracy, 15, 52, 95, 100, 101, 222, 255, 265–269, 271, 273, 277–279, 282–288, 295–298, 304–306, 321, 327, 328, 335, 348, 353, 370, 405, 410, 434
Autoimmune, 401, 402, 435

B
Bacon, F., 24, 128, 203, 396
Behavioral immune system, 60, 62–63, 66, 71–73, 75, 130, 174–178, 180, 188, 195, 197, 198, 205, 237, 260, 287, 296, 298, 356, 374, 375, 379, 389, 418, 420, 427, 430, 432, 433
Behavioral immunity, 63–66, 68, 74–76, 151, 196, 198, 202, 204, 205, 245, 254, 257, 260, 355, 356, 358–363, 376, 380, 382, 383, 388, 389, 413, 414, 418, 419, 434
Behavioral isolation of species, 364
Big-five, 196, 197, 203–205

R. Thornhill and C.L. Fincher, *The Parasite-Stress Theory of Values and Sociality: Infectious Disease, History and Human Values Worldwide*,
DOI 10.1007/978-3-319-08040-6, © Springer International Publishing Switzerland 2014

Bill of Rights, 295
Biodiversity gradient, 353, 354, 385
Black Codes, 379
Blank slate, 297
Blending cultural phylogenies, 386
Brain, 5, 23, 36, 39, 41, 84, 105, 149–151, 307, 308, 312–314, 318, 397, 402, 403, 417, 428, 432
Burying beetle, 94
By-product, 5, 28–29, 31, 38, 40, 49, 53, 203, 360, 381, 383, 414, 431

**C**

Camels, 47, 48
Capitalism, 267, 303, 348, 425
Caste, 3, 33, 38, 50, 62, 105, 146, 181, 267, 269, 271, 290, 298, 312, 343, 354, 378–379, 381, 388, 389, 399
Causation, 2, 3, 8–10, 21–23, 25, 27–28, 31, 36, 38, 44–46, 48, 50, 54, 83, 84, 96, 106, 124–128, 136, 155, 210, 230, 254, 271, 284, 285, 297, 298, 327, 333, 337, 344, 396–398, 404, 422
Centers for Disease Control (CDC), 123, 135, 161, 212
Child maltreatment, 60, 225–231, 233, 421
Child mortality, 60, 225, 403
China, 72, 140, 399
Civil conflicts, 2, 16, 85, 101, 185, 265, 342–349, 351, 435
Civil Rights Movement, 279, 288, 291, 293–294
Civil war, 2, 5, 12, 15, 101, 173, 294, 335–339, 342, 345–347, 350, 351, 379, 400
Classical fitness, 32
Classical immune response, 65, 70, 71
Classical immune system, 60, 61, 65, 71, 72, 151, 198, 205, 307, 308, 313, 327, 402, 432
Classism, 1, 22, 378, 400
Climate, 43, 59, 83, 84, 106, 147, 149, 197, 215, 217, 232, 258, 316, 318, 326, 336, 354, 365, 376, 381, 403–405, 409
Closed-mindedness, 14, 196, 197, 204, 271, 285, 322, 323, 334, 397, 404, 430
Cognitive ability, 13, 15, 24, 70, 84, 103, 149–151, 255, 303–328, 397, 414
Collectivism, 13, 26, 66, 85, 113, 172, 195, 207, 245, 269, 303, 334, 353, 398
*Combined Parasite Stress*, 132–134, 138, 139, 152, 175, 223, 224, 246, 248, 250, 256, 304, 321, 325, 326, 369, 401, 405–408, 411–416, 418
Common descent, 47, 148
Comparative methodology, 2, 21, 42–46, 113, 125–128, 155, 230, 315
Competitive exclusion, 364
Condition-dependent adaptation, 64, 66
Conditions-of-living model, 241, 242, 407, 410, 426
Confederacy, 378
Conformity, 1, 33, 46, 51, 72, 73, 87, 114, 119, 123, 179, 203, 240, 243, 256, 306, 334, 425, 431
Consanguineous marriages, 181–183, 190, 359
Consciousness, 5, 26, 27, 35, 36, 41, 149, 230
Consilience, 203, 327, 397, 398, 400
Contemporary Parasite Severity, 116, 117, 171, 275, 284, 293, 336, 339–342, 369–374
Cooperative breeding, 13, 33, 145–147, 154, 155, 355
Copying, 34–36, 43
Corruption, 15, 265, 293, 306, 307, 320–321, 327, 328, 348, 425
Coups, 15, 70, 101, 146, 202, 209, 220, 232, 286, 314, 333–351, 359–362
Crafoord Prize, 29
Creationists, 22, 125
Creativity, 7, 9, 14, 15, 23, 24, 28, 31, 33, 36–38, 40, 41, 44, 46, 49, 50, 53, 61, 69, 70, 99, 103, 115, 149–151, 175, 176, 178, 200, 224, 244, 254, 268, 270, 287, 306, 321, 322, 326–328, 334, 344, 348, 354, 360–364, 377, 379, 380, 385, 388, 389, 395–401, 410, 411, 422, 426, 428, 431, 432, 435
Criticisms of the parasite-stress theory, 16, 255, 395, 402–403, 426, 435
C-scale, 86, 87
Culture, 1–2, 30, 63–64, 84, 113, 171, 197, 209, 254, 272, 322, 334, 354, 399
  history, 2, 3, 7, 17, 44, 45, 47, 48, 50, 210, 217, 386, 410, 418
  of honor, 98, 99, 216–218, 272, 275, 289, 431
  inertia, 4, 5, 44, 217, 410
  learning, 35, 36, 271, 418
  of male honor, 1, 216–218
  self-reinforcement, 361
Culture–gene coevolution, 66
Currently adaptive, 29, 37, 271, 285, 346
Cytokine-interleukin-6, 70

**D**
Darwinian selection, 4, 22, 24, 28–30, 32, 36, 39, 40, 44, 59, 65, 70, 149, 151, 155, 187, 296, 382, 402, 419
Darwin's method, 22, 24, 54
Declaration of Independence, 42, 295
Declaration of the Rights of Man and of the Citizen, 295
Deep South, 1
Dehumanization, 87, 101, 186, 269
Democratization, 10, 15, 38, 52, 70, 85, 86, 95, 180, 208, 214, 265–298, 304–306, 313, 321, 324, 327, 328, 335, 336, 338, 345, 370–373, 379, 399–402, 405, 410, 426, 434, 435
Demographic transition, 140–141, 155
Despotism, 172, 222, 266
Developmental biology, 38
Dictatorship, 266
Diffusion, 42, 43, 84, 149, 268–270, 283, 285, 287–288, 305–307, 321–327, 435
Diffusion of innovations, 15, 149, 265, 270, 287, 288, 298, 306, 322, 328, 376, 428, 430
Direct selection, 28, 29, 37, 39, 40, 49, 344, 346, 360, 381, 382
Disease trap, 348
Disgust, 27, 60, 68, 71, 73, 99, 130, 177, 187–188, 190, 198, 257, 269, 286, 343, 421, 433
Disgust sensitivity, 68, 72, 73, 105, 130, 176, 177, 187, 188, 195, 260, 289, 413, 417, 421, 429
Dispersal, 12, 13, 62–64, 70, 76, 106, 141–146, 154, 204, 237, 355–359, 363, 374–376, 380, 384–386, 388, 418–420, 428, 432, 435
Divorce, 97, 102, 134, 140, 178–179, 229, 291
Dogs, 29, 74
Drift, 28, 30, 40

**E**
Ecological fallacy, 113, 129–130, 155
Ecological model of diversification, 363–364
Economic
- conservatism, 87, 88
- productivity, 15, 84, 95, 119, 231, 303, 305, 318–320, 322, 325, 328, 345, 400–402, 404, 405, 411, 428, 430, 435

Egalitarianism, 9, 12, 22, 40, 101, 147, 208, 209, 270, 274, 279, 296, 297, 343, 400, 401, 435
Emancipation Proclamation, 294
Emigration, 13, 144, 145, 186
Enculturation, 5, 17, 21, 34, 36, 42, 44, 45, 54, 67, 105, 174, 178, 294, 356, 396, 410, 418, 428
Energy consumption, 327, 400, 401
Enlightenment, 22, 53, 54, 86, 203, 295, 327, 396, 410
Estrus, 184, 211
Ethnocentrism, 10, 12, 15, 39, 62–66, 69, 70, 75, 76, 93, 141, 147, 152, 153, 155, 237, 255, 257, 269, 271, 283, 285, 293, 297, 306, 334, 335, 344, 350, 357, 414, 420, 426, 427, 433
Ethnogenesis, 12, 13, 16, 76, 354, 374, 377, 396, 411
Ethnologue, 365
Evolution, 21, 59, 83, 121, 174, 196, 208, 238, 267, 303, 333, 355, 395
Evolutionary medicine, 434
Evolutionary novelty, 37, 271, 309, 312, 402
Exapted learning mechanisms, 29
Exclusionism–universalism, 137, 138
Extra-pair mating, 14, 211, 431
Extra-parental nepotism, 31, 32, 146
Extraversion, 71, 196, 197, 199–201, 203, 204, 260
Extrinsic mortality, 139, 140, 345, 415, 416, 419, 422, 423
Extroversion, 14

**F**
Faith, 6, 53, 209, 431
False positives, 66, 188, 413
Family
- life, 13, 94, 140, 146, 155, 173, 334, 396, 399, 403, 428, 435
- ties, 13, 113–165, 173, 190, 245, 258, 287, 417, 419, 421, 430
- values, 70, 98, 216, 402, 403

Farming, 217, 323, 429
Fascism, 10, 22, 266
Fascist, 283
Female
- honor, 97, 209, 210, 213
- purity, 431

Fluoridation, 292
Flynn effect, 307, 313, 314
Foreign aid, 347, 348
Fossil, 53
Freedom House, 132, 133, 246, 255, 266, 272, 274, 276–281, 283, 321

Freedom in the media, 265, 283–284, 298
Freeland, B., 69
French Revolution, 85, 86, 295
Frontier spirit, 428, 429
F-scale, 282–283, 286, 298
Functional design, 8, 17, 29–31, 35, 37, 40, 41, 49, 53, 54, 178, 187, 188, 190, 271, 342, 343, 346, 383, 421, 427, 432
Future research directions, 6, 16, 31, 85, 178, 202–205, 224, 228, 231, 257, 385, 395–436

**G**
Galton's problem, 43, 44
GDP. *See* Gross domestic product (GDP)
GDP per capita, 118, 119, 132, 133, 135, 136, 145, 182, 241, 246, 248–250, 255, 256, 274, 277, 278, 286, 293, 303–306, 309, 311, 319, 320, 327, 328, 335, 336, 338, 345, 366–368, 371, 372, 401, 405–407
*Gelfand In-group Collectivism*, 115, 122, 138, 183, 275, 276, 278–280, 321, 337, 340–342
Gender
  equality, 15, 85, 97, 180, 208, 209, 272, 273, 275–276, 279–282, 288
  relations, 15, 40, 265, 272, 275, 276, 281, 288–289
Gene flow, 358, 360, 418
General theory of biology, 6, 8
Genetic
  correlation, 359, 360, 383, 388
  quality, 14, 67, 172, 174–176, 178, 189, 211, 431
  recombination, 360
Genetically deterministic, 38
Geographic barriers, 64, 70, 362, 377, 381, 384, 389
Ghosts, 4, 5
GIDEON. *See* Global Infectious Disease and Epidemiology Network (GIDEON)
Gini, 118, 135, 136, 213, 214, 217–219, 223, 224, 226, 227, 229, 230, 233, 250, 274, 277, 278, 286, 304, 306, 327, 328, 345, 371, 372, 407, 408
Global Infectious Disease and Epidemiology Network (GIDEON), 116, 117, 120, 121, 124, 132, 365
Global Leadership and Organizational Behavior Effectiveness (GLOBE), 115, 131, 171
Global Peace Index, 338, 339, 341, 342, 347
GLOBE. *See* Global Leadership and Organizational Behavior Effectiveness (GLOBE)
Glorious Revolution, 295
Governmental corruption, 15, 265, 293, 306, 320, 321, 348, 425
Gross domestic product (GDP), 84, 118, 182, 197, 241, 255, 274, 303, 319–320, 327, 335, 366, 409
Group-selection paradigm, 30, 31

**H**
Hamilton, W.D., 29, 32, 67, 69, 173, 180, 189, 356, 396
Healthcare, 37, 63, 231, 232, 268, 270, 285, 287, 293–296, 379, 400, 423, 430, 434
Helpers-at-the-nest, 33
Heuristic, 24, 32, 44, 70, 72, 73
*Historical Parasite Severity*, 116–119, 171, 181, 182, 197, 282, 283, 336, 337, 339–342
Historical–particularistic ideology, 5–7
*Hofstede Individualism*, 115, 116, 122, 138, 224, 275, 278–280, 304, 321, 340–342, 425
Hokkaido, 140, 428
Holistic, 51, 52, 95, 96, 419, 429
Homicide, 2, 14, 98, 99, 102, 185, 186, 207, 211–226, 228–233, 246, 280, 341, 353, 421, 422, 424, 433, 434
Homology, 47, 146, 204
Homosexuals, 60, 66, 73, 291
Honest signaling, 5, 29, 41, 100, 240, 254, 431–433
Honor
  culture, 1, 2, 98, 99, 210, 216–218, 228, 272, 275, 289, 431
  homicide, 215–218, 220–222, 224–225, 229, 233
Hookworm, 84, 217, 230–232, 294, 305, 308
Hormone markers, 176, 178
Host manipulation by parasites, 427
Human
  eye, 41
  inequality, 26, 88, 100, 214, 335
Humanitarian, 22, 101, 230, 232, 271, 297, 395, 410, 434
Human-specific parasites, 120, 122, 123, 180, 200, 281, 282

Human-unique cognitive abilities, 13, 149, 155, 318
Hybrid
corn, 322–324
disadvantage, 359, 360, 381–383
zones, 382, 386
Hybridization, 50, 62, 70, 381–383
Hygiene hypothesis, 401, 402, 431

## I
ID. *See* Index of Democratization (ID)
Immigration, 13, 86, 148, 186, 254, 411, 428
Immunity genes, 59, 177, 202, 297, 358–362, 380, 382, 387, 388, 427
Inbreeding, 14, 61, 62, 70, 97, 171–190, 258, 353, 359
Incest-avoidance, 34
Incidental effects, 29, 374, 414, 415
Inclusive fitness, 9, 32, 44, 54, 65, 75, 114, 146, 239, 271, 285, 343, 346, 356, 359, 396, 403, 418, 419, 423, 430, 431
Independence, 3, 4, 6, 7, 16, 25, 36, 42–51, 54, 64, 70, 87, 88, 93, 95, 96, 114–116, 125, 127, 133, 142, 146, 153, 172, 173, 179, 181, 199, 203, 204, 214, 216, 218–220, 227, 229, 230, 233, 238, 249, 286, 287, 290, 295, 311, 312, 317–320, 327, 335, 349, 396–398, 404, 406–409, 418, 419, 423, 429, 430, 433, 435
Index of Democratization (ID), 266, 273–274, 276–284, 296, 370, 371
Indian Removal Act, 379
Indigenous societies, 60, 67, 69, 113, 138, 139, 145, 148, 154, 171, 172, 183, 189, 190, 203, 265, 283, 353, 357, 359, 376, 405, 419, 420, 429, 430
Indirect reciprocity, 32, 33, 148, 346
Individual differences, 38, 65, 72–74, 86, 105, 176, 195, 196, 202, 204, 240, 433
Individualism, 23, 60, 83, 113, 172, 195, 208, 237, 265, 303, 333, 354, 397
Individual selection, 30–32
Infanticide, 60
Infant mortality, 60, 270, 290, 292, 293
*Infectious Disease DALY*, 117, 131–133, 246, 304, 309–312, 320
*In-Group Assortativeness USA*, 139, 245–248, 250–253, 357, 401, 406, 407, 414
Inheritance, 39
Innovations, 15, 84, 101, 140, 149, 221, 255, 258, 265, 268–271, 285, 287–288, 298, 306, 321–328, 349, 361, 376, 397, 398, 401, 404, 428, 430, 435
Intellectual autonomy, 100, 429
Intelligence, 34, 280, 295, 307–309, 312–314, 316, 318, 321, 341, 397, 402, 404, 435
Intelligence quotient (IQ), 15, 84, 103, 151, 306–320, 327, 397, 414, 423
Intercultural diffusion, 43
Interdependent, 6, 7, 17, 25, 47, 49–51, 54, 94, 115, 131, 183, 216, 334, 343, 350, 355, 398, 419, 429
Interpersonal violence, 14, 97, 102, 207–233, 280
Intrastate warfare, 70
Introversion, 14, 65, 196, 197, 201, 202, 204, 356
IQ. *See* Intelligence quotient (IQ)

## J
Jim Crow, 379

## K
*Kashima Collectivism*, 116, 122, 183
KKK. *See* Ku Klux Klan (KKK)
Ku Klux Klan (KKK), 294

## L
Language, 5, 13, 16, 34, 47, 70, 73, 94, 99–100, 115–116, 127, 152, 237, 238, 240, 254–255, 269, 336, 353–355, 364–370, 375–377, 386, 389, 411
Language diversity, 64, 364–366, 369, 371, 376, 377, 425
Latitude, 70, 114, 119, 135, 140, 142, 154, 197, 217–218, 241, 258, 273, 274, 294–296, 298, 309, 353, 354, 357, 358, 363–368, 370, 372–373, 376, 377, 380, 382, 384–387, 389, 409, 425
Laws, 36, 140, 173, 211, 212, 219, 232, 275, 289, 291, 322, 324, 379
*Leishmania braziliensis*, 61
Libertarian, 134
Life expectancy, 118, 133, 136, 197, 219–220, 250, 252, 290, 293, 404
Life-history theory, 139–140, 219, 220, 259, 261, 345, 396, 403, 415, 416, 419, 421–424
Lifespan, 133, 135, 136, 149, 219–220, 248, 404
Linguistics, 23, 29, 61, 73, 85, 95, 100, 106, 116, 149, 291, 336
Linkage disequilibrium, 359–362, 382, 388, 420

Livestock, 120, 123, 124
Lynching, 185, 186, 207

**M**
Marine Corps, 5
Marriage, 14, 63, 67, 97, 98, 102, 106, 134, 148, 171–190, 229, 346, 359, 431
Martin Luther, 23
Marxism, 51, 52
Mate choice, 14, 40, 41, 171–190, 289, 347, 364, 387
Mate guarding, 211
Matrilocal residence, 430
Matronyms, 102
Medical technologies, 322, 323, 325, 435
Medicare, 221, 323
Medicine, 24, 60, 63, 221, 260, 293, 434, 435
Modernization, 267, 268, 285, 291
Moral foundations, 105, 152–153, 155
Moral intuitions, 105, 259
Mormons, 172
Morning sickness, 68
Multihost parasites, 120, 122, 123, 154, 180, 200, 280–282
Murray, D., 27, 65, 70, 91, 92, 113, 179, 195–199, 204, 280, 288
Music, 104, 240
Mutually reinforcing, 177, 360, 361, 375, 380, 388

**N**
Naming of children, 102
Naturalistic fallacy, 8–9, 41, 51, 96, 129
Neophilia, 12–13, 25, 197
Neophobia, 12, 15, 25, 62–63, 75, 84, 182–183, 197, 237, 306
Nepotism, 13, 31–33, 61, 63, 114, 139, 145–148, 150, 182–183, 271, 289, 306, 321, 334, 345–347, 355–357, 377, 416, 422, 423
Neuroticism, 196, 197, 203, 204
Ngram, 94
Non-lethal physical aggression, 207–211
Non-state wars, 338–340, 342, 345, 350, 351
Nonzoonotic, 102, 120–124, 131–133, 136–137, 152, 154, 156, 190, 199–201, 204–205, 212–213, 215, 218, 219, 224, 227, 229, 230, 233, 245, 246, 252–254, 272, 276, 280, 282, 286, 288, 297, 304, 404, 405, 421, 422, 432
Norms, 1, 27, 33, 42, 47, 61, 68, 69, 75, 87, 97–99, 101, 106, 114–116, 123, 147, 179, 183, 187, 197, 211, 224, 225, 228, 229, 243, 257, 259, 269, 272, 334, 343, 355, 376, 377, 379, 399, 430
Nuclear family, 13, 33, 63, 94, 141, 146, 147, 271, 285, 334
Nursing, 69, 139

**O**
Old friends hypothesis, 402, 432
Old South, 1, 3, 16, 17, 46, 52, 173, 185, 186, 190, 269, 289, 293, 294, 354, 378, 379, 399, 400
Ontogeny, 27, 34–39, 54, 65, 72, 84, 100, 151, 174, 177, 196, 202, 238, 256–257, 261, 271, 290, 296, 297, 312–314, 318, 334, 344, 399, 402, 417, 418, 429, 432, 435
Openness, 12–14, 46, 51, 75, 114, 153, 184, 195, 197, 199–201, 221, 240, 258, 260, 270, 271, 285, 293, 296, 298, 306, 322, 323, 326, 334–335, 349, 397, 398, 405, 429, 435
Openness to experience, 14, 71, 87, 103, 114, 195–201, 203–205, 356, 428, 430
Origin of species, 354, 362, 363, 380–388
Out-group, 7, 11–15, 27, 60–66, 68, 69, 72–75, 86–88, 93, 97, 99, 103, 106, 114, 131, 141, 142, 144, 147–150, 153–155, 184–186, 190, 197, 199, 203, 237–242, 257–259, 268–271, 285, 288, 296, 320, 326–327, 333–335, 338, 343, 347, 350, 356, 358–363, 375, 380, 383, 386, 400, 404, 407, 411, 419, 420, 425, 426, 431–433

**P**
Parapatric diversification of parasites, 363
Parapatric speciation, 70, 362, 381–388
Parasite-driven-wedge, 64, 353–389
Parasite–host coevolutionary races, 12, 61, 62, 197, 355, 363, 379
Parasite richness, 121–124, 154, 173, 174, 180, 199–201, 204, 232, 280–282, 369, 371, 372, 376, 380
Parasite-stress hypothesis of economics, 305–307, 319
*Parasite Stress USA*, 135, 136, 144, 175, 186, 212–214, 218–221, 225, 227, 250, 252, 290, 315, 317, 323, 324, 350, 421–423, 425
Parasite theory of sexual selection, 67, 172–174, 189
Parental nepotism, 31, 32
Particularistic–historical approach, 3–7
Particularistic method, 2, 17

Patrilocal residence, 430
Patriotism, 95, 151–152, 155
Patronyms, 102
Perceived-vulnerability-to-disease (PVD), 65, 71–73, 105, 175, 184, 185, 195, 260, 413, 429
Personality, 4, 6, 9–11, 14, 24–27, 31, 40, 51, 52, 65, 70–73, 75, 84, 85, 87, 93–95, 97, 102, 103, 116, 151, 195–205, 209, 233, 241, 245, 260, 270, 271, 280, 283, 286, 297, 298, 306, 313, 320, 322, 348, 356, 396, 397, 402, 409, 427, 433–435
Pets, 73, 74
Phenotypic quality, 14, 27, 34, 35, 67, 172, 174–176, 178, 189, 211, 225, 226, 229, 233, 312, 431
Philopatry, 12, 13, 40, 62, 63, 66, 70, 75, 97, 102, 105, 113–165, 183, 186, 195, 216, 237, 255, 257, 258, 260, 306, 355–363, 374, 375, 380, 383–388, 411, 418, 420, 427, 429, 431
Philosophical aesthetics, 21–27
Philosophical method, 21, 53–54
Phylogenetic methodology, 50
Phylogenetic origin of the trait, 23
Phylogeny, 43–49, 54, 386, 396
Physical attractiveness, 14, 27, 34, 67, 72, 173, 174, 177, 178, 189, 195, 289, 290
Politic, 11, 15, 40, 42, 52, 83, 85, 86, 96, 101, 186, 202, 208, 222, 230–232, 265–276, 280, 281, 283–285, 288–291, 294–297, 321, 328, 333, 335, 336, 338, 340–347, 370, 377, 378, 410, 425
Political
- factionalism, 354, 377–378
- science, 11, 23, 24, 85, 255, 265–267, 274, 286, 296, 324, 337, 340, 396, 435

Polygynous marriage systems, 67, 172, 173, 346
Polygyny, 171–173, 189, 347
Population size, 117, 135, 142–144, 258, 336, 338, 345, 350, 366–368, 370–373
Pregnancy, 68, 69, 225
Prejudice, 1, 3, 10, 24, 40, 52, 60, 61, 66, 68, 72–74, 87, 88, 103, 106, 184–186, 190, 198, 239, 240, 269, 270, 288, 347, 379, 389, 400, 410, 413, 420
Primates, 36, 69, 149, 354, 358
Pronouns, 94, 95, 99, 100, 115, 116, 291
Property rights, 52, 70, 85, 101, 255, 265, 273, 275, 278–279, 283, 288, 297
Proportion of Believers, 242–244, 246, 248, 250, 406, 408, 412, 413, 415
*Proportion of Religionists USA*, 242–246, 248, 250, 252, 253
*Proportion of Religious Adherents*, 244, 245, 250, 252, 253
*Proportion Who Prayed Everyday*, 244, 246, 248
Protestant, 295
Provincialism, 40
Proximate, 8, 9, 39, 45, 65, 88, 96, 106, 151, 187, 201, 210, 268, 270, 285, 291, 379, 402, 410, 416
Proximate causes, 9, 21, 27, 29, 34, 38, 39, 44, 54, 83, 84, 96, 106, 124, 172, 195, 217, 220, 225, 226, 228, 231, 242, 255, 268, 270, 271, 284, 285, 294, 297, 303, 326, 328, 333, 344, 349, 399, 410, 427, 433
Psychological adaptation, 4, 5, 7, 11–13, 17, 26, 28, 29, 33, 34, 36–39, 41, 46, 53, 54, 64, 83, 96, 148, 149, 181, 187, 198, 210, 217, 226, 271, 285, 289, 292–293, 296, 297, 309, 333, 343, 344, 346, 356, 357, 374, 395, 413, 418, 419, 430, 432
Public goods, 36, 101, 222, 268, 335, 337, 345, 348
Puritans, 295
PVD. *See* Perceived-vulnerability-to-disease (PVD)

**R**

Racism, 1, 2, 22, 87, 185, 186, 293, 425
Rainfall, 47, 83, 142, 147, 345
Rape, 22, 96, 185, 220, 433
Rapoport's rule, 143–145, 154, 357, 385, 389
Reasoning styles, 95–96, 106, 428
Reciprocal altruism, 13, 113, 148, 346
Reciprocity, 13, 32, 33, 147–150, 153, 155, 322, 346, 355
Reconstruction, 45, 48, 173, 185, 269, 293–294, 378, 379
Reductionist, 51, 327
Religion, 10, 11, 16, 17, 22, 23, 51, 64, 70, 73, 85, 127, 181, 237–245, 248, 253, 254, 257–258, 261, 295, 336, 353, 354, 364, 370–376, 389, 407, 411, 425, 431
Religiosity, 1, 5, 9, 10, 15, 23, 28, 29, 40, 51, 52, 70, 73, 85–87, 101, 102, 106, 139, 153, 237–261, 269, 280, 291, 295, 304, 324, 353, 355, 361, 371, 374, 375, 396, 401–403, 406–414, 417, 419, 421, 424, 426, 428, 430–433, 435
Religious
- pluralism, 242
- ritual, 29

Religious participation and value, 242, 244–246, 248, 250, 252–254, 256, 406, 411, 412, 414, 417, 418
Reproductive isolation, 49, 50, 64, 70, 354, 362, 364, 381, 383, 384, 387
The Republic, 98, 228, 243, 265, 324
Republican, 98, 228, 324–325
Reputation, 32, 33, 98, 149, 216, 322, 346
Residential movement, 141, 144–145, 429
Resource distribution, 88, 99–101, 132, 133, 214, 246, 248, 249, 267, 268, 274, 277–280, 282, 285, 297, 401
Revolutions, 15, 32, 70, 101, 291–293, 295, 333–351, 396
*Rhagoletis*, 387
Right-wing Authoritarianism (RWA), 87, 88
Robbery, 219, 433
Romantic love, 97, 179, 276
RWA. *See* Right-wing Authoritarianism (RWA)

**S**

Sanitation, 60, 140, 149, 219, 231, 256, 268, 270, 287, 292, 294, 298, 348, 379, 400, 402, 410, 434
SCCS. *See* Standard cross-cultural sample (SCCS)
Schaller, M., 27, 44, 46, 60, 61, 65, 66, 68–72, 113, 116, 120, 152, 179, 181, 195–199, 204, 240, 259, 280, 288, 377, 413, 416, 421, 428, 433
Scientific aesthetics, 27
Scientific method, 21–23, 26, 27, 36, 53, 54, 125, 128
Scientific revolution, 22, 203, 396
SDO. *See* Social Dominance Orientation (SDO)
Secularism, 9, 22, 23, 40, 51, 125, 239, 241, 243, 255–256, 261, 267, 291, 293, 371, 410
Self-awareness, 5, 35
Self-concept, 6, 7, 17, 40, 93–95, 106, 216, 398, 435
Self-esteem, 94, 409
Self-evident truth, 41, 42
Self-reinforcing, 348, 360–362, 380, 382, 388, 389
Sex ratio, 208, 215, 220
Sexual
  restrictiveness, 70, 87, 97, 103, 179, 180, 190, 272, 276, 282, 288, 291, 297
  revolution, 140, 179, 180, 291, 298
  selection, 67, 69, 96, 172–174, 184, 189, 214, 343, 344, 346, 347, 360, 364
Sexual continence, 1, 14, 98, 179, 180, 190, 272, 282
Sexually transmitted diseases (STD), 421, 422
Skin darkness, 309, 313, 314
Slavery, 2, 3, 173, 217, 294, 349
Smith, J.M., 29
Social
  complexity, 14, 99, 203
  conservatism, 87, 88
  learning, 3, 34–36, 65, 271, 344, 358, 359, 361, 380
  navigation, 6, 35, 36, 40, 99, 148, 271, 298
  networks, 12, 32–34, 63, 75, 96–97, 102, 106, 141, 148, 150, 155, 196, 239, 268, 322, 334, 376, 409, 417, 425, 430, 432
  scientists, 13, 113
  welfare, 101, 270, 287, 296
Social Dominance Orientation (SDO), 87, 88, 172, 173
Socially-mediated speciation, 384, 387
Sociology, 2, 11, 23, 24, 37, 255, 286
Socio-relational model, 409
Sociosexual Orientation Inventory (SOI), 179, 180, 195, 276, 279, 280, 282
SOI. *See* Sociosexual Orientation Inventory (SOI)
Song dialects, 380
South, 1–3, 5, 6, 98, 139, 173, 185, 186, 216–218, 228, 230, 231, 274, 289, 293, 294, 305, 310, 311, 324, 349, 366, 378, 399
Southern democratic party, 98, 228, 378
Special-purpose adaptations, 40–41
Speciation, 50, 64, 70, 76, 354, 360, 362, 364, 378, 380–384, 387–389, 395, 396
Species, 13, 22, 59, 94, 113, 173, 201, 240, 271, 308, 333, 353, 399
Spices, 67, 68, 74
Spirits, 4, 5, 9–11, 41, 45
Spousal abuse, 14, 97
Standard cross-cultural sample (SCCS), 138, 142, 144, 145, 154, 171, 172, 183, 189, 283, 286, 298, 420
Statistical independence of cultures, 42, 45, 49
Statistically independent, 43, 45, 50, 290, 418
STD. *See* Sexually transmitted diseases (STD)
Step-parents, 225, 228, 229
Strength of family ties, 130–137, 139, 151, 152, 154, 245, 246, 254, 304, 319, 325, 339, 406, 408, 416–418, 425
Sub-Saharan Africa, 2, 311, 312, 314
*Suh individualism*, 115, 122, 124, 138, 275, 279, 280, 284, 321, 337, 340–342

Supermarket, 103, 322, 324, 328
Supernatural perspective on culture, 4
Symmetry, 27, 176, 178, 314
Sympatric speciation, 378, 381, 387–388

**T**
Team sports, 95, 333, 349–351
Technologies, 12, 25, 60, 84, 101, 114, 140, 149, 151, 255, 258, 267–269, 287, 288, 292, 298, 305–307, 321, 322, 325–327, 348, 349, 386, 397, 398, 401, 404, 430, 435
Temperature, 39, 67, 83, 126, 127, 142, 197, 215, 309, 311, 312, 316, 345, 384, 403
Territoriality, 141, 144, 146, 358, 366, 376, 385
Theory of pest pressure, 385
Third-party votes, 424, 425
*Total Index*, 403–405
Totalitarianism, 52, 266, 297
Tradition, 4–6, 9, 12, 14, 17, 22, 24–26, 37, 40, 45, 46, 51, 53, 54, 72, 83–87, 95, 97, 98, 101, 103, 104, 106, 114, 119, 137–139, 142, 152, 153, 173, 179, 180, 189, 190, 195, 203, 209, 217, 228, 240, 255, 267–268, 270, 272, 286, 287, 293, 295, 297, 306, 323, 325, 326, 334, 346, 357, 366, 389, 396, 399, 425, 430, 431, 434
Traditionalism, 1, 6, 7, 15, 46, 73, 173
Trait's maintenance, 23, 49
Tree of life, 27, 28, 30, 36, 39, 155, 344
Trial-and-error learning, 34, 35, 41
Trivers, R., 29, 32, 180, 346

**U**
Ultimate, 8, 9, 21, 25, 27–28, 31, 36, 38, 45, 54, 83, 96, 106, 124, 150, 151, 187, 195, 210, 220, 271, 284, 285, 296–298, 303, 327, 333, 344, 346, 349, 399, 410
Unit of analysis, 42
Unity of knowledge, 328

**V**
Vaccination, 74, 140, 231, 256, 292, 298
Vector, 65, 121, 140, 232, 256, 292, 294, 298, 400, 410
Venomous snakes, 28, 29
Violence, 13, 14, 95, 97, 98, 102, 106, 152, 185–187, 190, 207–233, 280, 284, 341, 396, 400, 433–435
Virulence, 62, 65, 121, 150, 151, 155, 357, 428

**W**
Warfare, 15, 70, 106, 185, 207, 233, 265, 286, 333, 335, 338, 343, 350, 400, 433
Wealth, 15, 83, 84, 101, 106, 118, 119, 132, 145, 182, 186, 213–215, 217, 218, 227, 229, 230, 232, 241, 246, 250, 255, 256, 258, 267, 270, 274, 277, 285, 287, 288, 303–306, 309, 311, 315–318, 325–328, 345, 366, 371, 403–410, 434
Whistling Dixie, 6
Wife-capture, 346
Wildlife, 120
Williams, G., 29, 30, 49, 396, 434
Women's liberation, 180
World Health Organization (WHO), 117, 293
World regions, 46, 48, 118, 119, 133, 134, 139, 142, 174, 182, 197, 204, 246, 286, 310, 311, 365, 366, 368, 369, 373, 374, 376, 415, 419, 426, 430, 431
World Values Survey, 130, 131, 151, 152, 244, 256, 272, 275, 276, 279, 287, 425

**X**
Xenophobia, 2, 10, 12, 15, 61–66, 68–70, 73, 75, 76, 93, 101, 146, 152, 155, 183, 237, 257, 285, 286, 293, 297, 306, 334, 335, 343, 344, 348, 350, 355–359, 361, 362, 375, 376, 388, 411, 413, 414, 419, 420, 427, 433

**Z**
Zoonotic, 120–124, 132, 136, 137, 152, 154, 161, 163, 179, 180, 190, 199–201, 204, 205, 212, 213, 215, 218, 219, 224, 227, 229, 230, 233, 245, 246, 252–254, 272, 276, 280–282, 286, 288, 304, 405, 421, 432

MIX
Papier aus verantwortungsvollen Quellen
Paper from responsible sources
FSC® C105338

If you have any concerns about our products,
you can contact us on
**ProductSafety@springernature.com**

In case Publisher is established outside the EU,
the EU authorized representative is:
**Springer Nature Customer Service Center GmbH**
**Europaplatz 3, 69115 Heidelberg, Germany**

Printed by Libri Plureos GmbH
in Hamburg, Germany